Polymer Interfaces
and Emulsions

Polymer Interfaces and Emulsions

edited by

Kunio Esumi

Science University of Tokyo
Tokyo, Japan

MARCEL DEKKER, INC.　　　　　NEW YORK · BASEL

Chemistry Library

ISBN: 0-8247-1975-1

This book is printed on acid-free paper.

Headquarters
Marcel Dekker, Inc.
270 Madison Avenue, New York, NY 10016
tel: 212-696-9000; fax: 212-685-4540

Eastern Hemisphere Distribution
Marcel Dekker AG
Hutgasse 4, Postfach 812, CH-4001 Basel, Switzerland
tel: 41-61-261-8482; fax: 41-61-261-8896

World Wide Web
http://www.dekker.com

The publisher offers discounts on this book when ordered in bulk quantities. For more information, write to Special Sales/Professional Marketing at the headquarters address above.

Current printing (last digit):
10 9 8 7 6 5 4 3 2 1

PRINTED IN THE UNITED STATES OF AMERICA

To Setsuko with thanks for her patience and love

Preface

Polymeric materials are used for many applications in a wide array of technological areas, and their surface/interface characteristics as well as solution properties are of great importance for their application. Considerable theoretical and experimental work has been carried out to understand the properties and performance of polymeric materials in solution and at various interfaces. In addition, polymer latexes produced by emulsion polymerization have a large and still growing importance as model colloids and industrial products. It is obvious that these polymer colloids will find further use in microdevices and micromachines.

The volume has 14 chapters. K. Ishizu outlines the structural ordering in polymer solutions such as the $(AB)_n$ star, core-shell polymer microsphere, and block copolymer in Chapter 1. In Chapter 2, S.-B. Lee deals with the influence of polymer surface structure on polymer surface behavior for polymer systems including block copolymers and graft copolymers. G. H. Ma reviews the advances in preparations of polymeric microspheres and gives examples for their application in Chapter 3. Some new methodologies for latex particle chemical heterogeneity as well as the fundamentals of each relevant methodology are discussed in Chapter 4, by F. Galembeck and E. Fátima de Souza. Chapter 5, by F. J. de las Nieves, A. Fernández-Barbero, and R. Hildago-Alvarez, outlines the electrokinetic behavior of polymer colloids and discusses the anomalous electrokinetic behavior. K. Furusawa focuses on the preparation of composite particles consisting of polymer latices and inorganic colloids and their physicochemical properties in Chapter 6.

Chapter 7 by H. D. Ou-Yang and M. M. Santore, reviews the current approaches used to understand polymer-bridging of colloids both thermodynamically and kinetically. T. Miyajima presents the polyelectrolytic nature and metal

v

complexation of polyelectrolyte systems in Chapter 8. In Chapter 9, by S. G. Dixit and A. K. Vanjara, the adsorption of quaternary ammonium compounds at polymer surfaces is described. In Chapter 10, W. Loh, J. R. Lopes, and A. C. S. Ramos review adsorption onto poly(tetrafluoroethylene) from aqueous solutions. Chapter 11, by Yu. S. Lipatov, T. T. Todosijchuk, and V. N. Chornaya, deals with the analysis of the data on the adsorption of polymer mixtures at solid interfaces from semidiluted and concentrated solutions. K. Esumi discusses polymer adsorption as well as the simultaneous adsorption of polymer and surfactant at oxide surfaces in Chapter 12. In Chapter 13, Y. Gushikem and E. A. Toledo describe the preparation of oxide-coated cellulose and some physicochemical characteristics of the oxide particles dispersed onto the polymer in the form of a fiber or membrane. M. Morra, C. D. Volpe, and S. Siboni outline the basic concepts and definitions related to acid–base properties and discuss some problems estimated with different methods.

I would like to thank Professor A. T. Hubbard for his advice and help, as well as the authors who participated in this effort. I am indebted to Mr. Joseph Stubenrauch and Ms. Anita Lekhwani of Marcel Dekker, Inc., for their assistance in preparing this volume.

Kunio Esumi

Contents

Contributors

Valentina N. Chornaya Physical Chemistry of Polymers, Institute of Macromolecular Chemistry, National Academy of Sciences of Ukraine, Kiev, Ukraine

F. Javier de las Nieves Department of Applied Physics, University of Almería, Almería, Spain

Elizabeth Fátima de Souza Biological Science and Chemistry Institute, Pontifícia Universidade Católica de Campinas, Campinas, São Paulo, Brazil

Sharad G. Dixit Applied Chemistry Division, Department of Chemical Technology, University of Mumbai, Mumbai, India

Kunio Esumi Department of Applied Chemistry and Institute of Colloid and Interface Science, Science University of Tokyo, Tokyo, Japan

A. Fernández-Barbero Department of Applied Physics, University of Almería, Almería, Spain

Kunio Furusawa Department of Chemistry, University of Tsukuba, Ibaraki, Japan

Fernando Galembeck Institute of Chemistry, Universidade Estadual de Campinas, Campinas, São Paulo, Brazil

Yoshitaka Gushikem Institute of Chemistry, Universidade Estadual de Campinas, Campinas, São Paulo, Brazil

R. Hildago-Alvarez   Department of Applied Physics, University of Granada, Granada, Spain

Koji Ishizu   Department of Polymer Science, Tokyo Institute of Technology, Tokyo, Japan

Soo-Bok Lee   Advanced Chemical Technology Division, Korea Research Institute of Chemical Technology, Taejon, Korea

Yuri S. Lipatov   Physical Chemistry of Polymers, Institute of Macromolecular Chemistry, National Academy of Sciences of Ukraine, Kiev, Ukraine

Watson Loh   Institute of Chemistry, Universidade Estadual de Campinas, Campinas, São Paulo, Brazil

Josias R. Lopes   Institute of Chemistry, Universidade Estadual de Campinas, Campinas, São Paulo, Brazil

Guang Hui Ma   Graduate School of Bio-Applications and Systems Engineering, Tokyo University of Agriculture and Technology, Tokyo, Japan

Tohru Miyajima   Department of Chemistry, Saga University, Saga, Japan

Marco Morra   Nobil Bio Ricerche, Villafranca d'Asti, Italy

H. Daniel Ou-Yang   Department of Physics and Polymer Interfaces Center, Lehigh University, Bethlehem, Pennsylvania

Antonio C. S. Ramos   Institute of Chemistry, Universidade Estadual de Campinas, Campinas, São Paulo, Brazil

Maria M. Santore   Department of Chemical Engineering and Polymer Interfaces Center, Lehigh University, Bethlehem, Pennsylvania

Stefano Siboni   Department of Materials Engineering, University of Trento, Trento, Italy

Tamara T. Todosijchuk   Physical Chemistry of Polymers, Institute of Macromolecular Chemistry, National Academy of Sciences of Ukraine, Kiev, Ukraine

Eduardo Aparecido Toledo   Department of Chemistry, Universidade Estadual de Maringá, Maringá, Paraná, Brazil

Ajay K. Vanjara   Applied Chemistry Division, Department of Chemical Technology, University of Mumbai, Mumbai, India

Claudio Della Volpe   Department of Materials Engineering, University of Trento, Trento, Italy

1
Structural Ordering in Polymer Solutions

Koji Ishizu
Tokyo Institute of Technology, Tokyo, Japan

I. INTRODUCTION

The star-branched, or radial, polymers have the structure of linked-together linear polymers with a low-molecular-weight core. Generally, the star polymer has smaller hydrodynamic dimensions than that of a linear polymer with an identical molecular weight. The interest in star polymers arises not only from the fact that they are model branched polymers but also from their enhanced segment densities. Zimm and Stockmayer were the first to study the conformation of star-shaped polymers [1]. Recently, Daoud and Cotton [2] have studied the conformation and dimension of a star polymer consisting of three regions: a central core, a shell with semidilute density in which the arms have unperturbed chain conformation, and an outer shell in which the arms of the star assume a self-avoiding conformation. Stars with multiarms (the critical number of arms is estimated to be of order 10^2) are expected to form a crystalline array near the overlap threshold (C^*) [3].

Leibler [4] has developed a Landau-type mean-field theory on the microphase-separation transition in diblock copolymers and has presented the phase diagram for the microdomain morphologies in the weak-segregation limits as a function of segregation power and composition of the block copolymer. Ohta and Kawasaki [5] have generalized Leibler's theory to the strong-segregation limit by taking into account a long-range interaction of the local order parameter on the basis of the Ginzburg–Landau-type mean-field theory. More recently, Fredrickson and Helfand [6] have corrected Leibler's mean-field theory to take into account the effect of composition fluctuations on the microphase-separation transition. These theories expect that the morphologies are changed with the segregation power. Because the segregation power is a function of temperature, morphology may be

1

reversibly controlled and, hence, the transition between the different kinds of morphology may occur by changing temperature.

The star-block copolymers have a molecular conformation similar to star polymers. De la Cruz and Sanchez [7] have calculated the phase-stability criteria and static structure factors in the weak-segregation limits for an n-arm diblock copolymer [(AB)$_n$ star]. According to their results, as the arm number (n) becomes large, the (AB)$_n$ star begins to develop a "core and shell"-type structure. This self-segregation or self-micellization tends to create significant concentration fluctuations at the core–shell interface.

Four ordered microphases for block copolymers are well known; they consist of alternating layers, cylinders on a hexagonal lattice, spheres on a body-centered-cubic (BCC) lattice, and a bicontinuous double-diamond structure [8,9]. The spherical, cylindrical, and lamellar structures are all stable in the strong-segregation limits [10]. The stable phases are considered to be spherical microdomains for the volume fraction of either diblock, $f = \sim 0.17$. If the structure of the self-assemblies can be fixed by cross-linking of the spherical parts (the spherical microdomains in the solid state and the core in solution), the cross-linked products can form core–shell polymer microspheres (see Fig. 1). In fact, polystyrene (PS)–block—polybutadiene(PB)–block–PS triblock copolymer micelles with cores of PB blocks in dilute solution were stabilized by cross-linking the chains in the micellar cores by ultraviolet (UV) irradiation in the presence of a photoinitiator and by fast electrons [11–13]. The irradiated, stabilized micelles, examined by light scattering, sedimentation, and gel permeation chromatography (GPC), did not decompose upon heating or when placed in good solvents for both blocks. Others have obtained the core–shell polymer microspheres by cross-linking of the core domains in micelles formed in selective solvents for several different block and graft copolymers [14–21]. It was concluded from these results that the reaction rate of intra micelle cross-linking and the lifetime of the polymer micelle were both important for successful cross-linking of the polymer micelle. A core–shell polymer microsphere could be synthesized when intra micelle cross-linking occurred before the breakup of the micelle.

The microphase-separated structures in bulk films are more stable than micelles in solution. The cross-linking of spherical microdomains in films is often a superior method for the preparation of core–shell polymer microspheres. A well-defined PS–block—poly(4-vinylpyridine) (P4VP) diblock copolymer (P4VP block, 24 wt%) was prepared by a sequential anionic polymerization [22]. The microdomain structure in this specimen showed a texture of discrete P4VP spheres in a PS matrix. These segregated P4VP chains in a sphere were cross-linked by using quaternization with 1,4-dibromobutane (DBB) vapor in the solid state. These microsphere particles had a narrow size distribution. The micelle of core–shell polymer microsphere moved like the pseudo-latex in solution. The particles of core–shell microspheres were well aligned in a hexagonal array on the

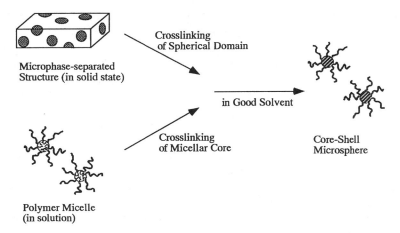

Figure 1 Schematic presentation of synthesis routes on core–shell polymer microspheres.

carbon substrate from two-dimensional observation [23,24]. It was expected that in the core–shell microspheres, such crystalline order in solution appeared close to the C^*.

In this chapter, we outline the structural ordering in polymer solutions such as the star, $(AB)_n$ star, core–shell polymer microsphere, and block copolymer. The ordering mechanisms of these polymers are also mentioned in detail. We also discuss the architecture of polymeric superstructures by locking cubic lattices formed by core–shell microspheres.

II. STRUCTURAL ORDERING IN STAR POLYMER SOLUTIONS

As mentioned in Sec. I, the stars with multiarms are expected to form a crystalline array near the C^*. Several groups [25–27] investigated the ordering phenomena of stars [8 < arm number(n) < 20] around the C^* by means of small-angle neutron scattering (SANS) and small-angle x-ray scattering (SAXS). They showed that ordering was very weak for 8- and 18-arm stars but became stronger with increasing arm numbers.

Star polymers are best prepared by coupling anionic living polymers with multifunctional electrophilic coupling agents. The coupling agents are either multifunctional chloromethylated benzene derivatives or multifunctional chlorosilane compounds. In general, it is difficult to extend the functionality of the stars with these compounds. Roovers et al. [28,29] have reported a new synthesis method of regular PB star molecules and their dilute-solution properties. These 64- and 128-arm stars were prepared by coupling of living PB–lithium with dendrimers having chlorosilane groups at the surface. At present, the most convenient way of preparing star polymers possessing more than 10 arms is by cross-linking monocarbanioic chains with divinylbenzene (DVB) [30–33]. More recently, we synthesized polyisoprene (PI) stars by the free-radical cross-linking of the vinylbenzyl-terminated PI macromonomers with DVB in n-heptane that dissolved PI but precipitated DVB [34]. The radical copolymerization of PI macromonomer with DVB led to microgelation in micelles formed by the primary copolymer radicals in the selective solvent (organized polymerization).

This idea could be applied to the PI star synthesis by anionic cross-linking of PI monoanions with DVB in n-heptane [35]. The anionic copolymerization of PI–lithium with DVB led to microgelation in micelles formed by the primary copolymer anions in the selective solvent. All of the PI stars prepared had an apparently narrow molecular-weight distribution ($\overline{M}_w/\overline{M}_n = 1.03$–$1.07$) from GPC profiles.

Table 1 lists the characteristics of PI stars. For example, the $(SI42)_{43}$ sample in Table 1, which shows the star structure having an arm molecular weight of 4.2×10^4 and arm number $n = 43$. All of the feed DVB was consumed in the core formation of star polymers. It was possible to estimate the core radius (R_0) of PI stars,

Table 1 Characteristics of PI Stars

| Code | Molecular weight | | Arm number (number/molecules) | $R_g{}^{c}$ (nm) | $R_h{}^{d}$ (nm) | $R_0{}^{e}$ (nm) | C^{*f} (wt%) |
	Arm[a] $(10^{-4} M_n)$	Star[b] $(10^{-6} M_w)$					
$(SI33)_4$	3.3	0.12	4	16.0	10.7	—	3.9
$(SI42)_{43}$	4.2	1.87	43	30.1	24.7	2.7	4.9
$(SI27)_{91}$	2.7	2.79	91	29.2	22.2	4.4	10.1
$(SI42)_{111}$	4.2	5.00	111	41.3	33.0	4.7	5.5
$(SI08)_{237}$	0.8	2.09	237	19.1	18.1	3.9	14.1

[a]Determined by GPC using PI standard samples.
[b]Determined by static light scattering (SLS).
[c]Radius of gyration (R_G) was determined by SLS.
[d]Hydrodynamic radius (R_H) was determined by dynamic light scattering (DLS; scattering angle = 90°).
[e]Core radius (R_0) was calculated from the equation $n = (4\pi/3P_E)R_0^3\rho N_A$, assuming that the core was of spherical form. ρ-density of DVB core; N_A-Avogadro's number; P_E-the number of monomer units (mixture of DVB and ethylstyrene) at the arm PI terminal end.
[f]Overlap threshold (C^*) was calculated from the equation $C^* = 3M_w/4\pi N_A R_H^3$.

because the yield of the star and the feed amounts of DVB were known. The values of R_0 were very small compared to the corresponding radii of gyration (R_G) for PI stars. For dendrimers used as coupling agents, R_0 was estimated to be 2.4–2.5 nm for the generation $m = 3$–4 [36]. Therefore, the core size of such PI stars seemed to correspond to the size of dendrimer with m somewhat larger than $m = 4$.

The hydrodynamic radius (R_H) was measured by dynamic light scattering (DLS) at 90° of scattering angles. The values of R_H are also listed in Table 1. The value of R_G/R_H is a sensitive fingerprint of the inner density profile of star and polymer micelle. The observed values of R_G/R_H approached unity as n became large. Even the stars with multiarm behaved not as neat hard spheres ($R_G/R_H = 0.775$) but as soft spheres that were penetrable near the outer edge in a good solvent.

We studied structural ordering of PI stars in cyclohexane [37]. According to the theoretical results of Witten et al. [3], a crystalline structure of the stars should appear near the C^*. The calculated C^* for each PI star is also listed in Table 1. Below C^*, the star polymers remain isolated, as any arrangement of stars in solution is expected near or above C^*. First, the SAXS intensity profiles of the $(SI42)_{43}$ star were measured at 6 and 13 wt% cyclohexane polymer solutions. Both polymer concentrations were higher than C^* (4.9 wt%). However, no regular scattering peaks appeared at these concentrations. As a matter of course, the structural ordering had never been observed in the $(SI33)_4$ star.

Figure 2 shows typical SAXS intensity profiles for the $(SI08)_{237}$ star in the small-angle region, where q [$= (4\pi/\lambda)\sin \theta$] is the magnitude of the scattering maxima and values in parentheses indicate the interplaner spacings (d_1/d_n) calculated from Bragg reflections. Below 8 wt% of polymer concentration ($C^* = 14.1$ wt%), no regular scattering peaks appeared due to disordering. At 11 wt% of polymer concentration (Fig. 2a), the first four peaks appear close together at the relative q positions of $1 : \sqrt{2} : \sqrt{3} : 2$ as shown in parentheses. The interplaner spacing (d_1/d_n) at the scattering angles is relative to the angle of the first maximum according to Bragg's equation: $2d \sin \theta = n\lambda$ (where θ is one-half the scattering angle and $\lambda = 1.5418$ Å). In general, this packing pattern appears in the lattice of not only simple cubic (SC) but also body-centered-cubic (BCC) structures. As mentioned, in the section of $(AB)_n$ stars, the $(AB)_n$ stars with a multiarm were packed in the lattice of a BCC structure near C^* [38,39]. The conformation of stars can be regarded as similar to one of $(AB)_n$ stars in solution. It is reasonable that these values correspond to the packing pattern of (110), (200), (211), and (220) planes in a BCC structure.

In the SAXS intensity profile at 33 wt% of the polymer concentration, the complicated scattering peaks appear as shown in Fig. 2b. The first five peaks appear at the relative q positions of $1 : \sqrt{4/3} : \sqrt{2} : \sqrt{8/3} : \sqrt{3}$. In general, the relative q positions of $1 : \sqrt{4/3} : \sqrt{8/3}$ correspond to a packing pattern of (111), (200), and (220) planes in a face-centered-cubic (FCC) structure. Therefore, it is concluded that the stars are packed in the mixed lattice of BCC and FCC structures

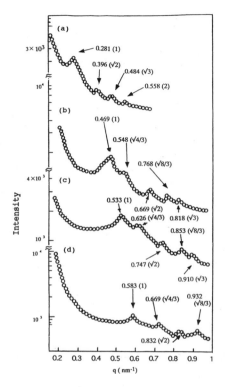

Figure 2 The SAXS intensity profiles for cyclohexane solutions of the $(SIO8)_{237}$ star: (a) 11 wt%; (b) 33 wt%; (c) 67 wt%; (d) bulk film.

at this polymer concentration. Similar mixed-lattice patterns are also observed in the SAXS profiles at 67 wt% of polymer concentration (Fig. 2c). It is noted that the first peak shifts to the side of the high-q position in the order of the increment of polymer concentrations. This fact means that Bragg spacing d_1 became shorter with increasing polymer concentration. It is also noted that the scattering intensity at the relative q position of $\sqrt{4/3}$ based on a FCC lattice increases with increasing polymer concentration. This means that the FCC lattice fraction in the mixed lattices increases during the concentration of polymer solutions.

Figure 2d shows the SAXS intensity profile for the $(SIO8)_{237}$ star film. In this measurement, the scattering peaks seem to originate from the DVB core of the PI star. It is found that the first four peaks appear at the relative q positions of 1 : $\sqrt{4/3}$; $\sqrt{2}$: $\sqrt{8/3}$, as shown in parentheses. The star is packed in the mixed lattice of BCC and FCC structures even in the bulk. However, the film specimen of the PI star did not take the state of thermal equilibrium, because the measurement temperature was 25°C. The precise lattice structure of the stars in the bulk could not be determined from the measurement method in this work. The results of SAXS data obtained for $(SI27)_{91}$ and $(SI42)_{118}$ stars were almost the same as those for the $(SI08)_{237}$ star.

One can consider spatial packing of the cubic lattice in solution. The measured Bragg spacing d_1 is related to the cell edge a_c of the cubic lattice and the nearest-neighbor distance of the spheres D_0:

$$D_0 = \left(\frac{\sqrt{3}}{2}\right)a_c = \left(\sqrt{\frac{3}{2}}\right)d_1 \quad \text{for bcc} \tag{1}$$

$$D_0 = \left(\frac{1}{\sqrt{2}}\right)a_c = \left(\sqrt{\frac{3}{2}}\right)d_1 \quad \text{for fcc} \tag{2}$$

Figure 3 shows the relationship between d_1 or D_0 and polymer concentration for the $(SI08)_{237}$ star. The PI star takes the disordered state below ~C^* (14.1 wt%). Near C^*, the star forms the lattice with a BCC structure. Beyond ~19 wt% of polymer concentration, this structure changes into a mixed lattice of BCC and FCC and leads to shrinkage of the spherical particles. A FCC lattice is the most efficient way of packing spheres. It is also found that both values of d_1 and D_0 decrease continuously with an exponential function increasing the polymer concentration. Figure 4

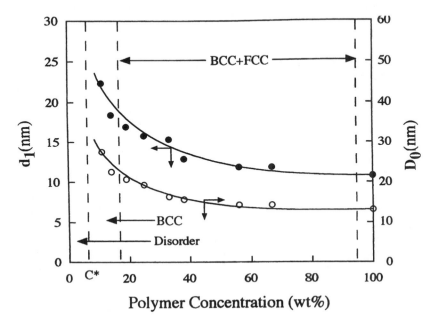

Figure 3 Relationship between d_1 or D_0 and polymer concentration for the $(SI08)_{237}$ star.

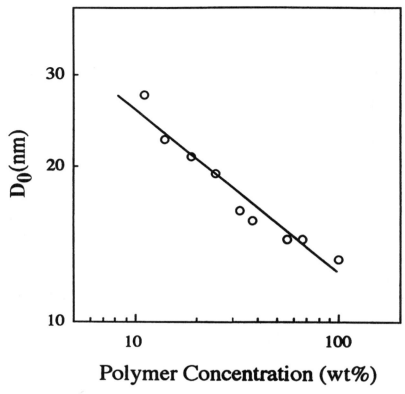

Figure 4   Double-logarithmic plot of D_0 as a function of polymer concentration for the $(SIO8)_{237}$ star.

shows the double-logarithmic plot of D_0 as a function of polymer concentration. It is found that the measured D_0 is proportional to the -0.32th power of the polymer concentration and fits well with the $-1/3$ power expected for a homogeneous system. This fact means that the spherical particles of $(SIO8)_{237}$ stars lead to isotropic shrinkage increasing the polymer concentration around a high polymer concentration.

Structural transformation of such star polymers can be explained by a thermal blob model (Fig. 5) according to Daoud and Cotton [2]. They attempted to predict the swelling properties of a star as a function of the quality of the solvent and the monomer concentration C. They suppose there is a spherical symmetry around the center: All the blobs at a given distance r from the central point have the same size $\xi(r)$. The size ξ of the region depends only on the monomer concen-

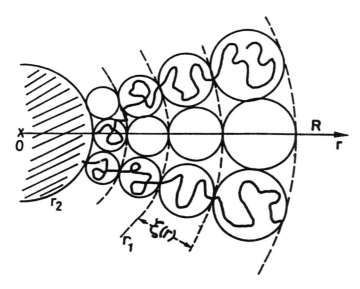

Figure 5 A representation of the Daoud and Cotton model: Every branch is made of a succession of blobs with a size ξ increasing from the center of the star to the outside. (From Ref. 2.)

tration C and is the same within the sample. The picture presented for a star polymer consists of three regions: a central core (A region), a shell with semidilute segment density in which the arms have their unperturbed chain conformation (B region), and an outer shell in which the arms of the star assume a self-avoiding conformation (C region). Then, the arm of the shell can be seen as a succession of growing blobs in the B and C regions. A BCC ordering of these stars appears close to the C^*. The inner cores do not interpenetrate with each other. The chain conformation of arm segments on the shell portions leads to shrinkage with increasing polymer concentration. The gravities of cores move to the arrangement of a FCC lattice. The star polymers lead to hierarchical structure transformation of the cubic lattice in the film formation to minimize the thermodynamic energy.

III. STRUCTURAL ORDERING IN (AB)$_n$ STAR COPOLYMER SOLUTIONS

De la Cruz and Sanchez [7] have derived the phase-stability criteria and static structure factors near the order–disorder transition (ODT) for n-arm star-block copolymers [(AB)$_n$ star]. Figure 6 shows the variation of $(\chi N)_s$ with composition f

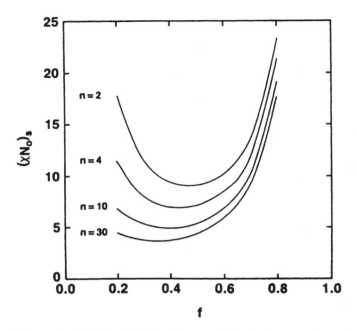

Figure 6   Variation of $(\chi N)_s$ with composition f and arm number n for $(AB)_n$ star copolymers. (From Ref. 7.)

and arm number n for $(AB)_n$ stars. The $(AB)_2$ star corresponds to a triblock copolymer. The $(\chi N)_s$ curves are asymmetric. As n gets large, the $(AB)_n$ star begins to develp a "core and shell"-type structure due to the small critical value of $(\chi N)_c$. The core will be rich in monomer A and the shell will be rich in monomer B, even in the disordered state. This self-segregation or self-micellization tends to create significant concentration fluctuations at the core–shell interface. Thus, this theory predicts that the phase separation is easier in star copolymers than in an analogous block copolymer.

In recent studies, the $(AB)_n$ star copolymers were synthesized by free-radical microgelation in micelles (organized polymerization) formed by a diblock macromonomer with ethylene glycol dimethacrylate as a linking agent [40], and the self-micellization of $(AB)_n$ star copolymers were studied as a parameter of n [41]. As a result, the microphase-separated structures of these stars ($n = 14$–30 and 16–19 wt% B blocks) were formed with the dimensions of a unimolecular micelle in strong segregation. The chain conformation of $(AB)_n$ stars is very similar to that of star-shaped polymers. Then, it can be expected that $(AB)_n$ stars also form the structural ordering near C^*.

The $(AB)_n$ star copolymers (arm number $n = 14$–41 and 16–19 wt% PI

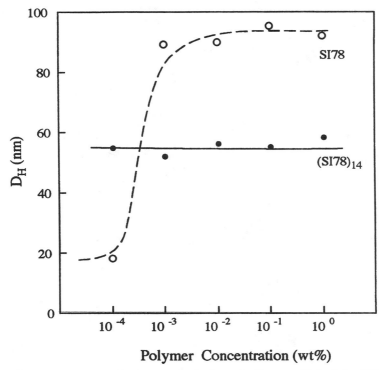

Figure 7   Relationship between hydrodynamic diameter (D_H) of the $(SI78)_{14}$ star and diblock arm SI78 and polymer concentration.

blocks) were prepared by anionic copolymerization of PS–*block*–PI diblock anions with DVB. The structural ordering of such $(SI)_n$ stars was investigated through SAXS and electron microscopy [38,39,42]. Figure 7 shows the relationship between the hydrodynamic diameter (D_H) and polymer concentration for the $(SI78)_{14}$ star (18 wt% PI block) and its diblock arm SI78 in N, N-dimethylformamide (DMF). The $(SI78)_{14}$ sample shows the star structure having a PS–*block*–PI arm molecular weight of 7.8×10^4 and an arm number of 14. DMF is a good solvent for PS but a poor one for PI. In diblock arm SI78, D_H increases suddenly beyond 2×10^{-4} wt% of polymer concentration. It indicates that diblock arm SI78 forms polymeric micelles, such as PI core–PS corona structure. In the $(SI78)_{14}$ star, however, D_H shows a constant value in the range 10^{-4}–1 wt% of polymer concentration. The inner PI chains of star copolymer are bonded radially with cross-linked DVB core. But a strong repulsion force works between PI and PS chains, because of incompatible segments. So, the inner PI parts of stars may behave as hard cores in a selective solvent such as DMF.

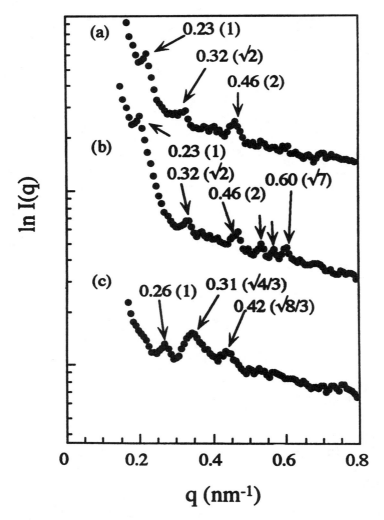

Figure 8   SAXS intensity profiles of the $(SI33)_{16}$ star copolymer: (a) 7.0 wt% in benzene; (b) 7.0 wt% in styrene; (c) bulk film.

As mentioned in the previously section, an ordering structure of the stars should appear near C^*. Below C^*, the $(AB)_n$ stars remain isolated, as any arrangement of stars in solution is expected near or above C^*. Figure 8 shows typical SAXS intensity profiles for the $(SI33)_{16}$ star (19 wt% PI block) solutions in the small-angle region. At 7 wt% of polymer concentration in benzene ($C^* = 6.5$

wt%), regular peaks appear at the relative q positions of $1: \sqrt{2} : \sqrt{3} : 2$, as shown in the parentheses (Fig. 8a). In general, this packing pattern corresponds to the lattice of SC or BCC structure. In order to determine such a packing form, styrene was employed as a solvent in the SAXS measurement. Figure 8b shows the SAXS intensity profile for the $(SI33)_{16}$ star in 7 wt% of styrene solution, which has a refractive index similar to PS corona chains of $(SI)_n$ stars. It is found that high-ordered scattering peaks appear clearly in the small-angle region. The sixth peak $(q = 0.6 \text{ nm}^{-1}, d_1/d_n = \sqrt{7})$ is the characteristic one for the lattice of BCC structure. Therefore, the $(SI33)_{16}$ star is packed with a BCC structure near C^*.

Figure 9 shows transmission electron microscopy (TEM) micrograph of the $(SI33)_{16}$ star specimen in the bulk film. The dark portions correspond to PI phases selectively stained with osmium tetroxide (OsO_4). The texture shows the morphology of discrete PI spheres (radius $R = 4.0$ nm) dispersed in a PS matrix. It is found also that these PI spheres are arranged regularly with long-range distance. This packing pattern can be determined by the SAXS measurement. Figure 8c shows the SAXS profile of the $(SI33)_{16}$ bulk film. The first four peaks appear at the relative q positions of $1 : \sqrt{4/3} : \sqrt{8/3} : 2$, as shown in parentheses. This packing pattern corresponds to the lattice of a FCC structure.

It is important to clarify the process of structural ordering for the $(SI)_n$ star solution through C^* into the bulk film. Figure 10 shows the SAXS intensity profiles of $(SI33)_{16}$ star at the concentrated polymer solutions (25 and 40 wt%). In the SAXS intensity profile at 25 wt% of polymer concentration, the first three peaks appear at the relative q positions of $1 : \sqrt{4/3} : \sqrt{3}$ (Fig. 10a). Moreover, the first four peaks at the relative q positions of $1 : \sqrt{4/3} : \sqrt{3} : \sqrt{11/3}$ in the SAXS intensity profile at 40 wt% of polymer concentration (Fig. 10b). The relative q positions of $1 : \sqrt{4/3} : \sqrt{11/3}$ correspond to a packing pattern of (111), (200), and (311)

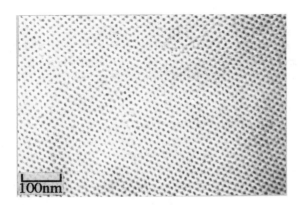

Figure 9 TEM micrograph of the $(SI33)_{16}$ star specimen in bulk film.

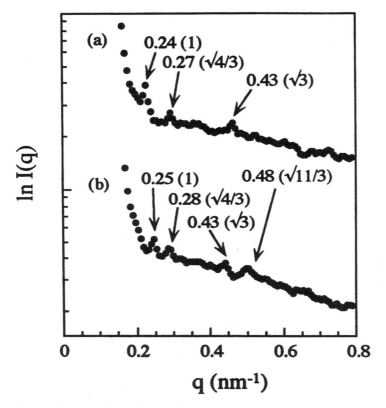

Figure 10 SAXS intensity profiles of the $(SI33)_{16}$ star copolymer: (a) 24 wt% in benzene; (b) 40 wt% in benzene.

planes in a FCC structure. Therefore, it is concluded that the $(SI33)_{16}$ star is packed with the mixed lattice of BCC and FCC structures at these polymer concentrations. It is noted also that the first peak shifts to the side of the high-q position in the order of the increment of polymer concentrations.

One can consider spatial packing of the cubic lattice in solution, as mentioned in Sec. II. Figure 11 shows the relationship between Bragg spacing d_1 or the nearest-neighbor distance of the spheres D_0 and the polymer concentration for the $(SI33)_{16}$ star. The $(SI33)_{16}$ star takes the disordered state below ~ C^* (7 wt%). Near C^*, this star copolymer forms a lattice with a BCC structure. Beyond ~ 20 wt% of polymer concentration, this structure changes into a mixed lattice of BCC and FCC and leads to shrinkage of the sphere particles. Both values of d_1 and D_0 decrease continuously with an exponential function increasing the polymer con-

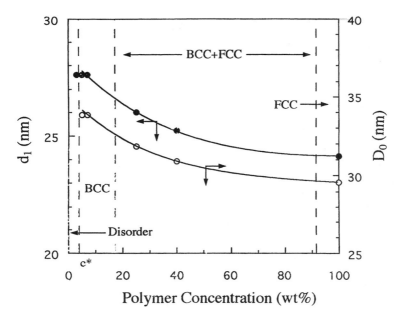

Figure 11 Relationship between d_1 or D_0 and polymer concentration for the $(SI33)_{16}$ star.

centration. Near the bulk film, these structures change into the lattice of a FCC structure. The structural ordering of $(AB)_n$ star copolymers is very similar to the hierarchical structure transformation of the cubic lattices observed on the star polymers.

In a $(SI)_n$ star copolymer, the PI spherical core can be regarded as the central core of region A in the Daoud–Cotton model shown in Fig. 5. The PS block of the arm can be seen as a succession of growing blobs in regions B and C. The PI spherical cores do not interpenetrate with each other and the $(SI)_n$ star copolymers are compatible with their PS blocks of the arm during solvent evaporation.

IV. STRUCTURAL ORDERING IN CORE–SHELL POLYMER MICROSPHERE SOLUTIONS

Ishizu and co-workers established a novel synthesis method of the core–shell polymer microspheres by cross-linking the segregated chains in spherical microdomains [22–24,43,44]. Figure 12 shows a typical TEM micrograph of the P4VP core–PS shell polymer microsphere SV1-M cast from 0.1 wt% 1,1,2-

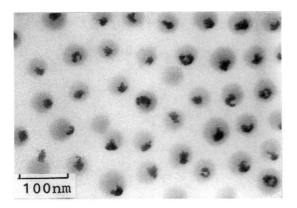

Figure 12   TEM micrograph of the SV1-M core–shell polymer microsphere.

trichloroethane (TCE)/nitrobenzene: 10/1 (v/v) solution [22]. This texture indicates a multimolecular micelle. In the preparation conditions of this cast specimen, multimolecular micelles are frozen with a structure like the core–shell microspheres. The dark, gray, and white portions indicate the cross-linked P4VP cores, the shell of PS chains, and the carbon substrate, respectively. This micrograph clearly shows the structure of the core–shell microspheres. It is also clear that these microsphere particles have only a narrow size distribution. These core–shell polymer microspheres stabilized in good solvent by highly branched arms. The morphological behavior of core–shell microspheres was studied from casting solution onto a carbon substrate [45]. The particles of microspheres were well aligned in a hexagonal array on the carbon substrate from two-dimensional observation. It was expected that such ordering in solution would appear close to the C^* in the core–shell microspheres. However, the three-dimensional ordered structure in solution was not made clear by SAXS, because the large-diameter microsphere particles were used. They determined the packing structure in bulk film formed by the core–shell microspheres using three-dimensional observation with the tilted method of TEM [46]. It was concluded that the packing structure of the core–shell microspheres was the lattice of a FCC structure. In recent reports, Ishizu et al. prepared the core–shell microspheres using small-diameter particles. Film formation of these microspheres was investigated through SAXS and TEM observations [47,48].

A well-defined PS–*block*–P4VP diblock copolymer SV was prepared by the sequential anionic addition polymerization using *n*-BuLi as an initiator. The morphological structure of the SV10 specimen cast from TCE showed dispersed P4VP spheres in a PS matrix. Table 2 lists the characteristics of the PS–*block*—P4VP

Table 2 Characteristics of Polystyrene–*block*–Poly(4-vinylpyridine) Diblock Copolymer

	Diblock copolymer			Domain size	
Code	$10^{-4}\bar{M}_n$[a]	$\bar{M}_w/\bar{M}_n$[b]	Content of P4VP block[c] (mol%)	$\bar{D}_n$[d] (nm)	$\bar{D}_w/\bar{D}_n$[d]
SV10	4.2	1.27	14	17	1.06

[a]Estimated by M_n of the precursor and composition of the P4VP block.
[b]Determined by gel permeation chromatograph.
[c]Determined by ^{1}H-NMR (nuclear magnetic resonance).
[d]D_n indicates the diameter of P4VP spheres determined by a TEM micrograph cast from TCE.
Source: Ref. 48.

diblock copolymer SV10. Core–shell polymer microspheres (SV-M) were synthesized by cross-linking the segregated chains in P4VP spherical microdomain with the vapor of 1,4-dibromobutane (DBB) in the solid state. The solubility of the core–shell microspheres depended strongly on the solubility of the block chains that construced shell parts. According to theoretical results of Witten et al. [3], a structure for the crystalline array of the star-shaped polymer should appear near C^*. Table 3 lists the physical values and C^* of core–shell polymer microsphere SV10-M.

The SAXS of microspheres was measured near C^*. The benzene solution of the SV10-M microsphere had a bluish tint at 7.0 wt%. Figure 13a shows the SAXS intensity distributions for this microsphere solution in the small-angle region; the arrows show the scattering maxima. It is shown that the first three peaks appear close at the relative angular positions of $1 : \sqrt{2} : \sqrt{3}$ (shown in parentheses). These values correspond to packing patterns of (110), (200), and (211) plane in a

Table 3 Physical Values and C^* Concentration of Core–Shell Polymer Microsphere

	Microsphere			
Code	N[a]	$10^{-8}\bar{M}_w$[b]	D_H[c] (nm)	C^*[d] (wt%)
SV10-M	1800	0.76	66	6.9

[a]Aggregation number.
[b]Molecular weight of core–shell polymer microsphere.
[c]Hydrodynamic diameter of microsphere determined by DLS.
[d]Overlap concentration.

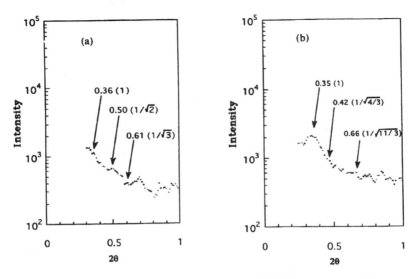

Figure 13 SAXS intensity distributions for the SV10-M microsphere. The arrows show the scattering maxima and the values of parentheses indicate the cubic packing (d_1/d_n): (a) 7 wt% benzene solution; (b) bulk film.

BCC structure. The microspheres are packed in the lattice of a BCC structure near C^*.

The three-dimensional arrangement of the microsphere film was determined by the SAXS measurement. Figure 13b shows the SAXS intensity distribution in the small-angle region for the SV10-M microsphere film. The maxima at $2\theta = 0.42$ and 0.66 occur at the angular positions of $\sqrt{4/3}$ and $\sqrt{11/3}$ of the first-order maximum. Therefore, the microspheres are packed in the lattice of a FCC structure in the bulk film.

In general, when a block copolymer forms a highly ordered microphase-separated structure, the spherical microdomains can be packed into one of the three following cubic forms; SC, BCC, or FCC. According to Ohta and Kawasaki [49], a BCC arrangement is only slightly more favored than the FCC arrangement. In fact, many BCC lattice formations have been reported for AB-type diblock copolymers and blends of block copolymer/homopolymer systems [50–53].

More recently, Chevalier et al. have investigated the coalescence of latex particles through SANS and TEM [54]. The latex particles arranged to polyhedral cells in dispersion and these structures changed to colloidal crystals with a FCC lattice at the condensed solution (74 wt%). The core-shell polymer microspheres lead to the hierarchical structure transformation of the cubuc lattice in the film formation to minimize the thermodynamic energy. Thus, the transformation of a

C^* solution into a continuous film for core–shell polymer microspheres is very similar to those for latex particles, star-shaped polymers, and $(AB)_n$ star copolymers.

V. ARCHITECTURE OF SUPERSTRUCTURAL POLYMERIC MATERIALS

The core–shell polymer microspheres can be considered the smallest units in the microphase-separated structure. However, the properties of the microsphere in the solvent are governed by the shell chains in a good solvent. Therefore, the core–shell microspheres can be used as composite materials in blends that allow the introduction of spherical microdomains into a matrix.

This study investigated the morphologies of blends of well-ordered core–shell microspheres with lamellar AB-type block copolymers [55]. The P4VP core–PS shell microsphere (SV05-M; $\overline{M}_w = 1.5 \times 10^8$, diameter of P4VP core = 49.7 nm, $D_H = 180.5$ nm, $C^* = 4.4$ wt% in benzene) was used because of its tendency toward hexagonal packing in two dimensions and FCC packing in three dimensions. A lamellar PS–*block*–P4VP diblock copolymer (SV100; $\overline{M}_n = 2.4 \times 10^4$, PS = 68 mol%, domain size of P4VP phases = 10.8 nm) was synthesized by anionic polymerization.

To investigate the arrangement of the chains in the novel ordered structure, the weight fraction f of SV100 in the blend was varied. Figure 14a shows a TEM micrograph of the binary blend SV05-M/SV100 ($f = 0.5$) cast from a 1 wt% benzene solution (below C^*) and stained with OsO$_4$. The dark regions represent spheres surrounded by a P4VP layer similar to a cell wall. The average number of microspheres in one domain surrounded by the P4VP layer (K) was 1.08 for $f = 0.5$, and each microsphere was surrounded by a P4VP layer. Figure 14b schematically shows the chain arrangement in this novel morphology with a cell wall structure. The cell of the microsphere has a hexagonal surface, and the AB-type diblock copolymer forms a bilayer between the microspheres. From the schematic rearrangement of the chains, the f value for a completely ordered morphology was calculated as 0.46. This value agrees well with the value $f = 0.5$, the point where the blend showed a completely ordered morphology. It was suggested that a microphase separation with three phases could be obtained by blending the microsphere and block copolymer. In fact, such three-phase separated structure in which diblock copolymers were arranged as a honeycomblike bilayer around the ordered microspheres appeared by blending the P4VP core–PS shell microsphere with the PS–*block*–PI diblock copolymer [56].

By adding the PS homopolymer in the solid state, the superlattice structures of P4VP core–PS shell polymer microspheres were hierarchically changed from a disordered state to a FCC via a BCC. The transition of the lattice structures was

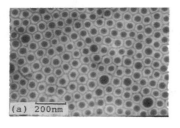

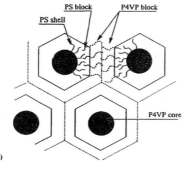

Figure 14 Microphase-separated structure of a binary blend of SV05-M/SV100 ($f =$ 0.5): (a) TEM micrograph; (b) schematic arrangement.

investigated from the viewpoint of the excluded volume of the microsphere, which was controlled by changing the molecular weights and blend ratios of PS homopolymers [57].

Figure 15 shows the phase diagram of the superlattice structures of the microspheres estimated by SAXS. It is found that the superlattice structures of the microsphere SV20-M ($\overline{M}_w = 3.2 \times 10^6$, arm number $n = 76$, $C^* = 10$ wt% in benzene) transited from the FCC to the BCC via the FCC/BCC mixture state by adding any type of PS. This hierarchical transition of the superlattice structure is peculiar to the core–shell polymer microspheres. When the molecular weight of PS increased, the critical PS concentration (ϕ_{PS}) between the FCC and the FCC/BCC mixture state decreased and the region of the FCC/BCC mixture state became narrower.

In order to give more detailed information of the lattice structure, the Bragg spacing d_1 values of the microsphere SV20-M are shown in Fig. 16. The d_1 value increases with the increase of ϕ_{PS} in the system. This tendency is clearer by decreasing of the molecular weight of PS. In other words, the d_1 was increased drastically when the molecular weight of the PS was low. This suggests that the microspheres are more widely expanded with the PS when the molecular weight of the PS is lower.

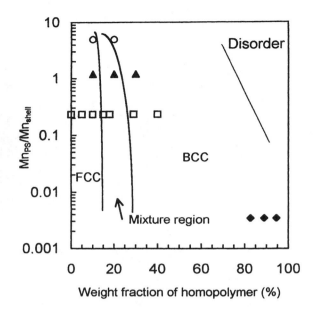

Figure 15 Phase diagram of the lattice structure of microsphere SV20-M in the blend films with homopolystyrene: (◆) monomer; (□) PS1 ($\overline{M}_n = 7.0 \times 10^3$); (▲) PS2 ($\overline{M}_n = 1.1 \times 10^4$); (○) PS3 ($\overline{M}_n = 1.5 \times 10^5$).

For the core–shell microspheres, because the cores were tightly cross-linked, the cores were not expanded with the homopolymers. Thus, the expansion of the microsphere is due to the expansion of the shell chains in the system. In order to quantitatively estimate the expansion of the shell chains by adding the PS, the shell chain length of the core–shell polymer microsphere, l_A, was simulated for the SV20-M system by using computer calculation techniques. The l_A was calculated by minimizing the total free energy of the micelle formation model of the AB-type diblock copolymer in a homopolymer as proposed by Whitomore and Noolandi [58]. The greatest difference between this case and the model of Whitomore and Noolandi was that the core of microsphere was chemically cross-linked. Therefore, it could be assumed that the homopolymer did not penetrate into the core and that the block copolymer chain did not exist in the system.

Figure 17 shows the l_A of the SV20-M with addition of the PS. When the molecular weight of the PS was less than 36,000, the l_A increased drastically when a small amount of the PS was added. Then, the l_A saturated when the ϕ_{PS} was larger than 10 wt%. The saturated values of the l_A were increased by the decrease of the molecular weight of the PS. By increasing the molecular weight of the PS, the expansion of shell chains and the penetration of the PS into the microsphere decreased. The wide expansion of the shell chains hierarchically optimized the

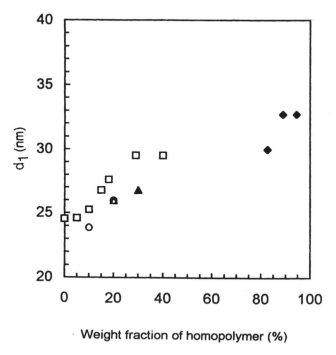

Figure 16   The d_1 distance of the lattice structure of microsphere SV20-M in the blend films with homopolystyrene: (◆) monomer; (□) PS1; (▲) PS2; (○) PS3.

superlattice structure of the microspheres. This understanding was supported by the d_1 values. Consequently, the hierarchical transition of the superlattice structure of the core–shell polymer microsphere was found to be a result of the wide, expanded shell chains in the blend system. Therefore, the FCC/BCC mixture region widened for the more expanded shell chains.

When the molecular weight of PS is larger than that of the shell chain, the shell chains will not be expanded by the PS. In fact, when PS3 with a molecular weight of 1.5×10^5 was added, the l_A was constant at any polymer concentration. This indicates that the shell chains of the SV20-M did not expand and the PS3 was excluded from the SV20-M. As a result, the superlattice structure was easily broken by the addition of PS3. However, the superlattice structure could not be transited by adding a small amount of the PS. This can be explained as follows. The PS can stabilize the superlattice formation of the core–shell microspheres in the blend system, not as a compatibilizer but as packing for the dead space.

On the other hand, Ishizu et al. reported the architecture of polymeric superstructure films formed by locking the cubic lattices of core–shell microspheres in

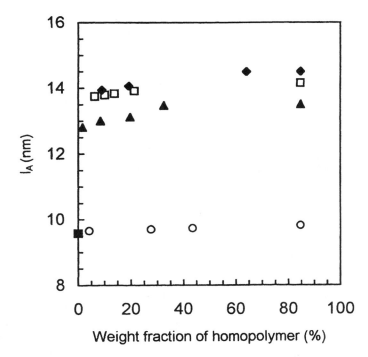

Figure 17 The calculated shell length of microsphere SV20-M in the blend films with homopolystyrene: (◆) monomer; (□) PS1; (▲) PS2; (○) PS3.

a poly(methyl methacrylate) (PMMA) matrix [59]. The possibility of structural ordering of a poly(α-methylstyrene) (PMS) core–PS shell microsphere (MSV1-M) was investigated in a methyl methacrylate (MMA) monomer by means of SAXS measurement, varying the microsphere concentration with reference to the C^* value in benzene. The characteristics of microsphere MSV1-M are listed in Table 4. The overlap threshold in polymer solutions of the MSV1-M microsphere in benzene was calculated to be 8.73 wt% from the D_H and the total molecular weight of the microsphere. Below a 7 wt% MMA monomer solution, no regular scattering peaks appeared due to disordering. At a 12 wt% microsphere concentration, the first five peaks appeared close at the relative q positions of $1 : \sqrt{2} : \sqrt{3} : 2 : \sqrt{5}$. These values correspond to a packing pattern of (110), (200), (211), and (310) planes in a BCC structure.

At first, the photopolymerization of a 12 wt% MMA solution of the MSV1-M microsphere containing 2,2′-azobisisobutyronitrile (AIBN) and benzoin methyl ether was carried out under ultraviolet (UV) irradiation (Method 1). Figure 18a

Table 4   Characteristics of Core–Shell Microsphere MSV1-M

Code	$R_c{}^a$ (nm)	$10^{-6}\overline{M}_w{}^b$ (g mol^{-1})	N^c	$D_H{}^d$ (nm)	C^{*e} (wt%)
MSV1-M	8.1	6.94	116	50.9	8.73

[a]Radius of P4VP cores determined by TEM micrographs.
[b]Total molecular weight of microsphere determined by SLS.
[c]Arm number (aggregation number of block copolymer), estimated from $\overline{M}_w$ of the microsphere and $\overline{M}_n$ of the arm block.
[d]Determined by DLS in benzene.
[e]Overlap threshold in benzene solution calculated from the D_H values in benzene.
Source: Ref. 59.

shows the TEM micrograph of the film specimen after locking treatment. Dark portions correspond to P4VP cores selectively stained with OsO$_4$. On the other hand, white portions correspond to PMMA or PMS shell parts. Large white domains appear at places in this micrograph independently of gathering parts of core–shell microspheres. It is reasonable to judge that these microdomains correspond to PMMA phases. This polymerization system seemed to lead polymerization-induced phase separation. Moreover, PMS chains on shell part were degraded by UV irradiation.

Next, free-radical polymerization of 12 wt% MMA solution of MSV1-M microsphere containing 2,2′-azobis(4-methoxy-2,4-dimethylvaleronitrile) (V-70) and CBr$_4$ was carried out in a sealed sample bottle (Method 2). The addition of CBr$_4$ as a chain-transfer reagent can be expected to shorten the propagation chain length of MMA with the intention of preventing the polymerization-induced macrophase separation. A transparent film was obtained by such a procedure. Figure 18b shows the TEM micrograph of the film specimen after locking treatment. The MMA monomers form a continuous matrix phase around the lattice of core–shell microsphere by polymerization. However, the microspheres are not maintained perfectly with a BCC packing, but they are packed with irregular arrangement at places.

Subsequently, free-radical polymerization of 12 wt% MMA solution of the MSV1-M microsphere containing V-70, CBr$_4$, and ethylene glycol dimethyl-methacrylate (EGDM) was carried out under the same conditions as Method 2 (Method 3). Figure 18c shows the TEM micrograph of the film specimen after locking treatment (MSV1-M-LA) by means of Method 3. It is indicated from this texture that P4VP cores are packed hexagonally in a two-dimensional aspect. Figure 19 shows the SAXS intensity profile of the solid film MSV1-M-LA. In this profile, the first three peaks appear at the relative q positions of $1 : \sqrt{3} : \sqrt{5}$, as shown in parentheses. This packing pattern is identical to the interplanar spacings

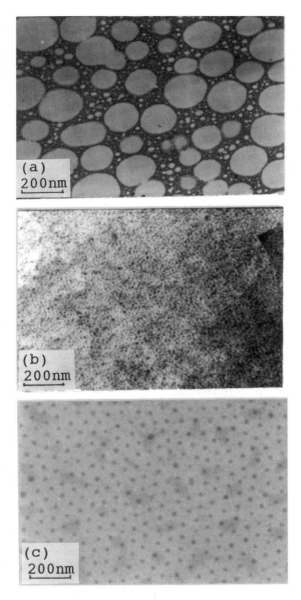

Figure 18 TEM micrographs of the film specimens after locking treatments: (a) locking by Method 1; (b) locking by Method 2; (c) locking by Method 3. (From Ref. 59.)

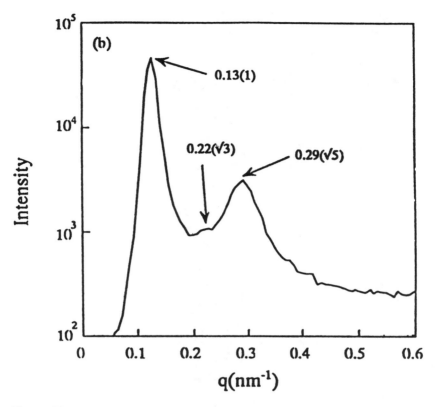

Figure 19 SAXS intensity profile for MSV1-M-LA film. (From Ref. 59.)

of a BCC structure. It is therefore concluded that the MSV1-M microspheres were locked with a BCC lattice in the PMMA network by free-radical polymerization. The addition of chain-transfer and cross-linking reagents was very effective in preventing the polymerization-induced macrophase separation of PMMA. The first peak ($q = 0.13$ nm^{-1}, $2\theta = 0, 18°$) of the MSV1-M-LA was identical to the q positions of an MMA solution of the microsphere (MSV1-M-LB; 12 wt% microsphere concentration). This means that the volume shrinkage of PMMA has never occurred during free-radical polymerization. Judging from the densities of MMA and PMMA, the volume shrinkage of PMMA is estimated to be 21.2 vol% after locking treatment. Then, the domain spacing of the nearest-neighbor microspheres should decrease by 8% in length compared to that of MSV1-M-LB.

In the differential scanning calorimetry (DSC) trace of MSV1-M-LA, three

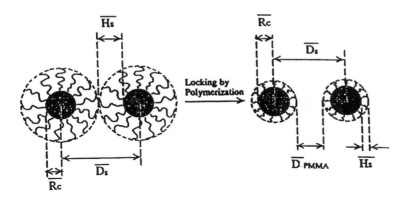

Figure 20 Illustration of space arrangement for the nearest-neighbor core–shell polymer microspheres before and after locking. (From Ref. 59.)

Table 5 Domain Spacings of Nearest-Neighbor Microspheres Before and After Locking

Code	State	D_s (nm)	$R_c{}^a$ (nm)	H_s (nm)	D_{PMMA} (nm)
MSV1-M-LB	MMA solution	60.0	8.5	21.5	—
MSV1-M-LA	Solid film	60.0	8.5	4.7	33.6

[a]Determined by SAXS data on the FCC lattice of MSV1-M film.
Source: Ref. 59.

endothermic peaks appeared at temperatures of 78°C, 105°C, and 156°C. Similar results were obtained in the DSC trace of binary blend of MSV1-M/PMMA. Then, the polymeric superstructure prepared by locking was composed of three phase-separated microdomains (i.e., the P4VP cross-linked core, the PMS shell, and the PMMA matrix). One can consider the space arrangement of the cubic lattice in polymeric superstructures using Eqs. (1) and (2). Figure 20 illustrates the space arrangement for the nearest-neighbor core–shell microspheres before and after locking; H_s, and D_{PMMA} are the thickness of shell part and the distance of PMMA matrix between nearest-neighbor microspheres, respectively. The stretched PMS arm chains stabilized the P4VP cross-linked cores sterically in the MMA solution. After the locking of the superlattice of microspheres, the PMS arm chains (shell thickness $H_s = 4.7$ nm) cover the surface of the P4VP cross-linked core ($R_c = 8.5$ nm, employed from the SAXS data) due to the immiscibility between PMS and PMMA chains, where the value of H_s was evaluated from the SAXS data (FCC lattice) of microsphere MSV1-M in the bulk film. Table 5 lists the domain spacings of nearest-neighbor microspheres before and after locking. The observed value of D_{PMMA} is 33.6 nm. Considering the effect of the volume shrinkage of the PMMA matrix, the D_{PMMA} value is speculated to be 30.9 nm.

The interplaner spacings of superstructures can be controlled by changing the monomer concentration and type of cubic lattices. These systems will have numerous applications in technology, such as optical materials.

VI. ORDER–ORDER TRANSITION OF BLOCK COPOLYMER MELTS

The theories concerning the microphase separation of block copolymers expected that morphologies are changed with the segregation power [4–6]. Because the segregation power is a function of temperature, the transition between the different kinds of morphology may occur by the changing temperature. Sakurai et al. found the thermally induced morphology transition from cylindrical to lamellar microdomains for a PS–*block*–polybutadiene (PB)–*block*–PS (SBS) triblock copolymer having a 0.56 weight fraction of PS blocks determined by SAXS and TEM measurements [60]. PB cylinders with hexagonal close packing were formed in a PS matrix when the SBS was cast from a methyl ethyl ketone solution. Such PB cylinders were transformed into lamellae on annealing at 150°C. This is a transition from thermodynamically quasi-stable to stable morphologies. On the basis of SAXS and TEM, the transition turned out to occur via coalescence of cylinders without their translational movements. It is proposed that an undulation of the interface induced by the instability of the interface might play an important role for the coalescence of the cylinders.

Thermoreversible morphological transition of a PS–*block*–PI (SI) diblock copolymer ($\overline{M}_n = 8.2 \times 10^4$, volume fraction of PS = 0.15) was also studied by SAXS and TEM [61]. The cylindrical and spherical microdomains of PS which are embedded in the PI matrix were thermoreversibly observed at 150°C and 200°C, respectively. Using experimentally determined interaction parameters χ, they compared SAXS and TEM results with the theory in the weak segregation limit presented by Leibler, which predicts the thermoreversible morphological transition between spheres and cylinders. Consequently, $(\chi N)_t < (\chi N)_{200} < (\chi N)_s < (\chi N)_l < (\chi N)_{150}$ was obtained, where $(\chi N)_t$, $(\chi N)_s$, and $(\chi N)_l$ denote the theoretical values of the product χN at the microphase-separation transition (MST), at the spinodal point of the MST, and at the morphological transition between spheres and cylinders, respectively, and N is the degree of polymerization of the copolymer. Further studies have been carried out in order to reveal a mechanism for the transition [62]. For this purpose, SAXS experiments were carried out for the solution of SI diblock copolymer in dioctyl phthalate. It was found that the angular position of the first-order peak in the SAXS profile did not change upon transition, indicating the invariance of the Bragg spacing, d_{110} (BCC) = d_{100}(hex). Based on this result, a possible model was presented for the mechanism of the morphological transition. The model consists of the coalescence of the spheres from the nearest-neighbor positions to form a cylinder whose axis is normal to a (111) plane of the BCC lattice of the original spheres.

Hajduk et al. recently reported the observation of a reversible thermotropic transition between the lamellar and cylindrical morphologies in a PS–*block*–poly(ethene-*co*-butene) diblock copolymer by SAXS and TEM [63]. Results showed that the transformation from the lamellar to the cylindrical morphology proceeded in two steps. Initially, fluctuations along the interface joining the components of the copolymer grow in amplitude until the lamella transforms into a sheet of evenly spaced cylinders; cylinders which form in adjacent lamellas are usually in poor register with one another. This intermediate structure subsequently anneals into the hexagonal packing of cylinders characteristic of the cylindrical morphology.

VII. CONCLUSION

Highly branched polymers such as star-shaped polymers, $(AB)_n$ star copolymers, and core–shell polymer microspheres lead to the hierarchical structure transformation of cubic lattices in the film formation to minimize the thermodynamic energy; that is to say, these branched polymers form a lattice with a BCC structure near C^*. In the bulk of the film; this structure changes to a FCC lattice. Thus, structural transformation can be explained by a thermal blob model proposed by Daoud and Cotton. It is possible to perform the architecture of polymeric superstructure films formed by locking the cubic lattices of core–shell microspheres in a PMMA matrix. The interplaner spacings of superstructures can be controlled by changing the MMA monomer concentration and the type of cubic lattices. This technique is one of the best methods of three phase-separated microdomains. These systems will have numerous applications in technology, such as optical and electronic devices.

REFERENCES

1. BH Zimm, WH Stockmayer. J Chem Phys 17:1301–1314, 1949.
2. M Daoud, JP Cotton. J Phys (Les Ulis Fr) 43:531–538, 1982.
3. TA Witten, PA Pincus. ME Cates, Europhys Lett 2:137–140, 1986.
4. L Leibler. Macromolecules 13:1602–1617, 1980.
5. T Ohta, K Kawasaki. Macromolecules 19:2621–2632, 1986.
6. GH Fredrickson, E Helfand. J Chem Phys 87:697–705, 1987.

7. MO de la Cruz, IC Sanchez. Macromolecules 19:2501–2508, 1986.
8. H Hasegawa, H Tanaka, K Yamasaki, T Hashimoto. Macromolecules 20:1651–1662, 1987.
9. EL Thomas, DB Alward, DL Kinning, DL Martin, DL Handlin, Jr, LJ Fetters. Macromolecules 19:2197–2202, 1986.
10. FS Bates. Science 251:898–905, 1991.
11. A Prochazka, MK Baloch, Z Tuzar. Makromol Chem. 180:2521–2523, 1979.
12. Z Tuzar, B Bednar, C Konak, M Kubin, S Svobodova, K Prochazka. Makromol Chem 183:399–408, 1982.
13. B Bednar, J Devaty, B Koupalova, J Kralicek. Polymer 25:1178–1184, 1984.
14. MH Park, R Saito, K Ishizu, T Fukutomi. Polym Commun 29:230–231, 1988.
15. MH Park, K Ishizu, T Fukutomi. Polymer 30:202–206, 1989.
16. R Saito, K Ishizu, T Nose, T Fukutomi. J Polym Sci, Polym Chem Ed 28:1793–1805, 1990.
17. R Saito, K Ishizu, T Fukutomi. Polymer 31:679–683, 1990.
18. R Saito, K Ishizu, T Fukutomi. Polymer 32:531–536, 1991.
19. R Saito, K Ishizu, T Fukutomi. Polymer 32:2258–2262, 1991.
20. R Saito, K Ishizu, T Fukutomi. J Appl Polym Sci 43:1103–1109, 1991.
21. R Saito, K Ishizu, T Fukutomi. Polymer 33:1712–1716, 1992.
22. K Ishizu, T Fukutomi. J Polym Sci, Polym Lett Ed 26:281–286, 1988.
23. K Ishizu. Polym Commun 30:209–211, 1989.
24. K Ishizu. Polymer 30:793–798, 1989.
25. L Willner, O Jucknischeke, D Richter, B Farago, LJ Fetters, JS Huang. Europhys Lett 19:297–303, 1992.
26. L Willner, O Jucknischeke, D Richter, J Roovers, L-L Zhou, PM Toporowski, LJ Fetters, JS Huang, MY Lin, N Hadjichristidis. Macromolecules 27:3821–3829, 1994.
27. K Ishizu, T Ono, S Uchida. Polym–Plast Technol Eng 36:461–471, 1997.
28. L-L Zhou, J Roovers. Macromolecules 26:963–968, 1993.
29. J Roovers, L-L Zhou, PM Toporowski, M van der Zwan, H Iatrou, N Hadjichritidis. Macromolecules 26:4324–4331, 1993.
30. LK Bi, LJ Fetters. Macromolecules 8:90–92, 1975.
31. LK Bi, LJ Fetters. Macromolecules 9:732–742, 1976.
32. DB Alward, DJ Kinning, EL Thomas, LJ Fetters. Macromolecules 19:215–224, 1986.
33. RN Young, LJ Fetters. Macromolecules 11:899–904, 1978.
34. K Ishizu, K Sunahara. Polymer 36:4155–4157, 1995.
35. K Ishizu, T Ono, S Uchida. Macromol Chem Phys 198:3255–3265, 1997.
36. RL Lascanec, M Muthukumar. Macromolecules 23:2280–2288, 1990.
37. K Ishizu, T Ono, S Uchida. J Colloid Interf Sci 192:189–193, 1997.
38. K Ishizu, S Uchida. J Colloid Interf Sci 175:293–296, 1995.
39. S Uchida, A Ichimura, K Ishizu. J Colloid Interf Sci 203:153–156, 1998.
40. K Ishizu, S Yukimasa, R Saito. J Polym Sci, Polym Chem Ed 31:3073–3080, 1993.
41. K Ishizu, S Yukimasa. Polymer 34:3753–3756, 1993.
42. K Ishizu, S Uchida. Polymer 35:4712–4716, 1994.
43. K Ishizu, A Onen. J Polym Sci, Polym Chem Ed 27:3721–3731, 1989.
44. K Ishizu, R Saito. Polym–Plast Technol Eng 31:607–633, 1992.
45. K Ishizu. J Colloid Interf Sci 156:299–304, 1993.

46. R Saito, H Kotsubo, K Ishizu. Polymer 35:1580–1585, 1994.
47. K Ishizu, M Sugita, H Kotsubo, R Saito. J Colloid Interf Sci 169:456–461, 1995.
48. K Ishizu. In JC Salamone, ed. Polymeric Materials Encyclopedia. Boca Raton, FL: CRC Press, 1996, vol 6, pp 4372–4375.
49. T Ohta, K Kawasaki. Macromolecules 19:2621–2632, 1986.
50. EL Thomas, DJ Kinning, DB Alward, CS Henkee. Macromolecules 20:2934–2939, 1987.
51. T Hashimoto, M Fujiura, H Kawai. Macromolecules 13:1660–1669, 1980.
52. RJ Roe, M Kishkis, VC Chang. Macromolecules 14:1091–1103, 1981.
53. RW Richards, JL Thomas. Polymer 22:581–589, 1981.
54. Y Chevalier, C Pichot, C Graillat, M Joanicot, K Wong, J Maqnet, P Lindner, B Cabane. Colloid Polym Sci 270:806–821, 1992.
55. R Saito, H Kotsubo, K Ishizu. Polymer 35:1580–1585, 1994.
56. K Ishizu. Polymer 38:2813–2817, 1997.
57. R Saito, M Sugita, K Ishizu. J Colloid Interf Sci 187:515–519, 1997.
58. MD Whitomore, J Noolandi. Macromolecules 18:657–665, 1985.
59. K Ishizu, T Ikemoto, A Ichimura. Polymer 39:449–454, 1998.
60. S Sakurai, T Monii, K Taie, M Shibayama, S Nomura, T Hashimoto. Macromolecules 26:5796–5802, 1993.
61. S Sakurai, H Kawada, T Hashimoto, LJ Fetters. Macromolecules 26:5796–5802, 1993.
62. S Sakurai, T Hashimoto, LJ Fetters. Macromolecules 29:740–749, 1996.
63. DA Hajduk, SM Gruner, P Rangarajan, RA Register, LJ Fetters, C Honeker, RJ Albalak, EL Thomas. Macromolecules 27:490–501, 1994.

2

Influence of Surface Structure on Polymer Surface Behavior

Soo-Bok Lee
Korea Research Institute of Chemical Technology, Taejon, Korea

I. INTRODUCTION

The surface layer plays an important role in the practical application of polymeric materials. The surface properties of polymers have recently received great attention because of its generic importance in diverse technological areas such as adhesion, biomedical materials, electronic materials, and coatings. Many desirable properties, such as compatibility, wettability, adhesiveness, permeability, solid-state morphology, and so forth can be influenced by polymer surface structure. Polymer surface properties are closely related to the functionality of materials (e.g., adhesion, fracture, environmental compatibility, biocompatibility, and electric properties); for example, lowering the surface tension of a film is desirable for formulating nonwettable surfaces. One of the most popular and successful strategies for lowering the surface tension of a film is the incorporation of fluorine into the polymer molecule comprising the coating. The fluorine can be incorporated into the main polymer chain [1–13]. Much research has been done to elucidate the influence of surface structure on their surface behavior for various polymeric systems. The polymeric systems, which are intensively investigated and substantially used in industry, can be classified into four groups as functional polymers, polymer blends, block copolymers, and graft copolymers.

The area of functional polymers is one of the most active in polymer sciences. Many desirable properties can be influenced by incorporating specific functional groups to polymers. These functional groups may be located anywhere in the polymer molecule. The functionalization of linear hydrocarbon polymers makes it possible to enhance the physical, mechanical, and rheological properties of the polymers. Functional homopolymers and copolymers have found wide applications for coating a diverse array of materials.

Polymer blends have become a very important subject for scientific investigation in recent years because of their growing commercial use. Blends are an inexpensive route to the modification of polymer properties. Examples of the properties of interest that may be altered upon blending are surface properties, impact resistance, fatigue behavior, heat distortion, and improved processability.

Multiphase polymers such as block and graft copolymers are known to exhibit unique surface properties due to their microphase separation at surface. The bulk and surface structures of multiphase polymers are usually different from each other. The significant surface accumulation of one polymer component at the surface is frequently noticed, producing a quite different surface structure than expected from that in bulk.

A multitude of experiments has been carried out to investigate polymer surface dynamics. In this chapter, the recent developments reported in the literature for the elucidation of the influence of polymer surface structure on polymer surface behavior are briefly reviewed and summarized by classifying the polymeric systems as functional polymers, polymer blends, block copolymers, and graft copolymers.

II. FUNCTIONAL POLYMERS

In recent years, the surface characterization of functional polymers having side chains of special functional groups has been performed extensively. The surface properties of polymers are significantly affected by the side chains of functional groups. To design polymer materials having specific surface properties, the effect of side chains should be clearly elucidated.

Rosilio et al. [14] assessed the miscibility of dimyristoylphosphatidylcholine (DMPC) and poly(ethylene oxide) (PEO) lipids through surface pressure and surface potential measurements at the water–air interface. The miscibility of PEO lipids with DMPC is essentially dependent on PEO chain length. The increase in the chain length of a hydrophilic PEO polymer makes the establishment of close contacts with DMPC molecules difficult, presumably because of the inclined position of the polymer chain toward the water–air interface. The differences in mixing behavior with DMPC of PEO lipids may be attributed both to the differences in the orientation of hydrophobic chains between PEO lipids, which depends on polymer chain length, and to the structuring of interfacial water, which is influenced by the volume of the random polymer coil.

Cai and Litt characterized a series of poly(N-acyl or N-aroyl ethylenimines) containing various pendent functional groups [15–18]. A series of decenyl(D)/heptyl(H) random copolymers (DH) were made by copolymerizing decenyl and heptyl oxazolines. In the DH copolymers with less than 30 mol% of D, the allyl pendent groups can rotate freely and interfere strongly with the crystallization of the polymers. The melting point and heat of fusion drop rapidly with the increase of D in this region. At the concentration of D higher than 30 mol%, more and more allyl groups are incorporated into the crystalline domain. The melting point and heat of fusion decrease more slowly with the increase of D than in the region below 30 mol% of D. Because the cross section of the allyl group is close to that of polymethylene, the average long spacing increases linearly in the whole composition range with the increase of D from 20.2 Å for the H homopolymer to 25.5 Å for the D homopolymer. Both D and H homopolymers have uniform crystalline domains, whereas the DH copolymers with 20–60 mol% of D have disordered crystalline structures because of the freely rotated allyl groups. The DH–OH copolymer molecules with high –OH concentrations in their side chain interact with each other strongly. Their alkyl side chains are not fully extended in the crystalline domain as in the DH copolymers. Most of the –OH groups in the copolymers are buried and the polymer surface is mainly composed of the methyl or methylene group. The –OH groups can bend back and form hydrogen-bonding with the carbonyl groups in the polymer backbone. Although the DH–OH polymers basically show a hydrocarbon surface in a normal environment, the polymer surfaces can reorganize when they are in contact with a polar material. The –OH groups are easily accessible and can be pulled out by a polar agent such as water or tape adhesive. The epoxy group of poly(N-acylethylenimines) is much less polar than a hydroxyl group. With 20–60 mol% of epoxy, the copolymers crystallize two dimensionally. However, the imidazole groups in the DH–imidazole polymers are very polar. They have a strong tendency to bend backward to meet the amide group in the polymer backbone. This is very similar to the DH–OH copolymers. In the DH–OH copolymers, the size of the –OH group is close to that of a methylene group. The side alkyl chains can pack very well and crystalline even with the –OH bending backward to the polymer backbone. However, the imidazole groups in the DH–imidazole polymers are much larger than the –OH groups. The strong interaction of imidazole groups and amide groups destroys the alkyl side chain packing and makes these polymers very difficult to crystallize.

Comb-shaped polymers occupy a special position among a large number of polymers. They differ from ordinary branched polymers because they have many side chains. Comb-shaped polymers with fluorocarbon segments in the side chains have attracted particular attention because of the very low surface free energy [5–12]. The aggregation states of a series of comb-shaped polymers with various types of main chain polymer having side chains of the heptadecafluorodecyl (C_8F_{17}) group [poly(perfluoroalkylfumarate) (PFAF), poly(perfluoroacrylate) (PFAA), and poly(perfluoromethacrylate) (PFMA)] were investigated by Volkov et al. [19]. PFAF was found to be an amorphous polymer, but PFAA and PFMA manifested a mesomorphic state at room temperature. Main-chain flexibility increases in the order PFAF, PFMA, and PFAA. Increasing the main-chain flexibility increases the probability of an ordered structure formation.

In the fluorosilicone case, the most favorable situation for the realization of polymers of very low surface energy can be obtained, as the most flexible polymer backbone, siloxane, $-(SiO)_n-$, is coupled with the groups having the lowest intermolecular interactions, fluorocarbons of structure $-(CF_2)_nCF_3$ [20]. At least two $-CH_2-$ groups are needed between the $-Si-$ and $-CF_2-$ groups in order to prevent the facile α- or β-elimination of the Si–F group as a result of the high S—F bond strength and the great electronegativity and strong inductive effect of fluorine. The length of the pendent fluorocarbon group for low surface energy is also important and should be sufficiently long. Surface energies decrease as this chain is lengthened.

Park et al. [21] performed surface characterization of PFMA [$CF_3(CF_2)_5$ $CH_2CH_2OCOCCH_3=CH_2$] by contact-angle measurements and electron spectroscopy for chemical analysis (ESCA). The critical surface tension of PFMA used was found to be about 9.0 m N/m. The outermost layer of the PFMA surface below the sampling depth of 4.7 Å from the polymer–air interface is composed mainly of perfluoroalkyl groups, and in the vicinity of 9.3 Å, methyl groups as well as perfluoroalkyl groups dominate the surface properties. The PFMA surface can be divided into three main sublayers consisting of the perfluoroalkyl layer, the perfluoroalkyl/methyl mixture layer, and the backbone layer. The thickness of the outermost layer that mostly affects the surface properties of PFMA is likely to be less than 20 Å. The outermost layer is composed of mainly perfluoroalkyl groups, and some ester groups existing in the outermost layer seem to be exposed to the air side. The structure of the outermost layer of PFMA surface is believed to be similar to Fig. 1b rather than Fig. 1a, which shows an irregular arrangement of perfluoroalkyl groups, most of which are not perpendicularly oriented to the air side. Annealing can arrange perfluoroalkyl groups to be more perpendicularly oriented to the polymer–air interface.

Thomas et al. [22] investigated a series of novel polymers prepared by free-radical solution polymerization of a variety of hydrocarbon monomers with the perfluoroalkylethyl methacrylate monomer ($H_2C=C(CH_3)OCO(CH_2)_2(CF_2)_nF$, $n = 7.7$). Through a judicious choice of reaction conditions, some control of the polymer architecture was exhibited. The resultant perfluoroalkylethyl methacrylate-containing acrylic polymers were shown to be quite surface active in solution and, even more importantly, in the solid state. The polymers were formulated into films and applied to a variety of surfaces. With incorporation of 1.5 wt% of perfluoroalkylethyl methacrylate into an acrylic polymer, water and oil-repellent surfaces were created. With the low levels of perfluoroalkylethyl methacrylate incorporated into the polymers, bulk properties are unaffected. In addition, these acrylic polymers can be formulated into coatings which offer unique repellency properties. The placement of fluorine along the polymer chain during copolymerization

(a) Air

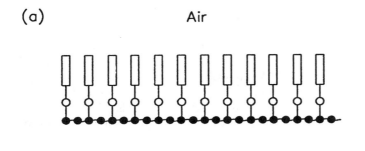

(b) Air

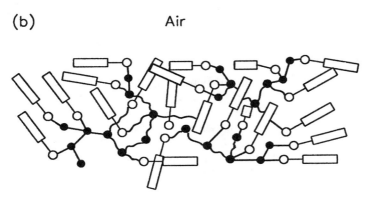

Figure 1 Arrangement of perfluoroalkyl groups of the PFMA surface at the polymer–air interface. (From Ref. 21.)

is critical in determining the propensity for fluorine enrichment at the air–polymer interface. The synthetic methodology employed permitted the control of polymer molecular weight and dispersity. This will have a substantial effect on the ability of the polymer chains to segregate to the air–polymer interface. The molecular weight will have a great negative impact on the ability of the perfluoroalkylethyl methacrylate-modified polymer chains to diffuse to the air–polymer interface. Polymer compatibility can play a substantial role in determining the distribution of material (specifically fluorine) in the film. Not only will the different types of polymer chain synthesized have a different chemical potential but they will also have different polymer–polymer interaction parameters which will determine the extent of their compatibility. Both molecular weight and compatibility become important in determining the distribution of material at the interface.

Although polymer surface dynamic behavior has been investigated extensively, a physical model would not only be useful in correlating data but also lead to a more thorough understanding of the interaction among polymer surfaces, interfaces, and the environmental factors before materials with specific surface properties are designed. The surface dynamic behavior of polymers cannot be predicted from bulk measurements, so a physical model is needed for describing the polymer surface dynamic behavior. Liu et al. [23] developed a hydrophobic/hydrophilic adsorption lattice (HHAL) model to characterize the polymer surface dynamic behavior by connecting the classical Young's equation and potential energy state with wetting kinetics and isothermal adsorption theories. Theoretical predictions of contact angles as a function of time or polymer surface functionality were compared with experimental data chosen from the literature. Fitting experimental data to the model indicates that the polymer surface dynamic behavior can be characterized by the HHAL model.

III. POLYMER BLENDS

In multicomponent polymer blends or copolymers, the homopolymer constituent or block having the lower surface energy will migrate to the surface of the material. The surface migration of the lower-surface-energy polymer is influenced by various important parameters, such as polymer functionality, polymer structure, molecular weight, miscibility, temperature, and casting solvent. The desirable low-surface-energy polymers should have a flexible linear backbone, sufficient pendent groups based on aliphatic hydrocarbon or fluorocarbon, linking groups between the backbone and pendent groups that introduce no additional adverse factors, and combinations of these entities that are adequately stable and have no undesirable interactions.

Numerous studies have indicated that there are differences between the composition of the surface and the bulk of a polymer blend. Surface tension measurements have suggested that the surfaces of various poly(dimethylsiloxane) (PDMS) solutions are enriched in the concentration of the lower-surface-energy component [24]. Similar conclusions were obtained as the results of surface tension measurements on miscible blends of low-molecular-weight poly(ethylene oxide) (PEO) and poly(propylene oxide) (PPO) [25]. In that system, the apparent increase in enrichment with molecular weight is attributed to the increased immiscibility of the components. This surface-enrichment phenomena has also been observed at the solid surface of immiscible polystyrene (PS)/PEO blends.

Pan and Prest [26] investigated the solid surface of an amorphous miscible polymer blend using PS/poly(vinyl methyl ether) (PVME) blends. A prominent example of amorphous polymer blends is the binary mixture of PS and PVME. The binary PS/PVME mixture is miscible in all proportions when it is cast from toluene, but it is phase separated when cast from trichloroethylene. The miscible binary mixture can be transformed into a phase-separated mixture consisting of PS-rich and PVME-rich phases by heating above its lower critical solution temperature

(LCST). It was found that the structure and properties of the free PS/PVME surface are strongly dependent on the bulk composition, the bulk-phase behavior, the degree of mixing, and the casting solvent of the PS/PVME blend. The free solid surface of all PS/PVME blends is significantly enriched in PVME, the component with the lower critical surface tension. The relative enrichment of the surface was found to be nearly independent of the bulk composition from 25 to 75 wt% of PVME. Near the air–polymer interface of a miscible 50/50 PS/PVME blend, the composition of PVME was found to be about 95 wt%. The surface enrichment effect is enhanced in phase-separated blends produced by either solvent casting or heating above the LCST. The extent of PVME enrichment in differently prepared blends follows the sequence: thermally induced phase-separated blend > solvent-cast phase-separated blend > miscible blend. The variation in the surface composition between the solvent-cast and thermally induced phase-separated blends reflects the difference in the degree of mixing in the bulk of the blends.

Bhatia et al. [27] extended the study of Pan and Prest [26] and reported the effects of blend composition and constituent molecular weights on surface enrichment in the miscible PS/PVME blends. Surface tension data and X-ray photoelectron spectroscopy (XPS) analysis demonstrated substantial surface enrichment of PVME for all of the blends examined and under all tested conditions. PVME exhibits strong surfactant behavior, causing a dramatic reduction in surface tension when even small quantities are added. A direct correspondence between the driving force for surface segregation (i.e., the surface energy difference between PS and PVME) and the resultant surface composition was shown. An empirical $M^{-2/3}_n$ (M_n: number of average molecular weight) power-law dependence of surface composition was observed, reflecting clearly the empirical $M^{-2/3}_n$ dependence of the surface energy difference. PVME content of the outermost surface layer increases with the bulk PVME content and attains values as high as 98% for certain blends.

Schmidt et al. [28] analyzed quantitatively the surface composition for solvent-cast films of poly(methyl methacrylate) (PMMA) and poly(vinyl chloride) (PVC) using angle-dependent ESCA and Attenuated total reflectance–Fourier transform infrared (ATR–FTIR). PMMA/PVC blends over the entire composition range were cast from tetrahydrofuran (THF) and methyl ethyl keton (MEK). Surface enrichment of PMMA was observed for all compositions of blends cast from THF, whereas blends cast from MEK exhibited surface compositions that were within error limits equivalent to the bulk compositions of the blends. Casting solvent systematically influences the surface composition as well as the bulk compatibility, showing that microphase segregation will be exaggerated at the air–film interface.

Schmidt et al. [29] also studied the surface composition for blends of bisphenol A polycarbonate (BPAC) and PDMS. The results show surface enrichment of the lower surface energy component of PDMS. All the blends of less than

11% bulk PDMS reach a surface concentration of approximately 85% PDMS over the topmost 50 Å. For the blends of BPAC/PDMS, the surface segregation of the PDMS component is much greater than that for the block copolymers of BPAC/PDMS.

The surface properties of PFMA/poly(*n*-alkyl methacrylate)s (PAMAs) blends were investigated by Park et al. [5]. The PAMAs used were PMMA, poly(ethyl methacrylate) (PEMA), and poly(butyl methacrylate) (PBMA). The addition of an extremely small amount of PFMA to PAMA as low as 0.01 wt% was found to be very effective in reducing the surface free energy of PAMA significantly. PFMA may preferentially concentrate at the outermost layer in PFMA/PAMA mixtures. The length of the side chain of PAMA affects the surface free energy for PFMA/PAMA blends by hindering the arrangement of perfluoroalkyl groups of PFMA to the air side. As the length of the side chain of PAMA increases, the arrangement of perfluoroalkyl groups to the outermost layer is more significantly interfered with due to the steric effect.

IV. BLOCK COPOLYMERS

Because block copolymers usually consist of more than two components with different surface free energies, a microphase-separation structure is generally formed at surfaces of block copolymers. Due to the microphase-separation morphology at the surface, block copolymers exhibit unique surface properties. For example, the incorporation of microphase-separated PDMS domains into a PMMA matrix allows the system to exhibit many of the desirable properties of both components. The PMMA matrix is known to possess very good optical clarity, good UV stability, high electrical resistivity, and hydrolytic stability. Some of the desirable properties of PDMS include a very low glass transition temperature, biocompatibility, low surface energy, high oxygen permeability, and resistance toward degradation by atomic oxygen and oxygen plasmas. Therefore, a heterophase material consisting of microphase-separated PDMS domains embedded in a PMMA matrix should illustrate the desirable properties of both components.

The relationship between surface and bulk properties of segmented polyurethanes with various polyol soft segments (SPUs) was investigated [30,31]. The polyols used in these studies were PEO, poly(tetramethylene oxide) (PTMO), poly(butadiene), hydrogenated poly(butadiene) (HPB), and PDMS. The hard segment of these segmented polyurethanes was composed of 4,4′ diphenylmethane diisocyanate and 1,4 butanediol and present at 50 wt%. The segmented polyurethanes based on the hydrophobic polyols such as PDMS and HPB exhibit distinct microphase separation between hard and soft segments. XPS revealed the surface enrichment of the hydrophobic component at the air–polymer interface. Dynamic contact-angle measurements indicated that the PDMS-based segmented polyurethane possesses a hydrophobic surface in water. Due to the large difference in surface free energy between their hard and soft segments, the polyol soft segment may be enriched at the air–polymer interface. However, after immersing the specimen in water, surface reorganization may occur in response to the system's requirement to minimize its interfacial free energy. The surface structure in an aqueous environment is different from that in the air. Because the surface free energy of water is higher than air, the component with a higher surface free energy is enriched at the water–polymer interface, whereas the component with lower surface free energy is enriched at the air–polymer interface. The size or hierarchy of structures ranging from blocks or domains to small side-chain functional groups is responsible for polymer surface structure and dynamics. XPS revealed that in the air-equilibrated state, lower surface free energy components are enriched at the air-polymer interface, whereas in the water-equilibrated state, higher surface free energy components are enriched at the water-polymer interface. The change in environment from air to water induces the surface reorganization in order to minimize interfacial free energy. Based on the results of both dynamic contact angle and XPS measurements, the surface structural models of SPUs were proposed. Figure 2 shows the surface structural

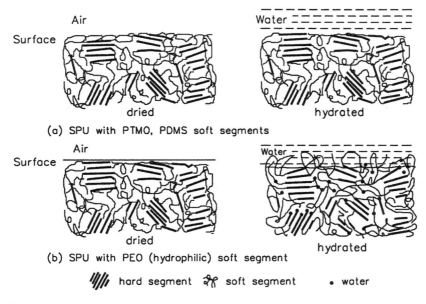

Figure 2 Schematic representation of surface structure of SPUs in the air- and water-equilibrated states: (a) SPU with hydrophobic PTMO or PDMS, (b) SPU with hydrophilic PEO. (From Ref. 33.)

models for SPU with a hydrophobic polyether soft segment (Fig. 2a) and a hydrophilic one (Fig. 2b) in the air-(dried) and water-equilibrated states. The hydrophilic component is dominant at the water–polymer interface, whereas the hydrophobic component is at the air–polymer interface. Because the glass transition temperature of the polyol in SPU employed in these studies is lower than room temperature, the surface layer may be mobile enough to reorganize its chemical composition upon immersion in water at room temperature.

Segmented poly(ether urethane)s (PEUs) and model polymers were examined by using static secondary ion mass spectrometry (SIMS) and XPS [32]. The PEUs were composed of a polyether [either poly(propylene glycol) (PPG) or poly(tetramethylene glycol) (PTMG)] capped with methylenebis (phenylene isocyanate) (MDI) and chain-extended with ethylenediamine. Model soft segments included poly(ethylene glycol), PTMG, and PPG of various molecular weights. Hard-segment models for the PTMG PEUs were based on MDI and butanediol, whereas those for the PPG PEUs were based on MDI and dipropylene glycol or tripropylene glycol. This study confirmed the enrichment in polyether at the PEU surface; however, it suggested that this surface layer of polyether is not pure but is interspersed in the uppermost 10–15 Å with small quantities of hard-segment components.

The properties, structure, and blood compatibility by platelet adhesion and deformation of segmented poly(urethaneureas) (SPUUs) with various aliphatic diamine chain extenders were investigated by Takahara et al. [33]. SPUU is a multiblock copolymer consisting of an alternating sequence of hard and soft segments. Due to the incompatibility of hard and soft segments, phase separation occurs. The domain sizes of the hard segments estimated from small-angle x-ray scattering are 7–20 nm. The hard-segment domain acts as a physical cross-link due to its strong intermolecular hydrogen-bonding between the urea groups. The SPUUs containing diamines with an odd number of methylene units show a remarkable degree of phase mixing between the hard and soft segments. Infrared (IR) spectra of SPUUs indicated that the state of hydrogen-bonding depends on the number of methylene units in the diamine. XPS measurements revealed that the surface concentration of the soft segment is independent of diamine structure, but the state of microphase separation strongly depends on the number of methylene units in the diamine. The SPUUs with an even number of methylene units in the diamines exhibit less platelet adhesion and deformation than those with an odd number of methylene units in the diamines.

Kajiyama et al. [34] investigated the surface properties of an A-B-type diblock copolymer composed of polystyrene (PS) and poly(methoxy poly(ethylene glycol) methacrylate) (PMPEGM). The end group of the hydrophilic chain has a large influence on surface chemical composition, depending on its environments. The diblock copolymer, the side chain of which is end-capped with a methoxy group, shows enrichment of the PMPEGM group at the air–polymer interface due to the hydrophobic character of the methoxy end group. On the other hand, the diblock copolymer, the side chain of which is end-capped with a –OH group, shows an enrichment of PS at the air–polymer interface. The dynamic contact-angle measurements revealed that the surface composition depends upon the environmental change; in the hydrated state, the surface is enriched with hydrophilic PEG side chains, whereas in the air, the surface is enriched with hydrophobic moiety. The driving force of surface structural reorganization is the difference in surface free energy between the polymer surface and the environment. They also compared the surface properties of the diblock copolymer composed of PS and PMPEGM with their corresponding random copolymers [35]. MPEGMs with different PEG chain lengths were used to change the hydrophilicity and molecular mobility of hydrophilic segments. The enrichment of PS components on the block copolymer surface become prominent with an increase in the PEG chain length. The block copolymer film cast from a good solvent of PS shows enrichment of PS on the surface. However, the remarkable effects of the PEG side-chain length and solvent on the surface concentration of PS or PMPEGM components were observed for random copolymers. The block copolymer with long PEG side chains shows a large-scale molecular rearrangement at the surface due to phase separation. The random copolymer does not show a dependence on the immersion cycle because of the small scale of the molecular rearrangement at the surface. This results clearly indicate the effect of copolymer structure on the surface molecular motion and the surface composition at the copolymer surface.

When triblock (PEO–PPO–PEO) copolymer molecules having one hydrophobic PPO block in the middle and two hydrophilic PEO blocks at the ends of the molecules absorb, the conformation of the block copolymer molecule will depend on the properties of the surface and the quality of the solvent [36]. For the triblock copolymer adsorption from water onto a hydrophobic surface, the PPO block acts as the anchor, being mainly situated near the surface because of its high surface affinity. The PEO segments do adsorb, as they have no affinity for the surface. A brush conformation is the result, as shown in Fig. 3. The thickness of a saturated brush layer depends mainly on the length of the ethylene oxide groups. For a hydrophilic substrate like silica, the situation is different. PEO has the highest surface affinity and will adsorb preferentially, thus forming a pancake conformation shown in Fig. 3.

The surface-induced orientation of symmetric diblock PS/PMMA copolymers near the copolymer–air and copolymer–substrate interfaces and in the bulk was investigated by Coulon et al. [37]. It was shown that films cast from toluene solutions, while microphase separated, are randomly oriented with respect to the surface of the film. Annealing the copolymers at 170°C for 24 h produces films with a strong orientation of the copolymer microdomains parallel to the surface. The effect of surface ordering was found to persist through the entire thickness of the film. PS and PMMA domains of approximately one-half the width of the

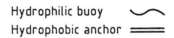

Hydrophilic buoy
Hydrophobic anchor

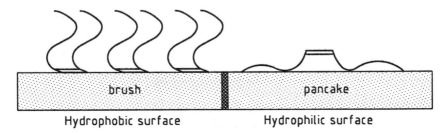

Figure 3 Schematic representation of block copolymers adsorbed at a hydrophobic and a hydrophilic surface. (From Ref. 36.)

domains in the film interior are located at the air–copolymer and copolymer–Si wafer interfaces, respectively. The ordering was found in films as thick as 5×10^3 Å.

The morphology of the microdomains of an A-B-type diblock polymer formed as a consequence of liquid–liquid microphase separation of the constituent polymers A and B at or near the air surface in contrast to the morphology in bulk was reported by Hasegawa and Hashimoto [38]. The morphology at the surface is dramatically affected by the surface free energy, and, therefore, is different from the morphology in bulk. A polystyrene (PS)/polyisoprene (PI) diblock polymer was synthesized and investigated. The electron micrographs of the PS/PI diblock polymer showed that in most areas at the free surface, the lamellae are parallel to the free surface of the film although the free surface of the film appears somewhat rough. The free surface of the film is always covered with the dark PI layer, which is usually thinner than the corresponding inner PI domains, and there is no evidence for the existence of an additional PI layer on the outermost PI layer. The outermost layer must consist of a single layer of PI block chains because the PI chain ends represented by the chemical junctions to PS chains cannot emerge from the air–polymer interface, whereas the inner layers are composed of bimolecular layers because the block chains can emerge from the two opposing interfaces of the domains. Therefore, the thickness of the outermost layer will be approximately half that of the inner domains if the density of the layer and the domains and the average distances between adjacent chemical junctions along the interfaces remain unaffected. The conformation of the PI block chains in the outermost layer should be considerably perturbed in comparison with that of the PI block chains in the PI microphases far from the interface because of the effect of surface tension.

Thomas and O'Malley [39] reported the studies of the surface properties of PS/PEO diblock copolymers. The results indicated that solvent-cast films have surface excesses of PS; that is, the concentration of PS in the surface region at the air–polymer interface is higher than the concentration of PS in the bulk. The concentration of PS at the surface of a given copolymer is solvent dependent and the trends follow the order ethylbenzene > chloroform > nitromethane. Furthermore, the surfaces of the block copolymer films were shown to be laterally inhomogeneous in PS and PEO, and isolated domains of each of these components are presents at the surface. These domains were found to extend more than 50 Å into the bulk.

O'Malley et al. [40] determined the surface compositions topographies of a series of PEO/PS/PEO triblock copolymer films cast from chloroform. The results indicated that the PS concentration at the air–polymer interface is substantially higher than the known bulk concentration of PS and that the copolymer surfaces are laterally inhomogeneous; that is, there are isolated domains of PS and PEO residing at the surface. Furthermore, the molar composition of the surface corresponds to the surface area occupied by each component. A comparison of these results on triblock copolymers with those found earlier on the PS/PEO diblock copolymers [39] indicated that these two systems have remarkably similar surface compositions and topographies. The results on the triblock copolymers clearly suggest that the components in the copolymers are partially miscible in the surface region and that this miscibility is a result of electronic interactions between the PEO and PS blocks in the copolymers.

Russell et al. [41] showed that the lamellar microdomain morphology of PS/PMMA symmetric diblock copolymers orients parallel to the surface of an Au or Si substrate. The nature of the substrate will, however, alter the component that is adsorbed preferentially. It was also shown that the domain orientation occurs at both the air–copolymer and copolymer–substrate interfaces and propagates from these interfaces throughout the film. Results on films prepared on a silicon substrate showed that for films with thickness less than the lamellar period, the microphase separation appears to be suppressed and only gradients in the composition at either interface are found. Finally, the use of solvents that preferentially solvate one of the blocks will, under slow-evaporation conditions, induce an oriented nonequilibrium morphology that is rapidly lost upon annealing, whereupon the equilibrium morphology is attained.

V. GRAFT COPOLYMERS

Graft copolymers are an important class of polymers that have the ability to exhibit the physical properties of both components. Generally, the bulk and surface structure of graft copolymers are not consistent with each other. The significant surface accumulation of one polymer component at the surface is frequently noticed, producing quite a different surface structure than expected from that in bulk. The control of the surface structure of polymer materials is an essential requirement for many industrial applications. Nevertheless, it is not yet clearly understood on the relationship between the bulk and the surface morphology of graft copolymers.

Graft copolymers of PDMA with organic polymers such as PMMA, poly(vinyl alcohol) (PVA), and PS have surfaces which are siloxane-rich, a consequence of the preferential migration to the polymer–air interface of the low-surface-energy, highly flexible dimethylsiloxane units [42–45].

Polymer ultrathin film self-assembly and organization on solid substrates was directed using grafted siloxane copolymers bearing mutually immiscible alkyl and perfluoroalkyl side chains [42]. Polysiloxanes grafted with both alkyl disulfide and perfluoroalkyl side chains as shown in Fig. 4 were synthesized and characterized. These terpolymer systems assemble spontaneously on gold surfaces, forming bound polymeric monolayers organized by intramolecular phase separation. Interfacially bound polymer monolayer fabrication is driven by chemisorption of multipoint alkyl disulfide side chains to gold surfaces from dilute organic solution. Immiscible perfluoroalkyl side chains of low interfacial energy enrich the ambient-exposed outer regions of these monolayers, yielding a novel bound polymer monolayer with an anisotropic layered structure and perfluorinated surface properties. Ellipsometry indicated that these polymer films have thicknesses ranging from 22 to 32 Å, depending on solution conditions and chemistry. Angle dependent XPS provided a depth profile of the bound polymer films, detailing the anisotropic composition resulting from perfluoroalkyl surface enrichment. SIMS measurements supported the enrichment of perfluoroalkyl groups in the outer atomic levels of these films. Polysiloxanes are highly flexible and mobile polymers, exhibiting very low glass transition temperatures and very limited miscibility with both hydrocarbon and fluorinated compounds. Because perfluorocarbon and hydrocarbon side-chain components phase separate from each other as well as from polysiloxane phases, the copolymer architecture unfolds once bound to the solid surface. Intrachain immiscibility compels the siloxane backbone to position itself upon an underlying the layer of chemisorbed alkyl disulfide side chains anchored on the gold surface. The lowest-energy perfluoroalkyl chains remain layered above a region enriched in siloxane backbone, exposed at the ambient interface by a thermodynamic drive to minimize film interfacial tension with air.

The surface structure and properties of PVA-g-PDMA graft copolymers with the controlled PDMS graft chain length as well as chain distribution were studied [43]. The surface of the graft copolymer was found to be covered with an essentially pure PDMS graft component even in only 5 mol% siloxane unit content. The significant surface accumulation of PDMS graft component was con-

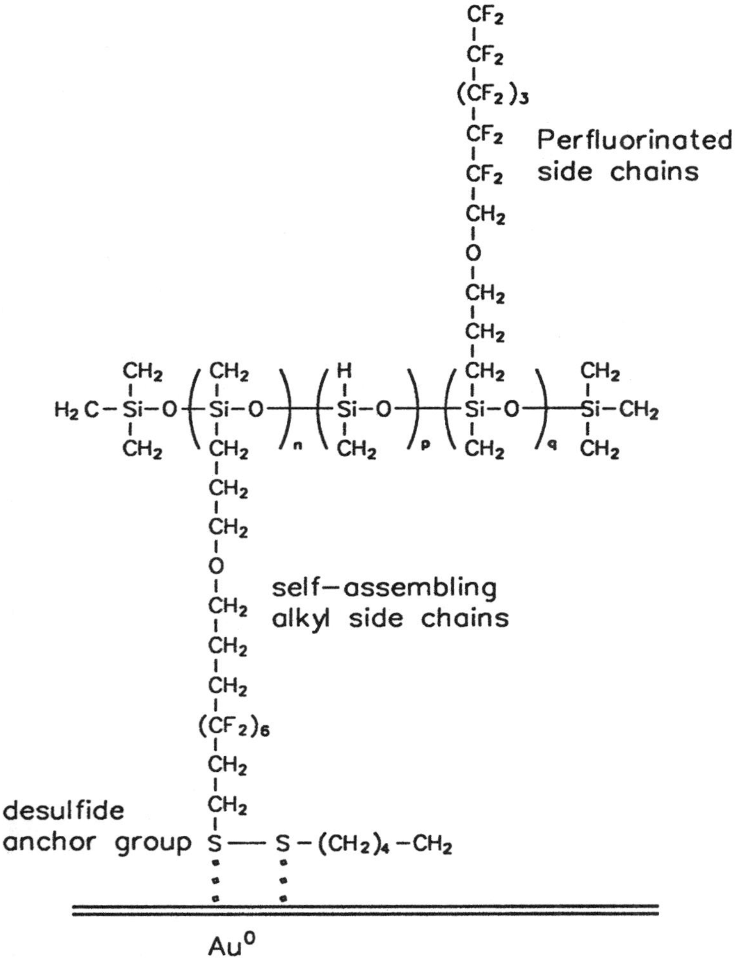

CF_2
|
CF_2
|
$(CF_2)_3$
|
CF_2 Perfluorinated
|
CF_2 side chains
|
CH_2
|
O
|
CH_2
|
CH_2
|
CH_2 CH_2
| |

$$H_2C-\underset{CH_2}{\overset{CH_2}{Si}}-O\left(\underset{CH_2}{\overset{CH_2}{Si}}-O\right)_n\left(\underset{CH_2}{\overset{H}{Si}}-O\right)_p\left(\underset{CH_2}{\overset{CH_2}{Si}}-O\right)_q\underset{CH_2}{\overset{CH_2}{Si}}-CH_2$$

CH_2
|
CH_2
|
O
|
CH_2 self−assembling
| alkyl side chains
CH_2
|
CH_2
|
$(CF_2)_6$
|
CH_2
|
des ulfide CH_2
anchor group S——S$-(CH_2)_4-CH_2$
              ·   ·
              ·   ·

Au^0

Figure 4 Chemical architecture for self-assembling polymers based on PDMA main chains, perfluoroalkyl side chains, and alkyl disulfide anchoring chains. (From Ref. 42.)

firmed with the graft copolymer/PVA blend of less than 1 mol% of siloxane unit content. In contact with water, the PVA-g-PDMS copolymer surface transforms its surface morphology remarkably, which was noticed by the contact-angle measurement with the air-in-water technique, where the contact angle of PVA-g-PDMS copolymer surface was different from that of pure PDMS-coated surface.

Smith et al. [44] investigated PMMA–g–PDMS copolymers. Because PDMS has a low surface energy and it exists in a microphase-separated system, PDMS graft dominates the air–polymer or vacuum–polymer interface in these copolymers. Water contact-angle measurements indicated a change in surface composition with molar mass and composition of the graft. Copolymers containing higher-molar-mass grafts, which exhibit a higher degree of phase separation, have higher contact angles, probably because of the presence of a more complete layer of the PDMS at the surface, as shown in Fig. 5. The lower molar mass graft does not dominate the surface at low siloxane compositions, indicating surface mixing.

The surface properties of poly(organophosphazenes) containing PDMS

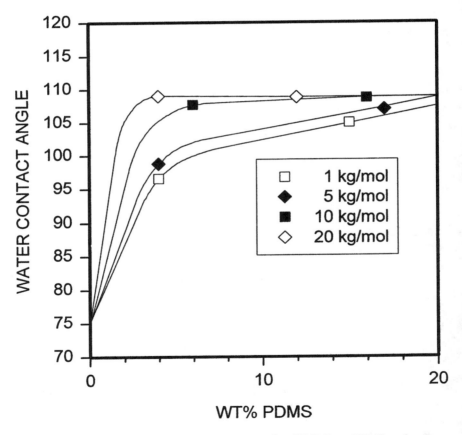

Figure 5 Variation of advancing water contact angle of PMMA–g–PDMS as functions of PDMS composition and PDMS graft molar mass. (From Ref. 44.)

grafts were investigated by Allcock and Smith [45]. A cosubstituent poly(organo-phosphazene) containing both trifluoroethoxy side groups and grafted dimethyl-siloxane side groups was found to have a surface rich in $-CF_3$ groups. Poly[bis(tri-fluoroethoxy)phosphazene] has a glass transition temperature of $-63°C$. As the glass transition temperature of a polymer reflects its surface mobility as well as its bulk mobility, this polymer should be sufficiently mobile at room temperature to allow the surface-active $-CF_3$ unit to be presented to the surface. The experimen-tally determined critical surface tension value of 16 mN/m is similar to that of other fluoroalkyl polymers [5–12,20], and this strongly suggests that $-CF_3$ groups do dominate at the surface. On the other hand, a poly(organophosphazene) with dimethylsiloxane grafts but phenoxy groups as the second substituent show surface enrichment of the dimethylsiloxane species. This demonstrates the relative surface-active natures of the three types of substituents: trifluoroethoxy, dimethylsiloxane, and phenoxy. Therefore, it is likely that the surface properties of multiple-sub-stituent polyphosphazene materials will be affected most significantly by the side group with the lowest surface energy, at least for those polymers which have the freedom to rearrange at the molecular level. The behavior of the surfaces of high-T_g polyphosphazenes with surface-active substituents would be of considerable interest in order to assess the effect of polymer mobility on surface rearrangement.

Takei et al. [46] investigated temperature-dependent surface properties for hydrophilic/hydrophobic changes using poly(N-isopropylacrylamide) (PIPAAm). Two types of PIPAAm of an end-functionalized PIPAAm with a carboxyl end group and a poly (IPAAm-co-acrylic acid) copolymer were used. By means of dynamic contact-angle measurements in water, the wettability of terminally poly-mer grafted surfaces prepared using end-functionalized PIPAAm with a carboxyl end group were compared with that of multipoint polymer grafted surfaces pre-pared using PIPAAm copolymers containing carboxyl groups along the polymer chain. The results are shown in Fig. 6. Both the terminal grafted surface and mul-tipoint grafted surface exhibit hydrophilic properties at temperatures below 20°C. Advancing contact angles of terminal grafted surfaces show a decrease in hydrophilicity at temperatures ranging from 20°C to 24°C. Above this tempera-ture, terminal graft surfaces exhibit hydrophobic properties. Terminal grafted sur-faces demonstrate hydrophilic/hydrophobic surface property changes at 24°C with a small temperature increase. The multipoint grafted surfaces also demon-strate surface property changes near 24°C, but the extent of decrease in the hydrophilic property is small compared to that of the terminal grafted surfaces. Temperature-responsive surface property changes for the terminal grafted sur-faces are more rapid and significant than that for the multipoint grafted surfaces. These features were suggested to be due to more effective restricted conforma-tional freedom for PIPAAm graft chains, which influence polymer dehydration and hydrogen bonding with water molecules.

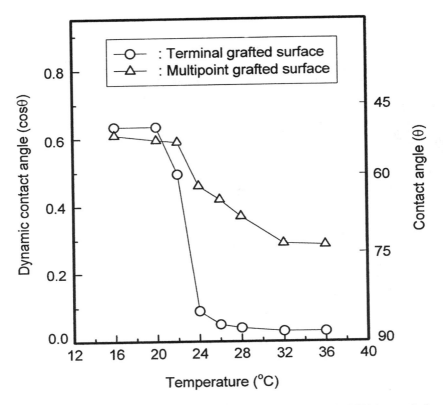

Figure 6 Temperature dependence of advancing contact angle for PIPAAm grafted surfaces. (From Ref. 46.)

Methacrylic acid (MAA) was grafted onto polyethylene (PE) surfaces by simultaneous irradiation with UV rays in the liquid phase to estimate hydrophilic and mechanical properties for MAA-grafted PE plates [47]. With an increase in grafting of MAA, the wettability from the contact angles of water is enhanced and the refractive index from the ellipsometry decreases. Although the contact angle remains constant above the grafted amount of 0.02 mmol/cm², the refractive index approaches the value of PMAA around 0.009 mmol/cm², indicating that the PE surfaces are fully covered with grafted PMAA chains. Surface properties depend on the density of carboxyl group at the surfaces of grafted layers, whereas adhesive properties depend on the structural properties of grafted chains as well as on the density of carboxyl group of the whole grafted layers.

Park et al. [48] synthesized fluorine-containing two-phase graft copolymers of PFMA-*g*-PMMA by using the macromonomer technique and emulsion polymerization. And their surface properties were investigated and compared with a PFMA-*r*-PMMA random copolymer and a PFMA homopolymer. The ability of PFMA-containing polymers to lower surface energy should be mainly determined by the mobility of PFMA to the outermost layer of the air–polymer interface. The mobility of PFMA in the PFMA/PMMA blend was found to be higher than the PFMA segment of PFMA-*g*-PMMA. The surface modification effect of the graft copolymer can be enhanced to nearly the same value as that of the PFMA homopolymer by annealing. This means that the PFMA segment of the graft copolymer has good mobility compared to that of the random copolymer. The PFMA segment of the random copolymer is more difficult to migrate into the outermost layer due to the hindrance of methylester group of PMMA segment. The graft copolymer may be more compatible with the PMMA homopolymer because of its PMMA segment.

VI. CONCLUSIONS

Various important studies on the influence of the polymer surface structure on polymer surface properties for functional polymers, polymer blends, block copolymers, and graft copolymers performed in recent years have been reviewed. The characteristics of side chains, such as composition, functionality, and length, significantly influence the surface properties of polymers. In multicomponent polymer blends or copolymers, the homopolymer constituent, block, or graft having the lower surface energy will migrate to the surface of material. The surface migration of the lower-surface-energy polymer is influenced by various important parameters such as polymer functionality, polymer structure, molecular weight, environment compatibility, temperature, and casting solvent.

Increasing the main-chain flexibility increases the probability of an ordered structure formation. The length of the pendent fluorocarbon group for low surface energy is important, and surface energies decrease as this chain is lengthened. The structure of the outermost layer of PFMA shows an irregular arrangement of perfluoroalkyl groups, most of which are not perpendicularly oriented to the air side. Therefore, the surface properties can be enhanced by making the side chains of perfluoroalkyl groups more perpendicularly oriented to the air side. The desirable low-surface-energy polymers should have a flexible linear backbone, sufficient pendent groups based on aliphatic hydrocarbon or fluorocarbon, linking groups between the backbone and pendent groups that introduce no additional adverse factors, and combinations of these entities that are adequately stable and have no undesirable interactions.

The extent of surface enrichment in differently prepared polymer blends follows the sequence: thermally induced phase-separated blend > solvent-cast phase-separated blend > miscible blend. The surface composition of the low-surface-energy polymer in polymer blends depends on the power law of $M^{-2/3}_n$. Casting solvent systematically affects the surface composition as well as the bulk compatibility, showing that microphase segregation will be exaggerated at the air–polymer interface. The surface segregation of the lower-surface-energy component for polymer blends is generally much greater than that for the block copolymers. The length of the side chain of base polymers affects the surface free energy for polymer blends by hindering the arrangement of functional groups exhibiting lower surface energy to the air side.

In block copolymer system exhibiting a large difference in surface free energy between their hard and soft segments, surface reorganization may occur in response to the system's requirement to minimize its interfacial free energy. The change in environment from air to water induces the surface reorganization in order to minimize interfacial free energy. The surface composition depends on the environmental change; in the hydrated state, the surface is enriched with hydrophilic side chains, whereas in the air, the surface is enriched with hydrophobic moiety. The conformation of a block copolymer molecule will depend on the properties of the surface and the quality of the solvent. The morphology at the surface is dramatically affected by the surface free energy.

Graft copolymers of PDMA with organic polymers such as PMMA, PVA, and PS have surfaces which are siloxane-rich, a consequence of the preferential migration to the polymer–air interface of the low-surface-energy, highly flexible dimethylsiloxane units. Copolymers containing higher-molar-mass grafts, which exhibit a higher degree of phase separation, have higher contact angles. The lower-molar-mass graft does not dominate the surface at low siloxane compositions, indicating surface mixing. The surface properties of graft copolymers having multiple side chains with different surface energies are affected most significantly by the side chain with the lowest surface energy. PIPAAms grafted surfaces exhibit temperature-dependent surface properties. The surface property for terminal grafted surface changes more abruptly and significantly with temperature than that for the multipoint grafted surface.

REFERENCES

1. RV Honeychuck, T Ho, KJ Wynne, RA Nissan. Chem Mater 5:1299–1306, 1993.
2. SC Yoon, BD Ratener, B Ivan, JP Kennedy. Macromolecules 27:1548–1554, 1994.
3. TM Chapman, KG Marra. Macromolecules 28:2081–2085, 1995.
4. YW Tang, JP Santerre, RS Labow, DG Taylor. J Appl Polym Sci 62:1133–1145, 1996.
5. IJ Park, S-B Lee, CK Choi. J Appl Polym Sci 54:1449–1454, 1994.
6. Y Kano, S Akiyama. Polymer 37:4497–4503, 1996.
7. JM DeSimone, Z Guan, CS Elsbernd. Science 257:945–947, 1992.

8. DL Schmidt, CE Coburn, BM DeKoven, GE Potter, GF Meyers, DA Fisher. Nature 368:39–41, 1994.
9. MGD Van der Grinten, AS Clough, TE Shearmur, R Bongiovanni, A Priola. J Colloid Interf Sci 182:511–515, 1996.
10. IJ Park, S-B Lee, CK Choi, KJ Kim. J Colloid Interf Sci 181:284–288, 1996.
11. CM Kassis, JK Steehler, DE Betts, Z Guan, TJ Romack, JM DeSimone, RW Linton. Macromolecules 29:3247–3254, 1996.
12. DL Schmidt, BM DeKoven, CE Coburn, GE Potter, GF Meyers, DA Fischer. Langmuir 12:518–529, 1996.
13. DR Iyengar, SM Perutz, C-A Dai, CK Ober, EJ Kramer. Macromolecules 29:1229–1234, 1996.
14. V Rosilio, G Albrecht, Y Okumura, J Sunamoto, A Baszkin. Langmuir 12:2544–2550, 1996.
15. G Cai, MH Litt. J Polym Sci Part A: Polym Chem 34:2629–2688, 1996.
16. G Cai, MH Litt. J Polym Sci Part A: Polym Chem 34:2689–2699, 1996.
17. G Cai, MH Litt. J Polym Sci Part A: Polym Chem 34:2701–2709, 1996.
18. G Cai, MH Litt. J Polym Sci Part A: Polym Chem 34:2711–2717, 1996.
19. VV Volkov, NA Plate, A Takahara, T Kajiyama, N Amaya, Y Murata. Polymer 33:1316–1320, 1992.
20. H Kobayashi, MJ Owen. TRIP 3:330–335, 1995.
21. IJ Park, S-B Lee, CK Choi, KJ Kim. J Colloid Interf Sci 181:284–288, 1996.
22. RR Thomas, DR Anton, WF Graham, MJ Darmon, BB Sauer, KM Stika, DG Swartzfager. Macromolecules 30:2883–2890, 1997.
23. FP Liu, DJ Gardner, MP Wolcott. Langmuir 11:2674–2681, 1995.
24. GL Gaines Jr. J Phys Chem 73:3143–3150, 1969.
25. AK Rastogi, LE St Pierre. J Colloid Interf Sci 31:168–175, 1969.
26. DH-K Pan, WM Prest Jr. J Appl Phys 58:2861–2870, 1985.
27. QS Bhatia, DH-K Pan, JT Koberstein. Macromolecules 21:2166–2175, 1988.
28. JJ Schmidt, JA Gardella Jr, L Salvati Jr. Macromolecules 22:4489–4495, 1989.
29. JJ Schmidt, JA Gardella Jr, L Salvati Jr. Macromolecules 19:648–651, 1986.
30. A Takahara, AZ Okkema, SL Cooper. Biomaterials 12:324–334, 1991.
31. MJ Hearn, BD Ratner, D Briggs. Macromolecules 21:2950–2959, 1988.
32. A Takahara, J Tashita, T Kajiyama, M Takayanagi, WJ MacKnight. Polymer 26:978–989, 1985.
33. A Takahara, H Takahashi, T Kajiyama. J Biomater Sci Polym Ed 5:183–196, 1993.
34. T Kajiyama, T Teraya, A Takahara. Polym Bull 24:333–340, 1990.
35. T Teraya, A Takahara, T Kajiyama. Polymer 31:1149–1153, 1990.
36. CGPH Schroen, MA Cohen Stuart, K van der Voor Maarschalk, A van der Padt, K van't Riet. Langmuir 11:3068–3074, 1995.
37. G Coulon, TP Russell, VR Deline, PF Green. Macromolecules 22:2581–2589, 1989.
38. H Hasegawa, T Hashimoto. Macromolecules 18:589–590, 1985.
39. HR Thomas, JJ O'Malley. Macromolecules 12:323–329, 1979.
40. JJ O'Malley, HR Thomas, GM Lee. Macromolecules 12:996–1001, 1979.
41. TP Russell, G Coulon, VR Deline, DC Miller. Macromolecules 22:4600–4606, 1989.
42. F Sun, DG Castner, G Mao, W Wang, P McKeown, DW Grainger. J Am Chem Soc 118:1856–1866, 1996.
43. Y Tezuka, A Fukushima, S Matsui, K Imai. J Colloid Interf Sci 14:16–25, 1986.

44. SD Smith, JM DeSimone, H Huang, GA York, DW Dwight, GL Wilkes, JE McGrath. Macromolecules 25:2575–2581, 1992.
45. HR Allcock, DE Smith. Chem Mater 7:1469–1474, 1995.
46. YG Takei, T Aoki, K Sanui, N Ogata, Y Sakurai, T Okano. Macromolecules 27:6163–6166, 1994.
47. K Yamada, T Kimura, H Tsutaya, M Hirata. J Appl Polym Sci 44:993–1001, 1992.
48. IJ Park, S-B Lee, CK Choi. Polymer 38:2523–2527, 1997.

3

Advances in Preparations and Applications of Polymeric Microspheres

Guang Hui Ma
Tokyo University of Agriculture and Technology, Tokyo, Japan

I. INTRODUCTION

The preparative techniques and systems of microspheres with diameters from 0.1 to above 100 μm have been well established. Especially in the last decade, remarkable progress has been made in the preparation of very small (several 10 nm) and very large (~100 μm) uniform microspheres and their applications. Microspheres can be prepared via a polymerization process from monomers and via a granulation process of preformed polymers. Most of researchers are concerned with the preparation of microspheres via a polymerization process, because various microspheres with required functionalities can be obtained economically at high concentration. This trend can be well understood from a large number of published papers. However, the preparative methods have also been developed for the preformed polymers such as natural, biodegradable, and other special polymers, because the microspheres composed of these polymers are difficult to prepare by a polymerization process. In this chapter, the preparative techniques via polymerization process are described, and a couple of examples on their advances are given. The preparative methods via a polymerization process are mainly divided to emulsion polymerization, soap-free emulsion polymerization, suspension polymerization, dispersion polymerization, precipitation polymerization, microemulsion polymerization, miniemulsion polymerization, and seeded polymerization. The characters of these methods, the size range, and the size distribution of obtained microspheres are shown in Fig. 1. Each polymerization system can be divided into a nucleation process and a growing process of the nuclei. These two

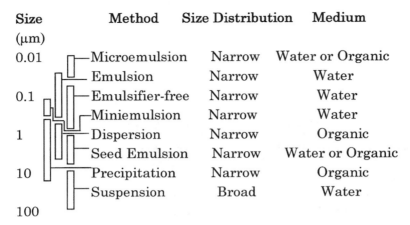

Figure 1   Size and size distribution of microspheres prepared by various polymerization methods.

factors are considered to be the key factors for controlling the size, size distribution, and polymer composition. A detailed discussion on polymerization mechanism will be presented in Sec. II.

Many advances in preparations and applications of microspheres in the last decade have been made (Sec. II–V): the dispersion polymerization has been applied to nonradical polymerization; the SPG (Shirasu Porous Glass) membrane emulsification technique has been introduced to the suspension polymerization, which can provide fairly uniform microspheres; a couple of special swelling methods now getting popular such as the activation swelling method, the dynamic swelling method, and the droplet swelling method have been devised for the seed polymerization; various techniques have been devised to prepare desired organic–inorganic microspheres; the optimum condition of microemulsion polymerization has been established; the nucleation loci in some polymerization process have been able to be detected by a special measurement technique.

Among various applications of microspheres, studies on using microspheres as carriers in bioreactors are most flourishing, such as the separation of DNAs and proteins, the immobilization of enzymes, and the refolding of denatured proteins. In these applications, a large specific surface area of microspheres and functionalities on the microsphere were utilized. Besides the above applications where the surface characteristics of microspheres were utilized, the applications of ordered or random aggregation of microspheres, where the internal characteristics of microspheres were utilized, are also getting much recognition. The concept of these applications is singularly different from the conventional one. A couple of examples and the detailed concept will be described in Sec. V.

II. GENERAL PREPARATIVE METHODS AND THEIR ADVANCES

A. Emulsion Polymerization

1. General Descriptions

Emulsion polymerization is a most typical and well-known polymerization method for preparing uniform polymeric microspheres composed of relatively hydrophobic monomers. The monodispersed microspheres with diameters from several tens to hundreds of nanometers can be obtained easily by this technique. The polymerization system usually consists of a hydrophobic monomer, water (medium), an emulsifier such as sodium salt of long-chain aliphatic acid, and a water-soluble initiator. The polymerization procedure is as follows. After the monomer is dispersed into the aqueous phase where the emulsifier is dissolved, nitrogen gas is introduced to replace oxygen in the reactor. Then, the temperature is elevated to a desired reaction temperature. Finally, the initiator is added to the system to start polymerization. The advantage of the emulsion polymerization is that the polymerization rate is fast and the size distribution of the obtained particles is very narrow. Usually, the polymerization is nearly complete within 1 h. When the concentration of the emulsifier is above the critical micelle concentration (CMC), the emulsion polymerization of the hydrophobic monomer is well described by the micellar nucleation mechanism [1–5] (Harkins' theory), as schematically presented in Fig. 2. Before the initiator is added, the system is com-

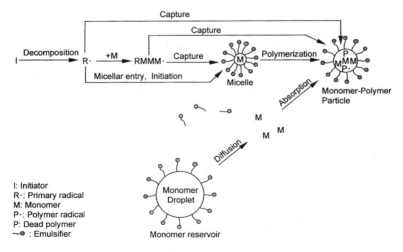

Figure 2   Schematic micellar nucleation mechanism (Harkins' theory) in emulsion polymerization.

posed of monomer-swollen micelles (~20 nm, ~10^{17-18}/cm^3), monomer droplets (1–10 μm, 10^{8-9}/cm^3) stabilized by an emulsifier, and an aqueous phase where free emulsifier molecules (~10^{18}/cm^3) and a small part of monomer are dissolved. After a water-soluble initiator such as potassium persulfate (KPS) is added in the system, it decomposes into primary radicals in the aqueous phase. The radical is either captured quickly by the monomer-swollen micelle or captured after growing to an oligomeric radical by adding several monomers dissolved in the aqueous phase. Then, the polymerization proceeds inside monomer-swollen micelles to form monomer–polymer particles. The polymerization does not occur inside the monomer droplets, although most of the monomer is present in the system as monomer droplets. This is because the number and the specific surface area of the monomer droplets are much smaller than those of the micelles. As a result, the possibility of primary radicals or oligomeric radicals entering the droplets is negligibly small. The monomer droplet serves only as a reservoir of the monomers from which the monomer diffuses into the monomer–polymer particles through the aqueous phase and thus the dynamic equilibrium is maintained among monomer droplets, aqueous phase, and monomer–polymer particles until the monomer droplets disappear. In parallel with the growth of the monomer–polymer particles, free emulsifiers adsorb on the surface of the particles to stabilize the growing polymer particles. After it is consumed completely, no more particles are formed and their number stabilizes at a constant value (at about 10% of monomer conversion). The polymerization then proceeds only in the monomer–polymer particles. On consumption of the monomer droplets (at about 50% of monomer conversion), the dynamic equilibrium is broken and the active centers of growing chains are fed only with monomers remaining in the polymer particles. Concluding the above description, the polymerization proceeds in the three stages: in the first stage (Interval I), the particles are generated from the monomer-swollen micelles by radicals entering the micelles from the aqueous phase. After the depletion of the micellar form of emulsifiers, the formation of new particles ends (at about 10% of conversion). In the second stage (Interval II), the particles grow by means of diffusion of the monomer from the monomer droplets until the monomer droplets disappear (at about 50% of conversion). In the last stage (Interval III), the particles consume the remaining monomer inside until the complete monomer conversion. Because the formation of the particles ends within a short period and the particles take a long time to grow, the emulsion polymerization mechanism yields a narrow size distribution of polymer particles.

Contrary to the above micellar nucleation theory, the homogeneous nucleation theory [6–8] is considered to be more applicable to relatively hydrophilic monomers such as methyl methacrylate (MMA) and vinyl acetate (VAc), especially when the emulsifier is below the CMC. As presented schematically in Fig. 3, this theory assumes that the initiation of polymerization starts in the aqueous phase through the decomposition of the initiator into primary radicals, followed by

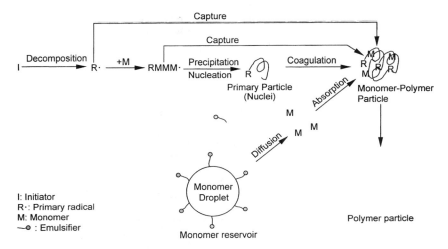

Figure 3 Schematic homogeneous nucleation mechanism in emulsion polymerization.

the addition of monomers dissolved in water. The growth of the oligomeric radical proceeds in the aqueous phase up to a particular critical chain length. On exceeding this critical length, the oligomeric radical precipitates from the aqueous phase and forms a water-insoluble polymer in the form of a spherical particle (primary particle, nucleus). The primary particles are unstable and they will coagulate to a large stable particle, which is stabilized by adsorbing emulsifiers or by the ionic charge of initiator fragments on the surface of the particle. The subsequent growing process of the particle is similar to that of the micellar nucleation process; that is, the growing particle absorbs monomers which diffuse through the aqueous phase from the monomer droplets to carry out polymerization. The monomer–polymer particle becomes the only polymerization locus.

2. A New Technique Distinguishing Nucleation Mechanisms

Recently, Napper and Kim [9] found that a fluorescence method was able to be used to distinguish between micellar nucleation and homogeneous nucleation mechanisms. The fluorescence of pyrene molecules is sensitive to changes in their local viscosity. Pyrene is located predominantly in the micelles and droplets because it is relatively water insoluble. Thus, the formation of a polymer in the micelles should induce a change in the fluorescence of the pyrene molecules contained therein, if micellar entry is operative. This change should be parallel with the onset of polymerization. In contrast, because the homogeneous nucleation is supposed to occur outside of the droplets or micelles containing the pyrene, initially no change should be observable in the fluorescence of pyrene even though polymerization occurs. Napper and Kim measured the fluorescence of pyrene for emulsion polymerization of the styrene (ST)/cationic emulsifier/V-50 [2,2'-azo-bis(2-methylpropionamidine)dihydrochloride] system and the methyl methacrylate (MMA)/V-50 system. The strong correlation observed between the onset of polymerization and the change in fluorescence is in good agreement with the prediction of a micellar nucleation mechanism for emulsion polymerization of ST. On the other hand, substantial polymerization occurred prior to any change in the fluorescence of pyrene for emulsion polymerization of MMA. This result conforms to the prediction of the homogeneous nucleation mechanism.

3. Advances in Emulsifiers and Their Advantages

Anionic, cationic, amphoteric, and nonionic types of emulsifier are usually used; for example, sodium dodecyl sulfate (SDS) or sodium dodecyl benzene sulfonate (SBS) is used as an anionic emulsifier, $CH_3(CH_2)_nN^+(CH_3)_3X^-$ is used as a cationic emulsifier, $CH_3(CH_2)(C_6H_4)O(CH_2CH_2O)_mH$ is used as a nonionic emulsifier, and $CH_3(CH_2)_nN^+(CH_3)_2CH_2SO_3^-$ is used as an amphoteric emulsifier. Because it is found that the desorption of emulsifier occurs and leads to an undesirable effect when the particles are applied to the film formation, used as carriers of drugs and biomaterials, the reactive emulsifiers (named as surfmer—acts as both of surfactant and monomer) are developed instead of conventional emulsifiers. The main method is to introduce a double bond to a conventional emulsifier. A few of examples of reactive emulsifiers [10–12] are shown as follows:

$$C_9H_{19}\overset{\text{CH=CHCH}_3}{\underset{}{\bigcirc}}\text{-}(CH_2CH_2O)_{10}SO_3NH_4 \qquad \text{(I)}$$

$$\begin{array}{c} H_2C\text{-}COOC_{12}H_{25} \\ | \\ H_4NO_3SCH\text{-}COOCH_2CHCH_2OCH_2CH=CH_2 \\ | \\ OH \end{array} \qquad \text{(II)}$$

$$\begin{array}{c} H_2C\text{-}COOC_{12}H_{25} \\ | \\ NaO_3SCH\text{-}COOCH_2CH=CH_2 \end{array} \qquad \text{(III)}$$

$$CH_2=CH\text{-}\underset{O(CH_2)_{12}SO_3Na}{\bigcirc} \qquad \text{(IV)}$$

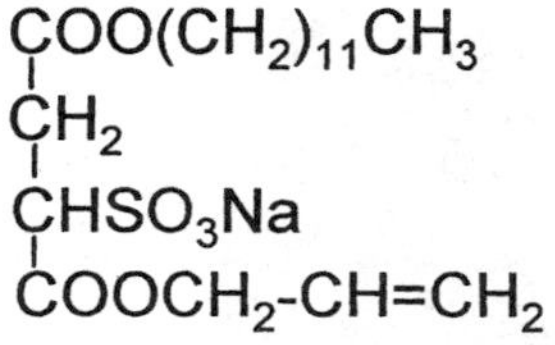

COO(CH2)11CH3
CH2
CHSO3Na
COOCH2-CH=CH2

(V)

Structures **(I)–(III)** have been used in the emulsion polymerization of acrylates [10]. It was found that at least 95% of these reactive emulsifiers were able to be copolymerized into particles. Films cast from the obtained lattices showed good water resistance, compared with those prepared by conventional emulsifiers. This is because the reactive emulsifier was distributed uniformly inside of the film, whereas the conventional emulsifier was usually localized on the surface of the film as water was evaporated in the film formation process.

Other special emulsifiers were also developed corresponding to respective purposes. Some examples are given as follows. The first example is a hydrophilic silicone emulsifier [13]; it can also be called a macromonomer. By using it in the emulsion polymerization of acrylates, the hydrophilic silicone macromolecule locates on the surface of the particles; the core–shell microspheres covered with the silicone can be obtained as shown in Fig. 4. The film formed from the emulsion of this kind of composite particle revealed the combined good properties of acrylic resin and silicone. For example, it showed water resistance, heat resistance, lubricity, and releasability of silicone, as well as the excellent adhesive power of acrylic resin. It has been widely used in the top coating on an ink ribbon where heat resistance and lubricity were required and as binders of heat-sensitive papers, and so forth.

Watanabe et al. [14] used reactive phospholipid **(VI)** as an emulsifier to prepare the microsphere with the phospholipid on the surface. This microsphere can be used as a model surface of the biomembrane and can be used to immobilize the enzyme.

CH3(CH2)14-C-O-CH2
O
O
CH2=CH-
-C-O-CH
O
CH2-O-P-O-CH2CH2N+(CH3)3
O-

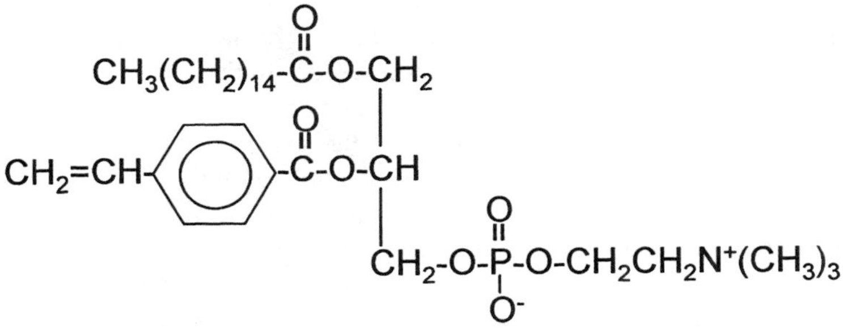

(VI)

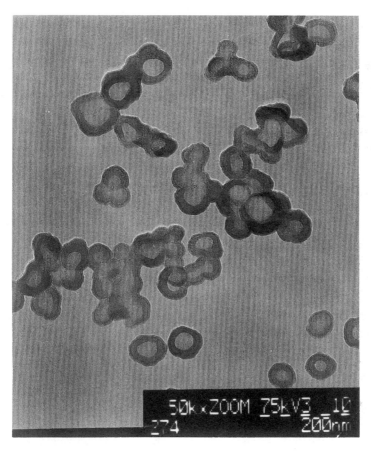

Figure 4 Transmission electron micrograph of poly(acrylate silicone) core–shell microsphere covered with silicone. (From Ref. 13.)

Recently, the emulsion polymerization using poly(vinyl alcohol) (PVA) with a –SH terminal (HS–PVA) as an emulsifier has been studied [15]. PVA without a –SH terminal is not suitable for emulsion polymerizations of styrene and acrylates, but is only effective for that of vinyl acetate and vinyl chloride. However, it was found that HS–PVA was suitable for emulsion polymerizations of styrene and acrylates because HS–PVA can be grafted on the particles to enhance their stability. Furthermore, it was found that the mechanical stability of particles obtained by using HS–PVA was much better than that of particles prepared by conventional emulsifiers. The film cast from such emulsions showed a large stress (about 40 kg/cm^2) and a lower strain (700%), whereas the stress and strain of the film obtained from the emulsions with conventional emulsifiers was 15 kg/cm^2 and 2000%, respectively; that is, the film obtained from the emulsions by using HS–PVA was a stronger film.

Kim et al. [16] used a special emulsifier—an alkali-soluble random poly (styrene-*co*-α-methylstyrene-*co*-acrylic acid) copolymer for the emulsion polymerization of ST and MMA in the alkaline condition. The feed amount of this special emulsifier was very high, up to 15–35 wt% based on the monomer. They obtained much smaller particles (about 40 nm) than those obtained in usual emulsion polymerization. They found the size distribution of the particles was narrow for the ST system but broader for the MMA system. They attributed this difference to the different nucleation process (micellar and homogeneous nucleation processes), as described above.

4. Advances in Initiators and Their Advantages

In emulsion polymerization, water-soluble persulfate initiators and azo initiators are usually used. Potassium persulfate (KPS), ammonium persulfate (APS), and 2,2'-azobis(2-methylpropionamidine)dihydrochloride (V-50) are typical examples. An initiator-surfactant (inisurf) which acts as both an initiator and emulsifier was developed and used in emulsion polymerization; for example, an azo initiator [17] containing emulsifier groups, as shown in **(VII)**. As the case of an reactive emulsifier, inisurf has the advantage that it can be incorporated in the microsphere by a covalent bond, so that it does not desorb from the microspheres. In the recent years, new initiators were developed to introduce a new functionality to microspheres, corresponding to the specific purpose. For example, Sugiyama et al. [18] used an azo initiator containing a phosphatidylcholine group **(VIII)**, which is present on the external surface of blood cells, in the emulsion polymerization of ST to obtain the microspheres with the phospholipid on the surface. The microspheres obtained showed excellent biocompatibility.

$$\left[N\text{-}C(CH_3)_2C\text{-}\overset{\overset{\displaystyle O}{\|}}{C}\text{-}O\text{-}R\text{-}SO_4^-\right]_2$$

R: (-CH$_2$-)$_n$, n=6-16, or (-CH$_2$CH$_2$O-)$_m$, m=3-12   (VII)

$$\left[N\text{-}\underset{\underset{\displaystyle CH_2CH_2\text{-}\overset{\overset{\displaystyle O}{\|}}{C}\text{-}O\text{-}CH_2CH_2\overset{\underset{\displaystyle CH_3}{|}}{\overset{|}{N}^+}CH_2CH_2\text{-}O\text{-}\overset{\underset{\displaystyle O^-}{|}}{\overset{\overset{\displaystyle O}{\|}}{P}}\text{-}O\text{-}CH_2CH_3}{}}{\overset{|}{\underset{|}{C}}\text{-}CH_3}\overset{\displaystyle CN}{}\right]_2$$

(VIII)

5. A Special Application of Emulsion Polymerization

Nagai et al. [19] developed an interesting method for limiting the emulsion polymerization on the silica gel particle to encapsulate a polymer shell on the silica gel
particle. They used reactive cationic emulsifiers and a KPS initiator in the polymerization. The reactive cationic emulsifiers were quaternary salts (C_8Br, $C_{12}Br$,
and $C_{16}Br$, mainly $C_{12}Br$) of dimethylaminoethylmethacrylate with alkyl bromide
having 8, 12, and 16 carbon atoms, respectively (**IX**). They found that the cationic
emulsifier adsorbed on the silica gel particle formed an oil-soluble salt with KPS,
which deposited on the adsorbed layer of the cationic emulsifier. This oil-soluble
salt formed on the surface of the silica gel particle can solubilize the hydrophobic
styrene monomer. As a result, the polymerization of styrene was limited in the
adsorbed layer of the cationic emulsifier to yield polymer-encapsulated composite
particles. It was found that the surface coverage of the resulting particles increased
with an increase of monomer concentration. The PST film impregnated with this
kind of composite particles showed an elongation and stress (kg/cm^2) about twice
that just impregnated by bare silica gel particles.

$$CH_2=\underset{\underset{\displaystyle CH_3}{|}}{C}-COOCH_2CH_2-\underset{\underset{\displaystyle CH_3}{|}}{\overset{\overset{\displaystyle CH_3}{|}}{N^+}}-(CH_2)_{n-1}-CH_3 \quad \overset{\displaystyle Br^-}{}$$

$$(n=8, 12 \text{ or } 16)$$ (IX)

Because the polymerization rate in emulsion polymerization is high, polymers with a high molecular weight ($\sim10^5$ g/mol) can be obtained easily, and the
water is used as the medium in this method, this method is utilized widely in the
industrial field, from the view of easy handling of the polymer, polymerization
equipment, and environmental protection.

B. Emulsifier-Free Emulsion Polymerization

1. General Descriptions

Emulsifier-free emulsion polymerization is a technique developed based on conventional emulsion polymerization, because the emulsifiers often give an unde-
sired effect when the particles are used in film formation or used as carriers in the
biomedical and bioseparation fields. It was found that polymerization also proceeded quickly by adding a small amount of a hydrophilic comonomer instead of
an emulsifier, even when the main monomer was fairly hydrophobic, such as
styrene. The mechanism of emulsifier-free polymerization is considered to obey

the homogeneous nucleation process; however, some supplementary remarks should be made. In the first stage of polymerization, the hydrophilic monomer copolymerizes with a small amount of hydrophobic monomer dissolved in the aqueous phase. When the copolymer chain reaches a critical length, it (amphiphilic oligomeric radical) precipitates from the aqueous phase to form a nucleus (primary particle). The primary particles coagulate to a large stable particle stabilized by hydrophilic monomer units and split fragments of hydrophilic initiator. In the second stage, the growing particles absorb monomers, which diffuse from monomer droplets through the aqueous phase, to carry out polymerization therein and become the only polymerization locus until the monomer droplets disappear (corresponding to about 50% of conversion). In the last stage, the hydrophobic monomers will be polymerized continuously in the monomer–polymer particles until the monomers are consumed completely. The hydrophilic monomer units and initiator segments usually locate at the surface of produced particles to stabilize them. By utilizing the emulsifier-free method, ingeniously, the functional groups of hydrophilic monomers can be concentrated on the surface of the particles, and core–shell-type particles can be obtained. However, the hydrophilic monomers may also locate in the core, dependent on the polymerization temperature, monomer composition, and copolymerization between hydrophilic and hydrophobic monomers. Detailed examples will be given below.

2. Some Examples of Emulsifier-Free Polymerization

Kawaguchi et al. [20–22] studied a series of copolymerizations between ST and various hydrophilic monomers with different hydrophilicities. The morphologies of obtained microspheres are schematically represented in Fig. 5. When acrylamide (AAm) was used, only a very small amount of AAm was introduced on the surface and about 90% of the AAm remained in the aqueous phase because hydrophilicity of AAm is very high and copolymerization with ST is poor. When N-acryloyl pyrrolidine (APr) with a lower hydrophilicity than AAm was used, the particle with the inner core and thin skin rich in APr was obtained at a lower polymerization temperature. When the polymerization temperature was lower, the nucleus rich in hydrophilic monomer was so hard that the hydrophobic monomers which diffuse from monomer droplets through the aqueous phase were not able to enter into the core completely and were polymerized near the surface of the particles. As a result, a APr-rich copolymer formed a core and a thin skin of the particles. When the feed fraction of APr was higher, the surface of microspheres became uneven because of apparent phase separation between ST-rich and APr-rich copolymers. On the other hand, when N-acryloyl piperidine (APp), the hydrophilicity of which was lower than APr and copolymerization with ST was more favored than APr, was used, APp was incorporated in microspheres uniformly. As a result, the apparent phase separation was not observed and the particle with a smooth surface was obtained.

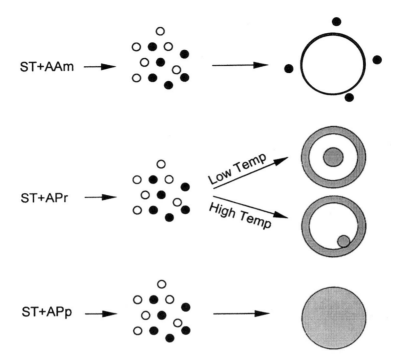

Figure 5 Schematic morphologies of microspheres obtained by emulsifier-free emulsion polymerization of styrene and various hydrophilic monomers with different hydrophilicities. Symbols: ○—styrene; ●—hydrophilic monomer; AAm—acrylamide; APr—*N*-acryloyl pyrrolidine; APp—*N*-acryloyl piperidine.

Many hydrophilic or relatively hydrophilic monomers have been used in copolymerizations with hydrophobic monomers to obtain particles with functional groups on the surface. For example, $-CONH_2$, $-N(CH_3)_2$, $-OH$, $-COOH$, a glycidyl group, a sulfonate group, a pyridine group, $-NO_2$ and so forth can be introduced on the surface of the particles by copolymering acrylamide (AAm), *N*-dimethylaminoethylmethacrylate (DMAEMA), 2-hydroxyethylmethacryle (HEMA), acrylic acid (AAc), or methacrylic acid (MAc), sodium styrene sulfonate (NaSS), 4-vinylpyridine (4VP), and 4-nitrostyrene with hydrophobic monomers such as styrene, butyl acrylate (BA), and so forth.

Ma and Fukutomi [23] developed a special method to prepare a poly(4-vinylpyridine) (P4VP) microsphere by using a partially quaternized reactive P4VP primary polymer as an emulsifier that served as a nucleus. Because the partially quaternized 4VP monomer cannot be copolymerized with 4VP, we first prepared the P4VP primary polymer and then quaternized it and introduced a double bond by using iodomethane (CH_3I) and *p*-chloromethylstyrene (CMS), respectively. Then, it was added into the emulsifier-free emulsion polymerization system. This meant that the emulsion polymerization started from the point where the primary particles were already present. The partially quaternized P4VP polymer behaved as a nucleus of polymerization. As the concentration of the partially quaternized P4VP polymer increased, the size of the particles decreased and the number of particles increased because the number of nuclei increased.

If a hydrophilic comonomer is not used and only the hydrophilic initiator is used in the emulsion polymerization of ST, the polymerization rate is very low; this occurs because the amount of ST monomer dissolved in the aqueous phase is very small and the formation of the nuclei needs a long time. Okubo et al. [24] tried to use acetone and water as the medium instead of pure water to enhance the solubility of ST in the medium. They found that the polymerization rate was raised steadily by adding acetone into water, and it attained to a maximum value when the volume fraction of acetone was 40%.

3. Special Application of Emulsifier-Free Polymerization

Yanase et al. [25,26] found that ferric particles, which were stabilized by the emulsifier in ferric fluid, were able to play the role of primary nuclei. The polymerization developed on the surfaces of the ferric particles. After coagulation of the primary nuclei and growth of the particles, the uniform microspheres containing ferric fluid were obtained. This discovery suggested that other functional substances can also be incorporated in microspheres by using them as the nuclei in the emulsifier-free polymerization process.

Another interesting application is the emulsifier-free emulsion polymerization of MMA in the presence of $CaSO_3$ using $K_2S_2O_8$ as an initiator [27]. MMA can be polymerized over the surface of $CaSO_3$ powder with chemical bonding at the interface of PMMA and $CaSO_3$ above the saturation concentration of $CaSO_3$, according to the following initiation mechanism in the presence of $CaSO_3$.

Redox reaction:

$$S_2O_8^{2-} + H_2SO_3 \rightarrow SO_4^{-}\cdot + HSO_3\cdot + HSO_4^{-}$$

$$S_2O_8^{2-} + HSO_3^{-} \rightarrow SO_4^{-}\cdot + HSO_3\cdot + SO_4^{2-}$$

$$S_2O_8^{2-} + SO_3^{2-} \rightarrow SO_4^{-}\cdot + SO_3^{-}\cdot + SO_4^{2-}$$

Hydrogen transfer:

$$CH_2=CCH_3CO_2CH_3 + H_2SO_3 \rightarrow CH_3-\dot{C}CH_3CO_2CH_3 + HSO_3\cdot$$

$$CH_2=CCH_3CO_2CH_3 + HSO_3^{-} \rightarrow CH_3-CCH_3CO_2CH_3 + SO_3^{-}\cdot$$

The larger and irregular particles consisting of the core of $CaSO_3$ and shell of PMMA were obtained. The bulk materials formed from this kind of composite microspheres showed higher tensile strength than those made by just blending $CaSO_3$ powder and pure PMMA latices.

Because the emulsifier-free polymerization does not require an emulsifier and can incorporate various functional groups by using various functional comonomers, the microspheres obtained can be used as carriers of biomaterials, besides conventional usage for coating, film formation, and so forth.

C. Miniemulsion Polymerization

1.   General Descriptions

As the mechanical mixing technique progresses, oil droplets can be stabilized to a small droplet with submicron size (<500 nm) by an emulsification process using a combination of an ionic emulsifier and a cosurfactant. As described in emulsion polymerization, the possibility of micron-sized monomer droplets capturing radicals from the aqueous phase is negligibly small and are unable to become nuclei; they just work as a monomer reservoir from which the monomers diffuse to monomer–polymer particles through the aqueous phase. When the droplets are very small and stable, however, a much larger amount of emulsifier will be adsorbed on the surfaces of monomer droplets because the specific surface area increases remarkably. As a result, the free emulsifier concentration in the aqueous phase can be adjusted below CMC level without the formation of micelles if the feed concentration of emulsifier is adequate. Therefore, the possibility of the small droplets capturing radicals becomes much higher, and the small droplets can become the nuclei and polymerization loci [28,29]. Furthermore, because the monomer droplets are very stable because of the combined effects of emulsifier and cosurfactant and there is no micelle present in the aqueous phase, the diffusion and solublization of the monomer into the aqueous phase can be suppressed. As a result, the possibility of nucleation in the aqueous phase can be avoided. The minidroplets (miniemulsion) are usually prepared by microfluidizer and sonication techniques [30], after the mixture of monomer and cosurfactant is added into the aqueous phase, dissolving the emulsifier. The size of monomer droplets prepared by microfluidizer is smaller than that prepared by the sonification technique. Either a hydrophilic or hydrophobic initiator can be used. If hydrophobic initiator is used, the initiator is present, predominantly inside the monomer droplet, the polymerization proceeds similar to the usual bulk polymerization inside an individual monomer droplet, but escape and the reentry of radicals are also take place during the polymerization. If a hydrophilic initiator is used, the minidroplets capture the primary radicals from the aqueous phase and then initiate polymerization inside the monomer droplets. The technique to enhance the stability of the miniemulsion is to introduce a "cosurfactant"—a water-insoluable and monomer-soluble compound. In conventional emulsions, smaller droplets have a higher vapor pressure than larger droplets, so the larger droplets will grow at the expense of the smaller droplets. In the miniemulsion, however, the presence of the cosurfactant acts to reduce substantially the rate of oil diffusion from smaller to larger droplets. Usually, a long-chain alkyl alcohol (dodecyl alcohol, cetyl alcohol, etc.) and a hydrophobe (hexadecane, dodecane, etc.) are used. Some researchers noted that the long-chain alcohol formed a liquid-crystal-like and electrically charged barrier with emulsifier molecules on the surface of the droplets to prevent droplets from coalescing. However, this mechanism of enhanced stability is not operative for a long-chain alkane (hexadecane, etc), because the long-chain alkane is certainly located in the interior of the miniemulsion droplets. Therefore, the low water solubility of a long-chain alcohol retarding the diffusion of a monomer to the aqueous phase (Ostwald ripening or diffusion degradation) is another important reason, as in the case of a long-chain alkane. Thus, the long-chain alkane, when used to stabilize the droplets against degrading by diffusion, should be referred to as a "hydrophobe" rather than a "cosurfactant." Miller et al. compared the stability of miniemulsions prepared in the presence of cetyl alcohol and hexadecane [31] and found that the miniemulsion prepared using hexadecane was more stable and smaller than that prepared using cetyl alcohol. This result implies that the finite solubility of cetyl alcohol (CA: $\sim 10^{-5}$ g/dm^3, HD: $\sim 10^{-6}$ g/dm^3) in the aqueous phase also allows it to undergo a slower degradation through the aqueous phase to deteriorate the stability of the miniemulsion, and that the effect of a hydrophobe retarding the diffusion of the monomer to the aqueous phase is a most important factor for stabilizing the miniemulsion.

In recent years, it is found that the preformed polymer can be also used as the hydrophobe. For example, PST and PMMA can be used as the hydrophobe in preparing miniemulsions of ST and MMA, respectively [32,33]. These polymers satisfy the requirement of a "hydrophobe" (cosurfactant): They are highly water insoluble and cannot diffuse out of the droplet. In addition, most of polymers are very soluble in their monomers. The most important advantage of using a polymer as a hydrophobe is that it is not innocuous in the final product, whereas cetyl alcohol or HD has to be removed from the particle after polymerization. Reimers and Schork [33] tried to use PMMA in the miniemulsion polymerization of MMA. They found that the stability of minidroplets depended on the molecular weight and concentration of PMMA, as well as the concentration of the emulsifier. PMMA with a higher or lower molecular weight is unable to stabilize the minidroplets; PMMA with a molecular weight of 350,000 (g/mol) showed the best effect as a hydrophobe. Increasing the concentration of PMMA can increase the stability of the miniemulsion, except for the case when PMMA has a large molecular weight (1,050,000 g/mol). In general, it is considered that the polymerization rate of miniemulsion polymerization is lower than that of emulsion polymerization when the same amount of emulsifier is used. Recently, however, Reimers and Schork [33] found that the polymerization rate of the PMMA miniemulsion system as described earlier was higher than that of emulsion polymerization when KPS was used as an initiator. They ascribed this result to the fact that the miniemulsion did not accompany a transport process of the monomer from monomer droplets to monomer–polymer particles through the aqueous phase.

2.   A Technique Distinguishing Nucleation Mechanisms

Because the size of particles obtained by miniemulsion polymerization is close to that of particles prepared by emulsion polymerization, it is difficult to determine which mechanism is dominant. Recently, it was found that the conductance measurement was a good method for this determination [33]. In the case of general emulsion polymerization, the conductance initially increases after KPS (initiator) is added into the system because KPS is an electrolyte. This is followed by a reduction in the conductance as monomer–polymer particles are formed and adsorb emulsifier because the concentration of free emulsifier decreases (Interval I). Then, a sharp rise in conductance appears because the monomer droplets disappear (end of Interval II). In the case of miniemulsion polymerization, however, this phenomenon becomes less pronounced because the miniemulsion polymerization occurs with very little change in surface characteristics and there are virtually no monomer droplets (minidroplets) to disappear.

3.   Advantages and Special Applications of Miniemulsion Polymerization

One important advantage of miniemulsion polymerization is that a hydrophobic macromonomer, a hydrophobic transfer agent, and other large functional hydrophobic substances can be used as a hydrophobe and incorporated into particles uniformly. It is very attractive for industries that a large hydrophobic molecule can be incorporated into such small particles. The large hydrophobic molecule is not capable of being incorporated into small particles by conventional emulsion or emulsifier-free emulsion polymerization, because it is difficult to migrate from monomer droplets into micelles through the aqueous phase. Ishitani et al. prepared poly(styrene-*co*-2-ethylhexylacrylate) composite microspheres containing silicone by using a silicone macromonomer as a hydrophobe [34]. They found that 30 wt% silicone was able to be incorporated into microspheres when the oil droplets were prepared to smaller than 500 nm, whereas it was difficult to incorporate them into microspheres and a large agglomerate was formed when the droplet size was larger than 500 nm. This result suggested that the emulsion polymerization process was predominant when the droplets were larger. The film prepared from an emulsion of this kind of composite microsphere showed lower film peel strength than those without silicone. Furthermore, they found that the peel strength increased as the content of silicone became larger. Yamamoto et al. [35]

prepared P(MMA-*co*-BA)-*g*-PMMA latex by using MMA and BA as monomers and a PMMA macromonomer as a hydrophobe. They found that the film cast from this latex showed good oil resistance after PMMA was grafted on a P(MMA-*co*-BA) copolymer by miniemulsion polymerization.

Another important advantage of miniemulsion polymerization is that it is far more robust to variations in the recipe or contaminants level than conventional emulsion polymerization. The particle number was found to be less sensitive, by at least an order of magnitude, to changes in the initiator, the water-phase retarder, and the oil-phase initiator concentration than emulsion polymerization. This is because, unlike conventional emulsion polymerization, there is no competition between particle nucleation and particle growth for the available emulsifier molecules. This property is an important advantage in the industrial production of microspheres.

D. Microemulsion Polymerization

1. General Descriptions

Microemulsion polymerization can provide very small particles. In most cases, the microspheres with diameters of 10–60 nm are available. Microemulsion covers oil/water (O/W), bicontinuous, and W/O (inverse type) microemulsion. These three types of microemulsion are formed corresponding to the relative amounts of oil phase and water phase, and HLB (hydrophilic–lipophilic balance) values of the emulsifier [36], as shown in Fig. 6. Compared to the conventional emulsion, the monomer droplets are not present in the microemulsion system; that is, all of the monomers are solubilized inside the micelle to form microdroplets or are dissolved in the continuous phase. Therefore, the microemulsion appears transparent or bluish, depending on the amount of monomer solubilized in the micelles. In order to highly stabilize monomer-solubilized micelles and extend the region of microemulsions, medium-chain alcohols (pentanol, etc.) are often used as a cosurfactant, besides a large amount of emulsifier. Medium-chain alcohol is localized in the interface between the monomer-swollen micelle and the continuous phase to decrease interfacial energy. It is convenient to describe a triangle phase diagram (for oil–water–emulsifier ter-components) to locate the microemulsion region, before carrying out polymerization. The O/W microemulsion is formed when the oil phase decreases and the HLB value of the emulsifier increases, whereas a W/O microemulsion is formed under the inverse condition [36]. The bicontinuous phase is an intermediate phase where the O/W emulsion transfers to the W/O emulsion; it is formed when the amount of oil and water are almost the same and the HLB value lies in the intermediate range of the above two cases. Hydrophobic and hydrophilic microspheres can be obtained by carrying out polymerization after O/W [37–43] and W/O [44–53] microemulsions are prepared using

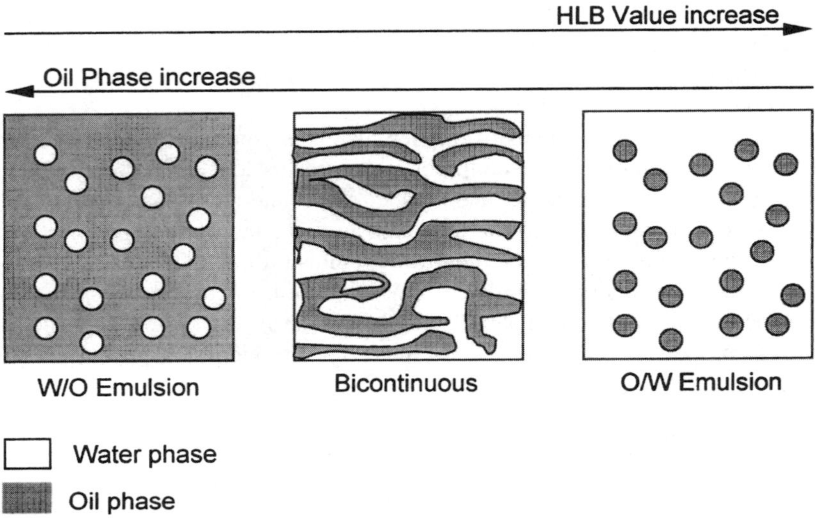

HLB Value increase ▶

◀ Oil Phase increase

W/O Emulsion   Bicontinuous   O/W Emulsion

☐ Water phase

▨ Oil phase

Figure 6 Relationship between microemulsion type (W/O, bicontinuous, O/W) and relative amounts of oil phase and water phase and the HLB values of the emulsifier.

hydrophobic and hydrophilic monomers, respectively. Usually, hydrophilic and hydrophobic initiators can be used in O/W or W/O microemulsion polymerization.

2.   Advances in Studies on Nucleation Mechanisms

The nucleation mechanism of emulsion polymerization can be applied to microemulsion polymerization after some modifications. It is considered that the monomer-swollen micelles capture the radicals or oligomeric radicals from the continuous phase, where the initiator decomposes to radicals and they grow to oligomeric radicals with monomers dissolved in the continuous phase, to polymerize monomers inside micelles to form nuclei. Because the monomer droplet is not present, the growing particle is provided with a monomer by other un-nucleated micelles through the diffusion process or collision [50,51]. Therefore, not all of the micelles convert to polymer particles. After the polymerization, a part of the micelles will remain in the continuous phase. Being different from conventional emulsion polymerization, it was found that the number of particles increased continuously until the polymerization was completed. Therefore, it is considered that the possibility of initiating free radicals in the continuous phase being captured by un-nucleated micelles, rather than by polymer-containing nuclei, is very high throughout the polymerization. Candau and Leung investigated the polymerization of acrylamide (AAm) in inverse microemulsion by using Aerosol OT [sodium bis(2-ethylhexyl)sulfosuccinate] as an emulsifier and AIBN or KPS as an initiator at 45°C [50]. The aqueous phase (AAm+water) was used as a dispersed phase and toluene was used as a continuous phase. They found that every final particle was composed of only one single-polymer molecule. This result suggested that once a micelle captured a radical and polymerized the monomer therein to form a nucleus, this nucleus no longer captured another radical. The newly decomposed radicals were likely to be captured by un-nucleated micelles. This was a very special case. Many authors found that the final particles consisted of 2–10 polymer molecules. Furthermore, Candau et al. found that adding AAm into the aqueous phase was able to extend the microemulsion region steadily when isoparaffinic oil (Isopar M) was used as a solvent [52]. A schematic phase diagram measured by them is shown in Fig. 7, reflecting the effect of AAm. This result suggested that AAm behaved like a cosurfactant and located in the interface between the microemulsions and the oil phase. Many authors also studied on O/W microemulsion polymerization of ST and MMA [37–43]. The emulsifiers with a high HLB value, such as SLS, cetyltrimethylammonium chloride (CTMA), and dodecyltrimethylammonium bromide (DTMA), were usually used. Gan et al. studied the microemulsion polymerization of MMA by using stearyltrimethylammonium chloride (STAC) as an emulsifier and KPS or AIBN [2,2'-azobis(isobutyronitrile)] as an initiator [42]. They changed the MMA concentration from 3 to 9 wt% and maintained the H_2O/STAC ratio at 8.1. They found that the polymerization rate was larger when using KPS than AIBN. This was because the solubility of AIBN in the aqueous phase (continuous phase) was lower than that for KPS, which resulted in the generating rate of radicals in the aqueous phase and, thus, the possibility of radicals entering into the micelles, was lower in the former case. It was concluded that the radicals from AIBN initially dissolved inside the micelles were terminated instantaneously by recombination and, hence, were not responsible for the initiation of polymerization. Antonietti et al. investigated the microemulsion polymerization of ST by using a CTMA/DTMA mixture as an emulsifier [39]. They found that adding 10 wt% of MMA or dimethylaminoethylmethacrylate (DMAEMA) polar monomer to styrene enhanced the stability of microemulsion and the smaller particles were obtained, compared with pure styrene. This result suggested that polar monomers were able to be used instead of a cosurfactant.

3. Advances in Microemulsion Polymerization

Many attractive advantages of microemulsion polymerization have been recognized. The molecular weight of the polymer obtained by microemulsion polymerization is higher (10^6–10^7 g/mol) than for conventional emulsion polymerization. Because the microsphere obtained by the microemulsion polymerization is very

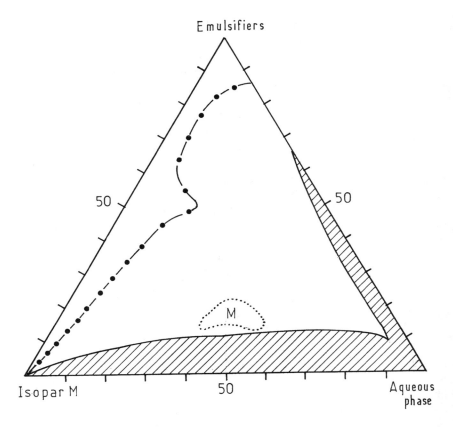

Figure 7 Pseudo-ternary phase diagram (wt%) reflecting the effects of monomers on the emulsion formation region. (–●–): aqueous phase composed only of water. (—): aqueous phase composed of water and monomers. In the aqueous phase, acrylamide/sodium acrylate = 7/3, (acrylamide + sodium acrylate)/water = 1.25. M area: polymerization system studied by Caudau et al. Emulsifier: mixture of sesquioleate sorbitan (Arlacel 83, HLB = 3.7) and polyoxyethylene sorbitol hexaoleate with 40 ethylene oxide residues (G1086, HLB = 10.2). Oil phase: isoparaffinic oil (Isopar M). (From Ref. 52.)

small, the film cast from such small microspheres shows high transparency and brightness; and because of its very small size, it can be used as drug carrier in venous injection of the drug delivery system. Another attractive advantage is that a uniform water-soluble polymer microsphere can be prepared by W/O microemulsion polymerization, which is difficult to provide by conventional emulsion polymerization.

Although microemulsion polymerization has many advantages, there are severe limitations concerning the emulsifier/monomer ratio in formulations (1–3) which is much higher than that used in conventional emulsion polymerization (0.01–0.03). This drawback restricts the potential uses of microemulsion polymerization, as high solid contents and low emulsifier levels are desirable for most applications. Candau et al. tried to start the W/O microemulsion polymerization of AAm in a phase-inversion region (bicontinuous region) from O/W to W/O microemulsion to enhance the monomer concentration, and they used a comonomer as a cosurfactant to lower the concentration of emulsifier. Furthermore, they tried to use a mixture of emulsifiers to adjust the HLB and to minimize the concentration of the emulsifier [52,53]. For example, they used a mixture of sesquioleate sorbitan (Arlacel 83, HLB = 3.7) and polyoxyethylene sorbitol hexaoleate with 40 ethylene oxide residues (G1086, HLB = 10.2) as the emulsifier in the isoparaffinic oil (Isopar M)/(AAm + sodium acrylate)–water system [52]. Both monomers acted as cosurfactants, leading to a considerable extension of the microemulsion region in the phase diagram, as shown in Fig. 7. They found that the optimum concentration of emulsifier was lowered when an optimum HLB value was realized by varying the mixing ratio of two emulsifiers. They also found that the optimum concentration of the emulsifier was lowered further by adding a comonomer (sodium acrylate) to the aqueous phase of AAm. They started the polymerization in the bicontinuous phase, where the concentration of monomer + water reached a maximum concentration (aqueous concentration: 50 wt%, AAm content: 25 wt%). During the polymerization, the bicontinuous phase transferred to the W/O microemulsion because the solubility of the obtained polymer in micelles became lower than that of the monomer, causing a change in the phase diagram. The size of the particles was found to decrease with increasing amount of sodium acrylate. More recently, microemulsion copolymerization of AAm and sodium 2-acrylamido-2-methylpropanesulfonate (NaAMPS) also was studied by Caudau et al. using the same emulsifier mixture [53].

E. Suspension Polymerization

1. General Descriptions

The microspheres in the size range from several to hundreds of micrometers can be obtained by suspension polymerization. The system usually consists of a hydrophobic monomer, water (medium), a stabilizer, and a hydrophobic initiator. The monomer droplets containing an initiator are formed by a mechanical stirring method. Because the size distribution of droplets is very broad, coalescence and redispersion of droplets will occur continuously during the polymerization. Stabilizer dissolved in the aqueous phase is adsorbed on the surface of monomer droplets to maintain their stability during polymerization. As a stabilizer, poly(vinyl alcohol) (PVA) and carboxymethylcelulose (CMC), and so forth are usually used.

In suspension polymerization, size and size distribution of droplets are controlled by stirring speed and the concentration of stabilizer. Therefore, the size distribution of obtained microspheres is very broad. However, because the process is simple and various functional substances can be incorporated into microspheres easily just by mixing them in the dispersed phase, this method is still a well-established method in the industrial field.

The drawback of broad size distribution of suspension polymerization was overcome to a certain extent by improving the dispersing equipment. Kamiyama et al. [54] devised a continuous-feed process in suspension polymerization to prepare composite particles from 2 to 20 μm. In their process, the monomer phase in which carbon black or magnetite was dispersed and the water phase were held in separate vessels. Then, they are fed to the disperser with a controlled rate. By this special dispersing technique, the particles with a relatively narrow size distribution were obtained.

2. SPG Emulsification Technique—A Special Suspension Polymerization Technique to Prepare a Uniform Microsphere

In recent years, Omi et al. [55,56] has developed a SPG (Shirasu Porous Glass) emulsification process combined with a subsequent polymerization process to prepare fairly uniform microspheres, instead of conventional suspension polymerization. The SPG membrane is a porous glass membrane with very uniform pores, consisting of the hydrophilic substance SiO_2–Al_2O_3 (Ise Chemical Co., Miyazaki Prefecture, Japan). Instead of the conventional mechanical stirring method, the droplets are prepared by permeating the oil phase containing a hydrophobic initiator through the uniform pores of a SPG membrane into the aqueous phase under a controlled pressure. A schematic diagram of equipment for preparing uniform droplet is shown in Fig. 8. Usually, a tubular membrane is used. The oil phase containing a monomer, a hydrophobe (hexadecane, lauryl alcohol, etc.), and a hydrophobic initiator stored in an oil tank is allowed to fill outside of the tubular SPG membrane by nitrogen pressure. After the pressure has attained a critical value, the oil phase passes through the uniform pores of the membrane to form uniform droplets. The stabilizer and emulsifier dissolved in the aqueous phase, which circulated continuously from the emulsion storage tank, will be adsorbed on the emerging droplets to stabilize them. After the emulsification process has been completed, the emulsion is moved to the reactor to be polymerized by elevating the temperature. Because the pores of SPG membrane are very uniform, the droplets obtained are fairly monodispersed (CV: around 10%) compared to the conventional suspension method. The size of droplets can be varied easily by changing the pore size of the membrane. At present, the SPG membrane with pore sizes from 0.5 to 18 μm is available. The size of droplets formed is about six times as large as the pore size of the membrane. In this method, the oil phase should be

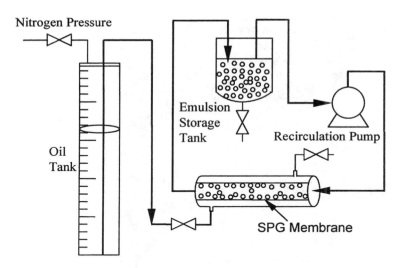

Figure 8 Schematic diagram of the SPG emulsification process. (From Ref. 55. J Appl Polym Sci. Copyright © 1994, John Wiley & Sons, Inc.)

very hydrophobic to obtain uniform droplets. It is because the SPG membrane is composed of a hydrophilic substance that it is easily wet by the hydrophilic monomer, generating polydispersed droplets. Usually, 5–10 wt% of a hydrohobe such as hexadecane is added to the oil phase to obtain the uniform droplets. The hexadecane has two roles. One is that it can enhance the hydrophobicity of the oil phase so that the pores of the membrane are difficult to wet. Another role is that it can prevent the monomer from diffusing into the aqueous phase (Ostwald ripening or degradative diffusion) to maintain the stability of the droplets, as in the case of a miniemulsion. The uniformity of droplets and the number of particles can be maintained during polymerization because the droplets are considerably stable due to their hydrophobicity and monodispersity. However, the emulsion polymerization (the secondary nucleation) is usually accompanied by or occurs dominantly in the aqueous phase because a small amount of initiator and monomer are soluble in the aqueous phase to enhance polymerization therein and form the secondary nuclei. Omi et al. found that it was effective to add a small amount of inhibitor to the aqueous phase to avoid the secondary nucleation. Hydroquinone (HQ) and $NaNO_2$ were found to be effective for the polymerization of ST and MMA, respectively. Some scanning electron micrographs (SEM) are shown in Fig. 9. When the inhibitor was not used in the polymerization of ST, the emulsion polymerization occurred and only small particles of submicron size were obtained. By adding 0.003% HQ (based on the aqueous phase), however, the large particles with almost the same size as the droplets were obtained and the conver-

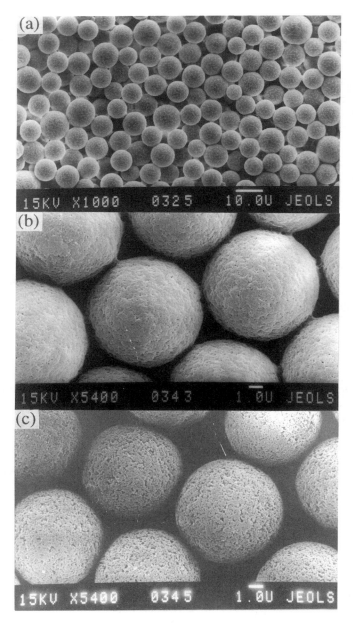

Figure 9 SEM micrographs of PST microspheres obtained by the SPG emulsification technique: (a) nonporous microsphere; (b, c) porous microsphere by using *n*-heptane as a diluent, and with ST/DVB/*n*-hexane ratios 40/10/50 and 25/25/50, respectively. (From Ref. 55. J Appl Polym Sci. Copyright © 1994, John Wiley & Sons, Inc.)

sion was maintained at a reasonably high value (around 90%), as shown in Fig. 9a. Furthermore, the uniform porous microspheres were obtained by adding a diluent and a cross-linking agent in the oil phase. Two examples are shown in Fig. 9b and 9c where *n*-hexane was used as the diluent, and the ST/DVB (divinylbenzene)/*n*-hexane ratio was 40/10/50 and 25/25/50, respectively. It is known from the SEM micrographs that the size of micropores can be adjusted by varying the concentration of the cross-linking agent.

As described above, fairly uniform large microspheres (diameter: from several micrometer to 100 μm; CV value: ~10%) can be obtained in water by a single-step SPG emulsification polymerization technique. It is important for its environmental protection. The uniform, large, porous microspheres can be used as carriers of biomaterials in the packing column because of their monodispersity and large specific surface area.

Another excellent advantage of the SPG membrane emulsification technique is that it can be used in the preparation of uniform particles composed of preformed polymer—for example, poly(lactide) [57], protein, and other polymers that cannot be prepared by radical polymerization. Furthermore, the functional substance also can be incorporated into the particles. We have tried to incorporate a polyimide prepolymer [58], a magnetite [59], into microspheres, and monodispersed microspheres containing these substances were obtained. However, this hydrophilic SPG membrane is not suitable for the preparation of microspheres composed of hydrophilic monomers such as MMA and 2-hydroxyethyl methacrylate (HEMA). Omi et al. also developed a special swelling method for droplets to prepare microspheres composed of hydrophilic monomers. The detailed technique will be introduced in Sec. II.H.

F. Dispersion Polymerization

1. General Descriptions

The size of microspheres obtained by dispersion polymerization ranges from submicron to several micrometers. This method can provide microspheres with a very narrow size distribution and can be used for various monomers from hydrophobic to hydrophilic monomers by choosing suitable solvent or a mixture of solvents. Because the monodispersed microspheres with several micrometers are difficult to obtain by emulsion, miniemulsion, microemulsion, and conventional suspension polymerization, this method is very attractive and was largely developed during the last decade. The polymerization mechanism is shown schematically in Fig. 10. Quite different from the above methods, the initial system is a homogeneous phase composed of a monomer, an initiator, an organic medium, which should be a good solvent for the monomer, and an initiator but a poor solvent for formed polymer, and a stabilizer which should show affinity with both the medium and the formed

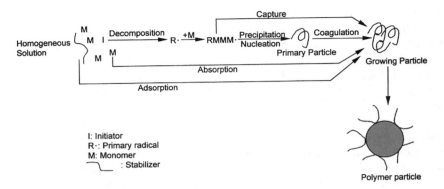

Figure 10 Schematic homogeneous nucleation mechanism in dispersion polymerization.

polymer. After the temperature is raised, the initiator decomposes to generate radicals which add to the monomer in the solvent. After the growing polymer chain reaches a critical length, it precipitates from the medium and becomes a primary nucleus. As in the case of the homogeneous nucleation mechanism in emulsion or emulsifier-free polymerization, the primary nuclei coagulate to a stable one (growing particle) which is stabilized by the stabilizer. Then the growing particles absorb monomers and capture oligomeric radicals and primary nuclei formed in the medium, and the polymerization site shifts to the growing particles from the continuous phase.

This method has some features in common with homogeneous nucleation of emulsion polymerization or emulsifier-free polymerization. All of them involve a nucleation process, a coagulation process of primary nuclei, and a growing process of the monomer–polymer particles. The nucleation process is completed within the initial stage (about 10 wt% of monomer conversion), and the nuclei are allowed to grow for a long period. However, the critical length of the oligomer at which the primary nuclei precipitate from the medium in dispersion polymerization is longer than that in emulsion polymerization, because organic solvent instead of water is used in the dispersion polymerization. Usually, the compatibility of an organic solvent with a polymer is better than water for allowing a longer oligomer to be dissolved in the solvent. This is also the reason why larger microspheres were obtained in the dispersion polymerization, compared to emulsion polymerization.

2. Advances in Stabilizers and Their Advantages

Poly(vinyl alcohol) (PVA) and poly(vinyl pyrrolidone) (PVP) are usually used as the stabilizer. These stabilizers are physically adsorbed on the surface of formed particles to stabilize them, although a small part of PVP is probably grafted on particles. However, it is found that because only a small amount of stabilizer added is adsorbed on the particles, it is necessary to add a large amount of stabilizer (above 10 wt% of monomer) to obtain a stable monodispersed microsphere. A block/graft copolymer, one sequence of which is compatible with formed polymer and another is compatible with the medium, is found to be a more effective stabilizer than the conventional stabilizers. Different from physical adsorption of usual stabilizers, one sequence of the block/graft copolymer can be anchored in the formed particles, so that the stabilizer can stabilize the formed particles more effectively. It is found that adding only one-tenth of a block/graft copolymer to conventional stabilizers is sufficient to stabilize the formed particle. For the same reason, macromonomers are also considered to be the effective stabilizers for the dispersion polymerization. Takahashi et al. [60] used a PST and a PST macromonomer (M-PST), and a PMMA and a PMMA macromonomer (M-PMMA) as the stabilizers in the dispersion polymerization of 2-hydroxyethylmethacrylate (HEMA) with a butanol/toluene (55/45 w/w) mixture as a medium. They found that the monodispersed particles were obtained only when macromonomers were used. Because the PHEMA-g-PMMA or PHEMA-g-PST graft copolymer is formed in situ when using the macromonomer as a stabilizer, the side chain of a graft copolymer is compatible with the medium and the main chain, which has the same monomer units with formed particles, can be anchored into the particles. Kobayashi et al. [61] used a poly(2-oxazoline) [poly(ROZO)] macromonomer with a styryl terminal (**X**) to prepare the PMMA microsphere in the methanol/H_2O (70/30 wt/wt) medium by using 2,2'-azobis(2,4-dimethylvaleronitrile) (ADVN) as an initiator. They found that the macromonomer with a higher molecular weight was more effective than that with a lower molecular weight, and the diameter of particles decreased with an increasing in amount of poly(ROZO). They also found that the poly(ROZO) macromonomer with a methyl substitute was most effective than other substitutes. This was because poly(ROZO) with a methyl substitute showed the best compatibility with the methanol/H_2O solvent. Furthermore, only adding 0.2 wt% of poly(ROZO) based on MMA was sufficient, whereas it was necessary to add at least 10 wt% of poly(2-ethyl-2-oxazoline) homopolymer without a terminal double bond to stabilize the obtained particles.

R: Methyl ($\overline{M}_w$=3900, 1200, 700), Ethyl ($\overline{M}_w$=3500, 1200), n-Propyl ($\overline{M}_w$=4200) (**X**)

Lacroix-Desmazes and Guyot [62] used several kinds of poly(ethylene oxide) macromonomers with mathacrylate (MA), styrene (ST), maleate (MAL), and methacryloyl isocyanate (MAI) terminals to prepare PST microspheres in an ethanol/water (70/30 wt/wt) medium by using AIBN as an initiator. Compared with PVP (K30), less stabilizer was needed for the production of the same particle size, less coagulum was formed, and the particles obtained were more uniform when a macromonomer was used.

Recently, Nakamura et al. [63] used the special macroazoinitiator (MAI) composed of successive poly(dimethylsiloxane) (PDMA) and azo group (**XI**) in dispersion polymerization of MMA. By using this initiator, poly(MMA-*b*-PDMS) was formed and served as an in situ formed dispersion stabilizer. They obtained a uniform PMMA-*g*-PDMA composite particle with a diameter of about 250 nm when using heptane/isopropyl alcohol mixture as a mixed medium. The microsphere showed a core–shell morphology with the PDMA-rich domain as the shell.

$$\left[CO(CH_2)_2\overset{CH_3}{\underset{}{C}}N=N\overset{CH_3}{\underset{}{C}}(CH_2)_2CONH(CH_2)_3(\underset{CH_3}{\overset{CH_3}{Si}O)_xSi}(CH_2)_3NH\right]_n$$

(X=70-80, n=7-8)   (**XI**)

As described above, the graft or block copolymer can be obtained by using a macromonomer or a macroinitiator. This advantage of dispersion polymerization was very attractive for researchers in the study of the preparation of graft/block copolymers.

3.   Advances in Preparation of Functional Polymeric Microspheres

(a)   Porous Microspheres.   Different than emulsion polymerization, the dispersion polymerization is very flexible method because many solvents or mixed solvents can be selected to correspond to the respective formed polymer. The microspheres composed of polar monomers can also be prepared, as well as nonpolar monomers. One interesting example is poly(acrylonitrile) (PACN) porous microspheres developed by Uyama et al. [64]. They selected a DMF/methanol mixture as a solvent and found that 70–85 wt% of DMF was most suitable for the preparation of uniform PACN particles when using PVP as a stabilizer. It is very interesting that the porous microspheres with a particular morphology were obtained as shown in Fig. 11. Because PACN cannot be swollen by ACN monomer inside growing polymer particles, the phase separation occurred between the PACN and ACN monomers to yield porous microspheres. Usually it is difficult to prepare porous microspheres by dispersion polymerization because the porogen, which is

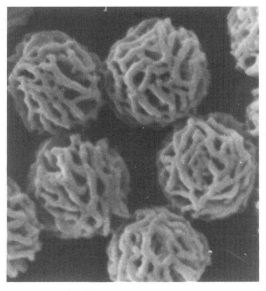

———————— 1 μm

Figure 11 Poly(acrylonitrile) porous microsphere prepared by dispersion polymerization using DMF/methanol mixture as a medium. (From Ref. 64.)

soluble in the initial homogeneous phase, will remain in the medium even after the polymerization. This is very important and is the first case reported.

(b) Highly Cross-Linked Microspheres. It is thought that highly cross-linked polymer particles are difficult to prepare directly by dispersion polymerization because of the instability of particles during polymerization, although there is a strong demand for highly cross-linked polymer particles with superior heat resistance and solvent resistance to serve as various spacers, slip property improvers for plastic films, and so forth. In recent years, however, El-Aasser et al. [65] successfully prepared poly(ST-*co*-DVB) particles in methanol medium by dispersion polymerization with AIBN as an initiator and PVP (K30) as a stabilizer. They observed the very surprising result that the uniform highly cross-linked poly(ST-*co*-DVB) particles were obtained only when the DVB content was higher than 11 wt% (based on total monomer weight). With less DVB, coagulation of the polymer resulted. More interestingly, they found that adding 4 mL of air into the reactor (1-oz. bottle) before polymerization was able to form a stable dispersion, whereas coagulation of the dispersion resulted when no air or less air was added. From these results, they proposed that oxygen probably promoted the formation of

DVB-*g*-PVP to form a smaller, stable particle. The detailed mechanism of PVP grafted on the particle proposed by them was as follows. First, AIBN decomposed to yield two primary radicals (2R·) and a nitrogen molecule. The oxygen reacted rapidly with a primary radical to form a peroxy radical (ROO·). Then, this radical abstracted hydrogen from PVP, creating an active site to which DVB started an addition. Because the growing particle did not absorb the DVB monomer because of its high cross-linking density, it was concluded that the growth of the particle occurred on the surface of the growing particle by adsorbing oligomers or primary nuclei formed in the medium. However, there are few reports on the preparation of other cross-linked polymeric microspheres.

(c) Microspheres Incorporating Functional Substances.   As in the case of porogen, functional substances are difficult to incorporate into microspheres by dispersion polymerization if they are soluble in the initial homogeneous system. Being different from a monomer, it is difficult for functional substances to become insoluble in the medium as the polymerization progresses. Recently, in order to incorporate a functional substance, a polyimide (PI) prepolymer, into polystyrene particles, Omi et al. [66] tried a dispersion polymerization system where PI and ST were dissolved in an isopropanol/2-methoxyethanol mixture and AIBN was used as an initiator and PVP (K30) as a stabilizer. Because only ST is a good solvent for PI, it is expected that PI can be incorporated into PST microspheres as the solubility of PI in continuous phase decreases gradually during the polymerization of ST. However, the obtained particle showed a broad size distribution, because of the negative effect of PI. In order to reduce the nucleation period so that we can obtain microspheres with narrow size distribution, we tried to add a small amount of L-ascorbic acid to remove oxygen soluble in the system. As a result, the very monodispersd (CV = 1.37%) and smaller particles were obtained as shown in Fig. 12. From the fact that the size and the size distribution of microspheres became much lower than the case without adding L-ascorbic acid, it was proposed that the oxidized form of L-ascorbic acid was also responsible for the abstraction of a hydrogen atom from the PVP chain to promote the formation of PVP–PST graft copolymers, which stabilized the precipitating chains in the nucleation period. The proposed stabilization mechanism of particles by L-ascorbic acid is shown in Fig. 13. The cross-linked microspheres can be obtained by post-heat-treatment after polymerization, and the cross-linking density can be adjusted easily by varying the feed amount of PI. The microsphere containing PI showed heat resistance and high mechanical strength, and can be used as a spacer of the liquid-crystal display and an adhesive agent of electron parts.

(d) Magnetic Polymer Microspheres.   Li and Sun [67] prepared a core–shell composite magnetic polymeric microsphere containing a magnetite core and a polymer shell by dispersion copolymerization of ST and HEMA in the presence of Fe_3O_4 powder. First, they dispersed Fe_3O_4 powder with ultrasonification in

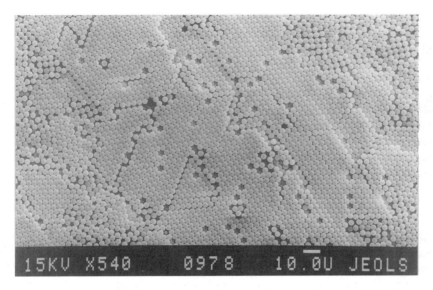

Figure 12 A monodispersed polystyrene microsphere containing a polyimide prepolymer by dispersion polymerization in the presence of L-ascorbic acid (diameter = 3.08 μm, CV = 1.37%).

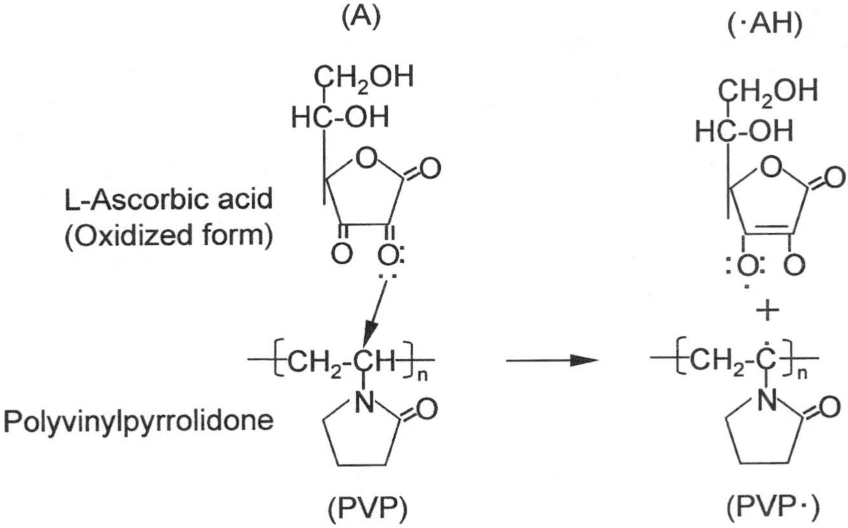

Figure 13 Proposed stabilization mechanism of particles prepared by dispersion polymerization in the presence of L-ascorbic acid. (From Ref. 66. J Appl Polym Sci. Copyright © 1998, John Wiley & Sons, Inc.)

poly(ethylene glycol) (PEG) solution to improve the affinity between the powder and the monomer and initiator. After this solution was fed into the H_2O/ethanol medium, dispersion copolymerization of ST and HEMA was carried out by using KPS as an initiator. By this process, the KPS initiator decomposed at a certain temperature to form radicals abstracting hydrogen atoms of PEG molecules on the surface of the Fe_3O_4 powder, and the Fe_3O_4 powder became the initiating site. Finally, core–shell composite particles with a narrow size distribution were obtained. They applied the magnetic polymer particles for immobilization of proteinase of *Balillus sublitis* by utilizing the hydroxyl group on the surface and the benzoquinone spacer. They found that the stability of the immobilized enzyme was obviously improved and the microspheres carrying the enzyme can be separated from the medium quickly in a magnetic field of 4200 G.

(e)   Microspheres Polymerized in Supercritical Carbon Dioxide.   DeSimone et al. [68] developed dispersion polymerization of MMA by using supercritical CO_2 as a dispersing medium. Supercritical fluids are prepared at a temperature and pressure above their critical values. They exhibit "liquidlike" densities, and thus liquidlike solvent power, while also revealing "gaslike" viscosities under the same conditions. The supercritical CO_2 is a very interesting solvent. It was found that the monomers which were soluble in hexane also were soluble in supercritical CO_2. However, supercritical CO_2 is a poor solvent for usual hydrophobic or hydrophilic polymers, whereas it is a good solvent for fluoropolymers and silicone polymers. Considering such distinguished properties of supercritical CO_2, Desimone et al. used poly(1,1-dihydroperfluorooctylacrylate) [poly(FOA)] (**XII**) as a stabilizer in the dispersion polymerization of MMA. The acrylic backbone will be preferentially adsorbed onto a PMMA colloidal particle, providing a surface anchor. The CO_2-philic fluorocarbon segment extends the poly(FOA) chain into the continuous phase, thereby preventing flocculation of particles through a steric stabilization mechanism. They carried out the polymerization by using AIBN or fluorinated AIBN (**XIII**) as an initiator at 65°C under a pressure of 204 ± 0.5 bars. The amount of MMA and initiator were 0.02 mol and 4×10^{-5} mol, respectively. Two stabilizers with different molecular weights ($M_n = 1.1 \times 10^4$ and 2×10^5 g/mol) were used. The conversion was very low and the resultant PMMA accumulated as a thick irregular film in the absence of stabilizer. As long as enough stabilizer (2–4 wt%/v) was added, the stable colloidal microspheres with a narrow size distribution were obtained with a higher conversion (above 90%). The diameter was varied from 0.9 to 2.7 μm. No major difference was observed in the use of AIBN versus fluorinated AIBN.

By using CO_2 as the medium, the problems in the posttreatment of organic medium used in dispersion polymerization can be solved, and the powder product can be obtained easily, just by releasing CO_2 below the critical point. Therefore, this process has a potential in industries if the supercritical CO_2 is supplied within a tolerable price range.

$$-\left[CH_2\text{-}CH\right]_n\left.\begin{array}{l}\\ \overset{|}{C}=O \\ \overset{|}{O} \\ \overset{|}{C}H_2 \end{array}\right\} \text{``PMMA-philic'' anchor}$$

$$\left.\begin{array}{l}(\overset{|}{C}F_2)_6 \\ \overset{|}{C}F_3\end{array}\right\} \text{``CO}_2\text{-philic'' steric stabilizing moiety}$$
$$\text{(XII)}$$

$$C_8F_{17}CH_2CH_2OOC(CH_2)_2\text{-}\overset{\overset{\displaystyle CH_3}{\displaystyle |}}{\underset{\underset{\displaystyle CN}{\displaystyle |}}{C}}\text{-}N\text{=}N\text{-}\overset{\overset{\displaystyle CH_3}{\displaystyle |}}{\underset{\underset{\displaystyle CN}{\displaystyle |}}{C}}\text{-}(CH_2)_2COOCH_2CH_2C_8F_{17}$$
$$\text{(XIII)}$$

4. Applications of Dispersion Polymerization on Nonradical Polymerization

Although the dispersion or precipitation polymerization method, which will be described in Sec. II.G, accompanies a problem of posttreatment of organic solvents, this method becomes more and more attractive in recent years because it has been developed for nonradical polymerization—for example, chemical oxidative polymerization, ring-opening polymerization of cyclic ester, enzymatic polymerization, and so forth. A couple of examples are given in the following subsections.

(a) Biodegradable Microspheres Prepared by Ring-Opening Polymerization. Sosnowski et al. [69] applied dispersion polymerization to anionic ring-opening polymerization of ε-caprolactone in heptane/1,4-dioxane (4/1, v/v) medium and successfully obtained monodispersed poly(L,L-lactide) microspheres. They used the poly(dodecylacrylate)-g-poly(ε-caprolactone) graft copolymer as a stabilizer and investigated the effects of the molecular weight of the copolymer, the number of grafted poly(ε-caprolactone) per copolymer, the molecular weight of poly(ε-caprolactone) on the size, and the size distribution of the obtained microspheres. The microsphere with the narrowest size distribution ($D_n = 3.44\ \mu m$, $D_v/D_n = 1.08$) was obtained when the average number of grafted poly(ε-caprolactone) per copolymer was 1.3 ($M_n = 4700$ g/mol), the concentration of copolymer was 1.6 (g/L), which was lower than the CMC (CMC = 5.1 g/L). By using the anionic polymerization method [catalyst: Sn(II)2-ethylhexanoate], the optical purity was maintained during polymerization; that is, the optical purity of poly(L,L-lactide) corresponded with that of the monomer. By dispersion polymerization, disadvantages related to the high viscosity of polymerizing mixtures, difficulties in con-

trolling the trans-esterification side reactions, and the molecular weight distribution which existed in conventional solution or bulk polymerization can be solved.

(b)   Electrically Conducting Microsphere Prepared by Oxidative Polymerization.   According to the definition of dispersion polymerization given earlier, it is possible to carry out the dispersion polymerization in water solvent if the monomer is soluble but the polymer is insoluble in water. Armes et al. [70] prepared polypyrrole particles by an aqueous dispersion polymerization technique at 20°C for 18–24 h using ferric chloride as an oxidant and PVA as a stabilizer. They obtained monodispersed particles in the approximate size range from 100 to 150 nm in diameter. The concentrations of the monomer and $FeCl_3 \cdot 6H_2O$ were maintained at 0.14 (mol/L) and 0.33 (mol/L) based on water. The PVA/pyrrole mass ratio was varied from about 0.1 to 0.6. The diameter and conductivity of polypyrrole particles were found to decrease with the increase of PVA concentration. At the optimum PVA/pyrrole mass ratio, the electrical conductivity of the particles was ~5Ω/cm. There are considerable advantages to use this route to prepare electrically conducting particles. First, no postreaction doping step is required because the Cl^- ion of $FeCl_3$ associates with pyrrole in the polymer chain. Second, the polymerization can be carried out under ambient conditions with no special precautions taken to exclude oxygen.

Masuda et al. [71] synthesized polyxylidine particles in an aqueous medium by oxidative dispersion polymerization. The system is composed of xylidine (0.14 g), ammonium persulfate (0.527 g) as an oxidant, PVA (0.014 g) as a stabilizer, and water (35 g) as a medium. They obtained needlelike particles with diameters from 100 to 200 nm and an average aspect ratio of 13.

The conducting polymer microspheres can be used in the adhesion of electron parts.

(c)   Polyphenol Microspheres Prepared by Enzymatic Polymerization.   Uyama et al. [72] synthesized polyphenol microspheres by enzymatic dispersion polymerization by using Western horseradish peroxidase (HRP) as an enzyme and 1,4-dioxane/phosphate buffer (pH = 7) (60/40 v/v) as a medium. They tried four kinds of stabilizer: poly(acrylic acid) (PAAc), poly(ethylene glycol) (PEG), PVA, and poly(vinyl methylether) (PVME). They found that the submicron-sized microspheres with a relatively narrow size distribution were obtained only when using PVME as a stabilizer. By changing from 5 to 50 wt% based on phenol, the diameter decreased with the increase of the stabilizer PVME. They also investigated the effect of the pH value of phosphate buffer. The relatively large particles were obtained at pH = 4–6 or 8, compared to pH = 7. Furthermore, they obtained the carbon particle without losing the sphericity of the particle by heat treating the polyphenol particles at 1000°C under nitrogen atmosphere.

(d)   Poly(amino acid) Microspheres Prepared by Catalytic Polymerization.   Yamagawa et al. [73] synthesized poly(amino acid) microspheres with the diame-

ter ranging from several micrometers to several tens of micrometers from γ-ben-zyl–gulutamate-N-carbonyl acid dehydrate (BLG–NCA) by using triethyl amine as a catalyst and water as a medium at room temperature for 5 h. The synthetic process is shown in structures (**XIV**). They tried to use nonionic, anionic, and cationic surfactants as the stabilizer and found that the nonionic surfactant with a high HLB value (HLB = 16–18) was most suitable for this system. They found that BLG–NCA did not hydrolyze even in the water if the high HLB nonionic surfactant was present in the water, although it hydrolyzed in the pure water instantaneously. The obtained particles were porous, probably because poly(amino acid) was insoluble in the BLG–NCA monomer, just as in the case of poly(acrylonitrile) described earlier. Because the poly(amino acid) microsphere shows biocompatibility and biodegradability, it can be used as a carrier of biomaterials, drugs, and so forth.

| BLG | BLG-NCA | PBLG | (**XIV**) |

G. Precipitation Polymerization

1. General Descriptions

The difference between precipitation and dispersion polymerization methods is that no stabilizer is used or that a stabilizer is formed in situ during precipitation polymerization by using a medium-compatible comonomer. This difference is similar to that between emulsion and emulsifier-free polymerization. For example, a poly(acrylamide-co-methacrylic acid) [poly(AAm-co-MAAc)] copolymer stabilizer rich in methacrylic acid units can be formed by adding a small amount of methacrylic acid (MAAc) when poly(acrylamide-co-methylenebisacrylamide) [P(AAm-co-MBAAm)] microspheres were prepared by precipitation polymerization in ethanol solvent [74]. MAAc units in the poly(AAm-MAAc) copolymer is compatible with ethanol, and AAm units can be anchored into particles. If MAAc was not added, a large coagulum was formed. On the other hand, a very uniform particle with a diameter about 1 μm was obtained when MAAc was added in the polymerization system.

As in the case of dispersion polymerization, it is also possible to carry out precipitation polymerization in water if the monomer is soluble but the polymer is insoluble in water. Kawaguchi et al. [75] prepared poly(N-isopropylacrylamide) (PNIPAM) microspheres from a NIPAM/MBAAm mixture at 70°C in water. This is because that PNIPAM shows hydrophobicity and becomes insoluble in water at 70°C, whereas a NIPAM monomer is soluble in water. A PNIPAM microsphere shows a very interesting characteristic. It swells below 32°C but shrinks above 32°C. Other characteristics such as surface potential and dispersion stability also change noncontinuously at around 32°C. Based on these interesting phenomena, PNIPAM microspheres have been applied to biomedical and bioseparation fields.

2. Advances in Preparation of Functional Polymeric Microspheres

(a)   Thermosensitive Magnetic Microspheres.   Kondo et al. [76] developed the precipitation polymerization for the preparation of thermosensitive magnetic microspheres and applied the microsphere obtained to the immunoaffinity purification of antibodies. Similar to the preparation of polymer magnetic microsphere in emulsifier-free polymerization, they used magnetites as the nuclei. First, the magnetite (ferrofluid) was coated with PST by the polymerization of ST for 6 h at 70°C using KPS as an initiator. Then, NIPAM, MAAc, and KPS were added, and the secondary polymerization was performed for 24 h. The obtained magnetic polymer microspheres showed thermosensitivity originated from PNIPAM, although the particles with higher magnetite content were less thermosensitive. Bovine serum albumin (BSA) was covalently immobilized on the thermosensitive magnetic polymer particles by the carbodiimide method. Thermosensitive polymer microspheres with immobilized BSA were thermoflocculated and separated quickly by applying a magnetic field. Moreover, they were successfully used for immunoaffinity purification of anti-BSA antibody from antiserum. Therefore, thermosensitive magnetic polymer microspheres are proved to be novel support materials for affinity purification of bioproducts.

(b)   Heat-Resistive Polyimide Microsphere.   As the case of dispersion polymerization, the precipitation polymerization method has also been developed for nonradical polymerization. An interesting example is the preparation of polyimide microspheres.

Nagata et al. [77] synthesized 1–10-μm polyimide particles with high crystallinity by precipitation polymerization from various acid anhydrides and diamines in N-methylpyrrolidone medium. Pyromellitic dianhydride (PMDA) or biphenyl tetracarboxylic dianhydride (BPDA) was used as acid anhydride, and p-phenylene diamine (PPD), m-phenylene diamine (MPD), or 4,4'-diaminodiphenylether (DPE) was used as the diamine. The structural formulas of the above compounds are shown in structures (XV)–(XIX), and synthetic process is shown in structure (XX). They first prepared poly(amide acid) by using acid anhydride and diamine, then heated this homogeneous solution at 150–200°C for 2–4 h. By heat treatment, the polyimide was formed and precipitated from the medium to form the particles. The amide acid residues played a role to stabilize the formed particles because the amide acid was soluble in the solvent. Then, they treated the particles at 400°C for 1 h to obtain particles with high crystallinity. The particles

Figure 14 SEM micrographs of polyimide microspheres prepared from (a) PMDA/PPD, (b) BPDA/PPD, (c) PMDA/DPE, (d) PMDA/MPD, and (e) BPDA/MPD. (From Ref. 77.)

prepared by this method showed higher crystallinity than general film products. These particles showed interesting morphologies as shown in Fig. 14. Furthermore, they obtained carbon particles without losing the spherical shape by treating the polyimide microspheres at 1000°C. The adsorption property and electric property of the carbon particles were noticed.

(PMDA)   (XV)

(BPDA)   (XVI)

(PPD)   (XVII)

(MPD)   (XVIII)

(DPE) or (ODA)   (XIX)

(Dianhydride)   (Diamine)   (Polyamic acid)

(Polyimide)   (XX)

Kawaguchi et al. [78] used a different precipitation technique to prepare monodispersed poly(imide) microspheres and studied the control of size and size distribution. First, they used pyromellic anhydride (PMDA) (**XV**) and oxydianiline (ODA) (**XIX**) to prepare poly(amide acid) particles at room temperature for 30 min in triethylamine (TEA)/acetone solvent, based on the fact that poly(amide acid) was not insoluble in acetone. Then, they carried out a posttreatment to modify poly(amide acid) particles to poly(imide). TEA worked not only as a catalyst of polymerization but also as a good solvent for poly(amide acid). Therefore, the obtained microspheres were stabilized by the affinity between poly(amide acid) and TEA. They also investigated the effect of triethylamine on the diameter of the poly(amide acid) microsphere.

Poly(imide) microspheres revealed novel properties such as heat resistance, high mechanical strength, electric insulation, and solvent resistance; it can be used in electron and electrical industrial fields.

As described above, an important drawback of dispersion or precipitation polymerization is that porous microspheres cannot be obtained and functional substances are difficult to be incorporated into microspheres, although monodispersed porous and functional microspheres are very important. Therefore, it is a challenging task to develop a dispersion polymerization process to overcome these problems.

H. Seeded Polymerization and Swelling Techniques

Seeded polymerization is a potential technique for preparing various functional polymeric microspheres. The system is composed of seed particles (or seed droplets), medium, an initiator, monomer droplets, and occasionally, a swelling agent and a stabilizer. After the swelling process is complete, the polymerization is performed. The seed particles can be of an origin from among the above methods; the medium and initiator should be selected according to purpose. By seeded polymerization, enlarged microspheres, composite, and irregular-shaped microspheres can be obtained by selecting a suitable monomer, initiator, and medium. Some detailed methods among seeded polymerization are described as follows.

1. Enlarged Microspheres

As described earlier, the diameter of the microsphere obtained in emulsion polymerization is from several tens to hundreds of nanometers and that obtained in dispersion polymerization is in the range of microns. However, the monodispersed micron-sized functional particles, such as porous particles and cross-linked particles, are difficult to prepare as described above. Furthermore, there were few methods to prepare relatively large particles around or more than 10 μm in diameter by one-step polymerization before the SPG emulsification polymerization was developed. Therefore, seeded polymerization was studied exhaustively by using particles obtained by emulsion and dispersion polymerization as the seeds for obtaining the desired large functional particles.

Generally, the seeded polymerization is carried out after the seed is swollen by the monomer until the swelling equilibrium is attained. The swelling process proceeds by monomer diffusion through the continuous phase from monomer droplets. Because the monomer absorption rate of every seed from the aqueous phase is almost the same due to the uniform size of the seed, the enlarged uniform particles can be obtained by seeded polymerization. However, the swelling degree of the seed polymer particle is usually limited, and the desired enlarged polymer particles cannot be attained only by one step of seeded polymerization. Some techniques have been developed in order to accomplish this purpose.

(a)   Activated Swelling Method.   The epoch-making swelling method, called the activated swelling method, was developed by Ugelstad et al. [79,80]. The detailed method is as follows. A water-insoluble low-molecular-weight compound or oligomer (Y compound) is incorporated into the seed particle at the first step; then the seed particle containing Y compound is swollen by monomer at the second step. After the seed-containing Y compound is mixed with a monomer droplet (Z), as shown in Fig. 15, an equation describing semiequilibrium distribution of the monomer (Z) can be written as follows:

$$\ln \phi_{Za} + (1 - m_{ZY})\phi_{Ya} + (1 - m_{ZP})\phi_{Pa} + \phi^2_{Ya}\chi_{ZY} + \phi^2_{Pa}\chi_{ZP} + \phi_{Ya}\phi_{Pa}(\chi_{ZY}$$
$$+ \chi_{ZP} - \chi_{YP} \, m_{ZY}) + (2\bar{V}_Z/RT) \, (\gamma_a/r_a - \gamma_b/r_b) = 0$$

where **a** denotes the seed phase containing polymer P, the Y compound, and the Z monomer, and **b** denotes monomer droplet phase. r_a and r_b are the radii of the swollen particles and the droplets of Z at equilibrium, respectively and γ_a and γ_b are the corresponding interfacial tension. ϕ_{ia} denotes the volume fraction of compound i in **a** phase, $\bar{V}_i$ is the partial molar volume of compound i, $m_{ij} = \bar{V}_i/\bar{V}_j$, and χ_{ij} is the interaction parameter between compounds i and j. In the case when no polymer is present, terms with index P should be omitted. Due to the very high molecular weight of the polymer, m_{ZP} may be set to zero. The swelling capacity can be calculated from the above equation. Let us compare two extreme cases where the seed is only composed of polymer P or compound Y, respectively. If Y is a compound with a chain length 5, m_{ZY} is equal to 0.2. Therefore, $(1 - m_{ZY})\phi_{Ya}$ = 0.8ϕ_{Ya} and $(1 - m_{ZP})\phi_{Pa} = 0$ when the seed is only composed of Y. $(1 - m_{ZY})\phi_{Ya}$ = 0 and $(1 - m_{ZP})\phi_{Pa} = \phi_{Pa}$ when the seed is only composed of a polymer. Comparing the left side of the above equation for the above two extreme cases, it can be known that ϕ_{Za} is larger when the seed is composed of Y than a polymer; that is, the seed can be swollen by a large amount of Z. Ugelstad et al. have calculated the effect of the amount (V_Y) and the chain length of the compound Y on the swelling capacity using a value of $\gamma_a/r_0 = 5 \times 10^{-4}$ N/m^2 (r_0: initial radius of the

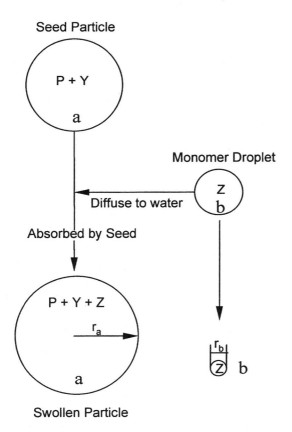

Figure 15 Schematic diagram of swelling process by using the seed particle containing the Y compound.

seed). The result is shown in Table 1. It can be seen that the swelling capacity increases with increasing m_{ZY} (decreasing chain length) and V_Y. The Y compound can be incorporated into seed particles by the swelling method. Because the Y compound is water insoluble, it is difficult to incorporate it into seed particles because the swelling process proceeds by diffusion from oil droplets through the aqueous phase. However, the purpose can be attained by finely dispersing droplets of Y or adding organic solvent to the aqueous phase to raise the diffusion rate or solubility of the Y compound. The oligomer can be incorporated into the seed by polymerization in the presence of the transfer agent after the seed has been swollen by the monomer and transfer agent. By this activated swelling method, the porogen, cross-linking agent, and various functional monomers can be introduced

Table 1   Swelling Capacity of Polymer Particle Containing the Y Compound, $V_Z/(V_Y + V_P)$

m_{ZY}	V_Y				
	1	0.75	0.5	0.2	0.1
0.2	390	250	135	34	13
0.1	135	85	48	13	9.3
0.05	48	31	17	7.3	5.7
0.02	13	10	7.3	5.3	4.9

Note: $\chi_{ZY} = \chi_{ZP} = 0.5$, $\chi_{YP} = 0$, $\gamma_a/\gamma_b = 5 \times 10^4$ N/m², $\overline{V}_Z = 10^{-4}$ m³/mol, $V_Y + V_P = 1$, $T = 323$ K. With $V_Y = 0$, the value $V_Z/(V_Y + V_P) = 4.5$. V_Y, V_Z, and V_P represent the volumes of the Y compound, monomer Z, and the polymer.
Source: Ref. 80.

into particles to produce porous, cross-linked particles and those with functional groups, as well as the enlarged particles. These monodispersed particles have been used as chromatographic supports in size-exclusion chromatography, high-performance affinity chromatography, protein recovery, ion-exchange chromatography, flow cytometric immunoassay, and so forth. The excellent performance of chromatography (uniform flow velocity, low back pressure, and high resolution) are obtained by using these very uniform particles, because the molecular diffusion condition over the whole column volume is very homogeneous.

(b)   Dynamic Swelling Method.   Okubo et al. [81] developed a dynamic swelling method in seeded dispersion polymerization by using a micron-sized particle obtained in dispersion polymerization as a seed. First, they dispersed the seed, monomer, and hydrophobic initiator into the methanol/water mixture, then continuously added water slowly into the system to decrease the solubilities of the monomer and initiator in the medium and allow them to be absorbed by the seed particles. By this method, a 1.8-μm-sized PST particle absorbed about 100 times the styrene monomer; finally, a 6.1-μm-sized monodispersed polystyrene particle was obtained by subsequent polymerization. Furthermore, they devised a cooling process after the dynamic swelling process [82] to further decrease the solubility of the monomer in the medium. As a result, a 7.7-μm-sized monodispersed particle was obtained.

(c)   Droplets Swelling Method.   Omi et al. [83–85] developed a special droplets swelling method to prepare very large particles up to above 100 μm by using uniform droplets with diameters up to ~ 30 μm. The uniform seed droplets composed of a hydrophobic substance are prepared by the SPG emulsification technique; a secondary finely dispersed emulsion composed of monomer or a monomer mixture, which hydrophilicity should be larger than that of seed emulsion, is prepared

by the homogenizer. By mixing these two emulsions, the secondary emulsion will be absorbed by the uniform seed emulsion by the diffusion process through the aqueous phase. The larger the difference of hydrophilicity of two emulsions is, the higher the swelling rate becomes. The hydrophobicity of the seed droplets can be enhanced by adding a long-chain hydrocarbon such as hexadecane into the oil phase before the SPG emulsification process. This effective swelling method has many advantages; for example, it involves only one step of polymerization, and the swelling rate and swelling capacity are very high. Compared to Ugelstad's activated swelling method, the seed droplets can be considered to be composed of 100% of a swelling agent (Y compound) because no polymer is present; it can be expected that much more of the monomer can be absorbed into seed droplets at swelling equilibrium. The porogen, cross-linking agent, or various functional monomers also can be introduced into droplets to obtain uniform porous, highly cross-linked, or other functional particles. If these agents are relatively hydrophobic, they can be introduced into droplets in the first SPG emulsification process. If these agents are relatively hydrophilic, they can be introduced into the droplets at the swelling process. Therefore, this swelling technique is a very flexible method. Omi et al. have also prepared uniform, porous, large PST particles from several to above 100 μm with –COOH functional groups on the surface by introducing porogen heptane into the seed droplets and used them in the immobilization of enzyme. It was found that the immobilized enzyme showed high relative activity and high stability [86]. By this swelling technique, PMMA microspheres composed of hydrophilic monomers also were successfully prepared by using benzene (Bz) to prepare the seed droplets, and MMA and ethyleneglycol dimethacrylate (EGDMA) to prepare the secondary emulsion. After polymerization, PMMA microspheres were obtained just by extracting Bz [83].

When using a submicron-sized microsphere obtained in emulsion or emulsifier-free emulsion polymerization as the seed, the secondary nucleation can be avoided easily when the concentration of emulsifier is lower even using a hydrophilic initiator. This is because the specific surface area of the seed particles is large enough because of the small diameter. As a result, the possibility of radicals entering into the seed particle is high and the polymerization locus is limited in or on the seed particles. When the micron-sized particle obtained in dispersion polymerization is used as the seed, it is necessary to use a hydrophobic initiator to limit the polymerization locus inside the seed particle [82]. However, when a very large droplet of 7–30 μm prepared by SPG emulsification is used as the seed, the secondary nucleation occurs even using a hydrophobic initiator because the specific surface area of the seed is so small. In this case, it is necessary to add a small amount of water-soluble inhibitor into the aqueous phase to inhibit the secondary nucleation [83–85].

(d)   Other Techniques.   Other techniques also were studied to enhance the swelling rate. Toshimatsu et al. [87] found that the rate of DVB absorbed by the

PST seed particles can be increased by mixing an adequate amount of isoamyl acetate into the monomer of the secondary emulsion of DVB. They attributed this result to the fact that the stability of the secondary emulsion decreased after isoamyl acetate diffused into the aqueous phase and was absorbed by the seed, because the size of the secondary droplets became smaller. Some researchers concluded that it was probably because the composition of the aqueous phase varied because of the first diffusion of isoamyl acetate into the aqueous phase. As a result, the solubility of DVB in the aqueous phase was enhanced. Because DVB is absorbed by the diffusion process through the aqueous phase, increasing solubility of DVB in the aqueous phase can raise the rate of absorption by the seed.

2. Composite Microspheres

The seeded polymerization was also carried out to obtain composite microspheres. Compared with emulsifier-free polymerization, this method has the advantage that a series of composite microspheres with the same diameter and size distribution but different surface characteristics can be obtained. Furthermore, the functional groups can be introduced on the particle surface with higher density, compared to emulsifier-free polymerization. For example, Inomata et al. found that a thicker shell of PGMA [poly(glycidyl methacrylate)] can be formed on the PST particle surface by polymerizing GMA using PST microspheres covered by a small amount of PGMA as a seed [88]. They applied the microspheres obtained in the immobilization of DNA and found that the microspheres with PGMA on the surface did not show nonspecific interaction with protein.

However, the core–shell morphology cannot always be obtained. The detailed morphology is determined by the relative hydrophilicity of the seed and monomer, the viscosity and cross-linking density inside the seed particle, and the location of the initiator. Many researchers studied the morphologies of microspheres based on thermodynamic and kinetic factors. Generally, if the hydrophilicity of the polymer formed at the second stage is higher than the seed polymer, the core–shell morphology will be obtained. If the case is different, the microspheres with an inverted core–shell morphology or microphase separation morphology, as shown in Fig. 16, will prevail. Utilizing the microphase separation, the microspheres with PST–PHEMA microphase separation on a surface similar to Fig. 16 were obtained [89–92]. Such particles with PST–PHEMA microphase separation showed excellent biocompatibility compared to pure PST or PHEMA particles and have been used in the immobilization of enzyme [89,90], immunoactivity [91], and protein separation [92]. By using such composite microspheres in the separation of bovine fibrinogen (BFb) and bovine serum albumin (BSA), the selective adsorption of BFb was observed for the composite microspheres, but not for homopolymer microspheres such as PST, PEMA [poly(ethyl methacrylate)], and PMMA [92].

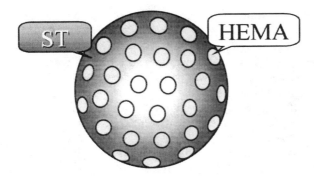

Figure 16 Schematic composite microsphere showing PST–PHEMA microphase separation morphology.

If the hydrophilicity of the added monomer is very much higher, only a low fraction can be incorporated into seed microspheres, a large part of them remaining in the aqueous phase. However, it can be improved by using the droplets swelling method because the mixing entropic change (ΔS) is larger when both the seed and the added monomer are of the monomer. Ma et al. [85] prepared the porous and nonporous microspheres containing HEMA up to 37 wt% with a diameter of up to 16 µm. Such very uniform hydrophilic functional microspheres are very useful in bioapplications. We found that the incorporated fraction of HEMA was enhanced when the secondary emulsion is composed of a HEMA/MMA mixture rather than a HEMA/ST mixture when using uniform ST monomer droplets as the seed. More interestingly, the microspheres with phase-separation morphologies categorized as snowman and popcorn were obtained, although the polymerization started after the monomer seed droplets have absorbed HEMA. This suggested that a part of the polymer rich in HEMA was also formed because copolymerization between ST and HEMA was poor.

3. Irregular-Shaped Microspheres

By seeded polymerization, many kinds of irregular-shaped particles were obtained, and some of them showed excellent functionality and have already been used in the industrial field. A couple of examples were introduced.

(a) Hollow Microspheres. Hollow microspheres have many attractive characteristics—for example, thermal resistance due to its very high cross-linking density, low density and thermal insulation due to its small air void, and optical opacity, which is used for paint formulation and resin compounds. They can be used in various fields such as paint, ink, and paper industries. Hollow particles have been

prepared by utilizing phase separation in the presence of a cross-linking agent by seeded polymerization.

Masukawa et al. [93] used submicron-sized PST microspheres of low molecular weight as a seed. The seed was obtained in the presence of mercaptan (radical transfer agent) by emulsion polymerization. After the seed microspheres were swollen by a mixture of MMA and DVB in the aqueous phase containing sodium dodecyl benzene sulfonate (SBS) emulsifier, a hydrophilic initiator was added to perform polymerization. Because the hydrophilic initiator and cross-linking agent DVB were used, a large part of DVB was copolymerized with MMA near the surface of particles and the outer diameter was fixed in the initial stage of polymerization. As the polymerization proceeded further, the phase separation occurred between the seed polymer (PST) and the subsequently polymerized polymer (PMMA). Then, complete phase separation occurred at the interface between PST and PMMA due to the shrinkage of the latter, and water filled the separated part. The size of the void became larger as the polymerization proceeded.

Okubo et al. [94,95] used their dynamic swelling technique to prepare hollow polymer microspheres with a diameter of several microns. First, the PST seed microsphere was dispersed into an ethanol/water (7/3, w/w) mixture, where DVB, solvent (toluene, etc.), BPO, and the PVA stabilizer were dissolved. After water was continuously added into the system to allow DVB, the solvent, and BPO to be absorbed by the seed, the polymerization was performed. As the polymerization proceeded, PST was exerted toward the outside because of cross-linking reaction of DVB, allowing toluene to separate in the center of particles, because toluene is very hydrophobic. As a result, hollow particles were obtained after toluene was removed. The size of the void can be controlled by varying the swelling degree or using benzene and xylene as the solvent instead of toluene.

Our group [96] prepared a hollow particle of diameter 10 μm by one-step polymerization after the uniform droplets mainly composed of a relatively hydrophilic monomer MMA, EGDMA, and hydrophobic solvent (mixture of heptane and benzene) were prepared by the SPG emulsification technique and the subsequent swelling process of the droplets as described above. As the polymerization proceeded, the phase separation occurred with heptane localized in the center and the PMMA–EGDMA network as the shell of the particles. After heptane was extracted, hollow particles were obtained. Furthermore, it was found that adding a small amount of 2-ethylhexyl acrylate into the oil phase can enhance the elasticity of the shell.

(b)   Fine-Particle Clustered Microspheres.   The fine-particle clustered (FPC) microspheres (about 1 μm) showing the specific morphology shown in Fig. 17 were developed by Hoshino et al. [97]. An alkaline swellable (AS) particle (0.32 μm) containing acrylic acid units was used as a seed. After the seed (100 parts)

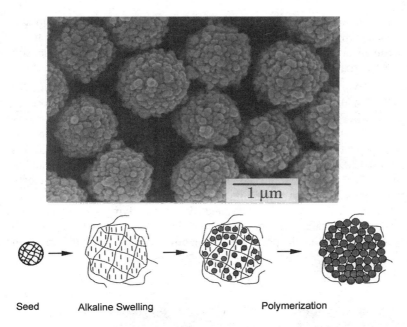

Seed   Alkaline Swelling   Polymerization

Figure 17 SEM micrograph and schematic formation process of a FPC microsphere by seeded polymerization using an alkaline swellable particle as a seed. (From Ref. 97.)

was dispersed in a pH 9.5 aqueous phase (1000 parts water) and the temperature was raised to 80°C, a mixture of ST (291 parts), 2-hydroxymethyl methacrylate (HEMA) (9 parts), and ammonium persulfate (APS) (1.5 parts) was added gradually throughout 3 h to allow the added monomer to be absorbed by the seed and to be polymerized inside the seed. Sodium dodecyl sulfate (SDS) (0.6 parts) was used as an emulsifier. By SEM and TEM observations, they found that a cluster of 0.1–0.2-μm fine polymer particles was formed throughout the whole seed particle, as shown in Fig. 17. Therefore, it was concluded that the polymerization mechanism was different than the general seed emulsion polymerization, in which a seed was considered as one locus. Probably, the nucleation at the seeded polymerization occurred all over the seed particle and progressed at every point, as shown schematically in Fig. 17. This supposition was proved by comparing the conversion-time curve of this special seeded polymerization with that of general seeded emulsion polymerization. The polymerization rate was much higher than the general seeded emulsion polymerization because the nucleation occurred at multiple sites in a seed. Because this FPC particle has a large specific surface area, and the fine particles inside the seed are not cross-linked, being different from general porous microspheres which were cross-linked by a large amount of cross-linking agent, it can absorb a large amount of heptane rapidly in the aqueous phase. It can be used as an absorbent of oil, a carrier in a drug delivery system, and a viscosity enhancer and flattening agent in the paint and paper coating industries.

(c) Flattened Nonspherical Microspheres. The flattened nonspherical micro-spheres (Fig. 18) also were developed by Hoshino et al. They used a PST micro-sphere with a lower molecular weight as a seed and the seeded polymerization was carried out in two stages at 80°C. The first stage was carried out for 3 h while con-tinuously adding an aqueous dispersion containing ST (85.5 parts), isooctane (OT) (18.0 parts), HEMA (2.7 parts), acrylic acid (1.8 parts), ammonium persulfate (APS), and SBS. The second stage was also carried out for 3 h while continuously adding an aqueous dispersion containing ST (82.8 parts), DVB (2.7 parts), HEMA (2.7 parts), acrylic acid (1.8 parts), APS, and SBS. By studying the effects of the polymerization temperature, the molecular weight of the seed polymer, and so forth, it was concluded that the phase separation proceeded as shown in Fig. 18. In

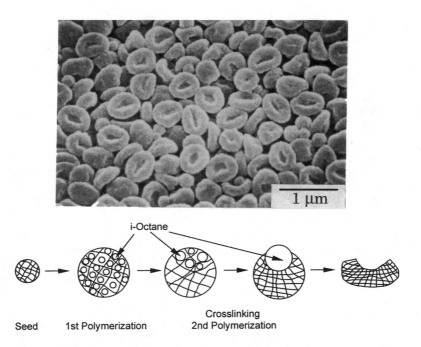

Figure 18 SEM micrograph and schematic formation process of a flattened nonspheri-cal microsphere by two-steps seeded polymerization. (From Ref. 98.)

the first stage, the spherical shape was maintained and OT was dispersed uniformly in the seed particle because no cross-linking agent was added. In the second stage, the particles shrank and the phase separation between the polymer and OT occurred because the cross-linked polymer was formed inside the particles. As a result, OT was driven out of the particles. This flattened particle showed excellent properties when used as a plastic pigment in paper coating. Its high-shear viscosity is lower than that of a spherical particle, and its sheet gloss, print gloss, and opacity are larger than those of spherical particles due to its flat shape.

III. OTHER PREPARATIVE METHODS OF MICROSPHERES AND THEIR ADVANCES

A. Evaporation of Solvent in the Aqueous Phase

All of the preparative techniques described earlier accompany a polymerization process. Sometimes, however, it is necessary to prepare microspheres from a preformed polymer such as a protein, poly(lactide), well-defined block/graft copolymers, polymer blends, and other specifically synthesized polymers. Especially, preparation of microspheres of biocompatible, biodegradable, and natural polymers is very important because these polymers show excellent properties when they are used as carriers in a drug delivery system (DDS), bioseparation, and biomedical use. Conventionally, a suspension method followed by a subsequent evaporation of solvent was used; that is, O/W emulsion or W/O/W emulsion was prepared by a mechanical stirring method at the first step. The polymer was usually dissolved in an organic solvent with a low boiling point, such as methylene chloride. If the drug to be incorporated into microspheres is water soluble, W/O/W double emulsion was usually prepared. Then, the solvent was evaporated to obtain a solid polymeric microsphere. Instead of the evaporation of the solvent, the extraction of the solvent, and cross-linking of the prepolymer were also employed.

The size distribution of microspheres prepared by a suspension method is very broad, however, the microspheres obtained are not suitable for the carriers in the DDS and the bioreactor without further size screening. By using the SPG emulsification technique, this problem can be overcome. We have prepared the uniform poly(lactide) (PLA) microsphere by the SPG emulsification technique with an evaporation process of the solvent [57]. First, the uniform droplet composed of PLA, lauryl alcohol (cosurfactant), and methylene chloride was prepared by the SPG emulsification technique. Then, methylene chloride was evaporated at room temperature for 24 h. It was found that the phase separation became apparent when the amount of lauryl alcohol (LOH) was increased, although the size distribution of droplets decreased with increasing LOH. As a result, the microsphere with the narrowest size distribution was obtained as shown in Fig. 19 when the LOH/CH_2Cl_2 ratio was 1/10 v/v.

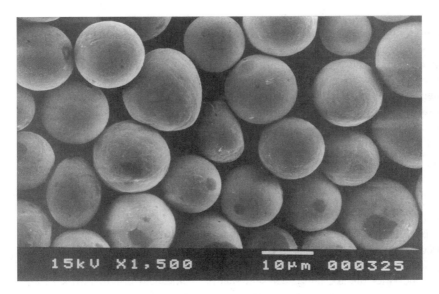

Figure 19 SEM micrograph of the PLA microsphere prepared by the SPG emulsification technique. PLA/(LOH + CH$_2$Cl$_2$) = 20 wt% by volume; LOH/CH$_2$Cl$_2$ = 1/10 v/v. (From Ref. 57. Reprinted from Colloids Surface A: Physicochem Eng Aspects. Copyright 1999, with permission from Elsevier Science.)

B. Precipitation from Homogeneous Polymer Solution

Hou and Lloyd developed a new technique for preparing a fairly uniform polymer particle of nylon [99]. They dissolved nylon in a theta solvent (formic acid/water) for 2 h at 70°C (above theta temperature) to form a clear polymer solution of 1 wt%. Then, they lowered the temperature of the solution quickly by using an ice bath to allow the polymer to precipitate out of the solution. They found that the nucleation occurred in a short period and no new nucleus continued to be formed when the cooling rate was higher (1.0°C/s). For this reason, the fairly uniform microspheres were obtained (the lowest CV value was 6.25%). On the other hand, the particles were nonspherical and tended to fuse together when the cooling rate was slower (0.05°C/s).

C. Casting of a Polystyrene Dilutent on the Water Surface

Kumaki [100,101] developed an interesting technique to prepare a very small PST microsphere (about 15 nm) composed of only a single PST molecule. First, he prepared a dilute PST/Bz solution, then dropped this solution on the water surface. With the rapid evaporation of Bz, the PST molecule became a particle to minimize the contact surface area with water. Kumaki and co-workers also used this technique to prepare a poly(ST-*b*-MMA) particle with a single PMMA hair on a PST particle by using a PST-*b*-PMMA block copolymer [102]. They used this special particle to study the conformation of a PMMA single-chain random coil by atomic force microscopy (AFM).

D. Cross-Linking of a Reactive Ionic Primary Polymer in the Aqueous Phase

Ma and co-workers [103] found that polymer particles can be formed by cross-linking reactive ionic polymers in the aqueous phase. For example, poly(4-vinylpyridine) (P4VP) microspheres can be prepared by cross-linking a partially quaternized P4VP primary polymer in the aqueous phase using diiodobutane as a cross-linking agent. When the partially quaternized P4VP was dispersed into the water, it behaved as a polymer micelle; that is, the nonquaternized units faced the inner part of the micelle and the quaternized units faced the aqueous phase. Thus, hydrohobic units of the inner part of the micelle can be cross-linked by diiodobutane to form a microsphere (microgel). It was found that the size of the microgel decreased with increasing quaternization degree, and the smallest microgel (20 nm) was obtained when the quaternization degree increased up to 70 mol%.

E. Cross-Linking of Block/Graft Polymer

It is well known that an A-B-type block/graft copolymer behaves like a micelle above the CMC in a solvent which is a good solvent for one sequence and a poor solvent for another. If double bonds or other reactive groups are introduced to the inner part of the micelle, a core–shell microgel can be obtained by cross-linking the inner part of the micelle [104–108]. Various microgels were obtained by utilizing this solution behavior of block/graft copolymers. For example, Saito et al. [106] carried out the cross-linking reaction of poly(methyl methacrylate) cores after preparing a poly(α-methylstyrene-*b*-methyl methacrylate) micelle in a mixed solvent of benzene and cyclohexane. They found that the core–corona-like microgels were obtained in a mixed solvent with a suitable composition of a block copolymer when the molecular weight of the block copolymer was sufficiently high and the α-methyl styrene content was > 39%. They also found that the particle size showed a minimum value at a benzene/cyclohexane ratio of 2/7–4/6 (v/v).

Recently, Saito et al. [109] used an A-B-A triblock copolymer to prepare a flower-type core–shell microgel. The concept of a flower-type microgel is shown

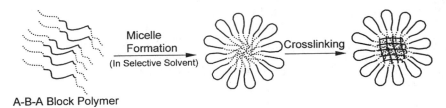

A-B-A Block Polymer

Figure 20   Concept of the formation of a flower-type microgel.

in Fig. 20. For example, they used poly(2-vinyl pyridine-*b*-styrene-*b*-2-vinyl pyridine) [P(2VP-*b*-ST-*b*-2VP)] to prepare a 1.0 wt% micelle solution in toluene. Then, the core part of micelle was cross-linked with 1,4-diiodobutane to form a flower-type microgel.

Morishima et al. [110] found that a random copolymer of sodium 2-(acrylamido)-2-methylpropanesulphonate (AMPS) and methacrylamide bearing bulky hydrophobic substituents as shown in structure (**XXI**) formed a unimicelle (a micelle formed by a single molecule) in the aqueous phase when the hydrophobic N-substitute group was the dodecyl, cyclododecyl, adamantyl, or 2-naphthymethyl group and the content of the hydrophobic group was higher than 40 mol%. In the unimicelle, the hydrophobic units aggregated by a strong hydrophobic bond and a hydrogen bond between the amide spacers connected with hydrophobic groups. The unimicelle was very compact; the diameter was about 11 nm when the polymer with a weight-average molecular weight of 5.2×10^5 (g/mol) was used. It was very interesting that the unimicelle was not formed when the amide spacer was replaced by the ester group.

$$\begin{array}{c} -\!\!\left[\!CH_2\text{-}CH\right]_{100\text{-}x}\!\!\left[\!CH_2\text{-}C\right]_x\!\!- \\ \quad | \qquad\qquad\quad | \\ \quad C\!\!=\!\!O \qquad\qquad C\!\!=\!\!O \\ \quad | \qquad\qquad\quad | \\ \quad NH \qquad\qquad NH \\ CH_3\text{-}\!\!\underset{|}{C}\text{-}CH_3 \qquad \boxed{Hydrophobe} \\ \quad CH_2 \\ \quad SO_3^-Na^+ \end{array}$$

-(CH$_2$)$_{11}$CH$_3$ Dodecyl

Cyclododecyl

Adamantyl

2-Naphthylmethyl

(**XXI**)

IV. SPECIAL PREPARATIVE METHODS OF ORGANIC–INORGANIC COMPOSITE MICROSPHERES

Interest in the preparation of organic–inorganic composite microspheres grew rapidly because composite microspheres show both the characteristics of polymers and those of inorganic materials—for example, flexibility, easy synthesis, biocompatibility, and so forth of polymers, and rigidity, magnetic, electric conductivity, and so forth of inorganic materials. Magnetic polymer microspheres were designed to be used in the drug delivery system (DDS) and bioreactors. By incorporating magnetite into a polymeric microsphere, for example, the separation between the product and the polymeric microsphere carrying an enzyme can be separated easily after the substrate is converted to the product; different cell can be separated by utilizing different magnetic forces between the composite microsphere and the cells; microspheres carrying drugs can be dragged to the target by the external magnetic field. By encapsulating alumina or silica particles with polymers, the inorganic particles can be dispersed into the polymer film uniformly, and the rigidity, releasability, and so forth of the film can be enhanced.

However, an organic–inorganic composite microsphere is difficult to produce because the affinity between the inorganic substance and the polymer is very poor. Furthermore, the control of the distribution of the inorganic substance in polymeric microspheres is also an important task. Besides the general polymerization method described earlier, such as emulsion, emulsifier-free, dispersion and suspension polymerization, many other special methods were devised to prepare organic–inorganic microspheres, according to the characteristics of the respective organic substances. A couple of examples follow.

A. Heterocoagulation Between Oppositely Charged Microspheres

Furusawa et al. [111,112] devised a heterocoagulation technique to prepare magnetite–polymer and alumina–polymer composite microspheres. In the preparation of the magnetite–polymer composite microsphere [111], $NiO–ZnO–Fe_2O_3$ fine magnetic particles with a diameter of 20 nm and PST particles with $-COO^-$ or $-SO_3^-$ functional groups were used. The magneitic particles and PST particles possessed positive and negative charges, respectively when the pH of the medium was 2.5. After these two microspheres were mixed, the smaller positive magnetic particles were adsorbed on the surface of negative larger particles by attractive force. They used four kinds of PST microsphere with diameters of 180, 530, 600, and 900 nm. Regular heterocoagulated composite particles were obtained when the PST microspheres larger than 530 nm were used. When magnetic composite microspheres are used in the bioreactor, the biomaterials will lose their bioactivities if bare magnetic particles are in contact with the biomaterials directly. There-fore, Furusawa et al. encapsulated the magnetic composite microspheres further with PST polymer by seeded polymerization.

In the preparation of alumina–polymer composite microspheres [112], a similar procedure was adopted. In this case, however, the system was comprised of larger positive α-alumina particles (1100, 850, and 400 nm) and smaller nega-tive PST microspheres. They found that the concentration of alumina particles dra-matically affected the formation of regular composite microspheres. The regular composite microsphere was obtained when the concentration was ~0.2 wt%, the pH of the medium was 3–4, and the particle number ratio of smaller to larger par-ticles was $\geq$ 300/1.

B.   Precipitation of Organic Substance on Polymeric Microspheres

Shimizu et al. [113] developed a surface swelling–precipitation technique to coat an electric-conducting polymer polypyrrole on the microspheres. In the first step, the surfactant (hydroxypropylmethylcellulose) was adsorbed on PST micro-spheres in an ethanol/water solvent for 24 h. Then, pyrrole was added to allow the surface of the microsphere to swell, where the surfactant was adsorbed. In the sec-ond step, after the unadsorbed surfactant and unabsorbed pyrrole were removed by filtration, ferric chloride ($FeCl_3$) (oxidant) was added to the system. Polypyrrole was formed on the surface of the microsphere by an oxidation reaction. They investigated the effects of water fraction in the medium and the amount of oxidant on the conductivity of the particles and found that the microsphere showed maxi-mum conductivity when the water fraction was 0.9 and the feed amount of oxidant was 7.68 mol/L.

C.   In Situ Formation of Magnetite in Polymeric Microspheres

Ugelstad et al. [80] prepared the magnetic polymer composite microspheres by in situ formation method in polymeric microspheres. First, they introduced a $-NO_2$ group throughout the whole volume of the polymeric microspheres by covalent coupling, then they dispersed the polymeric microspheres in an aqueous solution of Fe^{2+} salt. Under proper conditions, Fe^{2+} was continuously transported from the outer phase into the polymeric microspheres, where it was oxidized and precipi-tated as a magnetic hydrated iron compound. By subsequent heating, small grains of magnetite Fe_3O_4 and/or maghemite $\gamma\text{-}Fe_2O_3$ were formed evenly throughout the volume of polymeric microspheres. These uniform magnetic polymer composite microspheres can be applied in immunology, cancer research, transplantation, gene technology, clinical chemistry, virology, microbiology, clinical pathology, and so forth.

V. ADVANCES IN APPLICATIONS OF POLYMERIC MICROSPHERES

Polymeric microspheres have been applied in many fields by utilizing the various characteristics of microspheres described earlier. In this section, two kinds of application which utilize surface and domain characteristics of microspheres, respectively, are selected and introduced.

A. Utilization of Surface Characteristics of Microspheres

The most novel characteristics of microspheres are that the microsphere has large specific surface area and that various desired functional groups can be introduced on the surface of microspheres. Many studies utilizing such characteristics of microspheres have been undertaken. Several noticeable examples in recent years are discussed.

The first example is the refolding of protein using thiol-carrying microspheres, as studied by Fujimoto et al. [114]. They found that the microsphere could be used as a template to obtain the functional three-dimensional structures from denatured (not aggregated) polypeptide chains. They used the microsphere composed of a PST core and a poly(glycidyl methacrylate) (PGMA) surface layer, on which thiol groups were immobilized by coupling dithiothreitol with epoxy groups. Scrambled ribonuclease A (RNaseA) carrying incorrect disulfide pairings was chosen as a model protein for refolding. They found that the refolding was accelerated by the microspheres with thiol groups, although renaturation was not observed in the presence of unmodified microspheres without thiol group. Possibly, thiol groups on microspheres surface can preferentially attack disulfide bonds in the scrambled RNaseA. This attack can trigger protein refolding through thiol–disulfide exchange reactions at the microsphere–protein interface. Furthermore, they found that the scrambled RNaseA was bound on modified microspheres having refolding activity, whereas the native RNaseA was hardly adsorbed onto them. The extent of renaturation increased with incubation time, whereas the scrambled RNaseA adsorption showed a slight increase. This suggests that adsorption of misfolded protein and desorption of the refolded protein simultaneously proceed in this system.

Another attractive application field of polymeric microsphere is the one carrying a single-stranded DNA-fragment probe, which is used in the separation, purification, and detection of DNA which has complementary sequence with the probe DNA fragment. The DNA-fragment probe with 10–30 mers is thought to possess the highest sensitivity. Especially in the field of DNA diagnosis, polymeric microspheres carrying the DNA probe have been developed exhaustively. The general procedure is as follows: Prepare (synthesize or cleave a fragment with an enzyme and so forth) a DNA probe and immobilize it on a polymer particle with functional groups. Virus DNA from patients can be directly captured by microsphere-DNA (immobilized DNA fragment) with much higher selectivity than by a free DNA fragment. This kind of technique for DNA diagnosis has the advantages that the detection can be carried out instantaneously and directly with high sensitivity and high selectivity, compared with a conventional diagnosis technique. This technique has been applied to the detection of the HIV and HCV viruses.

Hatakeyama et al. [115] found that normal DNA could be separated almost completely from mutant DNA by using a PST–PGMA core–shell microsphere carrying a complementary DNA fragment (about 20 mers), whereas free DNA fragment not immobilized on the microsphere showed lower selectivity. That is, the normal DNA can be hybridized on microsphere-DNA, whereas few mutant DNA can be hybridized on the microsphere when the incubation time is shorter (less than 3 h). This is because that the mobility of DNA probe is suppressed after it is immobilized on the microsphere. As a result, the microsphere-DNA loses the flexibility to form a hybrid with a mutant DNA.

B. Utilization of the Domain Characteristics of Microspheres

Although the large specific surface area and functionalities on microspheres have been noticed and utilized for a long time (as described earlier), the idea to utilize the domain characteristics of a small microsphere is relatively new and was developed in recent years. In this area, the preparation of bi-continuous phase (also is called dual phase) polymer materials was most attractive.

Matsumoto et al. [116] prepared the dual-phase polymer electrolytes (DPE), which showed both high ionic conductivity ($>10^{-3}$ S/cm) and good mechanical strength by mixing poly(acrylonitrile-co-butadiene) (NBR) latex and poly (styrene-co-butadiene) (SBR) latex followed by swelling the membrane with lithium salt solution (1 M lithium perchlorate in γ-butyrolactone). They found that the NBR phase was polar and was impregnated selectively with polar lithium salt solution, whereas the SBR phase was nonpolar and formed a mechanically supportive matrix. The ionic conductivity increased dramatically in the NBR weight fraction range between 0.12 and 0.19. This result reveals that a percolation threshold for ion-conductive channels exists, and the NBR phase becomes continuous from this threshold. This membrane can be used in rechargeable lithium batteries.

Nakamura [117] prepared a bi-continuous mosaic charged membrane by mixing negatively charged poly(sodium styrenesulphonate) (PNaSS) latex and positively charged quaternized poly(4-vinylpyridine) (P4VP) latex in random under a suitable pH condition. When the volume ratio of the two latexes was between 3/7 and 7/3, the positively charged and negatively charged channels were arranged alternatively and each channel became continuous. This kind of membrane can be used in the deionization of seawater and in the separation of electrolytes and nonelectrolytes. Furthermore, by computer simulation, it was found that the possibility of percolation of each phase depended largely on the number of neighboring particles around one particle. This result implies that the possibility of percolation of each phase can be increased largely when particles form the ordered arrangement.

Ma et al. [118] found that the ordered arrangement was formed in film when the interaction between two compositions and the diameter ratio between them were proper. This study was based on the results of Hachisu and Yoshimura [119–121] and Hasaka et al. [122]. We prepared the films for three systems: the system composed of a P4VP microsphere and a P(ST-HEMA) microsphere with a HEMA surface layer; the system composed of two kinds of PST microspheres or P4VP microspheres with different sizes; and the system composed of a P4VP microsphere and a PST–P4VP (core–shell) microsphere with P4VP as a shell. We found that the ordered structure with a long range was formed only in the system composed of P(ST-HEMA) and P4VP microspheres, whereas a relatively ordered structure with a short range was formed in the system of P4VP and PST–P4VP (core–shell) microspheres, and no ordered structure was formed in the system composed of two kinds of PST microspheres or P4VP microspheres. Therefore, it was concluded that the weak attractive force between two compositions (for example the –OH group and the pyridine group) was necessary to maintain the ordered structure during the film formation. For the system where the ordered structure was obtained, we investigated further the effects of the mixing ratio and the diameter ratio of two particles on the formation of the ordered structure. When the diameter ratio of two compositions was 0.73, the ordered structure which is the same as that of $MgCu_2$ (Laves phase) was obtained (Fig. 21), and the fraction of the ordered structure attained a maximum value (92%) when the mixing ratio of large to small particles was 1 : 2. On the other hand, the ordered structure of the NaCl crystal was obtained when the diameter ratio of the two compositions was 0.43, and the fraction of the ordered structure attained a maximum value (53%) when the mixing ratio of the two particles was 1 : 1. Each phase in above two ordered structures is continuous and the possibility of percolation of each phase and the number of the channel are much higher than those of random mixing structure. Especially in $MgCu_2$ structure, the fraction of the ordered structure attained to a very high value, it can be used for bi-continuous phase polymer materials, for example, phase-transfer catalyst, mosaic charged membrane, and rechargeable lithium battery etc. The discovery of various new ordered structures will be expected by changing diameter ratio of two compositions. This method provides an alternative method to prepare polymer alloys, compared with conventional method from block/graft copolymers.

Fukutomi et al. [123] also prepared the bi-continuous phase by casting a mixture of PVA primary polymer and P(MMA–HEMA) microgel with a HEMA

Figure 21 SEM micrograph of a long-range ordered phase showing the $MgCu_2$ crystal structure.

surface layer, followed by a connection process of PMMA microgel using the seeded polymerization concept. They found that the homogeneous and regular arrangements of P(MMA–HEMA) microgel were formed in the PVA matrix when the weight ratio of the microgel and PVA polymer was 7 : 3; that is, the microgel was dispersed regularly in the PVA matrix to form a dispersed phase and the PVA matrix formed a continuous phase. Then, the microgels were connected to form another continuous phase by soaking the film in MMA monomer solution, followed by the seeded polymerization of PMMA microgel dispersed in the film. By such a method, a material of bi-continuous phase was obtained.

VI. CONCLUDING REMARKS

Various preparative techniques have been developed and the characteristics of microspheres have been well defined. By these techniques, the microspheres covering almost all sizes can be obtained. The microspheres prepared by these techniques showed many particular properties: (1) very uniform size; (2) large specific surface area; (3) various functional groups with desired surface density; (4) different properties of inner and outer parts or left and right parts; (5) various shapes and various morphologies; (6) porosity; (7) high mobility; and so forth. By utilizing these properties, the functional groups can be concentrated, and the reaction locus can be limited on or in a particular part of the microspheres (surface, inner part, and left or right side). A particular microenvironment can be prepared on or in the microsphere or between microspheres in the aggregation of microspheres. Therefore, it can be expected that the microspheres will be used as the precise microdevices and micromachines and will be called intelligent micromaterials in near future. In fact, it seems that the particular properties of microspheres have not been utilized thoroughly. Especially in the studies of morphologies, although many kinds of morphologies have been found and their formation conditions have been well defined, few applications of these microspheres have been reported except for core–shell, occlusion-type microspheres. Therefore, the tasks imposed on us are how to apply various well-defined microspheres already obtained to practical applications and how to design and modify these microspheres further based on the studies of practical applications. An accumulation of think tank researchers in various fields, and close cooperation between this area and other fields such as electron and electrical fields, medical, biochemical, architectural, textile fields, and so forth is urgent.

ACKNOWLEDGMENT

The author expresses her sincere appreciation to Professor Shinzo Omi for his support and his valuable advice while reading through this manuscript.

REFERENCES

1. WD Harkins. J Chem Phys 13:381–382, 1945.
2. WD Harkins. J Chem Phys 14:47–48, 1946.
3. WD Harkins. J Am Chem Soc 69:1428–1444, 1947.
4. WV Smith, RH Ewart. J Chem Phys 16:592–599, 1948.
5. WV Smith. J Am Chem Soc 70:3695–3702, 1948.
6. CP Roe. Ind Eng Chem 60 (9):20–33, 1968.
7. RM Fitch, CH Tsai. In: FM Fitch, ed. Polymer Colloids. New York: Plenum Press, 1971, pp 73–103.
8. RM Fitch, YK Kamath. J Colloid Interf Sci 54:6–12, 1976.
9. DH Napper, DR Kim. Experimental discrimination between competing nucleation theories. Preprints of International Symposium on Advanced Technology of Fine Particles, Yokohama, 1997, pp 8–9.

10. The Society of Polymer Science (Jpn), ed. Assessment of Polymer Production Process 17—Functionalization and Composition Techniques of Polymer Particles (Research Report), Tokyo, 1996, pp 253–257.
11. SL Tsaur, RM Fitch. J Colloid Interf Sci 115:450–462, 1987.
12. MB Urquiola, VL Dimonie, ED Sudol, MS El-Aasser. J Polym Sci, Polym Chem Ed 30:2631–2644, 1992.
13. K Maeda. Paint Coating Bus 6:37–37, 1996.
14. S Watanabe, H Ozaki, K Mitsuhashi, S Nakahama, K Yamaguchi. Makromol Chem 193:2781–2792, 1992.
15. K Yuki, T Sato, H Maruyama, J Yamaguchi, T Okaya. Polym Int 30:513–517, 1993.
16. DY Lee, JH Kim, TI Min. Role of alkali-soluble random copolymer in emulsion polymerization. Preprints of International Symposium on Advanced Technology of Fine Particles, Yokohama, 1997, pp 28–31.
17. K Tauer, S Kosmella. Polym Int 30:253–258, 1993.
18. K Sugiyama, K Ohga, K Kikukawa. Macromol Chem Phys 195:1341–1352, 1994.
19. K Nagai, H Kataoka, N Kuramoto. Kobunshi Ronbunshu 50:263–270, 1993.
20. Ohtsuka, H Kawaguchi, Y Sugi. J Appl Polym Sci 26:1637–1647, 1981.
21. F Hoshino, T Fujimoto, H Kawaguchi, Y Ohtsuka. Polym J 19:241–247, 1987.
22. F Hoshino, H Kawaguchi, Y Ohtsuka. Polym J 19:1157–1164, 1987.
23. GH MA, T Fukutomi. J Appl Polym Sci 43:1541–1547, 1991.
24. M Okubo, A Yamada, S Shibao, K Nakamae, T Matsumoto. J Appl Polym Sci 26:1675–1679, 1981.
25. N Yanase, H Noguchi, H Asakura, T Suzuta. J Appl Polym Sci 50:765–776, 1993.
26. H Noguchi, N Yanase, Y Uchida, T Suzuta. J Appl Polym Sci. 48:1539–1547, 1993.
27. CF Lee, WY Chiu. Polym Int 30:475–481, 1993.
28. J Ugelstad, MS El-Aasser, JW Vanderhoff. Polym Lett Ed 11:503–513, 1973.
29. ZR Pan, H Fan, ZX Weng, ZM Huang. Polym Int 30:259–264, 1993.
30. PL Tang, MS El-Aasser, CA Silebi, ED Sudol. J Appl Polym Sci 43:1059–1066, 1991.
31. CM Miller, J. Venkatesan, CA Silebi, ED Sudol, MS El-Aasser. J Colloid Interf Sci 162:11–18, 1994.
32. JL Reimers, FJ Schork. J Appl Polym Sci 59:1833–1841, 1996.
33. JL Reimers, FJ Schork. J Appl Polym Sci 60:251–262, 1996.
34. K Ishitani, WZ Zhang, T Noguchi. Graft copolymerization of miniemulsion using macromonomers. Preprints of 8th Polymeric Microspheres Symposium, Fukui, 1994, pp 67–68.
35. A Yamamoto, Y Hasegawa, F Yoshino. Development of macromonomer copolymerized emulsion polymers. Preprints of 8th Polymeric Microspheres Symposium, Fukui, 1994, pp 69–72.
36. Barton Cabek. In: J Barton, J Capek, eds. Radical Polymerization in Disperse Systems. London: Ellis Horwood, 1994, pp 186–215.
37. JS Guo, ED Sudol, JW Vanderhoff, MS El-Aasser. J Polym Sci, Polym Chem Ed 30:703–712, 1992.
38. J Santhanalakshmi, K Anandhi. J Appl Polym Sci 60:293–304, 1996.
39. M Antonietti, W Brenser, D Muschenborn, C Rosenauer, B Schupp, M Schmidt. Macromolecules 24:6636–6643, 1991.

40. LA Rodriguez-Guadarrama, E Mendizabal, JE Puig, EW Kaler. J Appl Polym Sci 48:775–786, 1993.
41. F Bleger, AK Murthy, F Pla, EW Kaler. Macromolecules 27:2559–2565, 1994.
42. LM Gan, CH Chew, KC Lee, SC Ng. Polymer 34:3064–3069, 1993.
43. M Antonietti, S Lohmann, CV Niel. Macromolecules 25:1139–1143, 1992.
44. JO, Stoffer, T Bone. J Polym Sci, Polym Chem Ed 18:2641–1648, 1980.
45. V Vaskova, V Juranicova, J Barton. Makromol Chem 191:717–723, 1990.
46. V Vaskova, V Juranicova, J Barton. Makromol Chem 192:989–997, 1991.
47. V Vaskova, V Juranicova, J Barton. Makromol Chem 192:1339–1347, 1991.
48. J Barton. Makromol Chem Rapid Commun 12:675–679, 1991.
49. F Candau, YS Leong, G Pouyet, S Candau. J Colloid Interf Sci. 101:167–181, 1984.
50. F Candau, YS Leong. J Polym Sci, Polym Chem Ed 23:193–214, 1985.
51. MT Carver, U Dreyer, R Knoesel, F Candau. J Polym Sci, Polym Chem Ed 27:2161–2177, 1989.
52. F Candau, Z Zekhnini, JP Durand. J Colloid Interf Sci 114:398–408, 1986.
53. F Candau, JY Anquetil. Polymerizable microemulsions: Some criteria to achieve an optimal formulation. Preprints of International Symposium on Advanced Technology of Fine Particles, Yokohama, 1997, pp 22–23.
54. M Kamiyama, K Koyama. Kobunshi Runbunshu 50:227–233, 1993.
55. S Omi, K Katami, A Yamamoto, M Iso. J Appl Polym Sci 51:1–11, 1994.
56. S Omi. Colloids Surface A: Physicochem Eng Aspects 109:97–107, 1996.
57. GH Ma, M Nagai, S Omi. Colloids Surface A: Physicochem Eng Aspects 1999.
58. S Omi, A Matsuta, M Nagai, GH Ma. Colloids Surface A: Physicochem Eng Aspects 1999.
59. A Kanetaka. Graduation thesis, Tokyo University of Agriculture and Technology, Tokyo, 1998.
60. K Takahashi, S Miyamori, H Uyama, S Kobayashi. J Polym Sci, Polym Chem Ed 34:175–182, 1996.
61. S Kobayashi, H Uyama, JH Choi, Y Matsumoto. Polym Int 30:265–270, 1993.
62. P Lacroix-Desmazes, A Guyot. Polym Bull 37:183–189, 1996.
63. K Nakamura, K Fujimito, H Kawaguchi. Dispersion polymerization using macroinitiator to prepare poly(dimethylsiloxane-b-poly(methyl methacrylate) microspheres. Preprints of International Symposium on Advanced Technology of Fine Particles, Yokohama, Japan, 1997, pp 116–116.
64. H Uyama, K Takahashi, S Kobayashi. Kobunshi Kako 45:492–496, 1996.
65. M Hattori, ED Sudol, MS EL-Aasser. J Appl Polym Sci 50:2027–2034, 1993.
66. S Omi, M Saito, T Hashimoto, M Nagai, GH Ma. J Appl Polym Sci 68:897–907, 1998.
67. XH Li, ZH Sun. J Appl Polym Sci 58:1991–1997, 1995.
68. JM DeSimone, EE Maury, YZ Menceloglu, JB McClain, TJ Romack, JR Combes. Science 265:356–359, 1994.
69. S Sosnowski, M Gadzinowski, S Slomkoski. Macromolecules 29:4556–4564, 1996.
70. SP Armes, JF Miller, B Vincent. J Colloid Interf Sci 118:410–416, 1987.
71. T Masuda, M Okubo, T Mukai. Synthesis of needle-like polymer particles by chemical oxidative dispersion polymerization of xylidine. Preprints of the First Asian

Symposium on Emulsion Polymerization and Functional Polymeric Microspheres, HangZhou, China, 1996, pp 62–63.

72. H Uyama, H Kurioka, S Kobayashi. Preparation of polyphenol particles by dispersion polymerization using enzyme as catalyst. Preprints of International Symposium on Advanced Technology of Fine Particles, Yokohama, 1997, p 115.

73. Y Yamakawa, K Kuroda, K Goto, T Hayashi. Preparation of poly(amino acid) particles by aqueous polymerization of amino acid NCA. Preprints of Polymeric Microspheres Symposium, Tukuba, 1996, pp 139–140.

74. H Kawaguchi, K Fujimoto, M Saito, T Kawasaki, Y Urakami. Polym Int 30:225–231, 1993.

75. H Kawaguchi, K Fujimoto, Y Kasuya. Colloid Polym Sci 270:53–57, 1992.

76. A Kondo, H Fukuda. Preparation of thermosensitive magnetic microspheres and their application to bio-processes. Preprints of International Symposium on Advanced Technology of Fine Particles, Yokohama, 1997, pp 144–144.

77. Y Nagata, Y Oonishi, C Kajiyama. Kobunshi Ronbunshu 53:63–69, 1996.

78. A Okamura, K Fujimoto, H Kawaguchi, H Nishizawa, O Hirai. Preparation and Strucural control of polyimide microsphere. Preprints of Polymeric Microspheres Symposium, Tukuba, Japan, 1996, pp 167–170.

79. J Ugelstad, PC Mørk, KH Kaggerud, T Ellingsen, A Berg. Adv Colloid Interf Sci 13:101–140, 1980.

80. J Ugelstad, PC Mørk, R Schmid, T Ellingsen, A Berg. Polym Int 30:157–168, 1993.

81. M Okubo, M Shiozaki, M Tsujihiro, Y Tsukuda. Colloid Polym Sci 269:222–226, 1991.

82. M Okubo, M Shiozaki. Polym Int 30:469–474, 1993.

83. S Omi, K Katami, T Taguchi, K Kaneko, M Iso. J Appl Polym Sci 57:1013–1024, 1995.

84. S Omi, T Taguchi, M Nagai, GH Ma. J Appl Polym Sci 63:931–942, 1997.

85. GH MA, M Nagai, S Omi. J Appl Polym Sci 66:1325–1341, 1997.

86. S Omi, K Kaneko, A Nakayama, K Katami, T Taguchi, M Iso, M Nagai, GH Ma. J Appl Polym Sci. 65:2655–2664, 1997.

87. A Toshimatsu, T Ito, A Kondo, R Tsushima. Kobunshi Ronbunshu 50:319–325, 1993.

88. Y Inomata, T Wada, H Handa, K Fujimoto, H Kawaguchi. J Biomater Sci, Polym Ed 5:293–302, 1994.

89. M Okubo, S Kamei, Y Tosaki, K Fukunaga, T Matsumoto. Colloid Polym Sci 265:957–964, 1987.

90. M Okubo, Y Yamamoto, S Kamei. Chem Express 6:145–148, 1991.

91. M Okubo, M Uno, Y Yamamoto, S Kamei, T Matsumoto. Kobunshi Ronbunshu 44:123–130, 1987.

92. M Okubo, H Hattori. Colloid Polym Sci 271:1157–1164, 1993.

93. T Masukawa, I Ozaki, M Hattori, N Itou, K Kasai. Cross-linked hollow polymer particles by emulsion polymerization. Preprints of International Symposium on Advanced Technology of Fine Particles, Yokohama, Japan, 1997, pp 125–125.

94. M Okubo, H Minami, T Yamashita. Macromol Symp 101:509–516, 1996.

95. M Okubo, H Minami. Colloid Polym Sci 274:433–438, 1996.

96. R Kanda. Graduation thesis, Tokyo University of Agriculture and Technology, Tokyo, 1998.

97. F Hoshino, M Nakano, T Yanagihara. Preparation and properties of fine particles clustered emulsion particles. Preprints of the Sixth Polymeric Microspheres Symposium, Fukui, 1990, pp 57–60.

98. F Hoshino, M Nakano, T Yanagihara. Preparation of flattened non-spherical emulsion particles. Preprints of the Seventh Polymeric Microspheres Symposium, Kobe, 1992, pp 197-200.

99. WH Hou, TB Lloyd. J Appl Polym Sci 45:1783–1788, 1992.

100. J Kumaki. Macromolecules 19:2258–2263, 1986.

101. J Kumaki. Macromolecules 21:749–755, 1988.

102. J. Kumaki, Y Nishikawa, T Hashimoto. J Am Chem Soc 118:3321–3322, 1996.

103. Y Koshiro, G-H Ma, T Fukutomi. Polym Gels Networks 2:29–47, 1994.

104. MH Park, K Ishizu, T Fukutomi. Polymer 30:202–206, 1989.

105. R Saito, K Ishizu, T Nose, T Fukutomi. J Polym Sci, Polym Chem Ed 28:1793–1805, 1990.

106. R Saito, K Ishizu, T Fukutomi. Polymer 31:679–683, 1990.

107. R Saito, K Ishizu, T Fukutomi. Polymer 32:531–536, 1991.

108. R Saito, K Ishizu, T Fukutomi. Polymer 32:2258–2262, 1991.

109. R Saito, Y Akiyama, M Tanaka, K Ishizu. Synthesis of flower-type microgels. Preprints of International Symposium on Advanced Technology of Fine Particles. Yokohama, 1997, pp 123–123.

110. Y Morishima, S Nomura, T Ikeda, M Seki M Kamachi. Macromolecules 28:2874–2881, 1995.

111. K Furusawa, K Nagashima, C Anzai. Kobunshi Ronbunshu 50:337–342, 1993.

112. K Furusawa, K Nagashima, C Anzai. Kobunshi Ronbunshu 50:343–347, 1993.

113. S Shimizu, N Saito, M Tanaka. Polypyrrole coating on microsphere. Preprints of 9th Polymeric Microspheres Symposium. Tukuba, Japan, 1996, pp 41–42.

114. K Fujimoto, H Shimizu, H Kawaguchi. Kagaku to Kogyo 50:1523–1525, 1997.

115. M Hatakeyama, S Iwato, K Nakamura, K Fujimoto, H Kawaguchi. Affinity latex for DNA diagnosis. Preprints of International Symposium on Advanced Technology of Fine Particles, Yokohama, 1997, pp 148–148.

116. M Matsumoto, T Ichino, JS Rutt, S Nishi. J Appl Polym Sci 32:2551–2558, 1994.

117. M Nakamura. Graduation master's thesis. Tokyo Institute of Technology, Tokyo, 1993.

118. GH Ma, T Fukutomi, N Morone. J Colloid Interf Sci 168:393–401, 1994.

119. S Hachisu, S Yoshimura. Nature 283:188–189, 1980.

120. S Yoshimura, S Hachisu. Hyomen 21:247–260, 1983.

121. S Hachisu. Phase Trans 21:243–249, 1990.

122. M Hasaka, H Nakamichi, K Oki. J Jpn Inst Met 45:347–353, 1981.

123. T Fukutomi, H Oomori, Y Sugito. J Polym Sci, Polym Chem Ed 34:2729–2735, 1996.

4

Latex Particle Heterogeneity: Origins, Detection, and Consequences

Fernando Galembeck
Universidade Estadual de Campinas, Campinas, São Paulo, Brazil

Elizabeth Fátima de Souza
Pontifícia Universidade Católica de Campinas, Campinas, São Paulo, Brazil

I. INTRODUCTION

A. Objective of This Chapter

Polymer latexes [1] produced by emulsion polymerization [2] have a large and still growing importance as model colloids and industrial products and intermediates. For this reason, we need detailed descriptions of the principal latex constituents, which are the polymer particles. Current analytical techniques yield much information on the polymer *chains* (dimensions and size distribution, tacticity, comonomer distribution, ionic group density and vicinity, ramifications, cross-linking) on a routine basis, but the only analogous information usually available for latex *particles* are the particle size distribution and zeta potentials. Much *average* analytical chemical information can thus be obtained, but we do not usually know how particles differ within a given population, considering their chemical compositions, polymer chain molecular weight (MW), and other characteristics, co-ion and counterion distributions, and so on.

The objective of this chapter is to present some new methodologies which are proving effective in providing information on latex particle chemical heterogeneity. The fundamentals of each relevant methodology are given, together with its main useful characteristics and some examples of their application, generally from the past experience of this laboratory.

The authors hope that this chapter will reveal, to those interested in polymer latexes, a number of new and unparalleled tools and answers to well-known but difficult problems in latex characterization. This new information on latex properties has been helpful in understanding some unexpected or anomalous latex behaviors, as well as in improving our predictive ability on latex properties and applications.

B.   Latexes as Model Colloids

Latexes have received much attention as model colloids [3,4], because many preparations have the following characteristics: paucidisperse or monodisperse particle sizes, uniformly spherical particle shape, and variable chemical composition depending on the monomers used and on the polymerization protocol. The surface composition may also change, depending on the initiator, comonomer, and surfactants used. Copolymer latex particles display intricate domain morphologies, the most common of which are core-and-shell, inverted core-and-shell, raspberry like, hamburger like, half-moon [5,6].

Since the earlier preparations of monodisperse latexes, many authors thought that these would be excellent model colloids: Following carefully designed procedures, spherical monodisperse particles are obtained [7]. Surface properties are also expected to be highly uniform, with polar groups tending to concentrate at the particle–water interface [8] but moving inward when the particles are dried.

Polystyrene latexes are particularly attractive, because the high-T_g polystyrene should make rigid and nondeformable particles at most temperatures of interest. For these reasons, polystyrene (PS) latexes have been widely used as model colloids for 30 or more years. There is now a vast number of publications in which PS latexes have been used in the study of colloid stability and aggregation, self-organization, and heterocoagulation [9], which are among the most attractive topics of current research on condensed-matter chemistry and physics.

1.   PS Latex in Coagulation and Aggregation Studies

The study of colloidal stability has been strongly influenced by the DLVO theory, for the past 50 years. During this time, many new factors were disclosed and studied in great detail [10]. We are now well aware of the importance of steric [11], hydration [12], and depletion [13] contributions to colloid stability, but these are often treated as perturbations to the fundamental DLVO behavior. A major challenge to this attitude comes from some authors led by Ise [14], who have emphasized the suitability of Sogami's potentials for the description of particle–medium–particle interactions in colloidal systems.

Many authors have published the results of careful experiments on the coagulation kinetics of PS latex particles [15,16]; in many cases, excellent agreement with the Smoluchowski kinetic theory is observed, considering that the DLVO interparticle interactions determine the energy barrier to the particle–particle approach. However, important deviations of experimental behavior from the DLVO predictions were also observed [17].

These difficulties led some authors to bring further considerations into this picture; for instance, Frens [17] has considered the effect of repeptization on coagulation rates, the Bristol group reported on the attainment of a monomer–multiplet equilibrium [18], and Midmore et al. [19] have considered that some latexes are "hairy" particles from which charged polymer chain ends protrude into the surrounding liquid, creating a sort of eletrosterically stabilizing polyelectrolytic chains grafted into the latex spheres.

Another idea has emerged from other authors, and it has been expressed using different key words: It is the effect of the relaxation rates of the colliding particles on the coagulation kinetics as compared to the relevant particle collision rates. Following this argument, the establishment of irreversible interparticle contact may be hindered by the slow relaxation of the electrical double layer [20], or by the poor plasticity of a layer of solvent and electrolyte, which prevents actual polymer–polymer contact. Of course, a slowly relaxing hydration layer might delay polymer chain diffusion from one to another particle, as well as particle–particle "neck" formation, which is a necessary intermediate step in the formation of irreversible aggregates.

Beyond the relaxation of the particle double layer (considering both the hydration layer and the ions contained therein as important factors), we have also to consider the role of the actual polymer relaxations also. Fast interdiffusing chains may bind two colliding particles irreversibly, whereas slowly diffusing chains may allow particles to maintain their identities during a collision, thus making an easily dissociable particle doublet. This is expressed by a general argument put forward by Israelachvili, which is based on the "Deborah number" D [particle relaxation time/particle perturbation (or collision) time]. Following this argument, the rate of energy dissipation in an irreversible process is maximum when D approaches unity, and it tends to zero when D approaches zero or infinity [21].

Recent experimental evidence in support of these ideas was recently published by this laboratory [22]: The exposure of some latexes to chloroform and toluene vapors has a strong effect in their colloidal stability toward high salt concentrations. Because the major particle characteristics relevant to electrostatic, electrosteric, depletion, and hydration stabilization mechanisms are not affected, we assume that the apolar solvent effect may be due to changes in polymer plasticity or in the double-layer relaxation rates.

A mechanism for particle plasticization is particle swelling with water, which reaches a large extent in the case of microgels. Model microgels are produced by incorporating significant amounts of a water-soluble monomer at the beginning of the polymerization [23]. Very interesting phenomena (e.g., temperature-dependent swelling and stability) are obtained using these model systems, which suggests that some unusual with latex behaviors may be due to microgel contributions.

2. Colloidal Crystal and Macrocrystal PS Latex Formation

Self-assembly [24,25] is an important and pervasive characteristic of molecular [26], colloidal [27], and biological systems [28]. It has been observed in proteins [29], Langmuir–Blodgett films [30] and other monolayers [31], micellar and mesophasic amphiphyle solutions, and latex dispersions [32]. Self-assembly systems now attract considerable attention because they are prospective candidates for preparing submicrosized and nanosized arrays, which display unique optical, electrical, and mechanical properties. Well-defined regularly patterned arrays will probably find use in the manufacture of devices for applications in various fields [33].

The self-organized assembly of colloidal latex particles into dry films with a highly ordered structure was first described by Alfrey and co-workers [34], using polyvinyltoluene latex. These iridescent, light-diffracting arrays are called macrocrystals. At first, the spontaneous formation of such ordered films was assigned solely to the high uniformity of particle sizes in the samples studied.

The ordering of latex particles in suspension and the consequent formation of the so-called colloidal crystals was reported only a decade after the pioneering work of Alfrey [35]. Considerable knowledge of the colloidal crystals has also been accumulated and sophisticated applications have developed, such as the thermally switchable diffracting films recently described by Asher and colleagues [36]. Such crystals are formed by unit cells similar to those of metallic alloy crystals and they exhibit crystalline-phase transitions [37].

As far as the formation of solid macrocrystalline films is concerned, the ordering of particles is not often observed, even in the case of monodisperse latex films. This shows that particle size uniformity is not the only relevant factor, and other particle characteristics may be decisive for the attainment of the required degree of cooperativity among the latex particles. Early in 1978, Distler and Koenig [38] found that heterogeneous latex particles made by two monomers of different hydrophilicity build up a continuous honeycomb surface structure during drying. Latex particles maintain their identity during film formation, and the resulting films are iridescent. These authors stressed the association between particle heterogeneity and light diffraction and concluded that diffraction was evidence of the regular distribution of particle heterogeneities within the films.

Recently, some research groups have developed new approaches for creating highly ordered patterned dry materials, especially those that exhibit iridescence. Ingenious ways of producing ordered films were proposed by Nagayama and co-workers [39], Michelleto et al. [40], Kim et al. [41], and Trau et al. [42]. These new routes have led to good results, but painstaking procedures are usually employed. These authors stress the importance of attractive lateral capillary forces [43] in creating a particle cluster, which then nucleates the formation of the ordered monolayer. On the other hand, the size of the ordered domains obtained by these authors is still limited. According to their micrographs, the number of correlated macrocrystalline planes is seldom greater than 10.

In a recent article, Burmeister et al. [44] state that the fabrication of close-packed monolayers of polystyrene lattices is now well established. These authors were quite successful in preparing and transferring latex monolayers to different substrates and using them as lithographic masks, but the two-dimensional (2D) structures presented by them are highly defective, and the number of ordered particles seldom reaches 30 in any row between grain boundaries. Picard [45] described recently a dynamic thin laminar flow method for 2D crystal preparation. He obtained highly ordered particle layers at high rates. However, its procedure is strongly dependent on the experimental conditions (e.g., the subphase pH).

All this effort required to make regular assemblies of spherical particles of regular size leads to the following hypothesis: There are some particle features (distribution of chemical constituents, symmetry of electrical charge distribution) beyond their diameters and surface hydrophilicity which have an important role in the making of self-assembled structures. Moreover, these features are often nonuniformly distributed among the particles so as to prevent these from interacting cooperatively and thus producing the desired self-assembly structures.

C. Latexes as Industrial Products

The importance of latexes as industrial products has been growing steadily for many decades. This has technical and environmental reasons: Emulsion polymerization benefits from the excellent heat exchange and stirring features allowed by a low-viscosity medium and it dispenses with the use of solvents, which are a source of many environmental and safety problems in polymer making and use. For this reason, it is quite likely that the importance of emulsion polymerization will continue to grow in the near future.

On the other hand, the number of process variables in latex fabrication is very large. For this reason, the number of different properties which can be found even in homopolymeric latexes is very high. An example is given in Table 1, which gives data for four different polystyrene latexes, prepared using different stabilizers and initiators, and exposed to the vapors of two water-immiscible solvents and turbidimetrically titrated for stability evaluation [46]. From these results, we observe two large differences among these polystyrene latexes:

1. One latex only (PS-M) shows a marked change in its effective diameter (as measured by PCS), under exposure to $CHCl_3$ vapor.
2. Two latexes (PS-LEV and PS-M) coagulate in the presence of 0.32 M

Table 1 Comparison of the Properties of Four Different Polystyrene Latexes, Prepared by Different Techniques: Effective Diameters,[a] Zeta Potentials,[a] and Coagulated Sediment Heights[b]

Exposure to vapor/ Solvent/ time (h)	Property	PS–LEV Renex 300/ Ascorbic acid–H_2O_2	PS–THS No surfactant/ $K_2S_2O_8$	PS-11 Renex 300/ $K_2S_2O_8$–NaHCO$_3$	PS-M Brij 35–SDS/ $K_2S_2O_8$
None	Effective diameter (nm)	405 ± 11	266 ± 3	94 ± 2	101 ± 2
	Zeta potential (mV)	-53 ± 2	-43 ± 1	-11 ± 4	-31 ± 2
	Sediment height (cm)	0.5	None	None	1.9
CHCl$_3$/48	Effective diameter (nm)	434 ± 9	283 ± 6	103 ± 2	220 ± 3
	Zeta potential (mV)	-51 ± 2	-23 ± 3	-15 ± 3	-58 ± 1
	Sediment height (cm)	None	None	None	0.6

(Sample: / Latex preparation: / Surfactant/Initiator:)

[a]Averages of triplicate measurements.
[b]In the presence of [NaCl] = 0.32 M.

NaCl, whereas the two others are stable in this high ionic strength. On the other hand, PS-LEV is stabilized by exposure to $CHCl_3$, but PS-M is not.

This and analogous examples show how much latex properties can change, depending on relatively small synthetic differences and, thus, on details of the bulk polymer and surface composition.

Thus, we may foresee a growing need for detailed information on latexes and their constituent particles, beyond all the usual sets of data available for polymers fabricated by other techniques. This specific information on particle characteristics should help the improvement of emulsion polymerization products and processes. In Secs. IV and V, we describe new techniques for latex analysis, emphasizing the contributions from our laboratory which allowed us to demonstrate some types of particle heterogeneities for the first time in the literature.

D. Emulsion Polymerization Kinetics and Mechanism

Emulsion polymerization is a heterogeneous reaction initiated by free radicals, used for the polymerization of styrene, vinyl chloride, acrylonitrile, vinyl acetate, methyl methacrylate, methyl acrylate, and other related monomers. The majority of these monomers have a limited water solubility, which is sufficiently low to prevent a substantial amount of polymerization from taking place in the aqueous phase.

According to the classical Harkins [47] model, a typical emulsion polymerization process is considered a three-phase reaction system. The reaction medium is generally a surfactant-stabilized aqueous emulsion of a poorly soluble monomer, which is partitioned among large droplets [1.5–3 μm in diameter], the surfactant micelles (~5 nm diameter), the aqueous phase (where it is dissolved together with the initiator), and the colloidal particles of a monomer-swollen polymer (usually 50–150 nm in diameter).

The initiator (often a persulfate or a redox couple) is present only in the aqueous phase. The free radicals generated in water meet the monomer molecules at the droplet or micelle surfaces, as well as in the solution. The final particle diameter (usually below 500 nm) is much smaller than the monomer droplet diameter, and the corresponding number of latex particles is greater than the number of micelles present at the beginning of the process, indicating that the preferential initiation sites are the monomer-swollen micelles [48].

Particle nucleation occurs in the beginning of a batch emulsion polymerization and, depending on the water solubility of the monomer, this step normally ends at a low (1–5%) conversion. The nucleation stops when all the micelles are consumed either by the formation of a polymer–monomer particle or by the adsorption of the surfactant into the growing particles.

The polymerization reaction proceeds to a great extent during a step characterized by the growth of the primary particles. This step begins when the particle number has stabilized and the volume of the particles starts to increase in proportion to the conversion. The monomer concentration in the particles remains approximately constant, whereas the monomer in the droplets is gradually transferred to the colloidal particles by dissolving in and diffusing through the aqueous phase. The particle size, the polymer–solvent interaction parameter, and the particle–water interfacial tension are factors which govern the equilibrium concentration of the monomer within the particles. The ionic strength, which can affect the surface potential and surfactant adsorption, influences the interfacial tension and thus the equilibrium swelling of particles.

By the end of the polymerization reaction, the separate monomer phase disappears; the monomer concentration in the particle decreases with a corresponding increase in polymer viscosity and a decrease of the particle volume. The increasing viscosity affects the intraparticle diffusion of both the monomer and free radical, and the polymerization stops when the monomer content within the particle is not large enough to plasticize the polymer.

The analysis of experimental kinetic data is usually based on the quantitative theory developed by Smith and Ewart [49] and is based on the Harkins model, relating the polymerization rates to the particle numbers. The particle number depends on the surfactant concentration to the 0.6 power and on the initiator concentration to the 0.4 power. Polymerization of relatively water-insoluble monomers (e.g., styrene) can be described by the Smith and Ewart theory below a particle concentration of 3×10^{14} particles/mL, but their model is not applicable to systems with water-soluble comonomers. Reactions of partially water-soluble monomers (e.g., vinyl acetate and methyl methacrylate), which have a water solubility higher than 0.04–0.07%, or those in which the monomer solubility is increased by the addition of some solvent to the system show large deviations from the theoretical predictions. Large deviations from the expected 0.6-power relationship were also found for such water-insoluble monomers as butyl acrylate and butyl methacrylate.

In order to explain these deviations from the Smith and Ewart particle nucleation model, a theory based on homogeneous nucleation was proposed [50]. According to this model, there is a homogeneous solution of monomer, surfactant, initiator, and water at the start of polymerization. The free radicals are generated in the water phase and polymerization proceeds in the continuous phase with the appearance of oligomeric radicals. The reaction proceeds in the solution until the polymeric chains are large enough to become insoluble and the polymer segments collapse upon themselves with the formation of the primary particles. The chance for an oligomeric radical to be captured by a particle increases with an increase in the number of particles. The rate of particle formation is thus proportional to the net rates of radical generation, oligomeric radical capture by particles, and particle

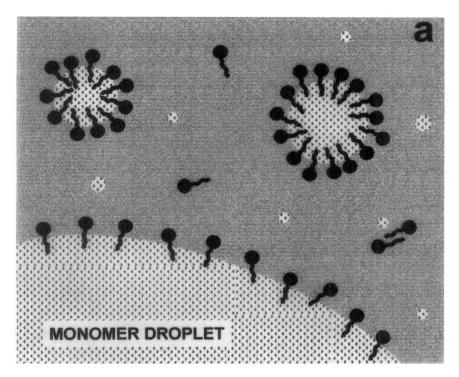

Figure 1 Schematic representation of emulsion polymerization (a), initial stage (b), interfacial and homogeneous initiation (c), final stage.

agglomeration. A schematic description of emulsion polymerization including homogeneous nucleation is presented in Fig. 1.

The ideas and facts mentioned in the previous paragraph suggest that different, concurring paths may lead to the formation of latex particles in any real case. Beyond that, the many diversifying factors affecting any chain polymerization (activity transfer, terminations) should help to create a great diversity within the polymer chains. Moreover, each of the different chain types thus generated would then concentrate, more or less, in specific particle subpopulations.

For these reasons, latex particles should display some degree of heterogeneity concerning their detailed chemical composition and properties. The impressive particle size uniformity which is often achieved should not be taken as evidence that all other particle characteristics are also uniformly distributed throughout the particles.

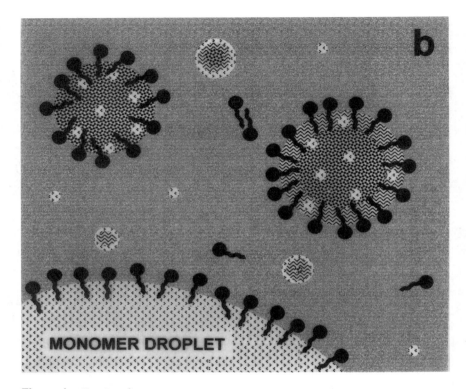

Figure 1 *Continued*

II. LATEX COLLOIDAL STABILITY

Latex colloidal stability problems arise in emulsion polymerization and in latex storage, use, and disposal. Latexes are often intentionally coagulated in the making of rubbers, thermoplastics, and their artifacts [51]. For this reason, a good control of latex stability and coagulation is essential for many products and processes. On the other hand, latexes have been widely used as particle models in colloidal chemistry. This gave a large amount of detailed information and revealed some important discrepancies with the current colloid stability theories, which have already been considered earlier in Sec. I.B.1.

There is common knowledge in industry about unexpected coagulation as well as excessive stability problems in latex polymerization and use [52]. Some difficult situations of both types arise that challenge even experienced professionals and researchers. This is evidence that relevant particle properties may change

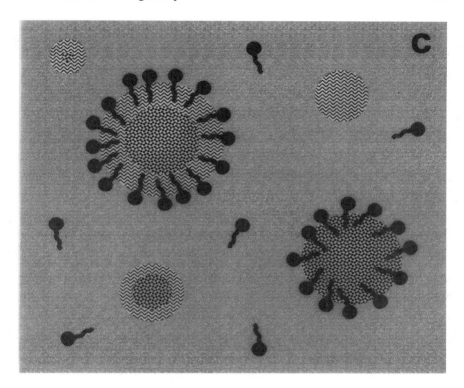

from one to another batch of the same product, and even within a continuous polymerization process.

Even in the scientific literature, we note a remarkable lack of information about the amount of coagulus and dispersed material obtained in a given preparation. All these observations bring us to a conclusion: There are serious, still partly unsolved problems in latex making and use which are perceived as a nonuniform behavior and should be related to nonuniform particle composition, morphology, plasticity, and surface properties, for which new evaluation techniques are required.

III. COAGULATION AND FILM FORMATION

Latexes are often used in one of the following ways: (1) they are coagulated, making a solid rubber or a thermoplastic, which will then be subjected to usual polymer processing techniques and (2) they are used as film-forming agents in paints, adhesives, condoms, and gloves. In this case, it is expected that the colloidal stable particles, which are prevented from aggregating in the dispersed state, change into a continuous apolar, preferably water-unswollen solid.

Many mechanisms and theoretical models have been proposed to describe the latex film formation. Considering many important contributions, it is now accepted that film formation develops through four sequential steps [53–55]:

1. The first phase is dominated by the water evaporation rate. This initiates at a high but decreasing rate as the particles approach each other and the aqueous phase becomes more concentrated in salt, surfactant, and other solutes. The particles within the latex sol are then reunited within coaguli or flocs, which percolate, transforming into a gel.

2. Contacting particles are now deformed under the action of interfacial tensions and the resulting capillary pressures. Hydrophilic layers surrounding the particles are exchanged with more hydrophobic polymer from the particle cores as external water is replaced by air. These powerful driving forces are opposed mainly by the polymer viscosity or by the changing viscosities of the drying particle domains.

3. The third step is characterized by densely packed, deformed, contacting, nearly dry particles subjected to the strong tensions due to capillary pressure. Voids decrease gradually as interparticle necking develops, driven by surface tension but limited by polymer diffusion.

4. The final step is film aging, characterized by polymer chain (segmental and translational) interdiffusion across the particle boundaries. This depends on polymer MW distribution, the degree of cross-linking, temperature, and plasticizers. The final state is determined by the phase diagram of the system and the related domain structure.

These four steps are schematically depicted in Fig. 2. The discussion of film formation has assumed that latex particles are uniform in most aspects. Of course, if the particles are not homogeneous with regard to their surface, viscoelastic, and thermal properties, important effects will be observed.

As a latex is transformed into a polymer solid, it displays marked changes in its properties. For instance, a finished wall paint should be highly water impermeable, but the drying film should be water permeable, to allow for water removal from the drying paint and often also from the paint substrate.

These seemingly contradictory sets of requirements are not easily met, and the particle characteristics which are desirable in one prospective application may be inconvenient in other cases. For instance, latex particles with thick hydrophilic outer layers form beautiful opalescent, densely packed macrocrystals easily, but these make brittle films.

On the other hand, the fractal nature of latex aggregates has been demon-

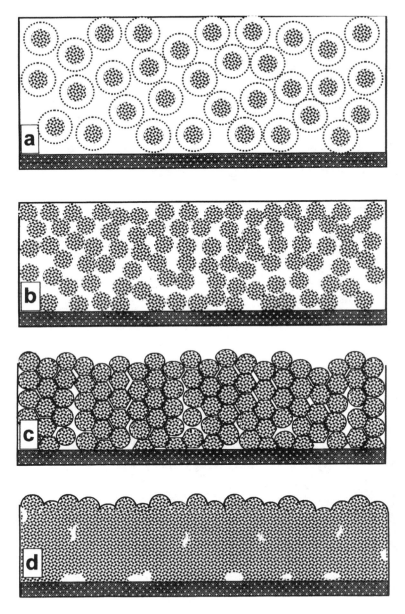

Figure 2 Main steps of latex film formation: (a) particles approach each other, due to water evaporation; (b) flocculation of the latex particles and percolation; (c) dense packing and deformation of the particles forming a honeycomblike structure; (d) coalescence and autoadhesion by interdiffusion of polymer chains of adjacent particles.

strated [56] and the fractal dimension is often low, which shows that the transformation of a wet, coagulated mass into a plastic solid requires overcoming a limited tendency to full interparticle void elimination, at least in some latexes. One example of incomplete film formation is shown in Fig. 3, in which particle individuality and film heterogeneities are easily perceived. When these films are subjected to temperatures more than 20°C or 30°C above their glass transition temperature, polymer diffusion occurs across the particle interfaces and particle structure finally disappears, as shown, for example by transmission electron microscopy (TEM) observation of latex particles [16].

This may seem perhaps too complicated to assure that high-quality final products will always be obtained. On the other hand, this complex behavior of latexes may be seen as a source for innumerable opportunities for making new materials with predesigned properties, using a set of a few monomers and initiators, but using different polymerization and coagulation protocols. Of course, achieving this objective requires the ability to obtain detailed information on latex particle heterogeneity, microchemistry, and topochemistry, using the techniques described in this chapter.

The formation of a solid polymer by latex coagulation and drying is affected by the same basic factors of film formation. Common knowledge within the industry tells that coagulation of the same latex but following different procedures leads to products with different properties, which is still poorly understood.

Figure 3 Atomic force microscopic picture of a dry poly(vinyl acetate) (PVAc) latex surface. Note the rugosity due to incomplete particle coalescence.

IV. LATEX PARTICLE HETEROGENEITY

A. Differences Among Particles

Particle size distributions are now determined using accurate techniques such as photon correlation spectroscopy (PCS), capillary chromatography, centrifugation, and ultracentrifugation, as well as field-flow fractionation [57]. However, other differences among latex particles are not detected as easily. For this reason, information is scarce about the uniformity of particle chemical composition, comonomer distribution and molecular-weight distribution within particles, surface composition, and electrical charge. Most researchers assume that these properties do not change significantly from one particle to another, but others have assigned an unexpected latex behavior to particle heterogeneity. For instance, Buscall and Ottewill [58] suggested that the anomalous coagulation behavior of some latexes under NaCl concentrations lower than the critical coagulation concentration could be due to particle heterogeneity, as discussed in Sec. II.

B. Centrifugation in Density Gradients

Excellent accounts on latex centrifugation have been presented by Mächtle [59,60] and Lange [61,62] using different techniques. In this section, we will be restricted to zonal centrifugation in density gradients, which is a powerful analytical and preparative technique for the study of colloidal particles [63]. Notwithstanding the many other centrifugation techniques, we have found that this one offers an excellent balance among the ease to perform experiments, their interpretation, and the amount of information thus produced.

Zonal centrifugation runs yield three results: (1) particle sedimentation coefficients, when the particle migration rates are determined; (2) particle densities, when the isopycnic equilibrium condition is reached; (3) fractionated samples (if the original latex preparation is heterogeneous), which are produced in amounts sufficient for further analytical work: infrared (IR) and nuclear magnetic resonance (NMR) spectroscopies, particle size and zeta potential determination, and so on.

One example of a fractionation according to particle sizes, in a low-speed centrifuge, is in a study on a partly coagulated PS latex sample. Band migration rates were measured in a zonal experiment within a density (and viscosity) gradient, which was accurately determined. Sedimentation coefficients and equivalent diameters for up to six aggregates were then calculated, from which the aggregate masses were obtained. This allowed the identification of latex singlets, doublets, and so on, up to hexaplets. These results are in Table 2 [64]. Using this kind of information, it is possible to obtain kinetic curves, showing the time evolution of monomer, dimer, trimer, and other aggregate concentrations [65] in a sample undergoing coagulation induced by salt (see Fig. 4).

Table 2 Determination of PS Latex Sedimentation Coefficients and Particle Diameters
by Sedimentation Rate Measurement in Density Gradients

Nominal diameter (μm)	Diameter of the equivalent sphere (μm)	Band number	Sedimentation coefficient (S)	Mass of particles in band i / Mass of particles in band 1
0.109	0.122	1	297	1
	0.163	2	470	2
	0.193	3	593	2.8
	0.225	4	723	3.8
	0.258	5	868	5.0
	0.275	6	947	5.7

Note: Gradient generated by osmocentrifugation of Ficoll, 8% w/w; starting density, 1.024 g/cm^3; centrifugation speed, 3000 rpm, T = 20°C. Sample: the supernatant of a partially coagulated 0.109-μm-diameter Sigma latex.
Source: Ref. 64.

Particle heterogeneity is thus easily detected from sedimentation coefficient data: A single zone is resolved in a number of bands, depending on the number of constituent particles or on the number of particle aggregates found.

On the other hand, if a run is prolonged until the particles reach the isopycnic condition, the band profile depends only on particle densities, which, in turn, depend on particle chemical composition, molar weight and tacticity, degree of hydration, and other more structural features. Some examples of isopycnic separation will be given in Sec. IV.D.

C. Performing Zonal Gradient Experiments

A zonal gradient experiment has three parts: (1) gradient preparation; (2) centrifugation run and data acquisition; (3) data analysis.

Concentration gradients are usually aqueous solutions of an inert, dense solute (cesium chloride, sucrose, Ficoll, metrizamide), the concentration of which increases from the top to the bottom of a centrifuge tube. They are prepared (1) by fast centrifugation of a solution of the gradient-forming solute, (2) premixing of a concentrated and a dilute solution of the gradient-forming solute, and (3) osmocentrifugation [66]. The first and third cases are suitable when the gradient former is macromolecular (Ficoll, dextran, albumin); the second case is adequate for micromolecular solutes (sucrose, CsCl).

A zonal centrifugation run is started by layering a small volume (typically 100 mL) on top of the concentration gradient and spinning the centrifugation tube,

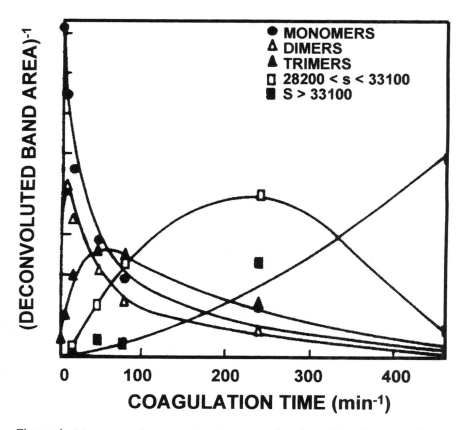

Figure 4 Monomer and aggregate band areas as a function of time, from a zonal centrifugation of partly coagulated latex. Polystyrene latex (Sigma, nominal diameter 0.46 μm) coagulation by the admixture of equal volumes of latex and NaCl stock solutions. Initial zone concentrations: 1.86×10^9 particles/cm^3 in NaCl 0.30 mol/dm^3. Concentrations in the density gradient: 0.30 mol/dm^3 NaCl ($\rho = 1009$ kg/m^3) at the top; the same plus Ficoll 70, 16% w/w at the bottom ($\rho = 1065$ kg/m^3). Centrifugation at 4500 rpm, 25 min, 25°C. (From Ref. 65.)

preferably in a fixed-angle rotor at a high speed. Current preparative centrifuges can be used with transparent plastic tubes at 1–30,000 rpm, which gives inertial fields as high as 10^5 times the normal gravity, and allows the completion of runs in short times. After spinning the sample for the desired time, the distribution of latex within the gradient is recorded by using a scattered-light densitometer or just by recording a digital tube picture, from which a line scan is obtained using image processing software. The result is a band profile as depicted in Fig. 5.

If the isopycnic equilibrium condition is met, each band corresponds to a particle subpopulation, the density of which is obtained from the known density gradient.

D. Density Heterogeneity of Latex Particles: Latex Fingerprinting in Density Gradients

Zonal centrifugation reveals that seemingly simple latexes may be very complex indeed. In one previous work [67], we found that a simple polystyrene latex is separated in many fractions, both in the transient sedimentation and under isopycnic equilibrium (see Fig. 6). The density differences are significant, even though these are just homopolymer particles from the same preparation. These three fractions in a polystyrene latex may be due to a number of factors, as discussed in Sec. IV.B.

A remarkable result is the sensitivity of particle isopycnic profiles to salt concentration. This is clearly seen in the intensity profiles recorded at the shorter times: In the lowest salt concentration (10^{-4} M), there is very little material at or close to the isopycnic position. On the other hand, at the highest salt concentration (10^{-2} M, still below the critical coagulation concentration), some bands move much faster than the main band, showing that some particles are much more sensitive to salt than others, which, in turn, is strong evidence for the heterogeneity of the particles' surface layers and thus of their colloidal stabilities.

Together, the time and salt concentration effects on the scattered-light-intensity profiles produce a particle-density two-dimensional fingerprint, which is unparallelled in latex particle heterogeneity characterization.

A detailed chemical characterization of a latex may be done by subjecting latex fractions to analysis by NMR, IR, and other suitable techniques. In our first work on this topic, we examined some latex industrial products, used as adhesives or paints [68]. At this time, we used only visual recording of the particle distribution profiles within the centrifugation tubes, as shown in Fig. 7. The density distribution for this PVA latex is broad, which allowed easy collection of the samples indicated by the letters A–D from which NMR spectra were taken, as shown in Fig. 8. The spectra show that the denser particles are also richer in the isotactic polymer sequences, which is as expected. On the other hand, the presence of CH_3CH_2O- groups is revealed by the 1.1-ppm triplet and the 3.6-ppm quartet, which appear only in the unfractionated sample and in the fraction with the highest density (D).

Copolymerization is often used to produce latexes. In this case, further information about the monomer distribution among polymer chains and particles is necessary. The question of monomer distribution can now be addressed by many techniques, including chromatographic separation combined with spectroscopic

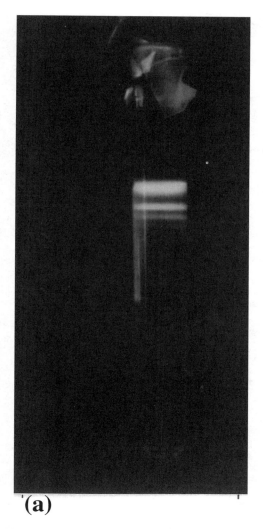

(a)

Figure 5 Bands formed by centrifugation of a partially coagulated latex in a density gradient. Polystyrene latex (Sigma, nominal diameter 0.885 μm) was added to 0.5 M NaCl for 12, 7, 3, and 1 min. Latex concentration: 1%; zone volume: 100 μL. Gradient starting solutions: 7% w/w aqueous Ficoll added to 0.5 M NaCl and aqueous 0.5 M NaCl. Zonal centrifugation for 20 min, 4500 rpm. (a) Tube picture; (b) line scan through the line drawn in (a).

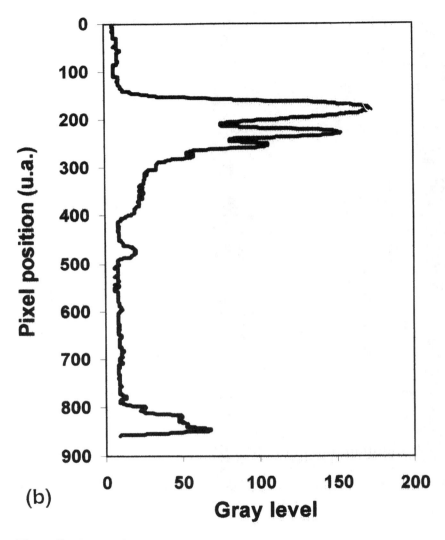

Figure 5 *Continued*

measurements [69,70]. On the other hand, monomer distribution among particles is often neglected in the literature. This may be approached by using the zonal centrifugation within density gradients in analytical or high-speed centrifuges, but only when there is a simple correlation between monomer content and particle density.

We have used centrifugation combined with NMR spectroscopy in the case

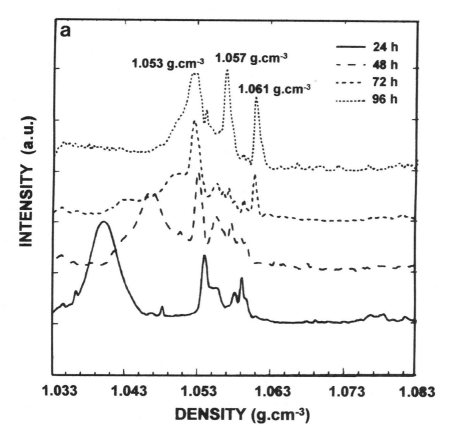

Figure 6   Scattered-light profiles of a fractionated 100-μL zone of PS latex (D_n = 79 nm, D_w = 84 nm), 1% solids. NaCl concentration in the gradient: (a) $10^{-2}\,M$; (b) $10^{-3}\,M$; (c) 10^{-4} M. Linear density gradients were made from aqueous salt–sucrose solutions with different sucrose concentrations. Centrifugation at 3000 rpm, 25 °C. Runs were stopped every 24 h, and scattered-light-intensity profiles were recorded. (From Ref. 67.)

of three poly[styrene-*co*-(butyl methacrylate)] latexes. This is a demanding system because the homopolymers have almost identical densities [71].

The three latexes were prepared following different protocoles: in one case, the styrene (S) and butyl methacrylate (BMA) monomers were added simultaneously (PSBMA); in the second case, styrene was added first (PS/PBMA); the third case was the opposite of the second. The scattered-light profiles of the centrifugation tubes are in Fig. 9. ¹H-NMR spectra of upper and lower fractions of these three latexes were obtained and are presented in Fig. 10. From these, the styrene

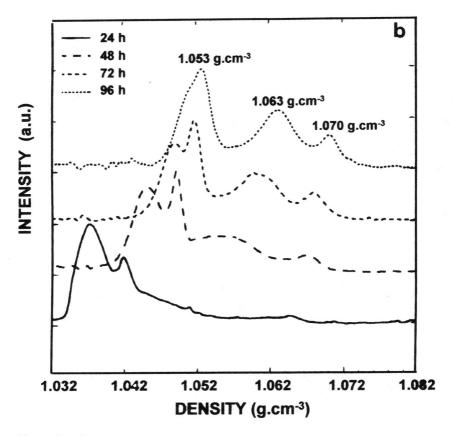

Figure 6 *Continued*

contents of the lower and upper density halves of the latexes are calculated, as shown in Table 3. The results show a significant difference between the monomer compositions of the latex fractions, together with the absence of a correlation between particle density and monomer composition, in this case. For this reason, great care should be exercised in calculating particle compositions from density data. This is only reliable in the case of monomers with highly different densities and uniform tacticity.

Another quantitative difference between upper and lower fractions spectra is observed in Fig. 10 at 1.3–1.8 ppm, which is a region dominated by peaks assigned to CH_2 and CH groups, from both styrene and butyl methacrylate. This difference stems from the polymer main chains, probably due to configurational or branching features.

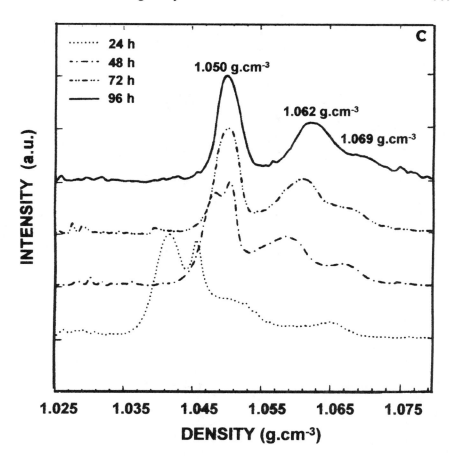

E. Osmocentrifugation in Density Gradients

An alternative to density gradient separations within preformed gradients in a high-speed centrifuge is osmocentrifugation in a density gradient. Osmocentrifugation is performed substituting the centrifuge tubes with dialysis cells fitted with a vertical membrane and mounted in a centrifuge swinging-bucket rotor [72–74]. This technique has two advantages: first, the gradients are produced in situ, which saves a time-consuming experimental preparation step; second, centrifugation times are shorter. On the other hand, there may be some loss of resolution and there is not yet a full description of mass transfer patterns within the osmocentrifugation cell, which makes a detailed analysis of some results difficult. A prac-

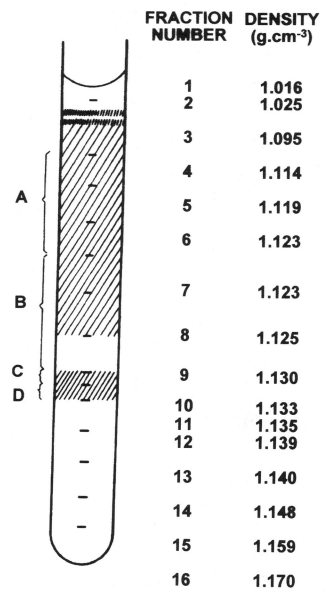

Figure 7   Isopycnic equilibrium of a PVAc latex-based adhesive in a Percoll (initial density: 1.129 g/cm^3) density gradient. Experimental conditions: a density gradient was self-generated by osmocentrifugation at 10 min, 1000 rpm, 20°C; then, a latex zone was layered and centrifuged; zonal centrifugation for 18 h at 1500 rpm, 20°C. (From Ref. 68.)

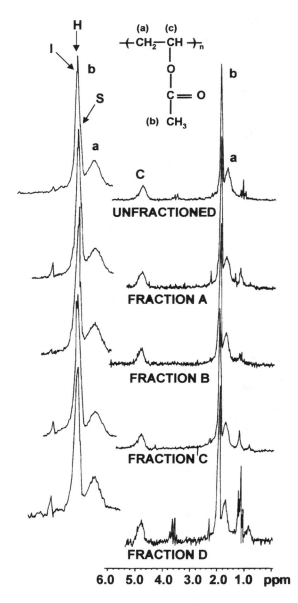

Figure 8 ¹H-NMR spectra of PVAc from a commercial adhesive and of four fractions separated by centrifugation in a Percoll (initial density; 1.129 g/cm³) density gradient. Solvent: CCl₄; Bruker AW-80 spectrometer. Concentrations: unfractioned, 8%; fractions A–C, 4.3%; fraction D, 1.4% (fractions are identified in Fig. 7). I, H, and S refer to iso-, hetero-, and syndiotactic sequences, respectively. (From Ref. 68.)

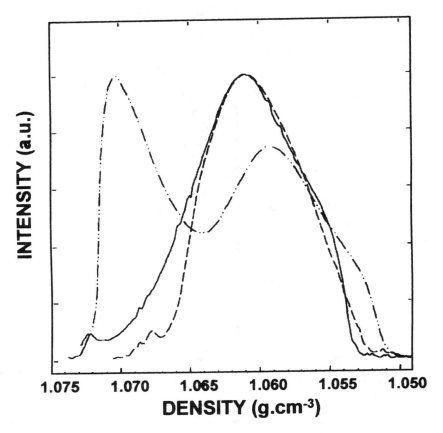

Figure 9 Scattered-light profiles obtained at isopycnic equilibrium of styrene–butyl methacrylate latexes, in aqueous sucrose density gradients:(—) P(SBMA),(----) PS/PBMA, (—··) PBMA/PS. The latexes were prepared by emulsion polymerization of styrene (S) and butyl methacrylate (BMA). Three procedures with different orders of monomers addition were carried out: (1) simultaneous addition of the monomers, labeled P(SBMA); (2) addition of S followed by acrylate (PS/PBMA); and (3) addition of BMA followed by the addition of S (PBMA/PS). Linear density gradients ranging from 1.030 to 1.090 g/cm^3 were prepared from sucrose aqueous solutions (8% and 22%). Centrifugation: 3500 rpm, 25°C (From Ref. 71.)

tical difficulty is that osmocentrifugation cells are not yet commercially available, which means that anyone wanting to use this technique may have to make his own cells.

PVA and PS latexes are easily separated in well-resolved bands in a short osmocentrifugation time, as shown in Fig. 11. Infrared spectra of the latex col-

Table 3   Styrene (S) Contents of the
Upper and Lower Isopycnic Fractions of
Three Styrene–butyl Methacrylate Latexes

	S (mol %)	
Sample	Upper	Lower
P(SBMA)	37.8	38.3
PS/PBMA	41.7	45.4
PBMA/PS	44.2	40.7

lected from each band shows that there is no detectable cross-contamination, as shown in Fig. 12.

The PS latex particles of different sizes are also well resolved by osmocentrifugation in a self-generated density gradient (Fig. 13).

The potential of this technique still has to be realized, but at this time, it depends on further progress in the modeling of mass transfer in the osmosedimentation cell.

F.   Particle Aging

Paint makers know that formulated latex paints "age" (i.e., their viscosity and other properties change with time). This may be related to any one of the paint components, including the latex. To approach this problem, we have applied the fingerprinting abilities of zonal centrifugation to demonstrate particle aging. We shall refer to the same PS latex described in Fig. 6, but considering runs made 1 year apart, as shown in Fig. 14 and Table 4 [75]. The latex changes upon aging, as follows:

1. The number of latex bands decrease and the particles densities also decrease. On the other hand, the particle density uniformity is increased.
2. The fractions coagulating at 10^{-2} M salt concentration disappear.

The observations made are consistent with the following hypothesis: During particle formation, solvated ions (from initiator and surfactant residues as well as the respective counterions) are trapped within the polymer domains, which are then frozen in a nonequilibrium situation. Ion rearrangement, redistribution and dissolution then proceed slowly, but their progress leads to more uniform polymer particles. Another aging factor could be the hydrolysis of terminal sulfate groups [76] in the polystyrene chains, arising from the persulfate initiator residues bound to the chains.

This demonstration of particle aging raises one question: Most data on poly-

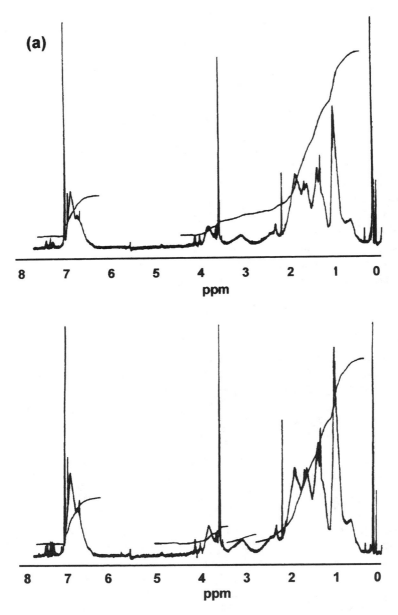

Figure 10   The 300.08-MHz-proton NMR spectra of the polymer latexes: (a) P(SBMA),
(b) PS/PBMA, and (c) PBMA/PS. Solvent CDCl$_3$ at room temperature. Varian Gemini 300
spectrometer, sample concentration 20 mg/mL, TMS standard. For each latex, two spectra
are given: one (upper) is for the lower-density fraction; the other for the high-density fraction. (From Ref. 71.)

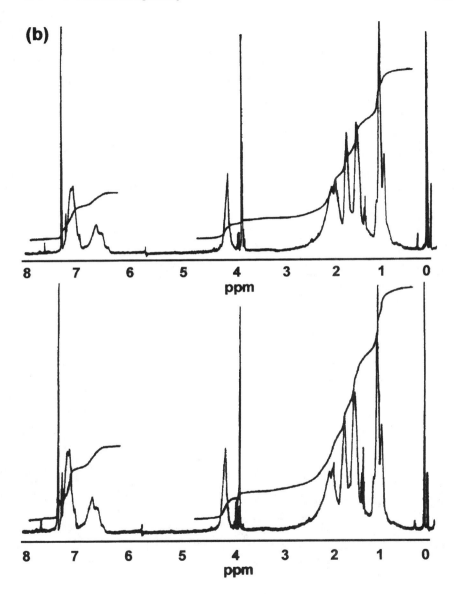

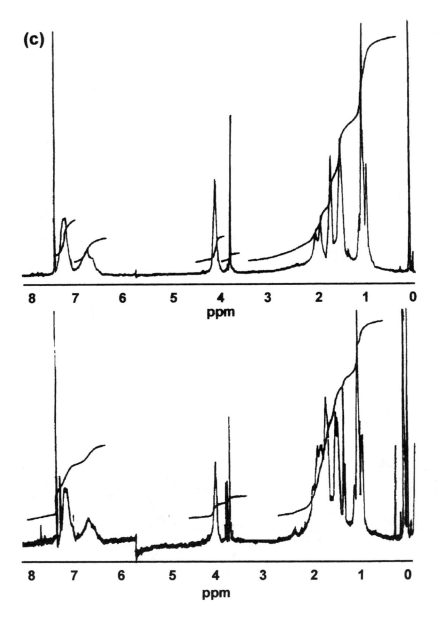

Figure 10 *Continued*

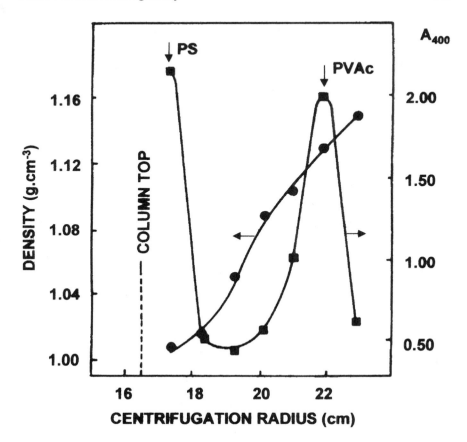

Figure 11   PS and PVAc latex particle fractionation in a preformed Percoll density gradient. Running conditions: 2000 rpm, 50 min osmocentrifugation (prior to zone layering) followed by 20 min zone fractionation. Polystyrene (Sigma, nominal diameter 0.460 μm) and poly(vinyl acetate) (Orniex, São Paulo) latexes were centrifuged in a Percoll gradient preformed by osmocentrifugation. Initial density: 1.129 g/cm³; concentrations: PS = 0.6% and PVAc = 8%. (From Ref. 64.)

mer latexes reported in the literature do not mention the time taken from latex preparation to the actual performance of the experiments in which the latex is used. This suggests that some difficulties of repeating some coagulation and self-arraying experiments, which are well known from many researchers, may just be assigned to intrinsic changes in the polymer due to aging.

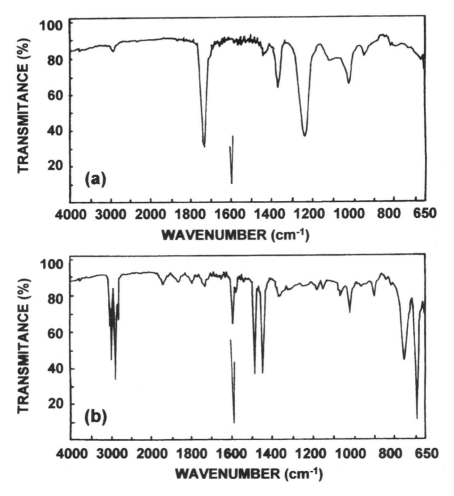

Figure 12 Infrared spectra of (a) PVAc and (b) PS obtained by latex particle fractionation in a Percoll density gradient. Each band was diluted with water, centrifuged in a glass vial (to separate the latex from Percoll); the solids collected at the bottom of the centrifugation vials were dried and extracted with toluene. The toluene–latex solutions were spread on IR NaCl windows and evaporated. Note the absence of band cross-contamination. (From Ref. 64.)

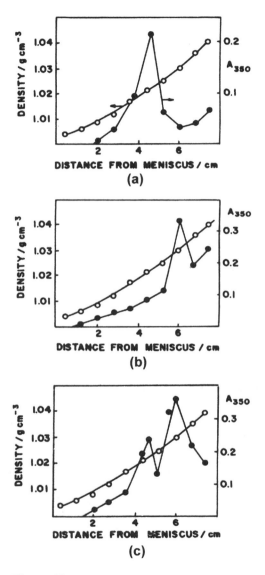

Figure 13   PS latex band formation in the osmocentrifugation cell. Latex particle diameters: (a) 0.305 μm; (b) 0.460 μm; (c) 0.305 μm plus 0.460 μm. Ficoll density gradients were preformed by osmocentrifugation of an aqueous solution (starting density 1.024 g/cm³) at 3000 rpm, 20°C for 2 h. A layer of PS latex (0.2 mL, 2% w/w) was placed over the density gradient column and the cell was spun for another 4 h. Cell content fractions were taken at various cell heights; their refractive indexes and turbidies were measured. Note the good band resolution. (From Ref. 64.)

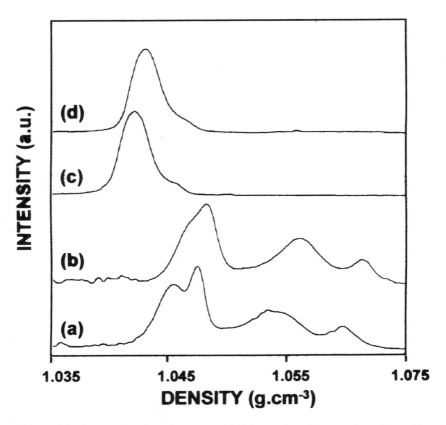

Figure 14 Latex aging detection: scattered-light profiles of isopycnic profiles of (a and b) fresh and (c and d) aged PS latex. Centrifugation running times: curves a and c, 72 h; curves b and d, 96 h. PS latex (D_n = 79 nm, D_w = 84 nm) was centrifuged in preformed linear density gradients made from aqueous salt–sucrose solutions. The volume of each latex zone was 100 μL (1% solids). Centrifugation: 3000 rpm, 25°C, the runs were stopped every 24 h. The PS latex was stored within tightly closed glass vials at a 6.3% (w/w) solids concentration. The storage time was 12 months within a cabinet in order to avoid lighting and mechanical disturbance. Just before their application to the top of the centrifugation gradients, the samples were diluted to a 1% solids concentration. The aged latex is more uniform and has a lower density than fresh latexes. (From Ref. 75.)

Table 4 PS Latex Zonal Centrifugation: Zonal Isopycnic Densities in NaCl–Sucrose Gradient Media

NaCl concentration added to sucrose density gradient (*M*)	Latex density[a] (g/cm^3)			
	Fresh			Aged
10^{-4}	1.053	Main band	65.5%	1.046
	1.063		28.0%	
	1.070		6.5%	
10^{-3}	1.050	Main band		1.050
	1.062			
	1.069			
10^{-2}	1.053	Band		1.050
	1.05 (7)	Coagulum		
	1.06 (1)	Coagulum		

Note: These data refer to the band profiles in Fig. 14 (from Ref. 75).
[a]±0.003 g/cm^3.

G. Solvent-Induced Latex Stability

A recent finding from this laboratory is the colloidal stabilization of some latexes by sorption of solvent (chloroform and toluene) vapors [46]. Latexes which coagulate in the presence of salt ($\sim 10^{-1}$ *M* NaCl) become stable under exposure to solvent vapor for 48 h. Under these conditions, the degree of swelling in small particle diameters change only by a few percent and the zeta potentials are unchanged. This phenomenon is not predicted by current colloidal stability theories, but it seems to be quite general (Figs. 15 and 16).

Since the amounts of swelling solvents absorbed are low and their effect on colloidal stability is important, we should now become aware of the possible consequences of uneven contaminant (solvents, residual monomers) distribution, in a latex population.

H. Particle Fractionation By SFFF

Sedimentation field-flow fractionation (SFFF) is a powerful technique for analytical and preparative particle fractionation, which can yield latex fractions for further analysis by other techniques [77]. This was used by Li et al. to fractionate a 10-component mixture of PS latex particles and to obtain the TEM images of fractions. However, these authors did not examine the fractions by other powerful ana-

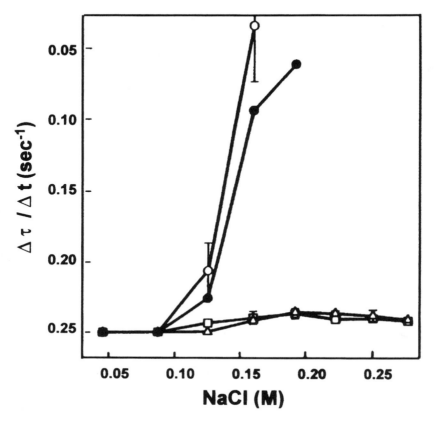

Figure 15 Apolar solvent effect on latex stability. Random styrene–butyl methacrylate copolymer P (SBMA) [D_n = 81 nm, ζ = –61 mV [within 0.01 M KCl], glass transition temperature T_g = 49.8°C, mol ratio s (S + BMA) = 0.41] was turbidimetrically titrated (100 μL, 1 M NaCl additions). (○) pristine latex; (●) the same but annealed at 80°C for 7 h; (□) latex exposed to toluene vapor; (△) latex exposed to chloroform vapor. (From Ref. 22.)

lytical techniques (size-exclusion chromatography, NMR, IR, Raman), perhaps due to the still insufficient size of the fractionated samples. For this reason, we cannot evaluate the merits of this technique for particle heterogeneity evaluation relative to centrifugation in density gradients.

I. Electrophoretic Heterogeneity

Latex electrophoresis has been the object of considerable activity, for many years. This technique has reached a high degree of sophistication as the "electrophoretic

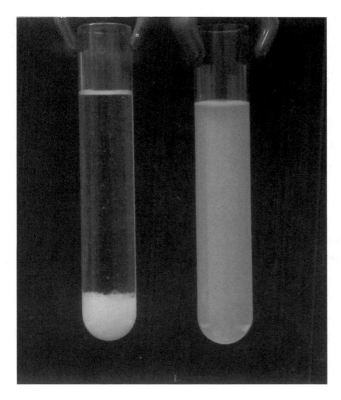

Figure 16 Photograph obtained 48 h after mixing 0.7 mL NaCl 1 M and 0.107% (w/w) P(SBMA) latex dispersion. Left: pristine latex; right: latex previously exposed to toluene vapor for 24 h. (From Ref. 22.)

fingerprinting" introduced by Rowell and colleagues [78]. Detailed comparisons of experimental data with existing models have been made; in some cases, good agreement with a simple model is obtained [79], but in other cases, anomalous behavior is observed, the interpretation of which requires the consideration of additional features. For instance, the anomalous electrophoretic behavior of a latex was examined by Verdegan Anderson [80], considering the effects of "hairiness" and of co-ion binding; the latter proved consistent with the experimental data. However, data in the literature always refer to average electrophoretic mobility data, suggesting that particles are electrophoretically uniform. Other techniques which can yield zeta-potential data (electrical conductivity, streaming potential, dielectric response measurements) are intrinsically macroscopic, which is why they cannot give data on particle heterogeneity.

However, a simple argument shows that large fluctuations are to be expected in the composition of particle surfaces: a typical (200-nm-diameter) polystyrene particle weighs 4×10^{-12} g; the outer 1-nm-thick layer weighs only 4×10^{-18} g, which is $\sim 4.10^{-20}$ mol of a 100-g/mol monomer unit, or 24×10^3 units of monomer. This is a relatively small number, for is why large fluctuations [81] are expected within a particle population, even if they have not yet been identified.

V. LATEX PARTICLE MICROCHEMISTRY AND TOPOCHEMISTRY

An important question which has challenged many researchers is the following: How are the chemical constituents of a latex distributed throughout each particle, and among the many particles? Moreover, which is the chemical composition of the particle surface layer, which is very important in determining particle characteristics? And still, how do these chemical constituents change positions, as the particles are moved from one environment to another, considering the presence or absence of salt, cosolvents, and surfactants in aqueous environments, or the transfer from an aqueous to a dry environment?

These questions have been approached in many ways, direct and indirect. An important example of an indirect way to evaluate a component distribution within a latex particle is the use of acid–base titration together with sulfur determination. Titration yields lower charged sulfate group concentration at the particle surfaces than expected from average sulfur concentration, evidencing the occlusion of a large fraction of sulfate groups (from persulfate initiator residues) within the particles. On the other hand, it has been shown that hydrogen ions are fully exchanged with bipositive metal ions (at least in one latex) if sufficient time is allowed [82]. Another, more general approach is the comparison of overall chemical analysis data with the surface data obtained (e.g., by ESCA). The problem here is that the depth of the surface layer sampled by ESCA is typically 2–10 nm, which means that the thickness of the latex particle surface may represent 10% or more of the particle diameter.

Electron and scanning probe microscopies underwent immense developments in the past 10 years. They are now revealing much newer and previously unsuspected information on materials, including latexes. Two specific types of microscopies which are particularly useful in this context will be examined in the following subsection.

A. Microchemical Spectroscopy: Energy-Loss Spectroscopy Imaging and X-Ray Probe Imaging

Two techniques for microchemical and topochemical analysis which found many important applications in materials science are energy-loss spectroscopy imaging

(ELSI) [83] and X-ray probe imaging [84]. The former has been associated mostly with the transmission electron microscope (TEM), whereas the second is used both in the scanning-transmission (STEM) and scanning electron microscopes (SEM).

X-ray probe microanalysis has not yet found application in the study of latexes, probably due to the low sensitivity of the predominating EDS (energy-dispersive spectroscopy) detection technique to light elements, to the limited availability of STEM instruments (which have the required resolution), and to the low resolution of the hot-filament SEM instruments. However, the relatively recent availability of field-emission, high-resolution scanning electron microscopes will perhaps change this picture, in the near future.

Only recently TEM–ELSI has also been applied to the microchemical study of latexes. Indeed, most TEM information about the distribution of chemical constituents within particles comes from standard bright-field experiments, performed using stained samples. This has been very important for demonstrating core-and-shell and other morphologies, but its potential is much more limited than that of actual elemental mapping, as done by TEM–ELSI. Because this technique has been accessible to polymer scientists for more than 10 years, it is odd that it has not been widely used for latex study. Perhaps, this is due to a misconception concerning the required sample thickness: many authors state that very thin samples should be used (50–100 nm), which is true for many elements. However, good elemental distribution maps from 500-nm-diameter latex particles are obtained, which is allowed by the long mean free electron paths within light-element matrices [85].

1. Elemental Distribution Using TEM–ELSI

When electrons impinge on a thin solid sample (10–500 nm, in the case of a sample made of light elements), a small fraction of the beam is inelastically scattered, as they lose energy to electrons from the inner levels of the constituent atoms. The inelastically scattered electrons now carry information about the distribution of the existing atoms throughout the sample. They are energy filtered, and those electrons within a given energy range are focused into an image. Combining images taken within different energy windows (to subtract the background contributions) gives an elemental distribution map.

The elemental distribution maps formed within the TEM optics do not have the same high quality of the bright-field images obtained with the same instrument but using unscattered electrons only. On the other hand, the maps contain spatially resolved microchemical information which is currently unmatched by any other technique. A few important examples of application of this technique to the study of polymers, or polymer blends and composites, are already seen in the literature [86–88]. However, its application to latexes is very recent [85].

The main advantages of the TEM–ELSI technique in the case of latex elemental mapping are as follows: (1) it is sensitive to light elements, such as carbon,

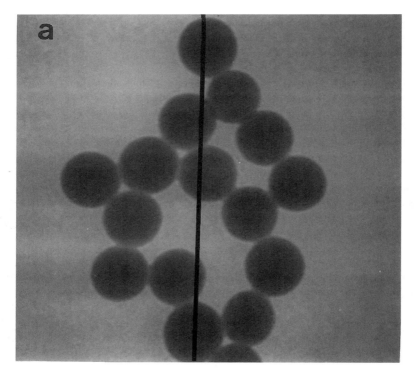

Figure 17   Bright-field (a) and ESI elemental distribution maps (b–e) of a PS latex [(b) sulfur, (c) potassium, (d) oxygen, and (e) carbon].

oxygen, nitrogen, sulfur, potassium, and others found in the polymer latex particles (but not to hydrogen); (2) small amounts of elements are detectable; (3) sample preparation is minimal, as whole particles up to ~500 nm can be observed and their images are interpreted in a straightforward manner.

Figure 17 displays a bright-field image together with C, S, O, and K elemental maps of a PS latex, prepared using a persulfate initiator. Line scans plotted in Fig. 18 (taken across the lines drawn in Fig. 17) indicated the relative amounts of each element in each image pixel. The results are as follows: (1) the minor component elements (S, K, O) are clearly mapped, as well as the major component (C); (2) sulfur (from the initiator residues) and O from the initiator as well as from an hydrolyzed initiator are distributed throughout the particles, but they are more concentrated in the particle interior than at the particle surfaces. This shows that initiator residues are actually trapped well away from the particle surfaces (at least in these dry particles). However, the S/O ratios grow from the particle surface to the particle center, indicating that there is increased sulfate hydrolysis closer to the

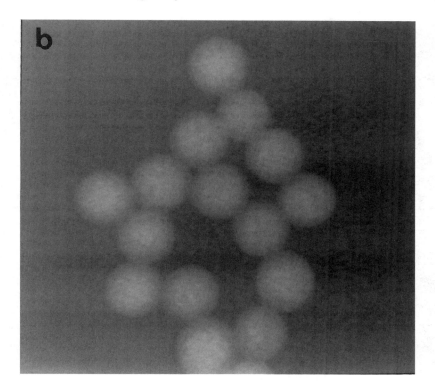

particle surfaces; (3) potassium counterions are accumulated at the polymer surfaces, but they are also found in significant amounts in the particle interior; (4) carbon is also accumulated at the surfaces, which is understood considering that less
polar chain segments should migrate to the surface when the particles are removed
from the aqueous environment and dried. This feature has not been observed in
every PS latex in this laboratory, suggesting that C surface accumulation in dry
particles depends on some particle characteristic associated with the apolar components' diffusion rates.

2. High-Resolution Sem X-Ray Microprobes

High-resolution scanning electron microscopes have been available for the past few
years, but we are unaware of their use in the elemental mapping of latex particles
at this time. However, it is quite probable that this will be done in the near future,
both using the x-ray probes and backscattered electron imaging (BEI) at a resolu-

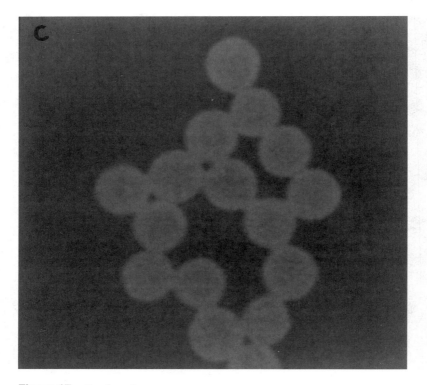

Figure 17 *Continued*

tion in the nanometer range. In this laboratory, we have already obtained some promising results using the second alternative, which should be published shortly.

It may be expected that results such as those presented here will be multiplied many-fold in the near future because of the increased availability of the analytical high-resolution microscopes of various kinds. Moreover, new scanning probe microscopies (AFM, electric force, electric potential, thermal, and so on) have been appearing at an impressive rate for the past 10 years, for which reason we may expect a surge of great progress in this area in the near future.

VI. CONCLUSIONS

Significant progress was made, recently allowing us to assess latex particle heterogeneity within a given particle population, as well as within individual particles. Progress was due to the introduction of an array of new techniques or new

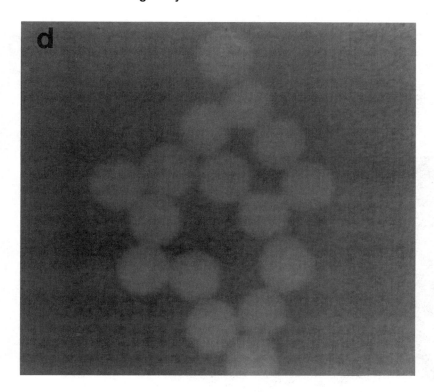

ways to use old techniques, such as centrifugal analysis, spectroscopies, and electron microscopies. New information is now available concerning the distribution of chemical constituents within a particle as well as among particles, and giving us a better understanding of their colloidal stability and film and self-assembly structure formation, which are decisive in many latex applications. This new information modifies some widespread pictures of polymer latexes and latex particles, revealing a more complex, richer spectrum of topochemical features, which may finally lead us into making better latexes, specially designed for both specialty and commodity applications.

ACKNOWLEDGMENTS

FG and EFS acknowledge continuing support of Fapesp, CNPq and Pronex/Finep/MCT. FG acknowledges also the contributions of AA Winkler-Hechenleit-

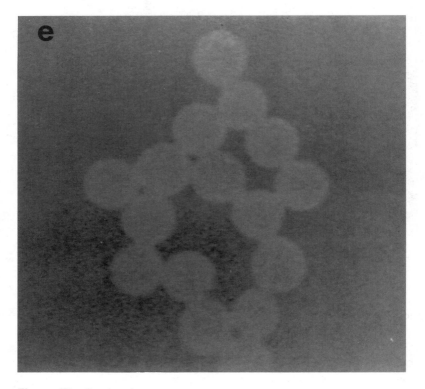

Figure 17 *Continued*

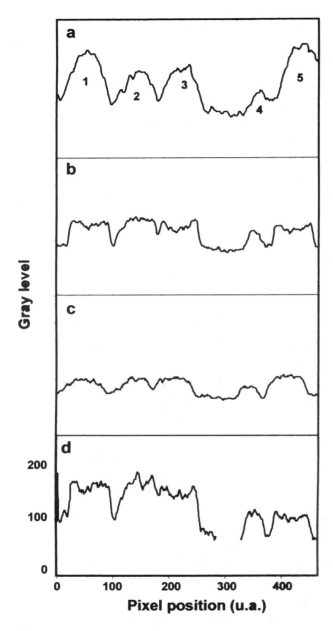

Figure 18 Line scans taken following the line indicated in Fig. 17a. Note the differences in elemental content in the various particles (indicated by numbers 1–5).

ner, SP Nunes, MCP Costa, MM Takayasu, JMM Moita-Neto, AH Cardoso, AO Cardoso, CAP Leite, and MCVM da Silva.

REFERENCES

1. R Buscall, RH Ottewill. Polymer Colloids. London: Elsevier, 1985, Chap 5.
2. DH Napper, RG Gilbert. In: A Ledwith, S Russo, P. Sigwalt, eds. Comprehensive Polymer Science. London: Pergamon, 1989, Vol 4, pp 171–218.
3. RH Ottewill. Ber Bunsenges Phys Chem 89:517–525, 1985.
4. RJ Hunter. Zeta Potential in Colloid Science. New York: Academic Press, 1981.
5. YC Chen, V Dimonie, MS El-Aasser. Macromolecules 24:3779–3787, 1991.
6. CL Winzor, DC Sundberg. Polymer 33:3797–3810, 1992
7. RH Ottewill. Specialist Periodical Reports on Colloid Science Vol 1. London: The Chemical Society, 1973, p 191.
8. B Vincent. In: E Pelizzetti ed. Fine Particles Science and Technology. Dordrecht: Kluwer, 1996, pp 791–814.
9. B Vincent, CA Young, ThF Tadros. Faraday Disc Chem Soc 65:296–305, 1978.
10. JN Israelachvili. Intermolecular and Surface Forces. 2nd ed. London: Academic Press, 1991.
11. DH Napper. Polymeric Stabilization of Colloidal Dispersions. London: Academic Press, 1983.
12. J Gregory. Crit Rev Environ Control 19:185–230, 1989.
13. RJ Hunter. Foundations of Colloid Science. Oxford: Oxford University Press, 1991, Vol 1.
14. H Kitano, S Iwai, T Okubo, N Ise. J Am Chem Soc 109:7608–7612, 1987.
15. PG Cummins, EJ Staples, LG Thompson. J Colloid Interf Sci 92:189–197, 1983.
16. MM Takayasu, F Galembeck. J Colloid Interf Sci 155:16–22, 1993, and references therein.
17. G Frens. Faraday Disc Chem Soc 65:146–155, 1978.
18. RM Cornell, JW Goodwin, RH Ottewill. Faraday Disc Chem Soc 65:182, 1978.
19. BR Midmore, D. Diggins, RJ Hunter. J Colloid Interf Sci 129:153–161, 1989.
20. SS Dukhin, J Lyklema. Faraday Discuss Chem Soc 90:261–269, 1990.
21. J Israelachvili, A Berman. Israel J Chem 35:85–91, 1995.
22. AO Cardoso, F Galembeck. J Colloid Interf Sci 182:614–616, 1996
23. RH Pelton, P Chibante. Colloids Surf 20:247–256, 1986.
24. G Whitesides, JP Mathias, CT Seto. Science 254:1312–1319, 1991.
25. H Ringsdorf, B Schlarb, J Venzmer. Angew Chem Int Ed Engl 27:113–158, 1988.
26. JM Lehn. Angew Chem Int Ed Engl 29:1304–1319, 1990.
27. J Fendler. Chem Mater 8:1616–1624, 1996.
28. JS Lindsey. New J Chem 15:153–180, 1991.
29. TE Creighton. Proteins, Structures and Molecular Principles. New York: Freeman, 1983
30. M Ahlers, W Müller, A Reichert, H Ringsdorf, J Venzmer. Angew Chem Int Ed Engl 29:1269–1285, 1990.

31. GM Whitesides, PE Laibinis. Langmuir 6:87–96, 1990.
32. N Ise. Ber Bunsenges Phys Chem 100:841–848, 1996.
33. E Kim, Y Xia, GM Whitesides. Nature 376:581–584, 1995.
34. T Alfrey Jr, EB Bradford, JW Vanderhoff, G Oster. J Opt Soc Am 44:603–609, 1954.
35. W Luck, M Klier, H Wesslau. Naturwissenschaften 50:485–494, 1963.
36. JM Weissman, HB Sunkara, AS Tse, SA Asher. Science 274:959–961, 1996.
37. S Hachisu, S Yoshimura. Nature 283:188–189, 1980.
38. D Distler, G Kanig. Colloid Polym Sci 256:1052–1060, 1978.
39. ND Denkov, H Yoshimura, K Nagayama, T Kouyama. Phys Rev Lett 76:2354–2357, 1996.
40. R Michelleto, H Fukuda, M Ohtsu. Langmuir 11:3333–3336, 1995.
41. E Kim, Y Xia, GM Whitesides. Adv Mater 8:245–247, 1996.
42. M Trau, DA Saville, IA Aksay. Science 272:706–709, 1996.
43. ND Denkov, OD Velev, PA Kralchevsky, IB Ivanov, H Yoshimura, K Nagayama. Nature 361:26–29, 1992.
44. F Burmeister, C Schäfle, T Matthes, M Böhmisch, J Boneberg, P Leiderer. Langmuir 13:1983–2987, 1997.
45. G Picard. Langmuir 13:3226–3234, 1997.
46. A Cardoso, F Galembeck. J Colloid Interf Sci 204:16–23, 1998.
47. WC Harkins. J Am Chem Soc 69:1428–1444, 1947.
48. DP Durbin, MS El-Aasser, GW Poehlein, JW Vanderhoff. J Appl Polym Sci 24:703–707, 1979.
49. WV Smith, RH Ewart. J Chem Phys 16:592–599, 1948.
50. RM Fitch, CH Tsai. Polymer Colloids. New York: Plenum, 1971.
51. HG Elias. Macromolecules, Vol 2. New York: Plenum, 1984.
52. G Fettis, personal communication.
53. RE Dillon, LA Matheson, EB Bradford. J Colloid Interf Sci 6:108–117, 1951.
54. DP Sheetz. J Appl Polym Sci 9:3759–3765, 1965.
55. F Dobler, T Pith, M Lambla. J Colloid Interf Sci 152:12, 1992.
56. MM Takayasu, F Galembeck. J Colloid Interf Sci 202:84–88, 1998.
57. RJ Hunter. Foundations of Colloid Science, Vol 1. Oxford: Oxford University Press, 1991, Chap 3.
58. R Buscall, RH Ottewill. In: R Buscall, T Corner, JF Stageman, eds. The Stability of Polymeric Latexes. Barking, U.K.: Elsevier, 1985, pp 141–217.
59. W Mächtle. Prog Colloid Polym Sci 86:111–118, 1991.
60. W Mächtle, G Ley, J. Rieger. Colloid Polym Sci 273:708–716, 1995.
61. H Lange. Colloid Polym Sci 258:1077–1085, 1980.
62. H Lange. In: DR Basser, AE Hamielec, eds. Emulsion Polymers and Emulsion Polymerization. Washington, DC: American Chemical Society, 1981.
63. CA Price. Centrifugation in Density Gradients. New York: Academic Press, 1982.
64. AAW Hechenleitner, MCP Costa, F Galembeck. J Appl Polym Sci 32:2369–2383, 1987.
65. MM Takayasu, F Galembeck. J Colloid Interf Sci 155:16–22, 1993.
66. SP Nunes, F Galembeck. Anal Biochem 146:48–51, 1985.
67. JM Moita Neto, ALH Cardoso, AP Testa, F Galembeck. Langmuir 10:2095–2099, 1994.

68. AA Winkler-Hechenleitner, F Galembeck. Separation Sci Technol 23:293–308, 1990.
69. GHJ van Doremaele, FHJM Geerts, LJ van de Meulen, AL German. Polymer 33:1512–1518, 1992.
70. GHJ van Doremaele, AM van Herk, AL German. Makromol Chem Symp 35/36:231–248, 1990.
71. AL Herzog-Cardoso, JM Moita-Neto, A Cardoso, F Galembeck. Colloid Polym Sci 275:244–253, 1997.
72. F Galembeck, PR Robilotta, EA Pinheiro, I Joekes, N Bernardes. J Phys Chem 84:112–119, 1980.
73. F Galembeck, ATN Pires Separ Purif Methods 15:97–126, 1986.
74. F Galembeck, MCP Costa. Braz J Med Biol Res 20:297–315, 1987.
75. JM Moita Neto, VAR Monteiro, F Galembeck. Colloids Surf A 108:83–89, 1996.
76. JE Seebergh, JC Berg. Colloids Surf A 121:89, 1997.
77. J Li, KD Caldwell, W Mächtle. J Chromatogr 517:361–376, 1990
78. RH Pelton, HM Pelton, A Morphesis, RL Rowell. Langmuir 5:816–818, 1989.
79. MR Gittings, DA Saville. Langmuir 11:798–800, 1995.
80. BM Verdegan, MA Anderson. J Colloid Interf Sci 158:372–381, 1993.
81. F Reif. Statistical Physics Berkeley Physics Course, Vol. 5. New York: McGraw-Hill, 1965, p 4.
82. W Mächtle, G Ley, J. Rieger. Colloid Polym Sci 273:708–716, 1995.
83. L Reimer. Energy-Filtering Transmission Electron Microscopy. Berlin: Springer-Verlag, 1995.
84. JI Goldstein, DE Newbury, P Echlin, DC Joy, AD Romig Jr, CE Lyman, C Fiori, E Lifshin. Scanning Electron Microscopy and X-Ray Microanalysis. New York: Plenum, 1992.
85. A Herzog-Cardoso, CAP Leite, F Galembeck, Langmuir (in press).
86. KF Silveira, IVP Yoshida, SP Nunes. Polymer 36:1425–1434, 1995.
87. G Lieser, SC Schmid, G Wegner. J Microsc 183:53–59, 1996.
88. HJ Cantow, M Kunz, S Klotz, M Möller. Makromol Chem Makromol Symp 6:191–196, 1989.

5
Electrokinetic Behavior of Polymer Colloids

F. Javier de las Nieves and A. Fernández-Barbero
University of Almería, Almería, Spain

R. Hildago-Alvarez
University of Granada, Granada, Spain

I. INTRODUCTION

Polymer colloids play an important role in many industrial processes such as paints, coatings, adhesives, sealants, latex foams, surface sizing, and many other products, and also in biomedical fields for immunoassays, biological cell-labeling, and drug delivery [1]. The rapid increase in the utilization of latexes over the last two decades is due to a number of factors: Water-based systems avoid many of the environmental problems associated with organic-solvent based systems; latexes can be designed to meet a wide range of application problems; emulsion polymerization on a large scale proceeds smoothly for a wide range of monomers.

The spherical shape of many polymer beads and their narrow size distribution make them most suitable for fundamental studies in colloid science requiring well-defined monodisperse systems. They provide valuable experimental systems for the study of many others colloidal phenomena and, recently, have been used as model systems for the simulation of molecular phenomena, including nucleation, crystallization, and the formation of glasses.

Polymer colloids are often treated as "model" particles because they are amorphous microspheres with smooth, uniform surfaces and rigidly attached, well-defined surface functional groups. In particular, their sphericity, monodispersity, and virtually zero dielectric constant in comparison to water made them particularly suitable for fundamental electrokinetic studies [2,3]. Electrokinetic processes are widely used to determine the electrical charge on the slipping plane of a polymer colloid. This electrical charge controls the colloidal behavior, the ζ-potential is the potential at the slipping plane, and electrokinetics is closely related to the slipping process [4,5].

The preparation of uniform polymer microspheres via emulsion polymerization, and techniques for their cleaning and characterization have been extensively reviewed by Ugelstad et al. [6]. Parameters such as particle size, surface charge density, and type of charge group can be controlled by varying the conditions of the polymerization, allowing latexes to be "designed" for specific end uses. For these reasons, polymer colloids have been widely used as model systems in investigations of electrokinetic and colloidal stability phenomena. However, it has been in the area of electrokinetics that the polymer colloids have failed to live up to much of the initial expectation. One particular disappointment has been the failure to find a convincing explanation for the behavior of the ζ-potential of polymer colloids as a function of electrolyte concentration [2,3], which has brought the ideality of the system into serious question. In fact, the controversy about the anomalous electrokinetic behavior of polymer colloids, especially with polystyrene latexes, continues giving experimental works and attempts [7–18] to give a general explanation.

Determination of the detailed structure of the electric double layer of polymer colloids is of primary importance in problems of stability and rheology of disperse systems, electrokinetic processes, filtration and electrofiltration, desalting of liquid on organic membranes, and so forth. Calculating the ζ-potential of polymer–solution interface is also important when looking for an accurate microscopic explanation of electrokinetic phenomena. The topic is advancing at the present time with the preparation of new materials and the controversy about the anomalous behavior of latexes is alive, with a new theoretical attempt to eliminate the controversy. From the point of view of the applications, the electrokinetic phenomena are useful tools in other fields, such as water detoxification of waste water by solar energy [19].

II. PURIFICATION AND SURFACE CHARACTERIZATION

Polymer colloids prepared by emulsion polymerization can have different electrokinetic and stability properties according to the type of monomers and procedure used during the synthesis. After the preparation, a cleaning procedure is required to remove salts, oxidation products, oligomeric materials, and any remaining monomers from the latex, which can modify the final electrokinetic properties of the colloidal system [9,20]. A worrying feature of the cleaning and characterization procedures is that different authors, using similar recipes for the latex preparation, have observed different surface groups. For example, Vanderhoff et al.

[21,22], using mixed-bed ion-exchange resins to remove ionic impurities, have concluded that only strong-acid groups (i.e., sulfate groups arising from the initiator fragments) are present on the surface; their conclusion was based on the single end point observed in the conductometric titration. Other authors [23,24], particularly those using dialysis as a cleaning procedure, have detected the presence on the surface of weak-acid groupings in addition to the strong-acid groupings. In fact, weak-acid groups could be produced by the hydrolysis of surface sulfate groups to alcohol (hydroxyl) groups followed by oxidation to carboxyl groups [9,23]. Moreover, Lerche and Kretzschmar [25] have shown that the surface charge density of several latex samples depended on the cleaning method used and that the ion-exchange and dialysis method was not able to remove charged oligomeric material from the particle surfaces completely. Thus, a very key question has arisen as to which set of results is correct and which cleaning procedure should be used for preference when preparing clean latexes for fundamental investigations. Serum replacement has been suggested as a reliable and easy method to clean latex suspensions [26]. Nevertheless, the ion form of the latexes cleaned by serum replacement only is achieved after an ion exchange with resins. A cycle of several centrifugation–redispersion might be the only method able to remove oligomer chains from the surface of polymer colloids [20,27,28]. The different methods for preparing clean latexes do provide polymer colloids with quite different electrokinetic properties. Cleaning of latexes is of paramount importance for electrokinetic and stability studies. The removal of polymeric impurities is essential in order to have both control over the surface charge and of the supporting electrolyte concentration. Also, the deionization of latex suspensions plays a decisive role in the formation of fluidlike, crystalline, or amorphous interparticle structure [29].

The surface structure and characteristics of polymer colloids are important because they determine the stability of the colloidal dispersion, the adsorption characteristics of surfactants or macromolecules, latex film formation mechanisms, and the properties of these films obtained from latex. The techniques suitable for surface analysis of polymer colloids are numerous. Conductometric and potentiometric titrations are often used and considered as a basic technique for surface charge determination [20,30–36]. Other classical methods include soap titration [37–39] and contact-angle measurements [40–42]. Moreover, surface topography of polymer beads is usually studied by electron microscopy (transmission electron microscopy) [42]. The problems associated with the surface characterization of polymer colloids have been studied by several authors [20,30,32,43–48]. Labib and Robertson [44] have shown that conductometric titration of polymer beads is more difficult to interpret than that of free acids and proposed a method to determine the stoichiometric end points in an appropriate and reliable manner. Hlavacek et al. [49] have demonstrated that the variation in the composition of the liquid phase which occurs during acid–base titrations of polymer colloids suspensions can be explained by a mechanism involving weak acid or base ion–exchange reactions coupled with surface ionization. Identification of the sites involved and their thermodynamics constants allows a good quantitative prediction of the experimental results and gives an explanation for the influence of ionic strength on the pH curves and an indication of the state of the solid surface. Also, Gilány [50] determined the surface charge density of polystyrene beads by using the concentration and activity of a binary electrolyte added to the latex dispersion. The distribution of ions was calculated by means of the non linearized Poisson–Boltzmann equation and the cell model. The effective charge of latex beads was found to be smaller than the analytical charge. It was concluded that a small effective charge cannot be explained with specific binding of counterions to the polymer colloids.

The hydrophobic or hydrophilic character of the polymer surface may have a certain influence on the surface structure of the latexes [42]. The hydrophobic surface of polymer colloids plays a crucial role in the ion distribution in the interfacial region. The contact angle is a measurement of the hydrophobicity of the polymer–solution interface and has been used to obtain information on the surface structure of core–shell and block polymer colloids [40–42,51]. To determine the water of hydration around charged polymer beads, Grygiel and Starzak [52] have studied the interfacial properties of carboxylated polymer beads using environment-sensitive laser excitation spectroscopy of the Eu^{3+} ion. This ion spectroscopy technique uses the changes in the electronic properties of the ion in different molecular environments to elucidate the structure and properties of those environments. Lifetime measurements show that when binding to a highly charged surface (32.3 $\mu C/cm^2$), the ion loses about half its waters of hydration while energy transfer from Eu^{3+} for these highly charged surfaces gives an ion separation (7.1 Å) that is consistent with the known average separation of the surface sites (7.1 Å). For lesser charged beads (15.1 and 2.6 $\mu C/cm^2$, respectively), the energy transfer separation distance is smaller than the surface site–site separation, indicating energy transfer between surface-bound and interfacial ions. For lesser charged beads, an ion separation of about 9.6–9.7 Å is found, indicating that bound ions retain most of their water of hydration. Using osmotic pressure measurements, Rymdén [47] has observed that the ion binding in aqueous polymer colloids depends on the surface charge density of carboxylated latex beads.

Recently, Moita Neto et al. have published an interesting article [53] in which they show that latex particle heterogeneity spontaneously decreases by aging. A latex 80 nm in diameter was examined by centrifugation employing NaCl–sucrose density gradients. Three fractions were found, with dissimilar densities. However, after 12 months of aging at room temperature, the latex presented a more homogeneous behavior, observing only a single band. Two possible slow events are taken into account for explaining the experimental results: (1) migration of hydrophilic groups trapped within the latex particles during the process of emulsion polymerization and slowly released to the particle surface and (2) migration of the most hydrophilic polymer chains and surfactant residues among the particles. These migrations could take place by dissolution and readsorption.

III. ELECTROKINETIC PHENOMENA

Electrokinetic phenomena is a generic term applied to effects associated with the movement of ionic solutions near charged interfaces. The determination of the detailed structure of the electric double layer (EDL) of latexes is the purpose of the electrokinetic studies, and calculating the ζ-potential of the polymer–solution interface is important when looking for an accurate microscopic explanation of electrokinetic phenomena.

To describe the structure of the EDL, information is needed on three potentials: the surface potential (Ψ_o), the potential of the Stern layer (Ψ_s), and the diffuse potential (Ψ_d). In the absence of organic impurities and polyelectrolytes adsorbed on latex surface, the Ψ_d-potential can be equated to the potential in the electrokinetic slipping plane (ζ-potential). In some cases, Ψ_o can be assumed approximately equal to Ψ_s with an indifferent electrolyte, and thus a detailed study of the structure of EDL only requires a knowledge of the Ψ_o and ζ-potentials. Extensive reviews [2,3,54–61] testify to the strong interest which has been shown in the electrokinetic phenomena during the past few decades.

Typical electrokinetic phenomena used to characterize polymer colloids are as follows:

- **Electrophoresis:** where a uniform electric field is applied and the particle velocity is measured [20,30–32,61–118].
- **Streaming potential:** where a liquid flux is allowed to pass through a porous medium and the resulting electric potential difference is measured [113,119–125].
- **Electro-osmosis:** where an electric field is applied to a porous medium and the resulting volumetric flow of fluid is measured [66,126,127].
- **Diffusiophoresis:** where a gradient of a solute in solution is applied and the migration of a suspended colloid particle is measured. Much of the early theoretical and experimental work on diffusiophoresis was on gaseous systems. Recent work, however, has focused on diffusiophoresis in liquid systems involving charged particles and electrolytes in solution.
- **Dielectric dispersion:** This technique involves the measurements of the dielectric response of a sol as a function of the frequency of an applied electric field. The complex dielectric constant [73,93,128–140] and/or electrical conductivity [131,141–146] of a suspension are measured as a function of frequency. The presence of dispersed particles generally

causes the conductivity of this dispersion to deviate from the conductivity of the equilibrium bulk electrolyte solution.

- **Electroacoustic phenomena:** where alternating pressure fields are applied and the resulting electrical fields are measured [95,147–153]. When an alternating voltage is applied to a colloidal dispersion, the particles move back and forth at a velocity that depends on their size and ζ-potential and the frequency of the applied field. As they move, the particles generate sound waves. This effect was predicted by Debye [154] in 1933.

- **Electroviscous effects:** where colloidal suspensions and electrolyte flows through electrically capillaries under a pressure gradient. The presence of an EDL exerts a pronounced effect on the flow behavior of a fluid. These effects are grouped together under the name of electroviscous effects [155–161].

In all cases, there is a relative motion between the charged surface and the fluid containing the diffuse double layer. There is a strong coupling between velocity, pressure, electric, and ion concentration fields. The literature pertaining to the study of electrokinetic properties of latexes has a long and confusing history. We note specifically: (a) experimental electrokinetic data performed in different laboratories on ostensibly identical systems often conflict; (b) minor changes (cleaning procedure, surface charge, and particle size) may result in major differences in the measured electrokinetic data; and (c) the ζ-potentials obtained using the various electrokinetic processes on the same dispersions are quite different in values. These studies are difficult due to the complex interactions involved.

IV. ELECTROPHORETIC MOBILITY

Recent development of laser-based instrumentation for electrophoretic mobility experiments has made it possible to determine the ζ-potential of particle suspended in liquid media. The new instrument use electrophoretic light scattering and allows direct velocity measurements for particles moving in an applied electric field by analyzing the Doppler shift of laser light scattered by the moving particles [76]. Recently Kontush et al.[118] designed a setup for studying nonlinear electrophoresis. By introducing the technique of electrophoretic fingerprinting (three-dimensional-profiles of electrophoresis), Marlow and Rowell [107] opened a new possibilities to describe the complex behavior of polymer colloids in aqueous media. In the electrophoresis cell, a controlled simultaneous variation of pH and ionic strength around the particles realized by a titration program is performed. As a result, the electrophoretic mobility as a function of pH and log λ (conductance) can be plotted. The method promises a fine differentiation between polymer colloids [18]. Using the technique of electrophoresis fingerprinting, Paulke et al.[162] have been able to distinguish their latex samples into two classes: (a) mobility is dominated by a salt dependence, and (b) pH dependence dominates the electrophoretic behavior. By this technique Donath et al. [18] have also investigated the consecutive layer-by-layer adsorption of anionic and cationic polyelectrolyte onto charged polystyrene latex particles. The main conclusion was that not only the top layer but also the underneath layers and the naked latex particle surface contribute to the particle mobility, and the thickness of the top adsorbed hairy layer is of the order of 1 nm. It is found that electrophoretic mobility curves pass through a minimum (anionic latexes) or a maximum (cationic latexes) as a function of increasing ionic strength. From a theoretical point of view, calculation of the ζ-potential from electrophoretic mobility data encounters a number of difficulties as a result of the polarization of EDL. The term "polarization" implies that the double layer around the particles is regarded as being distorted from its equilibrium shape by the motion of the particle. In general, for κa ≤ 30 it is necessary to account for EDL polarization when calculating ζ [62,93,117]. There are several theoretical treatments for converting electrophoretic mobility (μ_e) into ζ-potential under polarization conditions [2,3]. Monodisperse spherical polymer latexes have proved to be very useful model systems for testing the most recent theoretical approaches [2–18,62–63,79, 81,87,117]. As most of the theories deal with spherically shaped particles of identical size, the introduction of monodisperse latexes appeared to offer excellent chances for experimental verification of these theories. However, growing evidence of anomalous behavior of the ζ-potential as a function of 1 : 1 electrolyte concentration has appeared in the literature [7–20,30–31,33,62,79–82,90,93– 101]. The standard electrophoretic theories used for the conversion of mobility into ζ-potential give rise to a maximum in potential as well. This behavior contradicts the current double-layer models, which predicts a continuous decrease in ζ-potential. Various explanations for this maximum have been proposed [9–15,20,30,34,40,62,79,82,83,85,93,99,117,163–166], and some authors [122] have even pointed out that a maximum mobility value does not necessarily imply a maximum in ζ-potential, indicating that the conversion of mobility into the ζ-potential of the polystyrene microspheres–electrolyte solution interface should be done by means of a theoretical approach which takes into account all possible mechanisms of double-layer polarization. Other authors, on the contrary, have pointed out that the appearance of a minimum (or maximum) in the ζ-potential is unimportant because the EDL around polymer colloids, even with 1 : 1 electrolytes, cannot be explained on the basis of the Gouy–Chapman model. They proposed the use of a dynamic Stern layer [67] or an electric triple-layer model [151] instead.

Overbeek [114] and Booth [115] were the first to incorporate polarization of the EDL into the theoretical treatment. They assumed that the transfer and charge redistribution processes only involved the mobile part of the EDL. Also, O'Brien and White [69], starting with the same set of equations as Wiersema et al. [116], have more recently published a theoretical approach to electrophoresis, which takes into account any combination of ions in solution with the possibility of very high ζ-potentials (up to 250 mV), high enough against the values to be expected in most experimental conditions. In simple terms, the theory of O'Brien and White predicts the measured electrophoretic mobility of a colloidal particle in an applied electric field to be the sum of three forces: namely an electric force propelling the particle, due to the charged nature of the particle; a drag force due to hydrodynamic drag; and a relaxation force due to an electric field induced in the opposite direction to the applied field as a result of the induced polarization within the diffuse layer of ions surrounding the particle.

It predicts the electrical force propelling the particle to be proportional to ζ and the retarding forces to be proportional to ζ^2. A maximum in the conversion of mobility to ζ-potential is thus predicted for particles size and ionic strength conditions such that $5 < \kappa a \leq 100$.

The most striking features of O'Brien and White's theoretical treatment results are as follows:

1. For all values of $\kappa a \geq 3$, the mobility function has a maximum which becomes more pronounced at high κa values.
2. The maximum occurs at $\tilde{\zeta} = e\zeta/kT = 5–7$ (i.e., $\zeta \approx 125–175$ mV).
3. Their computer solution is much more rapid than the earlier theoretical treatments used in the conversion of mobility into ζ-potential (up to 250 mV).

A simplified version and analytical form for the mobility equation, accurate to order $1/\kappa a$ and valid for our purposes for $\kappa a > 10$, can be expressed as [58]

$$\tilde{\mu}_e = \frac{3\tilde{\zeta}}{2} - 6\left(\frac{\tilde{\zeta}}{2} - \frac{\ln 2}{z}[1 - \exp(z\tilde{\zeta})]\right)\left[2 + \frac{\dfrac{\kappa a}{1 + 3m}}{z^2}\exp\left(-\frac{z\tilde{\zeta}}{2}\right)\right]^{-1} \tag{1}$$

$$m \equiv \frac{2\varepsilon_0\varepsilon\, N_a^2\, kT}{3\eta z\, \Lambda_0} \tag{2}$$

m is the dimensionless ion drag coefficient, where Λ_0 is the limiting equivalent conductance.

All the above-cited theoretical approaches to convert mobility into ζ-potential assume the absence of ionic conduction inside the shear plane. In an attempt to account for this phenomenon, Semenikhin and Dukhin [72] developed an equation incorporating both the dimensionless ζ-potential and the dimensionless diffused Ψ_d-potential. The mobility μ_e for a spherical particle with a thin EDL ($\kappa a > 25$) in a 1 : 1 electrolyte presents a general dependence in the following variables:

$$\mu_e = f(\zeta, \kappa a, m, p, \Psi_d) \tag{3}$$

where p is the ratio of the counterion diffusion coefficient near the particle to its value in the bulk solution and m is the dimensionless ionic drag coefficient, previously defined.

If $\tilde{\zeta}$ and $\tilde{\Psi}_d$ are greater than 2, Eq. (3) can be expressed as

$$\tilde{\mu}_e = \frac{3}{2}\tilde{\zeta}\left[\frac{1 + \mathrm{Rel}\left(\dfrac{4\ln\cosh(\tilde{\zeta}/4)}{\tilde{\zeta}}\right)}{1 + 2\,\mathrm{Rel}}\right] \tag{4}$$

$$\mathrm{Rel} \equiv \frac{\exp(\tilde{\Psi}_d/2) + 3m\exp(\tilde{\zeta}/2)}{\kappa a} \tag{5}$$

Dukhin [61] introduced the dimensionless relaxation parameter Rel as a measure of the effect of surface conductance on electrokinetic phenomena. It is noted that Rel can be used with two meanings, namely in indicating the degree of EDL polarization (nonequilibrium degree) for curved surfaces and in indicating the relative contribution of surface conductance to the total conductance in nonpolarized systems (equilibrium states). An increase in the surface conductance and/or decrease in the radius results in an increase in Rel and thus in the polarization field in the direction of the induced electromigration current. We can distinguish two different mechanisms of electrical conduction: surface conduction associated with tangential charge transfer through the mobile portion of the EDL (normal conduction taken into account in the Overbeek–Booth–Wiersema theory), and anomalous surface conduction, which is related to the tangential charge transfer between the slipping plane and the particle surface. The Semenikhin–Dukhin theory [72] considers only a particular case of anomalous conduction associated with the presence of a boundary layer.

The induced tangential ionic flows near the surface have to be provided by radial ionic migration, diffusion and convection from beyond the EDL, where co-ion and counterion concentration can differ considerably from those near the surface. The concentration polarization results in an angular dependence of the ion concentration and the potential. Semenikhin and Dukhin [72] derived analytical formulas that express these dependencies for the cross section of the thin diffuse EDL of a spherical particle. Contrary to the mathematical procedures employed by Overbeek [114], Booth [115], and Wiersema et al. [116], the main advantage of the Dukhin–Semenikhin approach is the possibility of taking into account the effect of anomalous conduction on polarization.

The Semenikhin–Dukhin theory assumes that there is no contribution to the electrical conduction by any Stern layer ions, and that the Poisson–Boltzmann equation (PBE) applies up to the outer Helmholtz plane. The diffuse-layer ions between the shear plane and the outer Helmholtz plane do conduct a current and this anomalous surface conductance (inside the shear plane) dramatically reduces the mobility for a given value of ζ-potential. Consequently, for a description of electrophoresis under the conditions Rel > 1, not only must the effect of diffusion flows be taken into consideration but also the change in the polarization potential across the thin diffuse layer [61].

Comparative studies using different theoretical treatments of mobility data for ζ-potential determination have been accomplished by various authors [63,79,93,109,161]. This kind of study is closely related to the determination of the detailed structure of the EDL of colloidal particles. In this way, Baran et al. [79] have been successful in describing the structure of EDL for polystyrene latexes. They used potentiometric titration, conductometry, and electrophoresis to determine the surface charge, Ψ_o, Ψ_s, ζ-potentials, and the surface conductivity of monodisperse particles of a polystyrene latex in solutions of alkali metal chlorides. Until rather recently, only Ψ_o and ζ-potentials could be determined. Even here, the extensive data that are available on the ζ-potential cannot be regarded as a quantitative characteristic of the EDL, as any strict interpretation of electrokinetic data encounters at least three obstacles: polarization of the EDL by an external field, the possible existence of a boundary layer with reduced mobility, and the significant roughness of the surface of solid particles.

There are no direct methods for the measurement of Ψ_s. In most studies, this potential has been taken equal to the ζ-potential, on the assumption that the slip boundary coincides with the boundary of the Stern layer. In some cases, it has been assumed that $\Psi_s = \Psi_o$. Finally, some authors have attempted to estimate Ψ_s by using the Eversole and Boardman method. Baran et al. [79] have calculated Ψ_s from relative electrical conductivity measurements, and have shown that the dependencies of Ψ_o and Ψ_s on κa are identical. Hence, $\Psi_s = \Psi_o$ for the polystyrene latex particles. Baran et al. [79]. have accomplished this by using the tables of Wiersema et al. [116], together with Semenikhin and Dukhin's equation [72]. For comparison, they also calculated z using the relationship of Smoluchowskis, equation.

The ζ-potential calculated with allowance for EDL polarization, all the way up to $\kappa a \leq 60$, is substantially greater than the ζ-potential obtained from the Smoluchowski's equation. This effect was not taken into account in the second case of retarding action of the induced dipole created by polarization of the EDL. These differences are gradually smoothed out as the EDL becomes thinner, as would be expected. The greater values of ζ_{S-D} in comparison with ζ_{O-W} are readily explained on the basis that the Semenikhin–Dukhin (S–D) theory takes the contribution for polarization into account for all ions of the diffuse layer, whereas Wiersema et al. account only for the ions of the hydrodynamically mobile part of the EDL. As the EDL is compressed, the contribution from surface conduction drops substantially, and with $C \geq 5 \times 10^{-2}$ M, it becomes negligibly small. As a consequence, the difference between the values of ζ_{SM}, ζ_{O-W}, and ζ_{S-D} disappears. Figure 1 shows the values of diffuse potential and ζ-potentials as a function of

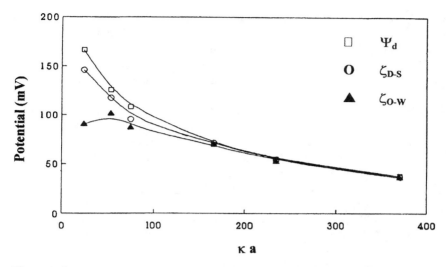

Figure 1 Diffuse potential (□) and ζ-potentials [Dukhin–Semenikhin (○), O'Brien–White (▲)] as a function of electrokinetic radius for an anionic polystyrene latex. (From Ref. 161.)

electrokinetic radius for anionic polystyrene latex particles [161]. In this case, the O'Brien and White, and Semenikhin and Dukhin theories were used.

The ζ-potentials obtained by applying the Smoluchowski, Henry, Overbeek, Booth, and O'Brien–White theoretical treatments of electrophoresis to experimental data of electrophoretic mobility of negatively charged polystyrene particles have been analyzed by several authors [167]. The results obtained applying the Overbeek and Booth theories are compared with those obtained by the O'Brien and White numerical methods. Important differences were found for nonsymmetrical electrolyte and small electrokinetic radius. Midmore and Hunter [93] have compared the ζ-potential values obtained from mobility data applying different theories (O'Brien–White, Semenikhin–Dukhin and Henry). Electrophoresis experiments were performed on two different negatively charged polystyrene latexes, using potassium fluoride, chloride, bromide, and iodide as indifferent electrolytes. They found little or no difference in the mobility data for the various co-ions, which seems to indicate that co-ions are not specifically adsorbed onto the polystyrene surface.

O'Brien and White's theory [69] assumes the absence of ionic conduction inside the plane of shear. Hence, in the opinion of Midmore and Hunter [93], this theory is not applicable to latex/electrolyte systems below electrolyte concentrations of about 0.01 M. Recently, however, Russell et al. [117] have concluded that the O'Brien and White theory correctly predicts the relaxation term in electrophoretic mobility measurements which implicates that deviations from theory observed in previous studies cannot be attributed to an incorrect relaxation term. By definition, it also negates the presence of surface conduction (ion mobility by the electrokinetic shear plane) for highly charged sulfonate/styrene latexes. This is an interesting observation in itself because highly charged latexes are generally viewed to be nonideal in this sense. Antonietti and Vorwerg [14] have very recently used a simple model based on the Gouy-Chapman theory and the O'Brien-White approach (see Eq. (1)) to calculate the dependence of the electrophoretic mobility on salt concentration. In this study they conclude that the theoretical and experimental curves are in good agreement in a number of qualitative features. A similar conclusion is reached by Midmore et al. [11], assuming that there is a constant electrokinetic charge and no anomalous surface conduction. The model used by Antonietti and Vorwerg [14] reveals that a monotonously increasing ζ-potential with falling electrolyte concentration results in a mobility minimum and that this so-called anomalous behavior is in accordance with the standard electrokinetic theory. According to these authors, no ion adsorption mechanism or the existence of a charged hairy layer, currently standard explanation for this anomaly, have to be invoked. They conclude that the so-called anomalous behavior of the electrophoretic mobility of polymer colloids is not atypical at all, but a direct consequence of the structure of the physics involved.

An important result from the combined measurements carried out by various authors [79,101,122,161] is that even if the ζ-potential is corrected for the EDL polarization, it is smaller than Ψ_d over a wide range of electrolyte concentration ($10^{-4} - 10^{-2}$ M), regardless of the method by which Ψ_d is determined. This may be due to the formation of a liquid layer with low hydrodynamic mobility on the particle surface, in which the ions retain high mobility. The thickness of this layer decreases with increasing ionic strength [2,79]. It is also possible that the relationship experimentally found by some authors is a consequence of the surface roughness of the latex particles due to the presence of strongly bound (chemisorbed) oligomer molecules [62,80,81,115].

An interesting article dealing on the behavior of the electrophoretic mobility with the pH for different copolymers at the particle surface is found in Ref. 168. This work tries to understand the copolymer composition of particle surfaces in relation to the bulk polymer and/or initial monomer composition. X-ray photoelectron spectroscopy (XPS) and time-of-flight secondary-ion mass spectrometry (ToF–SIM) were used to monitor changes in particle surface composition. These measurements indicate a substantial enrichment of the metacrylic acid component. The behavior of the mobility is consistent with the surface composition, showing a more significant decreasing of that as the pH decreases, in the case of particles with a higher content of methacrylic acid (MAA). This effect can be observed in Fig. 2.

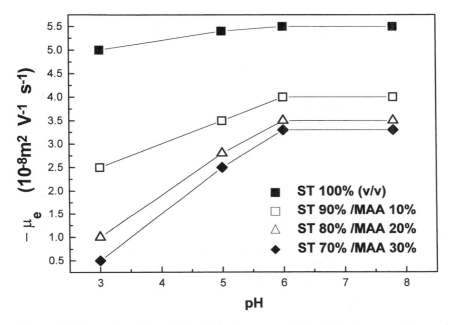

Figure 2 Electrophoretic mobility dependence on pH for different compositions of MAA at the particle surface. (From Ref. 168.)

In the last few years, some investigators have tried a direct determination of forces acting between two polymer colloids. One recently developed technique, based on "electrophoretic rotation or differential electrophoresis," permits the measurements of the electrophoretic rotation rates of colloidal doublets that were formed by Brownian coagulation of two different populations of polystyrene latex spheres in water [169,170]. The difference in ζ-potential of the two spheres was of the order KT/e. The rotation rate was proportional to the difference in the ζ-potentials of the two spheres and a geometric coefficient N. This coefficient can be calculated from the electrodynamic equations and depends on the kinematic boundary conditions imposed on the spheres. Velegol et al. [169,170] have shown that all the values of N for the colloidal doublets agree with the theoretical values for rigid-body rotation. This result is in disagreement with predictions of the DLVO theory.

Ohshima [171] has recently reviewed a theory of electrophoresis of spherical "soft" particles, that is, polyelectrolyte-coated spherical colloidal particles which unite two different electrophoresis theories for spherical hard colloidal particles and for spherical polyelectrolytes. For spherical soft particles, an interesting mobility behavior appears which may be used to classify the particles into soft or hard. This feature is that the electrophoretic mobility trend is completely different at high ionic concentration. For hard spheres, the mobility tends to zero when ionic concentration rises, whereas in the case of soft particles, the mobility tends asymptotically toward a limit value, different than zero. This limit value depends only on the charge density of the fixed charges in the polyelectrolyte layer and on the frictional coefficient of the polyelectrolyte. The interested reader should study a review about the electrokinetics of poly(N-isopropylacrylamide) hydrogel-coated particles and plates [172]. Electrophoresis measurements were used for particles, whereas electroosmosis measurements were employed for plate samples. The volume phase transition in the microgel particles is well reflected in the electrokinetic data which were found to be well described by a "soft" surface electrokinetic theory.

An extension of that theory for cylindrical soft particles is also reported by the same author in Ref. 173.

From a instrumental point of view, an application of tracking of colloidal particles using microscope image sequence analysis to microelectrophoresis and particle deposition may be found in Ref. 174. This method could be used to determine directly the particle velocity distribution in a fully observer fashion from microscopic images of colloidal particles moving in an applied electric field. In particle deposition onto collector surfaces, the method may be used to measure resident-time-dependent desorption of adhering colloidal particles.

V. STREAMING CURRENT AND POTENTIAL

According to the theory of Levine et al. [175], the streaming current in capillaries, I_s, is related to the ζ-potential by

$$\frac{I_s}{\Delta P} = \frac{\varepsilon \zeta}{\eta C} [1 - G(\kappa a, \zeta)] \tag{6}$$

where G is a correction function to the Helmholtz–Smoluchowski equation and C is the cell constant.

The streaming current data are actually obtained by multiplication of experimental streaming potential and A.C. conductance data. In this way, the effect of the polarization of the electrodes and their variable nature on electrokinetic signals is avoided, as is shown by van der Linde and Bijsterbosch [176]. Special attention has been paid to the preparation of homogeneous isotropic plugs of polystyrene spheres. The plugs have to be mechanically stable and completely wetted in order to avoid structural changes during electrokinetic investigations. The most successful preparation technique employs centrifugation, as also used for measurements of coagulation forces. During centrifugation, at a specified speed and ionic strength, the latex concentration at the bottom of the centrifuge tube increases. The lowest layers start to coagulate so that the noncoagulated sediment decreases in mass until the critical coagulation force is reached. In this case, streaming potential experiments are performed at pressure differences up to 30 cm Hg. Even at this pressure, no changes in the structure of the plug are observed, indicating a close-packing of the particles in the plugholder.

To determine the influence of surface conductance in concentrated dispersions, the cell constant C is determined according to Brigg's empirical method. The ratio between C and the electric resistance of the plug (R) provides the effective electric conductivity.

Van der Put and Bijsterbosch [141] have reported data on streaming potential and streaming current for plugs of monodisperse spherical polystyrene particles in aqueous solutions. Streaming current data calculated according to Smoluchowski's and Levine's theories as a function of electrolyte concentration were shown. Due to the constant charge in the system, a leveling off of I_s was predicted at low concentrations. It is worth mentioning that this effect is not present in the classical Smoluchowski approach. Nevertheless, the curves predicted by Levine's theory were not in agreement with the experimental streaming current data. It was particularly striking that at intermediate concentrations, all experimental curves passed through a maximum and that the nearly constant values of I_s at low ionic strength were far below the calculated values. Recently, Van der Linde and Bijsterbosch [176] have shown that streaming current measurements may lead to large errors depending on the type of electrode used. Therefore, ζ-potentials calculated by Levine's theory decrease monotonously with ionic strength when A.C. conductance is employed in the calculation of ζ-potentials. Thus, in order to obtain accurate data, they recommended measuring streaming potential and the conductance at sufficiently high frequency. In fact, the trends of variation in the experimental streaming potential data and the data calculated using Levine's theory are similar [118], although quantitative differences exist between both sets.

In a comparative study of the ζ-potential obtained from streaming potential (Levine theory) and electrophoretic mobility (Semenikhin–Dukhin theory), Hidalgo-Álvarez et al. [122] have shown that when A.C. conductance is used in the calculation of the ζ-potential, a better agreement between both methods is obtained. The results obtained with a cationic latex can be seen in Fig. 3.

VI. ELECTRO-OSMOSIS

O'Brien [126] has developed a theoretical treatment for the electroosmosis in a porous material composed of closely packed spheres immersed in an electrolyte solution. The equations obtained for the electro-osmotic flow are valid if the EDL

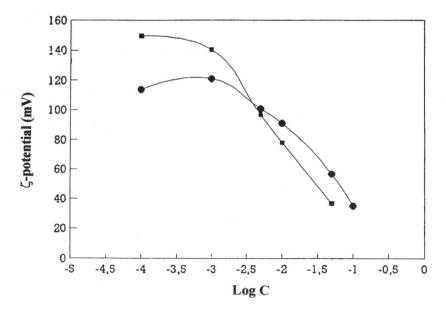

Figure 3 ζ-Potential from streaming potential (●) and electrophoretic mobility (■) of a cationic polystyrene latex. (From Ref. 122.)

thickness is much thinner than the particle radius. The total electro-osmotic flow as a function of the applied external electric field, E_f, is given by two contributions: one is identical to Smoluchowski's result and the second one is related to each component of the slip velocity; that is,

$$J_v = \frac{\varepsilon kT}{\eta e} \{\zeta[1 + 3\phi f(\phi)] - [\zeta - \gamma] g(\beta)\}E_f \tag{7}$$

where $f(\phi)$ is a function of the volume fraction and takes the values of −0.418, −0.384, and −0.378 for spheres in simple, body-centered-cubic and face-centered-cubic arrays, respectively. γ is a quantity which has the same sign as the ζ-potential but is independent of it. For a symmetric electrolyte, $\gamma = (2/z) \ln 2$.

The computed values for the nondimensional function $g(\beta)$ can be found in O'Brien's work [126]. β characterizes the relative importance of the tangential flux of counterions.

With the aid of Eq. (7), it is possible to calculate the electro-osmotic flow for any ζ-potential. For the simple cubic array, the error in applying Smoluchowski's result rises quite rapidly with ζ, from 15% at $\tilde{\zeta} = 3$–50% at $\tilde{\zeta} = 5.5$.

The ζ-potential obtained from Eq. (8) has been compared with those obtained using electrophoretic mobility and concentrated and dilute conductivity results. This was applied to Van der Put and Bijsterbosch's streaming current measurements [119].

Electro-osmosis and streaming current/streaming potential yield the same values for the ζ-potential if errors due to resistance measurements are avoided. This agreement, as well as the independence of the ζ-potential on the applied voltage (in the case of electro-osmosis) and on the hydrostatic pressure (in the case of streaming potential/streaming current), points out that the potential at the boundary immobile/diffuse layer can be determined. These theories call for more reliable experimental data sets.

VII. DIELECTRIC DISPERSION

When a colloidal dispersion is subjected to an harmonic alternating electric field, the macroscopic electric current which results has a phase and amplitude, which, in general, depend on the frequency ω of the applied field. The macroscopic electric current is of the form

$$\lambda\varepsilon \cos(\omega t) - \omega\varepsilon \sin(\omega t) \tag{8}$$

In general, λ and ε depend on the angular frequency of the applied field. The dielectric dispersion is related to both electrical conductivity and dielectric permittivity measurements. It is this frequency dependence that is referred here as dielectric dispersion [128]. The frequency dependence of the conductivity and the dielectric response occurs physically as a result of the inability of the polarized double layer to respond rapidly enough to the applied field at higher frequencies. Polymer colloid dispersions exhibit dielectric dispersion in two distinct frequency ranges: one around $\omega = D/a^2$, where D is the ion diffusivity, and the other around $\omega = \lambda^\infty/\varepsilon$, where λ^∞ and ε are the conductivity and permittivity of the background electrolyte. These frequencies typically lie in the KiloHertz (low-frequency) and megaHertz (high-frequency) ranges, respectively.

A. Dielectric Constant Measurements

The dielectric response of a colloidal dispersion at low frequency [177] and at high frequency [178] can be also a powerful tool to determine the ζ-potential of the polymer–liquid interfaces. In principle, this electrokinetic technique is applicable to dilute as well as concentrated dispersions. Both cases show very large dielectric dispersions at low frequency [172] which cannot be justified in terms of the Maxwell–Wagner theory [180]. From a theoretical point of view, the Schwarz–Schurr theory [181–183] has been successfully applied to aqueous polystyrene suspensions [183]. This theory, however, ignores the relaxation of the diffuse part, of the EDL. The Schwarz–Schurr theory is based on the polarization of the counterion cloud around the charge surface of the dispersed particles. It is the displacement of counterions in the EDL by an external electrical field that is responsible for the dielectric dispersion. In the Schwarz–Schurr theory, the dispersed particle with EDL is considered to be a sphere with a certain complex dielectric constant suspended in an electrolyte solution with a different complex dielectric constant. The substantial discrepancies between the polarization mechanisms of Schwarz and Schurr and that of Dukhin and Shilov [128] are due to the different kinetic model for the EDL, and so their respective qualities depend on the quality of the model assumptions. Lyklema et al. [184] have developed a new theoretical and experimental treatments of low-frequency dielectric dispersion (at frequencies of order D/a^2). According to these authors there are two ways in which a double layer can be polarized by an external field $E(\omega)$: by polarization of the diffuse part of the EDL and by polarization of the bound charge. They derived equations for the dielectric dispersion $\Delta\varepsilon(\omega)$ and the static permittivity $\varepsilon(0)$ for these two cases.

For the diffuse-layer polarization, the contribution of the particles to ε can be written as

$$\Delta\varepsilon(\omega) = \Delta\varepsilon(0) \, \frac{1 + W + W^2}{(1 + W)^2 + W^2 \, (1 + AW)^2} \tag{9}$$

$$W^2 \equiv \frac{\omega a^2}{2D_{eff}} \tag{10}$$

where D_{eff} is the effective coefficient of diffusion and $A = f(Z_+, Z_-, a, \kappa, \Psi_d, \zeta)$ [130]. The ζ-potential occurs because of convection currents due to electro-osmosis are accounted for, whereas Ψ_d is the diffuse double-layer potential determining conduction and diffusion in the diffuse EDL part.

The low-frequency dielectric increment $\Delta\varepsilon(0)$ can also be expressed as

$$\Delta\varepsilon(0) = \frac{9}{4} \, \phi\varepsilon_r \, (\kappa a)^2 B \tag{11}$$

where $B = f(Z_+, Z_-, a, \kappa, \Psi_d, \zeta)$.

The bound-layer mechanism is expected to apply only if there are bound counterions without significant exchange. In that case, a different polarization phenomenon applies, and the low-frequency dielectric dispersion obeys

$$\Delta\varepsilon(0) = \frac{9}{4} \, \phi\varepsilon_r \, (\kappa a) \left[\frac{\sigma_0\kappa}{4F_c} \, \sinh\left(\frac{F\Psi_d}{2N_A kT} \right) \right] M^{-1} \tag{12}$$

with

$$M = 1 + \frac{2}{\cosh\left(\dfrac{F\psi_d}{N_A kT}\right)} \left[\frac{\sigma_0 \kappa}{4F_c} \sinh\left(\frac{F\psi_d}{2N_A kT}\right)\right] \tag{13}$$

In the Schwarz–Schurr theory, $M = 1$, since there is no diffuse EDL part. However, introduction of the diffuse part is necessary to account for the screening of the bound charges. Therefore, ψ_d enters in the equations derived by Lyklema et al. [184]. These authors have compared the theoretical and experimental low-frequency dielectric increments. The first conclusion is that the high experimental values found for $\Delta\varepsilon(0)$ and its continuous rise with κa can never be quantitatively accounted for by the modified Schwarz–Schurr theory. This confirms that polarization of the diffuse EDL part is the leading feature. Quantitative discrepancies remain between theory and experiment, due to surface- and colloid-chemical peculiarities of the polystyrene latexes as pointed out by Springer et al. [130]. There is ample evidence that an anomalous surface conductance mechanism is responsible for the peculiar electrokinetic behavior of polystyrene latexes [132,185].

Originally, the low-frequency dielectric dispersion observed in dispersions was interpreted on the basis of the idea that the lateral transport of the counterions of the Stern layer plays the main role in forming the induced dipole moment of a dispersed particle [183]. Dukhin and Shilov [128] suggested an alternative mechanism. Theoretical [128,181] and, then, experimental [130,184] studies showed that the greatest dielectric increment $\Delta\varepsilon$ is caused by the concentration polarization of the EDL under the condition of free exchange between the double layer and the surrounding electrolyte concentration. One approach to study relaxation of double layers around charged particles is to apply dielectric spectroscopy to dilute colloidal dispersions. The dielectric response represents the conductive and the capacitive parts of the electric current flowing through the sol. The electric field will distort the ionic atmosphere around particles so that the double layer becomes polarized. It is relatively simple to relate the extent of this polarization to the dielectric response. The induced dipole moment itself is very sensitive to the ionic current flows around the particle, which are strongly dependent on the equilibrium double-layer structure. Therefore, dielectric spectroscopy in the proper frequency range enables us to study relaxation processes in the double layer as well as its equilibrium structure. Razilov et al. [133] have developed a quasiequilibrium theory for the concentration polarization in an EDL and the low-frequency dielectric dispersion, including the concentration polarization in the dense part of the double layer (Stern layer), under the condition of free ion exchange between the diffuse layer and the Stern layer. The application of this theory to low-frequency dielectric dispersion data obtained with polymer colloids might open new possibilities in the explanation of the electrokinetic behavior of these colloidal systems.

B. Electrical Conductivity Measurements

The frequency response of the polymer colloid dispersions is due to the sum of the polarizabilities of the EDL around the particles. The double-layer charge cloud is polarized due to the influence of the external field which is realized as the particle-dependent conductance increment and a very large effective dielectric response.

According to the value of volume fraction of solid particles in colloidal suspensions, we have concentrated or dilute suspensions of charged particles in a continuous phase (liquid medium). In both cases, the application of an electric field to such a suspension causes the ions and particles to migrate, giving rise to an electric current. Our aim is to determine the relationship between the applied field and the measured electric current. Electrical conductivity may be measured in a steady electric field (static conductivity) or in an alternating electric field (dielectric response measurements). This second possibility has been exhaustively reviewed by O'Brien [186]. In our case, we are especially interested in reviewing the conductivity of suspensions as this electrokinetic technique enables us to provide information on the electrical state of polymer colloid–liquid interfaces.

Precise measurements of conductivity are difficult to perform experimentally. The precision can easily be lost to many possible artifacts. Effects like electrode polarization have still not been thoroughly accounted for from a theoretical point of view. Electrode polarization is present at low frequency; at high frequency, there are stray capacitance and inductance effects. The modeling of these effects in order to extract the true sample signal is difficult, as the polarization and stray effects are up to orders of magnitude greater than the sample signal.

The theory describing the conductivity of suspensions of charged particles is based on the bulk or averaged properties at a finite, but small, particle volume fraction. In this way, it is very different from the theory of electrophoresis, which is concerned with the motion of a single particle at infinite dilution. The counterions which balance the charge on the particle are assumed to diffuse far away from the particle, hence not affecting the background ionic strength. In the theory of conductivity, the counterions which balance the charge on the particles do add to the background ionic strength and thus to the suspension conductivity. Because the theory of electrophoresis is based on single particles at infinite dilution and the theory of conductivity of suspensions is based on data obtained by these averaged properties at finite particle concentration, comparison of ζ-potentials calculated from two experimental techniques provides an excellent test for the completeness of the electrokinetic models employed [2,79,122,127,134,146]. Provided that the physical assumptions of the theory are valid, both processes should yield the same ζ-potential. Here, it is assumed that in the low volume fraction limit, the suspension conductivity, K^*, can be written [69,187]

$$K^* = K^\infty[1 + \phi\Delta K^*(\zeta, \kappa a)] \tag{14}$$

where k^∞ is the bulk electrolyte conductivity (outside the EDL), ϕ is the volume fraction of particles in suspension ($0 < \phi < 0.05$), κa is the electrokinetic radius and ΔK^* is the conductivity increment. As with the electrophoresis, conventional models relate the conductivity increment to a single interfacial property, the ζ-potential [134,145,146].

The experimental evaluation of ΔK^* is performed by measuring the conductivity as a function of the volume fraction. Voegtli and Zukoski [188] have found that at low volume fractions the conductivity of a latex suspension dialyzed against 10^{-4} M HCl becomes a linear function of the volume fraction.

The O'Brien conductivity model [189] provides values of ΔK^* much lower than those experimentally obtained by Watillon and Stone-Masui [190]. However, this is not the only case in which conductivity measurements yield different results compared to those obtained by other electrokinetic techniques. For instance, ζ-potentials calculated from electrophoretic mobility and conductivity measurements [163–166] have displayed significant differences. Various interpretation have been offered to justify those differences [79,166]. One of the difficulties is the dearth of data on well-defined systems where the tenets of theory can be tested.

The ζ-potential calculated from the conductivity increment approaches the ζ-potential calculated from the mobility both at high and low salt concentrations but is substantially larger at intermediate ionic strengths. It is apparent that the ζ-potentials calculated from ΔK^* are systematically higher than those from the electrophoretic mobility with the absolute magnitude of these deviations being at its largest (51–60 mV) at intermediate salt concentration. In agreement with Zukoski and Saville [163–166] these differences are due to inadequacies in electrokinetic theory applied to polymer colloids. The results obtained by these authors suggest a transport process occurring at the particle surface which is not taken into account in the extant electrophoretic theories. They proposed a model using a dynamic Stern layer. Using the model, they have shown that for a fixed ζ-potential, ionic transport behind the shear plane can increase the conductivity increment and depress the electrophoretic mobility [165].

In the opinion of Saville [191], the theories for the electrical conductivity of dilute dispersions fail to take proper account of the effects of nonspecific adsorption, which alters the concentration of ions in regions outside the EDL, and counterions derived from the particle charging processes. This might explain the poor agreement between theoretical and experimental data. Dunstan and White [145], however, have found that the volume fraction dependence of the supporting electrolyte concentration potentially explains the discrepancies between the mobility and the conductance-predicted ζ-potentials reported by other authors. The electrolyte concentration in latex dispersions changes markedly with the volume fraction of the particles and this explains the different ζ-potential values obtained using mobility and conductivity measurements. This work suggests that the observed electrokinetic behavior is general to colloidal dispersions and not intrinsic to polymer colloids.

Ideal electrokinetic behavior has been found by Gittings and Saville [192] of a polystyrene latex sample stabilized by sulfate charges (-1.06 μC/cm^2) with a diameter of 156 nm [measured by transmission electron microscopy (TEM)] and a hydrodynamic size of 160 nm (measured by Photocorrelation Spectroscopy [PCS]). Using the standard model of electrokinetics of electrophoretic mobility and the low-frequency dielectric response, these authors have found good agreement between the ζ-potentials of the O'Brien–White and Delacey–White theories, respectively. It seems that low charged polystyrene beads with diameters smaller than 200 nm are closer to the ideal colloidal system. There is a good agreement between the conclusions derived by Gittings and Saville [192] and Russell et al. [117].

C. High-Frequency Dielectric Dispersion

On the other hand, polymer colloids in the Smoluchowski limit ($\kappa a \gg 1$) can exhibit another type of dielectric dispersion at higher frequencies (of order k^2D). O'Brien [177] has described the high-frequency dielectric dispersion in terms of the complex conductivity $K^*(\omega)$ for concentrated and dilute colloidal dispersions. From an electrokinetic point of view, the relationship found between the surface conductance and z using the standard EDL model is very interesting.

The aim of the high-frequency conductivity and dielectric response experiments is to find the complex conductivity of the suspension as well as that of the background electrolyte in the frequency regime 0.1–40 MHz.

Experimentally, the complex conductivity of a suspension is obtained by measuring the admittance of a capacitor filled with the suspension in question. The admittance $Y(\omega)$ is converted into the complex conductance $K^*(\omega)$ [129] using

$$Y = C_c K^* \tag{15}$$

where C_c is the geometric cell constant.

The complex conductivity of the suspension can be expressed in terms of the conductivity $\Delta K^*(\omega)$ and the dielectric response $\Delta\varepsilon'(\omega) + i\Delta\varepsilon''(\omega)$ due to the presence of the polymer colloids, namely [129]

$$K^* = K_{el}^* + \phi\ [\Delta K^* - \frac{i\omega}{4\pi}\ (\Delta\varepsilon' + \Delta\varepsilon'')] \tag{16}$$

Alternatively, the complex conductivity of the suspension can be expressed in terms of the complex dipole strength $C_0(\omega)$ [129]

$$K^* = K_{el}^*\ [1 + 3\phi C_0\ (\zeta)] \tag{17}$$

Depending on the frequency range of interest, one of these expression is used. The choice is related to the structure of C_0, $\Delta\varepsilon'$, and $\Delta\varepsilon''$ when plotted against frequency. At low frequency, the real and imaginary parts of C_0 are approximately constant, whereas at high frequency, these curves possess greater structure, exhibiting well-defined maxima and minima for frequencies beyond 0.1 MHz. In the case of $\Delta\varepsilon'$ and $\Delta\varepsilon''$, all of the structure in the curves is observed at low frequency with a relaxation of $\Delta\varepsilon'$ occurring around 1 MHz and $\Delta e''$ decreases sharply to zero after reaching a maximum around 0.1 MHz. Very recently, Russell et al. [140] have studied the high-frequency dielectric response of highly charged sulfonate/stryrene latexes, they analyzed K^* in terms of the real and imaginary parts of the dipole strength using Eq. (17). Most early studies in this field were concerned with the low-frequency dielectric response and the experimental data were analyzed in terms of $\Delta\varepsilon'$ and $\Delta\varepsilon''$. Russell et al., however, used the high-frequency regime (0.1–40 MHz), and because the theoretical dipole strength is a function of the ζ-potential, a comparison of experimental and theoretical data will yield a ζ-potential. The experimental curves were compared to theoretical curves computed at various ζ-potentials using the numerical electrokinetic theory of Mangelsdorf and White [67]. The ζ-potentials derived from this study were compared to those obtained from electrophoretic measurements, and a very good correlation between the two electrokinetic processes was observed. This work shows that a high-frequency dielectric dispersion usually produces good fits to theory in contrast to dispersion at lower frequencies [165,188].

D. Dielectric Dispersion of Concentrated Dispersion

The problem of calculating the electrical conductivity of a porous plug is considerably more difficult than the apparently related problem of determining the conductivity of a granular material composed of purely conducting phases. The difficulty arises from the fact that the current in a porous plug (concentrated colloidal dispersion) is carried by electrolyte ions rather than electrons. In order to determine the current density at a point in the electrolyte, it is necessary to determine not only the local electric field but also the local ion density gradients and fluid velocity, because in addition to the component of current due to the electric field, there are also components due to the Brownian motion and convection of the ions with the neighboring fluid. To calculate the total current passing through a plug, it is therefore necessary to determine the distribution of ions, electrical potential, and flow field in the pores. This involves the solution of a set of partial differential equations. O'Brien and Perrin [193] have solved those equations and thereby theoretically determined the conductivity of a porous plug. The plug is assumed to be composed of closely packed spheres of uniform ζ-potential, and the Stern layer ions are assumed to be immobile. Besides, the particle radius a has to be much larger than the EDL thickness κ^{-1}. According to the procedure of O'Brien and Perrins, the electrical conductivity of a porous plug is given by

$$\frac{\lambda}{\lambda_0} = 1 + 3 \, \phi \left(f(0) + \frac{e^2 \, Z_i^2 \, u_i \, n_i^0}{\lambda_0}(f(\beta) - f(0)) \right) \tag{18}$$

where the subscript i refers to the counterion of highest charge. In this case, it is assumed that there is only one species of a highly charged counterion. β is a variable which relates the net tangential flux of ions entering a portion of the EDL to the flux passing out to the bulk electrolyte, and $f(\beta)$ is a complicated function of β which, in turn, depends on the geometrical distribution of particles constituting the porous plug (simple, body centered- and free-centered-cubic arrays). The computed values of the function $f(\beta)$ are shown in Ref. 193.

By using Eq. (18) together with the $f(\beta)$ values, we can compute the conductivity of the three cubic arrays for any electrolyte. In most practical applications, the conductivity is known through measurement and it is the ζ-potential which must be calculated. In this case, one would first calculate the conductivity of the plug over a range of ζ-potentials by the above procedure and then interpolate to find the ζ-potential corresponding to the measured conductivities. This method for ζ-potential calculation has been tested and compared with values experimentally obtained by Van der Put and Bijsterbosch [141]. The agreement between the theoretical and experimental ζ-potential values is quite satisfactory when conductivity data of concentrated and dilute polystyrene suspensions are used, but, again, ζ-potential from conductivity measurement are much larger than those from mobility measurements. Following a slightly different approach to the method used by O'Brien and Perrins [193], Midmore and O'Brien [194] have developed a cell-model formula for the low-frequency conductivity of a concentrated suspension of spheres with thin double layers. In this model, the effect of surrounding spheres on a reference sphere is approximated by taking the reference sphere to be at the center of a larger sphere. From the solution to this problem, we find

$$f(\beta) = \frac{2\beta - 1}{(1 - 2\beta) \, \phi + (2 + 2\beta)} \tag{19}$$

As the volume fraction approaches zero, Eq. (19) reduces to

$$f(\beta) = \frac{2\beta - 1}{2 + 2\beta} \tag{20}$$

So that, for a symmetric two-species electrolyte, in which the ionic diffusivities are equal, we find

$$\lim_{\phi \to 0} \frac{\left(\dfrac{\lambda}{\lambda^\infty} \right) - 1}{\phi} = -\frac{3}{2} \left(1 - \frac{3\beta}{2\beta + 2} \right) \tag{21}$$

This expression can be used for the unequivocal determination of β.

A new application of conductivity measurements is the experimental determination of the diffuse charge density (σ_d). If the diffuse-layer charge is high, which is generally the case for polystyrene latexes, the contribution to the charge by the negative adsorption will be small compared to that by the positive adsorption, so that

$$\sigma_d = \frac{\lambda_s F - \left(\dfrac{2\varepsilon NakT\sigma_{el}}{\eta z_+}\right)}{\Lambda_+} \tag{22}$$

σ_{el} may be determined from the ζ-potential and λ_s from conductivity measurements since

$$K = \frac{\lambda_s}{a\lambda_s^\infty} = \frac{\beta}{2} \tag{23}$$

when the electrolyte is symmetrical in terms of both charge and ionic diffusivity (e.g., KCl). Also, K can be calculated for any electrolyte from high-frequency and low-frequency conductivity measurements by means of more complicated formulas.

Midmore and O'Brien [194] found that the diffuse charge density for polystyrene latexes ranged from 4.1 and 4.4 $\mu C/cm^2$, whereas a value of 2.8 $\mu C/cm^2$ was obtained by titration. Once more, conductivity measurements yielded larger values for the electrokinetic parameters. Midmore et al. [195] have examined the effect of temperature and co-ion and counterion type on the diffuse-layer charge of monodisperse polystyrene latexes. After an extensive analysis, they concluded that conductivity measurements provide a better method for studying the EDL in latex systems.

Finally, conductance measurements have allowed the determination of the effective charge number of monodisperse polystyrene spheres [196]. The fraction of free macroions and/or gudgeons in deionized suspension is close to but smaller than unity for spheres having several and several tens analytical valency.

Dielectric spectroscopy is applicable to high-solids practical dispersions wherein there is a need for a fast and sensitive interfacial characterization technique. Chonde and Shabrang [197] have recently characterized the dependence of dielectric properties of styrene–butadiene latex dispersions on the latex concentration (up to 20% of solid content). These authors conclude that latex concentration seems to affect latex dielectric properties in a predictable manner, consistent with the understanding gained from the rheological measurements.

VIII. ELECTROACOUSTIC PHENOMENA

The electroacoustical methods for determining the ζ-potential of particles are based on the so-called Debye effect [154]. In an electrolytic solution, ultrasonic waves produce alternating potentials between points separated by a phase distance other than an integral multiple of the wavelength. This effect occurs when cations and anions of the electrolyte have different effective masses and frictional coefficients and is a consequence of the resulting differences in the amplitudes and phases of the displacements of the cations and anions. When the anion is lighter than the cation, the former is further displaced by the ultrasonic pressure amplitude than the latter. Thus, a region (A) will be charged positively with respect to the another region (B).

If inert metal probes are placed in regions A and B, an alternating potential difference will be observed with the same frequency as the sound waves. The frequency of the alternating ultrasonic vibrational potential (UVP) corresponds to that of the sound field. A similar effect exits for colloidal particles caused by the distortion of the ionic atmosphere. It was first predicted by Rutgers [198] and has been treated theoretically by others [199,200].

Babchin et al. [201] have reviewed the use of electroacoustical methods in the determination of electrokinetic properties. The main advantage of the electroacoustical methods is that it is perfectly capable of providing electrokinetic data in nontransparent and nonpolar media along with the ability to monitor coagulation/coalescence processes [202]. Enderby [200] obtained for the colloidal vibration potential an analytical expression. This vibrational potential was first measured by Yeager et al. [203,204] and by Rutgers and Vidts [205]. More recently, Beck et al. [206] and Babchin et al. [201] have developed new electroacoustical techniques that, usually, yield different ζ-potential values than other electrokinetic techniques. However, the cases studied by Babchin et al. were often in reasonable agreement.

The main disadvantage of the ultrasonic potential lies in the fact that it is a complex combination of the colloidal and the ionic vibrational potential. At low free-salts concentration (less than $10^{-2}M$), the ionic contribution can be neglected. Above this limit, the influence of the retention aid on the particle potential is no longer detectable.

The use of electroacoustical methods to obtain electrokinetic properties of polymer colloids is very recent [151,207–209].

Electroacoustics, when compared to microelectrophoresis and streaming potential, shows considerable advantages: (1) measurements on particle sizes up to 3 mm can be performed using low frequency (100 K Hz) and a wide electrode distance (13 mm). (2) The concentration of the suspension may vary within wide ranges (>>1% or more) depending on the electrolyte concentration. (3) The measurement is almost instantaneous (milliseconds range). (4) Continuous monitoring is, in principle, possible.

Shubin et al. [151] have performed an interesting comparative study on the electroacoustic and dielectric responses of carboxylated polystyrene beads. The ζ-potentials are calculated from dynamic mobility using the measured surface conductances obtained from the complex conductivity of the polymer colloids. These ζ values are found to be smaller than those calculated from conductivity data. A triple-layer model (TLM) is used to interpret measurements of electrophoretic mobilities at zero and high (1 MHz) frequencies and the complex conductivities over the frequency range from 1 to 20 MHz. In order to reconcile these two sets of data, these authors considered two alternative electrokinetic mechanisms in the TLM: (1) the ζ-potential is measured some distance (from 1 to 3 nm) from the outer Helmholtz plane (OHP) and the anomalous conductivity occurs in the diffuse layer or (2) the ζ-potential is measured in the OHP and the anomalous conductivity occurs in the Stern layer. The second model gives a reasonable description of the ζ-potentials and surface conductance for the entire pH range. According to these authors, the anomalous conductivity (the extra surface conductivity not accounted for in the standard theory) is always presented in the highly charged ($\sigma_0 > 10$ μC/cm^2) polymer–water interface, which is not generally accepted by other authors [117,134,146]. The mechanisms of sound attenuation have also been studied for concentrated polystyrene colloids with low-density contrast (neoprene latex) [210].

Hunter and O'Brien [211] have published an excellent article recently in which an AcoustoSizer (Mactec Applied Science) has been also used to follow the electroacoustic behavior of oil-in-water emulsions stabilized by an anionic surfactant. From the particle dynamic mobility (magnitude and phase angle), it is possible to estimate the ζ-potential and the size of particles. In this article, a comparison is made between the ζ-potential values obtained on the emulsions and measurements using D.C. methods. They conclude that the D.C. measurements are ambiguous electroacoustic measurements are unequivocal in ascribing a high ζ-potential to the data. They also display an unusual electroacoustic phase behavior by 4-μm semiconductor particles (silicon powder dispersed in water at low ion concentration) due to the very large values of the effective particle dielectric permittivity, which is proportional to the particle size. The phase lag is large at low frequencies and becomes smaller as frequency rises.

IX. ELECTROVISCOUS EFFECT

The presence of an EDL exerts a pronounced influence on the flow behavior of a fluid and leads to an increase in the suspension viscosity due to energy dissipation within the EDL. All such influences are grouped together under the name of electroviscous effects. These kind of effects appear both in concentrated colloidal dispersions (porous plugs or membranes) [212] and in dilute ones [213]. There are three distinct effects, called respectively the primary, secondary, and tertiary electroviscous effect. The primary electroviscous effect is due to an increase of the viscous drag forces on the particles as their EDL are distorted by the shear field.

The resulting contribution to the viscosity is, in first order, proportional to the volume fraction of the suspended particles. The primary electroviscous effect is a useful tool for investigating EDL at the polymer–liquid interfaces. This effect can be interpreted on the basis of several theoretical treatments for spherical particles.

The primary electroviscous effect occurs in a suspension, in which the particles are electrically charged [158]. The first theory for the primary electroviscous effect was presented without proof for the limiting case of a thin EDL by Smoluchowski [214]. Later, Krasny-Ergen [215] calculated the viscous dissipation in the same limit to obtain a result similar to Smoluchowski's, only different by a numerical factor. Booth [155] performed a definitive analysis in the low shear limit for arbitrary EDL thickness, obtaining a Newtonian viscosity with an $f(\phi)$ coefficient which increased with increasing surface charge and EDL thickness.

The magnitude of the primary electroviscous effect is proportional to the first power of the particle concentration, and for a suspension of spherical particles, it appears as a correction p to the Einstein equation:

$$\frac{\eta}{\eta_0} = 1 + 2.5\,(1 + p)\phi \tag{24}$$

The predicted enhancement of the coefficient over the Einstein value of 2.5 is generally of the same magnitude as or smaller than 2.5 and hence difficult to measure accurately; nevertheless, some experiments with polystyrene latexes show similar effects [156].

The calculation of p for any κa value can be carried out by a numerical analysis of the equation developed independently by Hinch and Sherwood [216] and Watterson and White [157]. The various equations describing the primary electroviscous effect of dilute monodisperse suspensions with spherical particles can all be put in the form

$$\frac{\eta}{\eta_0} = 1 + 2.5\,\phi\left(1 + \frac{3Q^2e^4N_a}{2\pi\eta_0\varepsilon kT_a^2}\,F(\kappa a)G(\lambda_i)\right) \tag{25}$$

where η is the viscosity of the suspension and η_0 that of the solvent, ϕ is the volume fraction, a is the radius of the suspended particles, Q is the number of elementary charges on each particle, N_a is the Avogadro number, ε is the dielectric constant of the dispersion, and λ_i are the conductivities of the various ionic species in the dispersion. e, k, and T have their usual significance. The functions $F(\kappa a)$ and $G(\lambda_i)$ can be found in Ref. 217.

For simple electrolytes, the concentrations of the ionic species enter only through the function $F(\kappa a)$. According to the Debye–Hückel (D–H) approximation:

$$Q = \frac{a\varepsilon}{e}\,(1 + \kappa a)\,\zeta \tag{26}$$

Hence, Eq. (26) can be utilized in low ζ-potential determinations.

For the primary electroviscous effect, all approximate equations are based on the following assumptions:

1. Laminar flow, neglect of inertia terms, no slip at the surface of the particles
2. Small volume fraction of particles, which are nonconducting and spherical
3. Uniformity of ionic conductivities, dielectric constant, and solvent viscosity throughout the electrolyte solution
4. Uniform immobile charge density on the surface of the particles
5. No overlap of double layers of neighboring particles
6. The electroviscous effect must be less than the Einstein coefficient (2.5ϕ) (i.e., low surface charge or low ζ-potential)

Honing et al. [218] have determined the dependence of the viscosity of silica (Ludox) sols on the electrolyte concentration of various aqueous solutions. For volume fractions below 5% silica, the viscosity was found to be linearly dependent on the volume fraction of silica. The sols showed Newtonian behavior at these low volume fractions. The electrolyte concentration had a marked effect on the viscosity. Very reasonable fits were obtained with the Booth equations by assuming the charge of the particles to be constant. Delgado et al. [160] could not fit their data to the Booth or the Watterson and White equations. The source of the discrepancy may be the fact that their particles (anionic polystyrene latex) were not smooth, whereas the theory assumes smooth particles with an uniform surface charge. In the same way, McDonogh and Hunter [158] indicated that the results obtained with polystyrene latexes lead to the conclusion that either polystyrene latexes prepared by emulsion polymerization and purified by dialysis or ion exchange are good examples of smooth, spherical hydrophobic particles or they are not. If they are smooth spheres, then there are serious experimental discrepancies between experimental results and the current theories of the primary electroviscous effect. This would be a matter of real concern because the mathematical model assumptions used for developing the theory are the same as those employed in the theoretical description of the electrophoretic mobility of a colloidal particle. If so, this could jeopardize the use of the ζ-potential as the main characterizing parameter of the EDL.

In the opinion of McDonogh and Hunter [158], it would be more comfortable to believe that this discrepancy is caused by some peculiar feature of the polystyrene latex system and, certainly, there seems to be some evidence that this system when prepared by the usual procedures (with presence of emulsifiers) is not as well behaved as would be expected from its appearance in electron micrographs. The results obtained by Honing et al. [218] with silica sols seem to support that second possibility, because silica sols are indeed smooth. Ali and Sengupta [159] have measured the primary electroviscous effect in two different series of polystyrene latexes of increasing particle size, prepared without the addition of surfactants.

They found that whereas for the larger particle size latexes, the measured electrical contribution [indicated by Eq. (27)] to the primary electroviscous effect decreases

$$\frac{\eta_{el}}{\eta_0 \, \phi} = \frac{\eta \eta_0}{2.5 \eta_0 \, \phi} \tag{27}$$

with increasing counterion concentration (quaternary ammonium of different ionic sizes) as required by the theory, it either remains (more or less) constant or increases slowly in the case of smaller latexes. This might possibly be due to the simultaneous occurrence of a slow feeble agglomeration of latex particles in the presence of increasing counterion. Also, Yamanaka et al. [217] found a satisfactory agreement between experimental primary electroviscous effect and the Booth's theory for relatively high salt concentrations. Furthermore, they used ionic polymer latexes prepared without surfactants. Recently, Garcia-Salinas and de las Nieves [219] using carboxylate polystyrene latexes found good agreement between the theoretical and the experimental values of the primary electroviscous coefficient (p), when the p values were calculated by the Booth's theory [155] and the ζ-potentials were calculated by the O'Brien and White's theory [69]. Finally, Quadrat et al. [220] have demonstrated that the dissociation of carboxylic groups on the surface of polymer colloids induces changes in the effective hydrodynamic volume due to an increased electroviscous effect and swelling.

The secondary electroviscous effect results from the overlap of the EDL of neighboring particles, resulting in repulsion. This repulsive force leads to a larger effective volume of the particles and, hence, to an increase in viscosity. The leading contribution to the viscosity is proportional to the square of the volume fraction, because at least two particles are involved. Although the secondary electroviscous effect depends on the square of particle concentration and the primary effect is only proportional, in most practical cases the secondary effect is much larger than the primary effect, even at low concentrations. Actually, the primary effect is important only if the thickness of the EDL is of the same order of magnitude as the size of the particles. This implies that the particles must be very small for the primary electroviscous effect to be dominant [218]. With regard to the secondary electroviscous effect, Stone-Masui and Watillon [156] have obtained results which indicate the treatment of Chan. et al. [221], although it predicts to high values, represents a more realistic approach than Street's equation, which yields too low values for the viscosity.

X. ANOMALOUS ELECTROKINETIC BEHAVIOR OF POLYMER COLLOIDS

Marlow and Rowell [107] have noted five possible explanations for the anomalous electrokinetics behavior of polymer colloids:

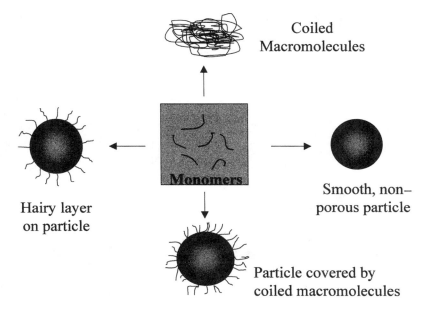

Figure 4 Types of layer of flexible polymer chains formed during emulsion polymerization.

1. Shear plane expansion (the hairy-layer model) [10,15,20,30,80,81,83, 85,91,93,113,119–122,141]
2. Preferential ion adsorption [9,31,44,64,74,82,142,145,146,163–166, 185]
3. Osmotic swelling or core–shell redistribution [50]
4. Crossing of the mobility/ζ-potential minimum [11,14,192]
5. Anomalous surface conductance [33,42,62,68,79,80,87,93,109,122, 132,161,184]

Explanations 1, 2, and 3 are qualitative explanations, whereas explanations 4 and 5 try to explain quantitatively the electrokinetic behavior of polymer colloids.

Two very different models have been proposed to account for the nonclassical behavior of polymer colloids. Recently, Seebergh and Berg [222] have published a most interesting article in which both models are extensively discussed.

The hairy-layer model postulates the presence of a layer of flexible polymer chains, or hairs, at the surface [119]. Zimehl and Lagaly [223] have suggested that the hairy layer is just one of several types which may be formed during emulsion polymerization, as shown in Fig. 4.

Electrophoretic Mobility

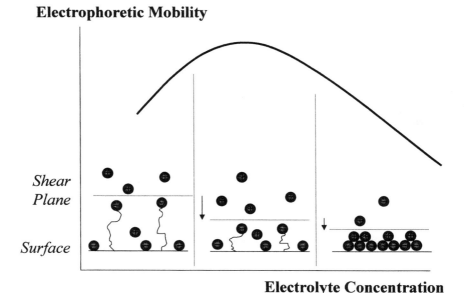

Electrolyte Concentration

Figure 5   Surface scheme for a "hairy" layer.

The hairs extend into the bulk solution, owing to the electrostatic repulsion between the ionic groups terminating the hairs and the ionic groups anchored at the surface. The distance of hair extension (i.e., the thickness of the hairy layer) will also be influenced by the extent of solvation of the polymer chains and the ionic end groups, as elucidated by some authors [224]. At low ionic strength, the hairs are in an extended conformation, so that some fraction of the total fixed charge of the particle is now located in the Stem layer and/or the diffuse layer. The shear plane has shifted away from the surface, so the ζ-potential is less than it would be for a nonhairy particle with the same fixed charge. The shear-plane shift may also allow for electrical conduction between the shear plane and the surface, further reducing the electrophoretic mobility or streaming potential. As the ionic strength increases, the hairs collapse back toward the surface, owing to charge shielding. Thus, the shear plane moves closer to the surface and the ζ-potential increases. At high ionic strength, the hairs have completely collapsed and the ζ-potential decrease with increasing electrolyte is due to the compression of the electrical double layer. This mechanism is shown in Fig. 5.

Seebergh and Berg [222] have published an article in which they verify the presence of a hairy layer via simultaneous measurement of both the size and electrokinetic characteristics of a given latex over a range of conditions. The presence of a hairy layer was investigated by comparing size, mobility, critical coagulation concentration (ccc), and surface charge density measurements of three different surfactant-free latexes before and after heat treatment. Most of the literature has focused on trying to explain the anomalous electrokinetic behavior of polymer colloids. In fact, polymer colloids exhibit another behavior which does not agree to the predictions of theory. Several studies have reported decreases in the hydrodynamic diameter as a function of increasing ionic strength, based on photon correlation spectroscopy measurements [224–227]. Seebergh and Berg [228] also have studied very recently the effect of organic cosolvent on the aggregation stability of an aqueous polystyrene latex dispersion. The counterion displacement mechanism suggested by Vincent [229] may be operative, although experimental evidence regarding the occurrence of specific adsorption does not always support this mechanism. Changes in the conformation of the hairy layer are not responsible for the anomalous behavior, as demonstrated by hydrodynamic size measurements. Any proposed model of the polymer colloid–fluid interface must account for these observations as well as for the electrokinetic behavior. According to the hairy-layer model, with increasing ionic strength, the hairs move gradually closer to the surface and the shear plane shifts inward. The hydrodynamic diameter is thought to reflect the existence of bound surface layers, so a decrease in the hairy-layer thickness should result in a decrease in the hydrodynamic diameter. The observed hydrodynamic diameter behavior is therefore consistent with the hairy-layer model of the interface [230]. An alternative way to determine the hairy-layer thickness is to measure the force acting between two latex spheres. Wu and van de Ven [10] have obtained the layer thickness by interpreting the force–distance profile [using an ultrasensitive force measurement method, colloidal particle scattering (CPS)] with the aid of a theoretical model. Two latex samples were used in the experiments: 5-mm divinylbenzene–cross-linked polystyrene latex (denoted as Duke latex) and 4.7-mm surfactant-free polystyrene latex (denoted as IDC latex). The CPS method is based on the analysis of particle collision trajectories between an immobile particle stuck to a wall and a freely mobile particle subjected to a wall shear flow of known shear rate. These collision trajectories are recorded and analyzed with an image processing board. Collision trajectories can also be calculated by solving the trajectory equations. These theoretical trajectories are fitted to the experimentally observed ones varying the parameters in the colloidal force equations which enter in the trajectory equations. The best fit yields the optimum parameters with which the force–distance profile can be calculated. The two parameters to be fitted are the Hamaker constant, A, and the hairy-layer thickness, L_h. From the experimental data, these authors obtained that $L_h = 11$ nm and $A = 1.1 \times 10^{-20}$ J for the Duke latex, whereas for the IDC latex the parameters are 3 nm and 8×10^{-21} J, respectively. According to these authors, the stabilizing mechanism in this case is related to the hairy layer but cannot be ascribed to steric repulsion. Instead, a layer of immobilized water in the hairy

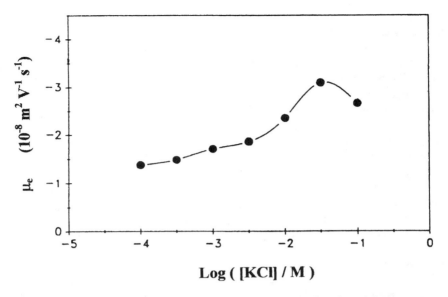

Figure 6   Electrophoretic mobility for an anionic latex as a function of the ion concentration in the presence of 10^{-4} M LaCl$_3$ as the background electrolyte. (From Ref. 82.)

layer effectively lowers the van der Waals interactions, thus increasing the relative importance of electrostatic repulsion.

An alternative explanation of the size behavior of the polymer colloids was offered by Saaki [226], who suggested that the reduction of the hydrodynamic radius with increasing electrolyte is due to the ion-initiated destruction of the rigid water layer at the polymer–bead surface.

The hairy model has been criticized for several reasons. A number of studies have reported a secondary maximum in the mobility of anionic latex at a very low electrolyte concentration which cannot readily be explained by the hairy-layer mechanism [82,165,225]. Elimelech and O'Melia [82] reported that the addition of lanthanum ions (La^{3+}) to an ionic latex dispersion did not eliminate the minimum in mobility as a function of potassium chloride concentration, even though La^{3+} ions should strongly adsorb onto the negatively charged sites and collapse a hairy layer through charge shielding (see Fig. 6).

Also, Midmore and Hunter [93] considered it unlikely, based on energetic considerations, that hydrocarbon chains would extend into the aqueous medium. Owing to these apparent weaknesses, some investigators have invoked models based on ion adsorption to explain their results.

The second model is based on ion adsorption. This model postulates that the electrokinetic behavior of the polymer colloids is a consequence of the extent of counterion and co-ion adsorption at the surface [9,31,44,64,74,82,142,145,146, 163–166,185]. At very low electrolyte concentrations, counterions adsorb, causing a decrease in the ζ-potential with increasing electrolyte concentration. At somewhat higher electrolyte concentrations, co-ion adsorption occurs as counterion sites become filled. This causes an increase in the ζ-potential with increasing electrolyte concentration. At relatively high electrolyte concentrations, there is no further ion adsorption, and the ζ-potential decreases as the electric double layer is compressed.

Very recently, a new qualitative explanation for the minimum in the mobility–ionic strength curves of anionic polymer colloids has been proposed [230]. According to it, three competing processes are involved in determining the shape of the mobility curve in the presence of electrolytes; for the case of anionic polystyrene particles, they are as follows: (1) neutralization of negative charge on the surface by adsorption of counterions, causing an increase in the electrokinetic potential (less negative); (2) approach of co-ions close to the hydrophobic surface of the particles, causing a decrease in the electrokinetic potential (more negative); (3) compression of the diffuse double layer due to high bulk concentration of electrolyte, causing an increase in electrokinetic potential (less negative). Thus, the effect of each process along the mobility curve determines the shape of the electrokinetic potential as a function of the electrolyte concentration. Figure 7 shows this process.

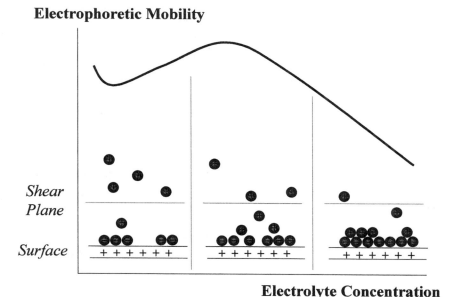

Figure 7 Ion adsorption mechanism.

The issue of co-ion adsorption has recently gained greater significance because of the works of Zukoski and Saville [164,165] and Elimelech and O'Melia [82]. They interpret their electrokinetic data by invoking specific co-ion adsorption. This ion-adsorption model postulates that the anomalous behavior of polymer colloids is due to counterion and co-ion adsorption at the surface [82, 164,165,188].

The ion-adsorption model has also been criticized. In particular, the driving force for co-ion adsorption of simple, inert electrolytes onto polymer colloids is not well understood [165,230]. Some authors have suggested that the driving force is hydrophobic in nature [146], where the co-ions have an affinity for the relatively hydrophobic surface of polymer colloids. This hypothesis is plausible for anionic polymer colloids, where the co-ions are relatively hydrophobic ions; however, it is difficult to accept in the case of cationic polymer colloids, where the co-ions are strongly hydrated cations. Several experimental studies have demonstrated that cationic polystyrene latexes exhibit a mobility maximum as a function of electrolyte concentration [62,89,113,122], which implies that co-ion adsorption is not hydrophobically driven. In any case, although some conductivity studies indicate that co-ions adsorb at the surface of the polymer colloids [165], there have been no direct measurements of such adsorption [93,230]. In fact, adsorption studies have shown that neither chloride nor sulfate ions adsorb at the surface of anionic polystyrene colloids dispersed in water [231].

Another criticism of the ion-adsorption model is that it cannot account for the observed insensitivity of the mobility to the co-ion species. Midmore and Hunter [93] measured the mobility of an anionic polystyrene latex as a function of KF, KCl, KBr, and KI concentration and found that the mobility behavior was virtually identical for each electrolyte. These results cast additional doubt on the mechanism of co-ion adsorption, as it seems likely that the adsorption affinity would differ according to the ionic species. Nevertheless, the hydrated size of these ions are quite similar, and this might explain the insensitivity of the mobility behavior to these ions. A slightly different ion-adsorption model has been proposed by Goff and Luner [64], "the ion-exchange" model. In this model, they assumed the original negative surface charge (i.e., in the absence of additional electrolyte) to be fully compensated by H^+ counterions. It is assumed that these protons are mainly inside the slipping plane and so do not contribute to the diffuse charge which is responsible for the ζ-potential. Thus, both the electrokinetic charge density and the ζ-potential are also less negative at low ionic strength. As electrolyte is added (e.g., NaCl), the H^+ ions are gradually replaced by Na^+ ions which takes place in the diffuse layer outside the slipping plane and thus act to decrease the ζ-potential (more negative) until compression of EDL at still higher ionic strengths causes the latter to increase (less negative) again. The transition occurs with an electrolyte concentration of 10^{-3}–10^{-2} M and this gives rise to a minimum for the ζ-potential in that region. This model is supported by observations made in connection with the conductometric titration of latexes. However, Goff and Luner [64] pointed out that their model is only valid for latexes whose charge is due to strong acid, so a different explanation must be sought to cover the behavior of latexes whose charge is due weak acids. Also, this "ion-exchange" model has been criticized by some authors. Van den Hoven and Bijsterbosch [120,121] obtained electrokinetic data where the particles were strictly in K^+ form only, so that any ion exchange was out of question. Under these experimental conditions, electrophoretic mobility as well as streaming potential techniques used on the same particles showed the disputed minimum. Also, de las Nieves et al. [20] working with highly sulfonated polystyrene latex beads obtained no appreciable difference in the electrophoretic mobilities versus NaCl concentration where the sulfonated particles were in H^+ or Na^+ forms. Figure 8 shows the m_e values for a sulfonated latex in Na^+ and H^+ forms. From these results, the ion-exchange theory does not seem to be the final answer to the controversial electrokinetic behavior of polymer colloids. Also, Ma. et al. [74] found that the minimum in mobility with increasing ionic strength was not due to the titratable surface charge. In their

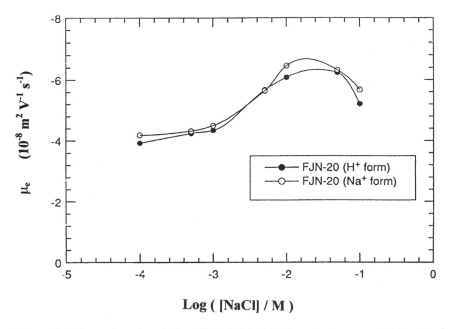

Figure 8 Electrophoretic mobility of latex FJN-20 (248 nm in diameter and 139 mC/m² of surface charge density) in sodium and hydrogen forms versus electrolyte (NaCl) concentration. (From Ref. 20.)

experiments, the titratable surface charge was varied by hydrolysis of the sulfate surface groups such that the resulting latex had no measurable titratable surface charge. However, the electrophoretic properties of these hydrolyzed latexes were virtually identical to the original sample over a wide salt-concentration range. These authors suggest that their results can be understood if the origin of the latex electrokinetic charge were primarily due to the adsorption of anions onto the hydrophobic parts of the latex surface.

Although the hairy-layer model and the ion-adsorption model are capable of explaining many of the observed phenomena, neither is able to account for all of them. Thus, some authors suggest that both mechanisms are operative [185,227]. The primary weakness of each model is a lack of direct experimental verification of the underlying mechanism [227]. Independent measurements of co-ion adsorption and clarification of the hydrophobic adsorption mechanism, for cations in particular, are necessary to validate the ion-adsorption model. This is a very challenging experimental problem, as ion adsorption is usually inferred via ζ-potential determinations. Likewise, additional nonelectrokinetic proof of hairs at the polymer surface is necessary to validate the hairy-layer model.

The appearance of a minimum in ζ-potential is closely related to the theoretical treatment used to convert mobility into ζ-potential. The ζ-potential values obtained from the Dukhin–Semenikhin equation [72] agree reasonably with those theoretically expected. Hence, the most likely explanation for the mobility/concentration minimum seems to be a movement of the shear plane away from the surface with decreasing electrolyte concentration. This phenomenon results in two effects. First, it lowers the ζ-potential in the usual way and, second, it lowers the mobility by a much greater percentage by introducing ionic conduction in the diffuse layer but inside the shear plane [93]. Henry's equation, with the surface conductance correction, and the Dukhin–Semenikhin equation seem to give ζ-potentials that agree remarkably well, and both remove the ζ-potential maximum. Also, this explanation agrees with the model proposed by Midmore and Hunter [93] for the electrolyte–polymer interface. According to this model, the Stern layer is empty when the surface charge density is less than 6 $\mu C/cm^2$ and the Poisson–Boltzmann equation applies to all the solution side of the double layer. At high electrolyte concentration, the surface is smooth and the shear plane corresponds with the outer Helmholtz plane. At lower electrolyte concentrations, however, the surface becomes either rough or hairy, causing the shear plane to move some slight distance (~1 nm, this is the diameter of a hydrated sulfate ion) away from the surface. This leads to some lowering of the ζ-potential in the normal way but to a much greater reduction in the electrophoretic mobility because of the surface conduction effect. The postulated shift in the shear plane has the effect of immobilizing a layer of water of 1 nm thickness but does not influence the mobility of the diffuse-layer ions. Thus, these ions are able to conduct in this region and thereby reduce the mobility of the particle.

Surface conductance has been very useful to explain why the ζ-potentials obtained from electrophoretic mobility data are usually less negative than the ζ-potentials obtained from low-frequency dielectric dispersion [130,132,184]. Nevertheless, Russell et al. [117] have also cast doubt on this explanation. These authors claim that the theory of O'Brien and White [69] correctly predicts the electrokinetic potential without the need to invoke the surface conductance. They studied the high-frequency dielectric response of sulfonate latexes, and the ζ-potentials obtained were compared to ζ-potentials from electrophoresis measurements. They found that the differences encountered between both techniques may not be related to theory or anomalous effects such a Stern-layer conduction but merely to experimental limitations. Thus, and according to Russell, only one technique is needed to characterize the electrokinetic properties of polymer colloids and it is highly unlikely that their latex sample is unique. The shot-growth polymerization process employed to prepare these polymer colloids will lead to a highly charged "hairy" surface [16,20,30,80,87] with high concentrations of specifically adsorbed counterions. Extant hypothesis would make it a prime candidate for Stern-layer conduction [67]. Despite this postulate, they did not find evidence of surface conduction and only 1 out of 10 systems studied showed any dielectric response anomalies. They also claimed that the method of sample preparation may be of importance and that this could be the reason for the inconsistencies observed by Zukoski and Saville [163–166]. Dunstan and White [145] found reasonable agreement between static conductivities and electrophoretic mobilities with an amphoteric latex of 176 nm and 62 μC/cm^2 of surface charge density when dialysis was used to bring the samples to an equilibrium electrolyte concentration. Other methods of sample preparation, such as centrifugation and dilution, were found to be totally unsuitable. They also showed that very small errors in the electrolyte concentration caused large discrepancies between the two measurement techniques. In a second article, Dunstan and White [134] reported on the dielectric response of amphoteric latexes and suggested that surface conductance is not present in these systems or, if present, that the ions in the Stem layer have the same conductivity as those in the bulk. However, they claimed that the latexes used in their study did not display a maximum in the mobility versus KCl concentration.

An alternative approach to the explanation of the electrokinetic behavior of negatively charged polymer colloids is the possible crossing of the mobility/ζ-potential minimum (or maximum for cationic polymer colloids). The phenomenon of the mobility/ζ-potential minimum is caused by the distortion of the double layer at high ζ-potentials resulting in a reduction of the mobility, μ_e, compared with that expected from Smoluchowski's equation which predicts a linear relation between z and μ_e. If z is high enough, the mobility begins to drop with increasing z, giving rise to a minimum. There exist, therefore, two possible ζ-potentials as solution to any given mobility, one low and the other high. Thus, it is possible that an apparent drop in z with a decrease in concentration is caused by taking the wrong solution for a given mobility. Very recently, Midmore et al. [11] have shown that the maximum in the mobility–ionic strength plot is predicted by classical theory and that it is a reflection of the theoretical mobility versus ζ-potential plot. A linear relationship between z and KCl concentration was found when the high branch of the mobility versus z plot was selected, a phenomenon not often encountered. Gittings and Saville [192] have also published an article in which they measured the dielectric response and electrophoretic mobility to test the classical theory of electrokinetics. Good agreement between the ζ-potentials derived from the two complementary measurements was found using the standard model of electrokinetics and the ζ-potential from the upper branch of the mobility–ζ-potential relation.

A curious hypothesis on the latex particle surface has been introduced by Chow and Takamura [81]. According to these authors, the ζ-potential maximum may be attributed to surface roughness, which results in a more distant location of the shear plane, and thus in a smaller value for the ζ-potential. In order to check this hypothesis, they modified the surface roughness of the latex by heating it above its glass transition temperature. Under these conditions, the polymer chains will acquire mobility and will leave or collapse onto the surface of the particles. Such alterations in the structure of the interfacial region should be revealed in the surface properties and electrokinetic behavior. Their experimental results showed a decrease in the particle radius and an increase in the measured mobilities (and, therefore, ζ-potential) of the heat-treated latex for various electrolytes below 10^{-4} M. They suggested that surface roughness is one of the most important factors in the determining ζ-potentials from electrophoresis and the O'Brien and White computer solution [69] overestimates the electrokinetic relaxation effect when κa < 100.

Other works have been published recently that use the heat treatment to verify if a "hairy layer" exists on the surface of the polymer beads [7,12,142,146, 176,185,222]. Van der Linde and Bijsterbosch [176], using a negatively charged latex, obtained an apparent decrease of the ζ-potential calculated with the O'Brien–White theory, upon prolonged exposure at high temperature, whereas the maximum disappeared. The two effects were accompanied by a notorious decrease in the experimentally obtained (surface) conductivity. The particle diameter remained unchanged, contrary to the observation of Chow and Takamura. But in this article, none surface charge density for the heat treated latex was presented. In an early article Rosen and Saville [142] compared results obtained for an amphoteric and anionic latexes before and after heat treatment by making low-frequency dielectric measurements. Their results indicated that the mean particle diameter and uniformity were essentially unaffected by the heat-treatment procedure. For the amphoteric latex, the dielectric constant and conductivity decreased with heat-treatment time, and the characteristic relaxation frequency increased. They suggested that the heat treatment smoothes the surface, creating particles which more closely conform to the assumptions of the classical theory. Experiments with the anionic latex suggested the presence of surface structure, but discrepancies between experiment and theory persisted. They concluded that factors other than hairiness contribute to these discrepancies. In a later article [185], they used low-frequency dielectric spectroscopy and electrophoretic light scattering to test the applicability of the classical electrokinetic theory to latex particles with and without a surface-grafted water–soluble polymer. They employed two surfactant-free anionic latexes: a bare poly(methyl methacrylate)/acrolein particle, and a hairy particle, prepared by chemically grafting a layer of water-soluble poly(acrylamide) onto the bare particle surface. Their results showed that the mean particle diameter and uniformity were not significantly altered by heat treatment, but that the surface charge densities had increased after the heat treatment. Dielectric spectroscopy response of the heat-treated bare latex agreed with theoretical predictions based on the ζ-potential measured electrophoretically. For the heat-treated hairy latex, the agreement between theory and experiment improved. They concluded that the "beneficial" effects of heat treatment are attributed to smoothing of the molecular surface structure.

Seebergh and Berg have published an article [222] in which they confirmed the presence of a hairy layer on the surface of some polymer colloids via simultaneous measurement of both size and electrokinetic characteristics of a given latex for a range of conditions. The presence of a hairy layer was investigated by comparing size, mobility, critical coagulation concentration, and surface charge density measurements on three different surfactant-free latexes before and after heat treatment. Light-scattering measurements of size as a function of electrolyte concentration indicated that the thickness of the layer may be greater than 7 nm. The results showed that pronounced mobility and ζ-potential minima were no longer observed after heat treatment, but these parameters were smaller than the originals ones (contrary to predictions of the hairy-layer model). This result was attributed to a loss of surface charge density after heat treatment. The high c.c.c. values of untreated latexes were consistent with the presence of a hairy layer, which would provide an additional steric barrier to coagulation. After the heat treatment, there was a decrease in c.c.c. values and a better agreement was found between theoretical DLVO predictions and experimental c.c.c. results. They conclude that their results strongly support the hypothesis that latex particles are covered with a layer of polymer hairs, although they note that the new results do not rule out ion adsorption as a comechanism.

Dunstan [146] presented experimental data from electrophoretic mobility, conductivity, and dielectric response measurements on polystyrene latexes and colloidal alkane particles. Measurements had been made on both normal (raw) and heat-treated latexes. Figure 9 shows the measured electrophoretic mobility values for the raw sulfate latex, the heat-treated sulfate latex, and the docosane particles as a function of KCl concentration. Two significant features are apparent; the

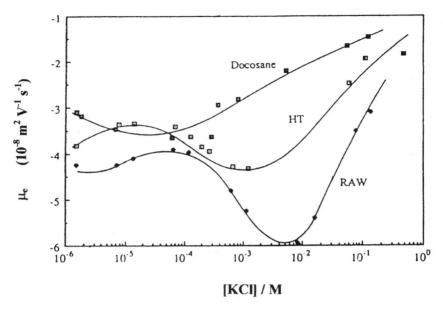

Figure 9 Electrophoretic mobility of 0.30-μm-diameter sulfate latexes versus KCl concentration. The data sets are for the heat-treated (HT), raw latexes (RAW), and 1.0-μm-diameter docosane particles. Measurement correspond to 25°C and pH = 5.8. (From Ref. 146.)

mobility minima are present both before and after heat treatment and the colloidal alkane shows the mobility minimum. From these results, he concluded that the mobility minima are not due to surface hairiness. The docosane particles do not have polymeric hairs protruding into the solution. The reduced mobility for the heat-treated particles compared with the raw particles arises from the hydrolysis of some sulfate groups on the surface of the latexes. The data of the concentration of excluded electrolyte as a function of the electrolyte concentration at which the particle were exposed showed that no specific co-ion adsorption occurs on the surface of the particle. From the results obtained with dielectric and conductivity measurements, he suggested that the hydrophobic surface disturbs the water structure, causing preferential solvation of the ions of one charge in the interfacial region, combined with overall exclusion of those ions from interfacial region. As a conclusion, he claimed that the interpretation of electrokinetic data from different measurements do not yield the same values for the ζ-potentials for the same suspension. The perturbation in the ion distribution due to the hydrophobic surface of the latexes is postulated as the reason for the inadequacy of the theory, as such polystyrene latexes are not ideal classical electrokinetic model colloids. To describe these systems electrokinetically, a spatially varying electrostatic and chemical potential should be used in the Poisson–Boltzman description of the interface. The spatially varying chemical potential term arises from the fact that the interfacial water is entropically different from the bulk water.

Bastos et al. [7,12] studied the effect of heat treatment (115°C for 17 h) on four surfactant-free polystyrene latexes with different functional groups on their surface (sulfate, sulfonate, aldehyde, and carboxyl). The latex with and without heat treatment were analyzed by PCS, infrared spectroscopy, conductometric and potentiometric titrations, adsorption of non-ionic surfactant, and measurement of their electrophoretic mobilities as a function of pH and electrolyte concentration. It was observed that the mean particle diameter of four latexes did not change after heat treatment. This result is similar to those obtained by Rosen and Saville, and van der Linde and Bijterbosch but contrary to that obtained by Chow and Takamura, and Seebergh and Berg. In relation to the surface charge densities, it was found that the strong-acid charge was converted into weak acid and that there was an increase in the total surface charge density of the heat-treated sulfate and aldehyde latexes. They justified these changes by taking into account that sulfate groups oxidized in carboxyl groups and that part of the surface groups were hydroxyls, products of Kolthoff reaction, and therefore impossible to detect through titrations. As result of heat treatment, these groups could be oxidized in carboxyl groups and become measurable with titration. Similar results were observed by Rosen and Saville when they used potassium persulfate as the initiator in the synthesis of their latexes. The sulfonate and carboxyl latexes showed similar electrokinetic behavior before and after heat treatment. They concluded that heat treatment could affect latexes with sulfate groups on the surface due to hydrolysis of these groups but that changes observed could not be attributed to the presence of a hairy layer.

The effect of heat treatment on the electrokinetic behavior of polymer colloids is another controversial aspect of these colloidal systems. However, there appears to be a consensus that the heat-treated polymer colloids are closer to the "ideal colloid," as predicted by standard electrokinetic theories.

The behavior of pronounced "hairy" character latexes are tackled In ref. 232. Dynamic light scattering and electrophoretic light scattering were used to examine the surface characteristic of two polystyrene latexes—one with sulfate and the other with sulfate and aromatic amino groups. These systems are also compared with styrene–butadine copolymer latex particles. Zhao et al. conclude that the surface polymer composition and the surface charge density influence the adsorption of inorganic electrolyte ions. The existence of ionizable amino/carboxyl groups leads to a strong pH dependence of the surface properties and complicated changes in both the hydrodynamic radius and the mobility with pH. This effect is plotted in Figure 10.

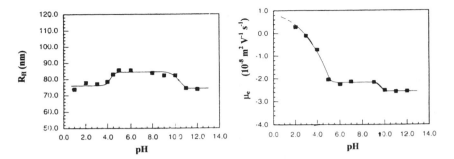

Figure 10 Hydrodynamic radius (R_H) and electrophoretic mobility (μ_e) as a function of the pH of an aqueous solution of latex C with no salt at 25°C. (From Ref. 232.)

The stability of particles with a "hairy" surface in the presence of free polymers is also of both theoretical and experimental interest [233–236]. An interesting topic related to this mechanism is the depletion flocculation beyond a critical free-polymer concentration [237]. Depletion of polymers takes place if the loss of configurational entropy near the interface is not balanced by positive interaction energy. Flocculation can occur when the osmotic attraction energy due to depletion outweighs the loss in configurational entropy [238]. In the so-called depletion zone, the polymer segment density is lower than that in the bulk, resulting in a reduced viscosity in comparison with the bulk polymer solution. Electrophoretic studies on charged smooth particles in polymer solutions have been performed and generally revealed an increase in mobility compared with that predicted from the Smoluchowski equation. These findings were attributed to the presence of a depletion layer. Krabi and Donath [239] calculated the apparent thickness of depletion layers from electrophoretic data for human red blood cells and liposomes in dextran and poly(ethylene)glycol, assuming an exponential viscosity profile in a linear approximation. This was possible for smooth and hairy surfaces at high ionic strength. The calculated thickness were in reasonable agreement with the radii of gyration of the polymers [240]. It is challenging to study the electrophoretic mobility of hairy surfaces together with a depletion layer, as little is known about the structure of the depletion layer in front of the hairy surface. Also, it cannot be ruled out that the free polymer can influence the density, as well the equilibrium distribution, of the terminally anchored polymer of the hairy layer.

ACKNOWLEDGMENTS

This work was supported by the "Comisión Interministerial de Ciencia y Tecnología" under project MAT 96-1035-C03-03 and by Junta de Andalucía (Grupo de Investigación FQM 0230).

REFERENCES

1. ES Daniels, ED Sudol, MS El-Aasser. Polymer Latexes—Preparation, Characterization and Application. Washington DC: American Chemical Society, 1992.
2. R Hidalgo-Alvarez. Adv Colloid Interf Sci 34:217–342, 1991.
3. R. Hidalgo-Alvarez, A Martin, A Fernandez-Barbero, D Bastos, F Martinez, FJ de las Nieves. Adv Colloid Interf Sci 67:1–118, 1996.
4. J Lyklema. Colloids Surf A 92:41–49, 1994.
5. J. Lyklema. Fundamentals of Interface and Colloid Science. Vol. I. London: Academic Press, 1993.
6. J Ugelstad, PC Mork, KH Kaggerud, T Ellingsen, A Berge. Adv Colloid Interf Sci 13:101, 1980.
7. D Bastos, R Hidalgo-Alvarez, FJ de las Nieves. J Colloid Interf Sci 177:372–379, 1996.
8. J Zhao, W Brown. J Colloid Interf Sci 179:255–260, 1996.
9. G Tuin, JHJE Senders, HN Stein. J Colloid Interf Sci 179:522–531, 1996.
10. X Wu, TGM van de Ven. Langmuir 12:3859–3865, 1996.
11. BR Midmore, GV Pratt, TM Herrington. J Colloid Interf Sci 184:170–174, 1996.
12. D Bastos, R Hidalgo-Alvarez, FJ de las Nieves. Prog Colloid Polym Sci 100:217–220, 1996.
13. D Bastos, FJ de las Nieves. Colloid Polym Sci 274:1081–1088, 1996.
14. M Antonietti, L Vorwerg. Colloid Polym. Sci 275:883–887, 1997.
15. JE Seeberg, JC Berg. Colloids Surf A 121:89–98, 1997
16. JM Peula, R Hidalgo-Alvarez, FJ de las Nieves. Colloids Surf A 127:19–24, 1997.
17. JM Peula, R Hidalgo-Alvarez, FJ de las Nieves. Langmuir 13:3939–3942, 1997.
18. E Donath, D Walther, VN Shilov, E Knippel, A Budde, K Lowack, CA Helm, H Möhwald. Langmuir 13:5294–5305, 1997.
19. A Fernández-Nieves, F. J. de las Nieves. J Non-Equilib Thermodynam 26:45, 1998.
20. FJ de las Nieves, ES Daniels, MS El-Aasser. Colloids Surf 60:107–126, 1991.
21. HJ Van den Hul, JW Vanderhoff. Br Polym J 2:121, 1970.
22. JW Vanderhoff, HJ Van den Hul, RJM Tausk, JThG Overbeek. In: G Goldfinger. Clean Surfaces: Their Preparation and Characterization for Interfacial Studies. New York: Marcel Dekker, Inc., 1970.
23. JW Goodwin, J Hearn, CC Ho, RH Ottewill. Br Polym J 5:347, 1973.
24. DE Yates, RH Ottewill, JW Goodwin. J Colloid Interf Sci 62:356, 1977.
25. KH Lerche, G Kretzschmar. Mater Sci Forum 25–26:355, 1988.
26. SM Ahmed, MS El-Aasser, GH Pauli, GW Poehlein, JW Vanderhoff. J Colloid Interf Sci 73:388, 1980.
27. Y Chonde, I Krieger. J Colloid Interf Sci 77:138, 1980.
28. KH Van Streum, WJ Belt, P Piet, AL German. Eur Polym J 27:931, 1990.
29. T Palberg, W Härl, U Witting, H Versmold, M Würth, E Simnacher. J Phys Chem 96:8180, 1992.
30. H Tamai, K Niino, T Suzawa. J Colloid Interf Sci 131:1, 1989.
31. WC Wu, MS El-Aasser, FJ Micale, JW Vanderhoff. In: P Becher, MW Judenfreund, eds. Emulsions, Latices and Dispersions. New York: Marcel Dekker, Inc., 1978.

32. SF Schulz, T Gisler, M Borkovec, H Sticher. J Colloid Interf Sci 164:88, 1994.
33. D Bastos, JL Ortega, FJ de las Nieves, R HIdalgo-Alvarez. J Colloid Interf Sci 176:232–239, 1995.
34. AA Kamel, MS El-Aasser, JW Vanderhoff. J Dispers Sci Technol 2:183, 1981.
35. WT McCarvill, RM Fitch. J Colloid Interf Sci 64:403, 1978.
36. TW Healy, LR White. Adv Colloid Interf Sci 9:303, 1978.
37. M Okubo, A Yamada, T Matsumoto. J Polym Sci, Polym Chem Ed 16:3219, 1980.
38. SI Ali, JC Steach, RL Zollars. Colloids Surf 26:1, 1987.
39. BR Vijayendran, T Bone, C Gajiria. J Appl Polym Sci 26:1351, 1981.
40. WM Brouwer. Colloids Surf 40:235, 1989.
41. JA Holgado, A Martín, F Martínez, MA Cabrerizo. Solid/Fluid Interfaces: Capillarity and Wetting, Dynamical Phenomena. Arhem, The Netherlands, 1992.
42. A Martín, MA Cabrerizo, R Hidalgo-Alvarez. Colloids Surf A 125:263, 1996.
43. DH Everett, ME Gültepe, MC Wilkinson. J Colloid Interf Sci 71:336, 1979.
44. ME Labib, AA Robertson. J Colloid Interf Sci 77:151, 1980.
45. WT McCarvill, RM Fitch. J Colloid Interf Sci 67:204, 1978.
46. J Hen. J Colloid Interf Sci 49:425, 1974.
47. R Rymdén. J Colloid Interf Sci 124:396, 1988.
48. K Sakota, T Okaya. J Appl Polym Sci 21:1009, 1977.
49. M Hlavacek, P Chenevière, M Sardin, J Dodds. Colloids Surf A 95:101, 1995.
50. T Gilány. Prog Colloid Polym Sci 93:45, 1993.
51. F Dobler, S Affrossman, Y Holl. Colloids Surf A 89:23, 1994.
52. W Grygiel, M Starzak. Luminiscence 63:47, 1995.
53. JM Moita Neto, VA do Rego Monteiro, F Galembeck. Colloids Surf A: Physicochem Eng Aspects 108:83, 1996.
54. DA Saville. Ann Rev Fluid Mech 9:321, 1977.
55. BH Bijsterbosch, J Lyklema. Adv Colloid Interf Sci 9:147, 1978.
56. SS Dukhin, VN Shilov. Adv Colloid Interf Sci 13:153, 1980.
57. M Fixman. J Chem Phys 72:5177, 1980.
58. RH Hunter. In: RH Ottewill, RL Rowell, eds. Zeta Potential in Colloid Science. New York: Academic Press, 1981.
59. DC Prieve. Adv Colloid Interf Sci 16:321, 1982.
60. A Voigt, E Donath, G Kretzschmar. Colloids Surf 47:23, 1990.
61. SS Dukhin. Adv Colloid Interf Sci 61:17, 1995.
62. JA Moleón, FJ Rubio, FJ de las Nieves, R Hidalgo-Álvarez. J Non-Equilib Thermodynam 16:187, 1991.
63. A Delgado, F González-Caballero, G Pardo. J Non-Equilib Thermodynam 10:251, 1985.
64. JR Goff, P Luner. J Colloid Interf Sci 99:468, 1984.
65. RW O'Brien. J Colloid Interf Sci 171:495, 1995.
66. S Levine, GH Neale. J Colloid Interf Sci 47:520, 1974.
67. CS Mangelsdorf, LR White. J Chem Soc Faraday Trans 86:2859, 1990.
68. CS Mangelsdorf, LR White. J Chem Soc Faraday Trans 88:3567, 1992.
69. RW O'Brien, LR White. J Chem Soc Faraday Trans II 74:1607, 1978.
70. H Ohshima, T Kondo. J Colloid Interf Sci 130:281, 1989.

71. JThG Overbeek, BH Bijsterbosch. In: PG Righetti, CJ van Oss and JW Vanderhoff, eds. Electrokinetic Separation Methods. Amsterdam: Elsevier, 1979.
72. NM Semenikhin, SS Dukhin. Kolloid Z 37:1123, 1975.
73. M Fixman. J Chem Phys 78:1483, 1983.
74. CM Ma, FJ Micale, MS El-Aasser, JW Vanderhoff. In: DR Basset, AE Hamielec, eds. Emulsion Polymer and Emulsion Polymerization. ACS Symposium Series 165, Washington DC: American Chemical Society, 1981.
75. P Stenius, B Kromberg. In: GW Poehlein, RH Ottewill, JW Goodwin, eds. Science and Technology of Polymer Colloids. Boston: Martinus Nijhoff, 1983.
76. R Xu. Langmuir 9:2955, 1993.
77. RH Ottewil, JN Shaw. J Electroanal Chem 37:133, 1972.
78. RH Ottewil, JN Shaw. J Colloid Interf Sci 26:110, 1968.
79. AA Baran, LM Dukhina, NM Soboleva, OS Chechik. Kolloid Z 43:211, 1981.
80. D Bastos, FJ de las Nieves. Colloid Polym Sci 271:860, 1993.
81. RS Chow, K Takamura. J Colloid Interf Sci 125:226, 1988.
82. M Elimelech, CR O'Melia. Colloids Surf 44:165, 1990.
83. R Hidalgo-Alvarez, FJ de las Nieves, AJ Van der Linde, BH Bijsterbosch. Colloid Polym Sci 267:853, 1989.
84. AA Kamel, CM Ma, MS El-Aasser, FJ Micale, JW Vanderhoff. J Dispers Sci Technol 2:315, 1981.
85. JJ Spitzer, CA Midgley, HSG Slooten, KP Lok. Colloids Surf 39:273, 1989.
86. T Okubo. Ber Bunsenges. Phys Chem 91:1064, 1987.
87. D Bastos, FJ de las Nieves. Prog Colloid Polym Sci 93:37, 1993.
88. E Donath, P Kuzmin, A Krabi, A Voigt. Colloid Polym Sci 271:930, 1993.
89. F Galisteo, FJ de las Nieves, M Cabrerizo, R Hidalgo-Alvarez. Prog Colloid Polym Sci 82:313, 1989.
90. A Fernández-Barbero, R Martínez, MA Cabrerizo, R Hidalgo-Alvarez. Colloids Surf A 92:121, 1994.
91. WM Brouwer, RLJ Zsom. Colloids Surf 24:195, 1987.
92. H Higashitani, H Iseri, K Okuhara, A Kage, S Hatade. J Colloid Polym Sci 172:383, 1995.
93. BR Midmore, RJ Hunter. J Colloid Interf Sci 122:521, 1988.
94. H Tamai, H Hasegawa, T Suzawa. J Appl Polym Sci 38:403, 1989.
95. RT Klingbiel, H Coll, RO James, J Texter. Colloids Surf 68:103, 1992.
96. IH Harding, TW Healy. J Colloid Interf Sci 107:382, 1985.
97. M Deggelmann, T Palberg, M Hagenbüchle, EE Maier, R Krause, C Graf, R Weber. J Colloid Interf Sci 143:318, 1991.
98. M Dittgen, B Zosel. Colloid Polym Sci 269:259, 1991.
99. K Ito, N Ise, T Okubo. J Chem Phys 82:5732, 1985.
100. K Makino, S Yamamoto, K Fujimoto, H Kawaguchi, H Ohshima. J Colloid Interf Sci 166:251, 1994.
101. A Fernández-Barbero, A Martín, J Callejas, R Hidalgo-Álvarez. J Colloid Interf Sci 162:257, 1994.
102. E Donath, P Kuzmin, A Krabi, A Voigt. Colloid Polym Sci 271:930, 1993.
103. H Ohshima. J Colloid Interf Sci 168:269, 1994.
104. H Ohshima. Adv Colloid Interf Sci 62:189, 1995.

105. SS Dukhin. Adv Colloid Interf Sci 36:219, 1991.
106. D Bastos, R Santos, J Forcada, FJ de las Nieves, R Hidalgo-Álvarez. Colloids Surf A 92:137, 1994.
107. BJ Marlow, RL Rowell. Langmuir 7:2970, 1991.
108. JH Prescott, SJ Shiau, RL Rowell. Langmuir 9:2071, 1995.
109. A Martín, MA Cabrerizo, R Hidalgo-Alvarez. Anales Fis 91:100, 1995.
110. SL Tsaur, RM Fitch. J Colloid Interf Sci 115:450, 1987.
111. SL Tsaur, RM Fitch. J Colloid Interf Sci 115:463, 1987.
112. HJ Van den Hul, JW Vanderhoff. J Electroanal Chem 37:161, 1972.
113. R Hidalgo-Alvarez, FJ de las Nieves, AJ Van der Linde, BH Bijsterbosch. Colloids Surf 21:259, 1986.
114. JThG Overbeek. Adv Colloid Interf Sci 3:97, 1950.
115. F Booth. Proc Roy Soc London Ser A 203:514, 1950.
116. PH Wiersema, AL Loeb, JThG Overbeek. J Colloid Interf Sci 22:78, 1966.
117. AS Russell, PJ Scales, Ch Mangeldorf, SM Underwood. Langmuir 11:1112, 1995.
118. SM Kontush, SS Dukhin, OI Vidov. Colloid J 56:579, 1994.
119. AG van der Put, BH Bijsterbosch. J Colloid Interf Sci 92:499, 1983.
120. ThJJ van den Hoven, BH Bijsterbosch. J Colloid Interf Sci 115:559, 1987.
121. ThJJ van den Hoven, BH Bijsterbosch. Colloids Surf 22:187, 1987.
122. R Hidalgo-Álvarez, JA Moleón, FJ de las Nieves, BH Bijsterbosch. J Colloid Interf Sci 149:23, 1992.
123. FJ Rubio. J Non-Equilib Thermodynam 17:333, 1992.
124. R de Backer, Q Watillon. Berichte VI. International Kongress grenzflächenakt. München: Stoffe, Carl Hanser Verlag, 1972, p 651.
125. FJ Rubio, R Hidalgo-Álvarez, FJ de las Nieves, BH Bijsterbosch, AJ van der Linde. Prog Colloid Polym Sci 93:341, 1993.
126. RW O'Brien. J Colloid Interf Sci 110:477, 1986.
127. A Vernhet, MN Bellon-Fontaine, A Doren. J Chim Phys 91:1728, 1994.
128. SS Dukhin, VN Shilov. Dielectric Phenomena and the Double Layer in Disperse Systems and Polyelectrolytes. New York: Wiley, 1974.
129. EHB Delacey, LR White. J Chem Soc Faraday Trans II 77:2007, 1981.
130. MM Springer, A Korteweg, J Lyklema. J Electroanal Chem 153:55, 1983.
131. BR Midmore, RJ Hunter, RW O'Brien. J Colloid Interf Sci 120:210, 1987.
132. J Kijlstra, HP van Leeuwen, J Lyklema. J Chem Soc Faraday Trans 88:3441, 1992.
133. IA Razilov, G Pendze, SS Dukhin. Colloid J 56:612, 1994.
134. DE Dunstan, LR White. J Colloid Interf Sci 152:308, 1992.
135. LA Rosen, JC Baygents, DA Saville. J Chem Phys 98:4283, 1993.
136. R Barchini, DA Saville. J Colloid Interf Sci 173:86, 1995.
137. F Carrique, L Zurita, AV Delgado. Colloids Surf A 92:9, 1994.
138. DA Saville. Colloids Surf A 92:29, 1994.
139. JC Baygents. Colloids Surf A 92:67, 1994.
140. AS Russell, PJ Scales, CS Mangelsdorf, LR White. Langmuir 11:1553, 1995.
141. AG van der Put, BH Bijsterbosch. J Colloid Interf Sci 75:512, 1980.
142. LA Rosen, DA Saville. J Colloid Interf Sci 140:82, 1990.
143. BR Midmore, RW O'Brien. J Colloid Interf Sci 123:486, 1988.
144. BR Midmore, D Diggins, RJ Hunter. J Colloid Interf Sci 129:153, 1989.

145. DE Dunstan, LR White. J Colloid Interf Sci 152:297, 1992.
146. DE Dunstan. J Chem Soc Faraday Trans 89:521, 1993.
147. RW O'Brien. J Fluid Mech 190:71, 1988.
148. RW O'Brien, BR Midmore, A Lamb, RJ Hunter. Faraday Discuss Chem Soc 90:1, 1990.
149. RO James, J Texter, PJ Scales. Langmuir 7:1993, 1991.
150. M Loewenberg, RW O'Brien. J Colloid Inter Sci 150:158, 1992.
151. VE Shubin, RJ Hunter, RW O'Brien. J Colloid Interf Sci 159:174, 1993.
152. RW O'Brien, DW Cannon, WN Rowlands. J Colloid Interf Sci 173:406, 1995.
153. FN Desai, HR Hammad, KF Hayes. Langmuir 9:2888, 1993.
154. PJ Debye. J Chem Phys 1:13, 1933.
155. F Booth. Proc Roy Soc London A 203:533, 1950.
156. Stone-Masui, A Watillon. J Colloid Interf Sci 28:187, 1968; 34:327, 1970.
157. IG Watterson, LR White. J Chem Soc Faraday II 77:1115, 1981.
158. RW McDonogh, RJ Hunter. J Rheol 27:189, 1983.
159. SA Ali, M Sengupta. J Colloid Interf Sci 113:172, 1986.
160. A Delgado, F González-Caballero, MA Cabrerizo, I Alados. Acta Polym 38:66, 1987.
161. A Chabalgoity, A Martín, F Gallisteo, R Hidalgo-Álvarez. Prog Colloid Polym Sci 84:416, 1991.
162. BR Paulke, PM Möglich, E Knippel, A Budde, R Nitzsche, RH Müller. Langmuir 11:70, 1995.
163. CF Zukoski, DA Saville. J Colloid Interf Sci 107:322, 1985.
164. CF Zukoski, DA Saville. J Colloid Interf Sci 114:32, 1986.
165. CF Zukoski, DA Saville. J Colloid Interf Sci 114:45, 1986.
166. CF Zukoski, DA Saville. J Colloid Interf Sci 132:320, 1989.
167. A Delgado, F González-Caballero, J Salcedo. Acta Polym 37:361, 1987.
168. MC Davis, RAP Lynn, J Hearn, AJ Paul, JC Vickerman, JF Watts, Langmuir 12:3866, 1996.
169. D Velegol, JL Anderson, S Garoff. Langmuir 12:675, 1996.
170. D Velegol, JL Anderson, S Garoff. Langmuir 12:4103, 1996.
171. H Ohshima, Colloids Surf A: Physicochem Eng Aspects 103:249, 1995.
172. K Makino, K Suzuki, Y Sakurai, T Okano, H Ohshima, Colloids Surf: Physicochem Eng 103: 221, 1995.
173. H Ohshima, Colloid Polym Sci 275:480, 1997.
174. PJ Wit, J Noordmans, HJ Busscher, Colloids Surf: Physicochem Eng 125:85, 1997.
175. S Levine, JR Marriott, G Neale, GN Neale. J Colloid Interface Sci 52:136, 1975.
176. A Van der Linde, BH Bijsterbosch. Colloids Surf 41:345, 1989.
177. RW O'Brien. J Colloid Interf Sci 113:81, 1986.
178. J Lyklema, SS Dukhin, VN Shilov. J Electroanal Chem 143:1, 1983.
179. C Ballario, A Bonincontro, C Cametti. J Colloid Interf Sci 54:415, 1976; 72:304, 1979.
180. LKH van Beek. Prog Dielectr 7:69, 1967.
181. G Schwarz. J Phys Chem 66:2636, 1962.
182. JM Schurr. J Phys Chem 68:2407, 1964.
183. HP Schwan, G Schwarz, J Mackurk, H Pauly. J Phys Chem 66:2626, 1962.

184. J Lyklema, MM Springer, VN Shilov, SS Dukhin. J Electroanal Chem 198:19, 1986.
185. LA Rosen, DA Saville. J Colloid Interf Sci 149:542, 1992.
186. RW O'Brien. Adv Colloid Interf Sci 16:281, 1982.
187. PA Saville. J Colloid Interf Sci 91:34, 1983.
188. LP Voegtli, CF Zukoski. J Colloid Interf Sci 141:92, 1991.
189. RW O'Brien. J Colloid Interf Sci 81:234, 1981.
190. A Watillon, J Stone-Masui. J Electroanal Chem 37:143, 1972.
191. DA Saville. J Colloid Interf Sci 91:34, 1983.
192. MR Gittings, DA Saville Langmuir 11:798, 1995.
193. RW O'Brien, WT Perrins. J Colloid Interf Sci 99:10, 1984.
194. BR Midmore, RW O'Brien. J Colloid Interf Sci 123:486, 1988.
195. BR Midmore, D Diggins, RJ Hunter. J Colloid Interf Sci 129:153, 1989.
196. T Okubo. J Colloid Interf Sci 125:380, 1988.
197. Y Chonde, M Shabrang. J Colloid Interf Sci 186:248, 1997.
198. A Rutgers. Physica 5:13, 1938.
199. J Hermans. Phil Mag 25:426, 1938.
200. J Enderby. Proc Royal Soc (London) A 207:329, 1951.
201. AJ Babchin, RS Chow, RP Sawatzky. Adv Colloid Interf Sci 30:111, 1989.
202. FN Desai, HR Hammad, KF Haynes. Langmuir 9:2888, 1993.
203. E Yeager, J Bugosh, F Hovorka, J McCarthy. J Chem Phys 17:411, 1949.
204. E Yeager, H Dietrick, F Hovorka. J Acoust Soc Am 25:456, 1953.
205. A Rutgers, J Vidts. Nature 165:109, 1950.
206. U Beck, R Zana, E Rohloff. Tappi 61:63, 1978.
207. RO James, J Texter, PJ Scales. Langmuir 7:1993, 1991.
208. RW O'Brien, DW Cannon, WN Rowlands. J Colloid Interf Sci 173:406, 1995.
209. M James, RH Hunter, RW O'Brien. Langmuir 8:420, 1992.
210. AS Duhkin, PhJ Goetz, ChW Hamlet. Langmuir 12:4998, 1996.
211. RJ Hunter, RW O'Brien. Colloids Surf A: Physicochem Eng Aspects 126:123, 1997.
212. FJ Rubio-Hernández. J Non-Equilib Thermodynam 18:195, 1993.
213. ThF Tadros. Colloids Surf 18:137, 1986.
214. M von Smoluchowski. Kolloidzeit 18:190, 1916.
215. W Krasny-Ergen. Kolloidzeit 74:172, 1936.
216. EJ Hinch, JD Sherwood. J Fluid Mech 132:337, 1983.
217. J Yamanaka, H Matsuoka, H Kitano, N Ise. J Colloid Interf Sci 134:92, 1990.
218. EP Honing, WFJ Pünt, PHG Offermans. J Colloid Interf Sci 134:169, 1990.
219. MJ Garcia-Salinas, FJ de las Nieves. Prog Colloid Polym Sci 110:134, 1998.
220. O Quadrat, L Markvickova, J Snuparek. J Colloid Interf Sci 123:353, 1987.
221. FS Chan, J Blanchford, D Goring. J Colloid Interf Sci 22:378, 1966.
222. JE Seebergh, JC Berg. Colloids Surf A 100:139, 1995.
223. R Zimhel, G Lagaly. Colloids Surf 22:225, 1987.
224. JWS Goossens, A Zembrod. Colloid Polym Sci 257:437, 1979.
225. JH Prescott, S Shiau, RL Rowell. Langmuir 9:2071, 1993.
226. S Sasaki. Colloid Polym Sci 262:406, 1984.
227. JWS Goossens, A Zembrod. J Dispers Sci Tech 2:255, 1981.
228. JE Seebergh, JC Berg. Colloids Surf: Physicochem Eng 121:89, 1997.
229. B Vincent, Adv. Colloid Interf Sci 42:279, 1992.

230. BM Verdegan, MA Anderson. J Colloid Interf Sci 158:372, 1993.
231. HJ van den Hul. J Colloid Interf Sci 92:217, 1983.
232. J Zhao, W Brown, J. Colloid Interf Sci., 179:255, 1996.
233. DH Napper. Polymeric Stabilization of Colloidal Dispersion. New York: Academic Press, 1983.
234. GJ Fleer, JMHM Scheutjens, Colloids Surf 51:281, 1990.
235. B Vincent, PF Luckham, FA Waite, J. Colloid Interf Sci., 73:508, 1980.
236. C Cowell B Vincent, J Chem Soc Faraday Trans I 74:337, 1978.
237. B Vincent, J Edwards, S Emmett, R Croot, Colloids Surf 31:267, 1988.
238. GJ Fleer, JMHM Scheutjens, MA Cohen Stuart, Colloids Surf 31:1, 1988.
239. A Krabi, E Donath, Colloid Surf A: Physicochem Eng. 92:175, 1994.
240. E Donath, L Pratsch, H Bäumler, A Voigt, M Taeger, Stud Byophys 130:117, 1989.

6

Interaction of Polymer Latices with Other Inorganic Colloids

Kunio Furusawa
University of Tsukuba, Ibaraki, Japan

I. INTRODUCTION

The exceptional uniform shape and wide diameter range of latex particles are of interest because their combination with other inorganic (or metallic) colloid particles should extend their application in the field of academic research as well as industrial material development. This is especially true of the synthesis of composite particles comprised of organic and inorganic substances, in which the adoption of monodispersed polymer latices as their one component largely expands the variety of composite particles because many techniques to prepare latex samples with different sizes and structures have already been developed; see Ref 1 for example.

Core–shell and composite-type colloid particles have been widely investigated recently, as these could find important applications in practical fields such as biology, medicine, phamacology, electronics, and so forth [2]. Interest is also shown in the fabrication and characterization of porous particles and aggregates [3].

These composite particles are usually obtained by attaching a coat of small particles onto the surface of a larger one. Such composites could be obtained by heterocoagulation when the core and the shell particles are of opposite surface charge [4,5]. Other techniques for the formation of composites include polymerization of organic monomer in presence of inorganic particles [6] and dry mixing [7] or embedding [8] of two-component particles. The composites usually incorporate a combination of metallic/organic [5], oxide/organic [9], or two types of organic particle. In some publications, the successful combination of heterocoagulation, polymerization, and encapsulation is reported [10].

In the synthesis of organic–inorganic composite particles, the establishment of a method to control the orientation of each element should be regarded as an important factor, because the characteristics of the composite particles largely depend on the orientation of the component particles. Simultaneously, because the function of the composite particles is frequently determined by their size and uniformity, the development of a method to fix the particle diameter in a narrow size distribution is also indispensable. In practice, for example, when some magnetic composite particles are applied for the separation of T cells from B cells, it can be considered that the further out the magnetic particles are in the composite particles, the more efficiently the separation proceeds. Moreover, when immuno latices, on the surface of which some antibodies are placed, are applied for diagnosis of pathological symptoms, their shape uniformity and particle size dramatically effect the accuracy [11].

Aiming at the systematic resolution of these problems, the synthesis method of magnetic composite particles was investigated, adopting the concept of heterocoagulation theory; heterocoagulates were prepared from magnetic ultrafine powder and monodispersed negative latices with different particle diameters. In addition, the optimum condition for encapsulating the heterocoagulates by a polystyrene layer via emulsion polymerization was surveyed [10]. The size of multilayer composite particles can be controlled by the diameter of the core latices which are applied primarily. This method realizes the localization of the magnetic particles on the outside of the composite particles covered with a thin polystyrene layer; therefore, they can give full function to the surrounding system.

Finally, the technique and the data on the assembly of ordered, empty spherical aggregates from latex particles are presented [12]. This technique is based on using emulsion droplets as templates, onto whose surfaces the latex particles are gathered, assembled into ordered structures, and then fixed together. The particles are then extracted from the emulsion droplets by dissolving the droplets in the surrounding environment. The different possibilities of the emulsion-based assembly scheme are also discussed [13]. On this point, we describe the data on using the emulsion-based technique for the formation of ball-like latex aggregates in the ball of the emulsion droplets. These aggregates are formed in the bulk of the emulsion droplets. These ball-like aggregates are used further as a core for the fabrication of composite particles.

II. TYPICAL EXAMPLES OF THE SYNTHESIS OF COMPOSITE PARTICLES

A. Synthesis of Composite Particles By Using Heterocoagulation Technique

In recent articles, Matijevic and other workers [14,15] have investigated various behaviors of mixed colloid particles obtained by mixing differently preformed particles or precipitating one kind of solid in the presence of already existing particles. In the former, they showed that the nature of a mixing system depends on the conditions of preparation and concluded that the important parameters to control are the concentration of the particles, the relative numbers of the component particles, and their contrast to their relative surface charges. In some cases, pH, temperature, and the nature of counterions will affect the products in the most sensitive way.

Taking this into consideration, we report here on the syntheses and properties of heterocoagulate generated from mixtures of amphoteric polymer latices with a constant particle size ($2a$ = 250 nm) and monodispersed spherical silica with a wide range of particle diameters (240–1590 nm) [9,16]. Because this combination is a typical standard system and includes a wide applicability to other systems. The characterization data for these single dispersions are shown in Table 1. In this combination of components, the contrast between surface charge and the particle size ratio of the components can be changed arbitrarily within their wide range of particle size. Furthermore, the heterocoagulates produced are composed of organic and inorganic materials, whose surface charge in both sign and magnitude can be controlled by changing the pH of the medium. We can expect that these new characteristics will serve to extend its usefulness in a wide variety of ways as a functional fine particle.

1. Parameters in Controlling the Morphology of the Heterocoagulate

It was confirmed that the most important parameter in controlling the morphology of composite particles is the surface charge of the component particles, especially the contrast between the surface charges of the two component particles. A stable system consisting of a regular composite particle could be prepared only in a medium controlled at pH 4–6, where the two components were charged with

Table 1 Particle Diameters of Spherical Silica and Latex Samples

Sample	Diameter (nm)	D_w/D_n[a]	Particle size ratio (silica/latex)
Silica-L	1590	1.03	6.36
Silica-M	960	1.01	3.84
Silica-S	460	1.03	1.84
Silica-SS	240	1.04	0.96
Latex	250	1.02	—

[a]Polydispersity of the latex particle, where D_w and D_n are the weight-average and number-average particle diameters.

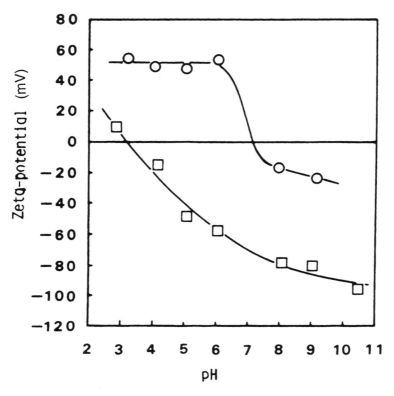

Figure 1   Zeta-potentials of silica (□) and latex particles (○) as a function of pH at 5×10^{-3} mol/dm^3 KCl.

opposite signs, as seen in Fig. 1. In Fig. 1, ζ-potentials of silica and latex particles employed here are plotted as a function of medium pH at 25°C.

The next important parameter to control is the particle number ratio (PNR) of the component particles when they are mixed in the vessel. To investigate this parameter, the stability of the mixed systems prepared under the various PNRs was examined under a constant ionic medium (pH 5.2, 1×10^{-5} mol/dm^3 KCl), where the numbers of silica particles were kept constant at 30 mg/mL. Figure 2 shows some typical transmittances of the systems versus elapsed time curves obtained from the mixtures of different PNRs. Figure 3 shows the relation between the transmittance of the system and its PNR at 1 h and 24 h after mixing the two single components. Under the systems with relatively low numbers of latex particles (PNR = 110/1–120/1), the systems coagulated irregularly and settled quickly because of their large mass, as indicated by the high transmittance values. With increasing particle numbers of latices (PNR = 150/1–500/1), however, the mixed

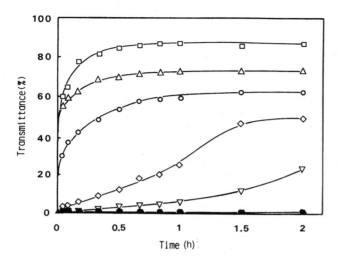

Figure 2 Relationship between transmittance and elapsed time at the various PNRs: (□) 20/1, (△) 80/1; (○) 100/1; (◇) 120/1; (▽) 140/1; (●) 0/1; (▲) 160/1–500/1.

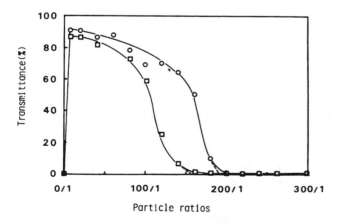

Figure 3 Transmittance versus particle number ratio curves at 1 h (□) and 24 h (○) after mixing two-component particles.

systems became stable and exhibited a transmittance comparable with those of the initial single components. The transition in the curves depicting the unstable and stable binary system is rather abrupt. The critical PNR is obtained by extrapolating the steep portion of the transmittance versus PNR curves to the abscissa. The critical value obtained from this analysis enabled calculation of the number of latex particles adsorbed per silica particle. The numbers determined from the two series are 160 and 195, respectively. The average of these two numbers agrees well with the value of 180, calculated by following the method of Harding and other workers [4,14].

Figure 4 shows optical micrographs of the heterocoagulates generated under conditions of lower PNR (Fig. 4A) and of higher PNR (Fig. 4B) relative to the critical value. It may be seen that at a PNR higher than the critical value, the suspension is comprised of uniform composite particles with white edges, which are composed of many adsorbed latex particles, and each composite particle is still undergoing Brownian motion as an isolated unit. Figure 5 shows a scanning electron micrograph of one composite particle prepared at PNR = 300/1, which defines, in detail, the structure of the composite particle. It is apparent that the composite takes a raspberry shape with one silica in the core. In contrast to this, the heterocoagulates generated under the conditions of PNR lower than the critical value are composed of large, irregular aggregates, and regular coagulates were rarely formed at any medium pH and particle number ratio investigated.

It is interesting to analyze the different heterocoagulation behaviors from the concept of the adhesion isotherm for the amphoteric latices on the different silica samples. Figure 6 shows the adhesion isotherms for the amphoteric latices on the four kinds of silica sample with different particle sizes which were obtained under the same medium conditions (pH 5.2). The amount adhered, A, is expressed in the form of fraction coverage, ϑ, as given by $\vartheta = A/A_s$, where A_s is the maximum number of latex particles calculated by Bleier and Matijevic [14]. A remarkable feature is the dependence of the ϑ value on the particle size of the silica sample

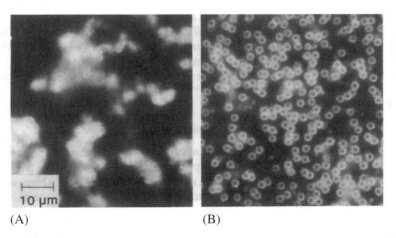

(A)   (B)

Figure 4   Optical micrographs showing the heterocoagulates prepared at different PNRs: (A) 100/1, (B) 400/1.

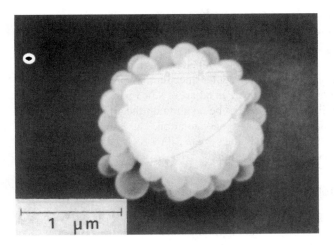

Figure 5 Scanning electron micrograph of one composite particle prepared at PNR = 300/1.

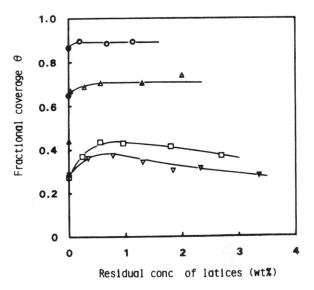

Figure 6 Adsorption isotherms of amphoteric latices onto silica particles with different sizes: (○) $2a$ = 1590 nm, (△) $2a$ = 960 nm; (□) $2a$ = 460 nm; (▽) $2a$ = 240 nm.

(i.e., the ϑ value decreases with decreasing particle size), and roundish isotherms appeared in the systems Silica-S + latices and Silica-SS + latices. This behavior may be explained as being due to the coagulation state of the component particles, as shown in Fig. 7, where the morphologies of heterocoagulate are indicated by the schematic picture (i.e., in the state of irregular flocculation, one latex particle may be adhering simultaneously to two or three adjacent silica particles by interparticle bridging) and the limited area able to be adhered on the silica surface is thus saturated more easily than in regular heterocoagulation. Therefore, a small ϑ value and a marked isotherm are observed in the system Silica-SS + latices, especially in the high latex particle concentration region.

2. Effects of Electrolyte and Water-Soluble Polymer

Figure 8 shows some typical isotherms for the latices on Silica-L at various K_2SO_4 concentrations, where all the systems were controlled at pH 5.2. It is evident that the isotherms are all of the very high-affinity type with a well-defined plateau and the plateau value increases with increasing K_2SO_4 concentration within the K_2SO_4 concentration range from 10^{-5} to 10^{-2} mol/dm^3 with a well-defined plateau. This means that in this concentration range, adhesion proceeds in the manner characteristic of monolayer adhesion. This may be due to the strong blocking effect of adhering particles. However, in K_2SO_4 aqueous solutions more concentrated than 2×10^{-2} mol/dm^3, no reproducible isotherm could be obtained under any condi-

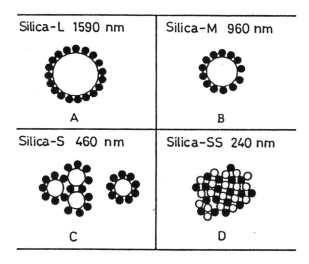

Figure 7 Schematics showing the morphology of heterocoagulate particles formed from different silica samples.

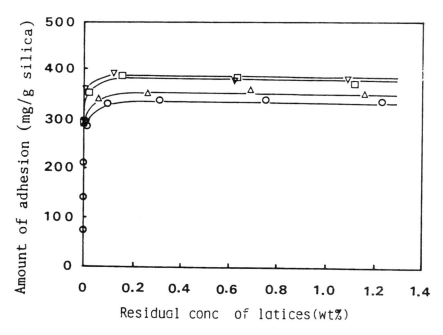

Figure 8 Adhesion isotherms of amphoteric latices onto silica at various K_2SO_4 concentrations: ($\triangledown$) $1.46 \times 10^{-2} M$; ($\square$) $1.46 \times 10^{-3} M$; ($\triangle$) $1.46 \times 10^{-4} M$; ($\bigcirc$) $0 M$.

tions tested, and only some irregular aggregates were generated in the course of the experiment.

In Fig. 9, the ζ-potentials of the heterocoagulates prepared at the different electrolyte concentrations are presented as a function of the medium pH, as well as the data on the amphoteric latices. As may be seen, a reversal of charge is observed in all the samples, and the isoelectric point (IEP) in the heterocoagulated systems occurs at about pH 8, which is not so different from the IEP of the single latices. Moreover, the fact that the limiting, net positive ζ-potential attained at pH 6–3 increases with increasing electrolyte concentration is also in line with the increase in latex adhesion with increasing electrolyte concentration.

The effect of a water-soluble polymer has been studied using hydroxyl propyl cellulose (HPC), which has a lower critical solution temperature (LCST) at about 45°C. The molecular structure and the data on HPC are shown in Table 2. First, HPC adsorption on the single latex and silica particles has been examined at 45°C. As may be seen in Fig. 10, HPC-M adsorbed fairly well on the silica surface and showed a maximum adsorption at 0.5×10^{-1}wt% residual concentration of HPC. However, no adsorption of HPC can be detected on the amphoteric latex sur-

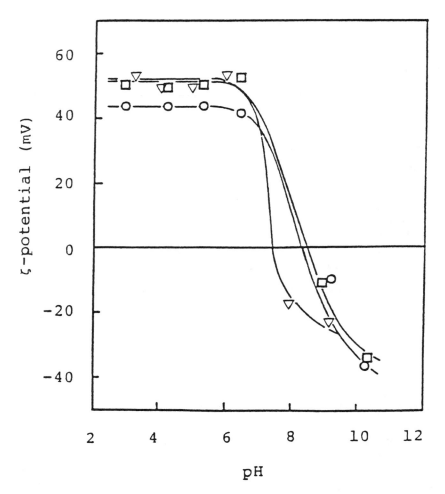

Figure 9 Zeta-potentials versus pH curves for heterocoagulate particles prepared at different electrolytes: (∇) heterocoagulates prepared at 2.48×10^{-1} M MgCl$_2$; ($\bigcirc$) heterocoagulates prepared at 1×10^{-5} M KCl; ($\square$) amphoteric latices.

faces. So, hereafter, adhesion behavior for latex particles on the silica surfaces covered by HPC layers will be examined. Figure 11 shows adhesion isotherms of amphoteric lattices on Silica-L in a solution containing 0.6 mg/mL HPC with different molecular weights. It is apparent that the isotherm obtained in the HPC-L aqueous solutions is of the high-affinity type and the adsorption layer of HPC-L has no effect on the latex adhesion. In contrast to this, the isotherms obtained in the

Table 2 Molecular-Weight Characterization of HPC Polymer

Sample	[η]	$M_w{}^a$	$M_w/M_n{}^b$	Degree of ether substitution per monomer
HPC-L	1.03	5.3×10^4	2.76	2.4
HPC-M	3.06	30.3×10^4	1.84	2.4
HPC-H	6.16	92.5×10^4	2.49	2.4

[a]M_w was calculated from [η].
[b]M_w/M_n was determined from the gel-permeation chromatography elution curve.

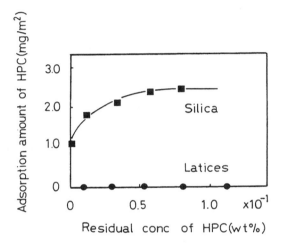

Figure 10 Adsorption isotherms of HPC-M onto silica and latex particles.

HPC-M and HPC-H aqueous solutions show curved shapes with low maximum adhesions; the effect of an HPC-H layer is especially remarkable.

All these results indicate that these additives (electrolyte and polymer) are useful in controlling the composition of organic and inorganic compounds in the regular composite particles prepared by the heterocoagulation technique.

3. Removal Behavior of Adsorbed Particles

To check the structural stability of the heterocoagulates, the removal behavior of the latex particles from the silica surface was investigated. In Fig. 12, the adhered amounts of latex particles left on the silica after washing with pure water of various pH values are compared with the initial value of adhesion as a function of the

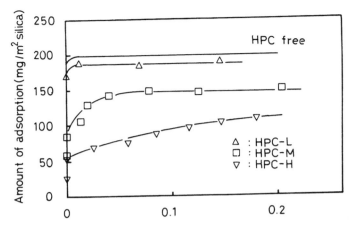

Figure 11   Adsorption isotherms of latex particles onto silica-L; effect of molecular weight of HPC (HPC concentration 0.6 mg/mL).

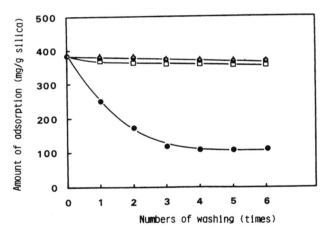

Figure 12   Effect of pH of washing medium on the desorption efficiency: (○) pH = 5.2; (△) pH = 9.1; (□) pH = 10.4; (●) pH = 11.8

number of washing times, where the initial equilibration was performed in a medium held at pH 5.2. As may be seen, repeated washing has no significant effect on the amounts left on the silica, except at pH 11.8; that is, it is clear that under usual washing conditions, the adhesion of the present system behaves as an irreversible process.

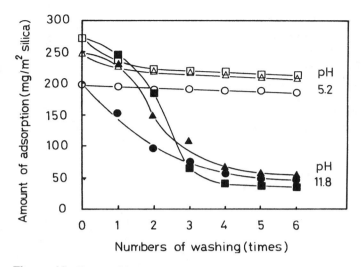

Figure 13 Removal behavior of amphoteric latices from the heterocoagulate particles prepared in concentrated salt solutions, where two different washing media were used: ($\square$, $\blacksquare$) 2.48×10^{-1} M MgCl$_2$; ($\triangle$, $\blacktriangle$) 7.24×10^{-2} M K$_2$SO$_4$; ($\bigcirc$, $\bullet$), 1×10^{-5} M KCl.

Figure 13 shows the removal behavior for the heterocoagulates prepared in concentrated salt solutions, where the removal efficiency is compared for two different washing media (pH 5.2 and pH 11.8). The behavior observed is similar to the results seen in Fig. 12, but it is important to notice that the residual amount on the silica, after washing six times with pure water at pH 5.2, increases with increasing electrolyte concentration.

4. Analyses of Heterocoagulation and Desorption Behavior Using Heterocoagulation Theory

The heterocoagulation between amphoteric latices and silica particles will proceed at the same time as the homocoagulation between latex particles and silica particles themselves. The interaction potential energy between these identical particles (two equal spheres of radius a) is given by

$$V = \frac{8R^2T^2\gamma^2\varepsilon a}{Z^2F^2} \exp(-2\kappa h_0) - \frac{A}{6}\left[\frac{a^2}{2h_0(h_0 + 2a)} + \frac{a^2}{2(h_0 + a)^2} + \ln\left(\frac{h_0h_0 + 2a)}{(h_0 + a)^2}\right)\right] \tag{1}$$

where

$$\kappa = \sqrt{8\pi n^0 Z^2 e^2/\varepsilon kT}, \qquad \gamma = \frac{\exp(Ze\psi/2kT) - 1}{\exp(Ze\psi/2kT) + 1}$$

In the above equations, R is the gas constant, $2h_0$ is the shortest distance of separation between two particles, ε is the dielectric constant of the medium (78.5 for water at 20°C), T is the absolute temperature of the suspension, Z is the valency of the ion, n^0 is the ionic density, e is the elementary charge, ψ is the surface potential of the particle (here, the ζ-potential measured at the appropriate salt concentration), and A is the Hamaker constant, which is reported as 0.3×10^{-20} J for silica in water and 1×10^{-20} J for latices in water [15].

The potential energy of interaction between two dissimilar spherical particles is given [16] as

$$V = \frac{\varepsilon a_1 a_2 (\psi_1^2 + \psi_2^2)}{4(a_1 + a_2)} \left[\frac{2\psi_1 \psi_2}{\psi_1^2 + \psi_2^2} \ln\left(\frac{1 + \exp(-2\kappa h_0)}{1 - \exp(-2\kappa h_0)} \right) + \ln\left[1 - \exp(-4\kappa h_0) \right] \right] - \frac{A_{12/3} a_1 a_2}{6 h_0 (a_1 + a_2)}$$

(2)

where a_1 and a_2, and ψ_1 and ψ_2 are radii and surface potentials, respectively, of the respective dissimilar particles, and for ψ_1 and ψ_2, the ζ-potentials obtained in the same salt solutions were used. $A_{12/3}$ in Eq. (2) has been calculated from the following equation, using $A_{33} = 4.83 \times 10^{-20}$ J for water in vacuum, $A_{11} = 14.8 \times 10^{-20}$ J for silica in vacuum, and $A_{22} = 11.5 \times 10^{-20}$ J for latices in vacuum:

$$A_{12/3} = (A_{11} - A_{33})(A_{22} - A_{33})$$

(3)

In Figs. 14 and 15, the total interaction potential curves between silica–silica, latex–latex, and silica–latex particles in 1×10^{-3} mol/dm³ and 1×10^{-2} mol/dm³ K_2SO_4 aqueous solutions are shown respectively. From Fig. 13, it is clear that in 1×10^{-3} mol/dm³ K_2SO_4 solutions, strong attraction energy between latex and silica particles appears and heterocoagulation will occur predominantly in comparison with the other two homocoagulations. In this salt concentration, the interaction potential between latex–latex particles shows a high enough barrier to prevent mutual coagulation. However, in 1×10^{-2} mol/dm³ K_2SO_4 solution, the height of the potential barrier between latex–latex particles becomes about 10 kT which is the lower limit needed to keep the colloid stable. This analysis indicates that the fractional coverage of latex particles on the large silica surface increases with increasing K_2SO_4 concentration, but the concentration of 1×10^{-2} mol/dm³ K_2SO_4 will be an upper limit to make a stable heterocoagulation, because above 1×10^{-2} mol/dm³ K_2SO_4, some aggregates of the latex particles will be generated in the course of the experiments. It is shown in Fig. 7 that the same behavior could be observed in the present system.

Figure 16 shows the total interaction potential curves between silica and latex particles at the various pH values in 5×10^{-3} mol/dm³ KCl aqueous solutions. These curves are useful for analyzing the removal behavior shown in Figs. 12 and

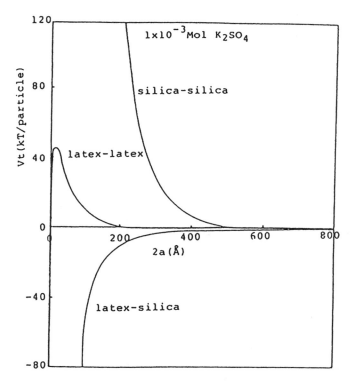

Figure 14 The total interaction potential curves of silica–silica, latex–latex, and latex–silica particles in 1×10^{-3} M K_2SO_4.

13. According to the usual consideration of the interparticle potential curve, heterocoagulation of silica–latex particles hardly exists at pH 9 and 10, because high potential barriers exist between the particles. However, it is considered that in the removal process of the adhering particles, the particles overcome the potential barrier from the inside. So, the existence of potential barriers tends to make the latex particles adhere on the large silica particles. However, a further increase of the medium pH causes removal of adhering latices. This different removal behavior of the latex particles from the silica surfaces can be explained by the different interaction forces between latex and silica surfaces. It is expected that the interaction force between these two kinds of particles changes from lateral electrostatic attraction to repulsion in the two media of high pH. Hydration forces [17] on the surface must be invoked to explain why the system becomes redispersed or why desorption occurs at the high pH values.

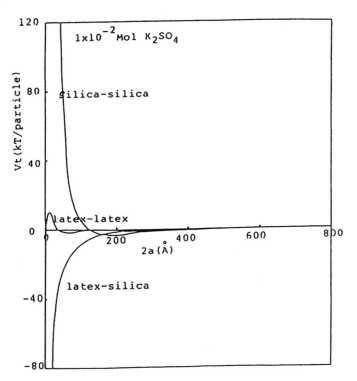

Figure 15 The total interaction potential curves of silica–silica, latex–latex, and latex–silica particles in $1 \times 10^{-2} M \, K_2SO_4$.

B.   Preparation of Composite Particles by Polymerization of Monomer with Inorganic Particles

Polymer colloids, more commonly known as latices, are important in the manufacture of synthetic elastomers, commodity polymers, surface coatings, adhesives, and numerous speciality products. In addition to these ordinary uses, recently some special needs for the latices have been gradually increasing in fields such as biology, medicine, phamacology, electronics, and so forth. They are, for example, carriers in antibody–enzyme diagnostic tests [11,12], conductive paints, plastisols for display devices [18], and so forth. For these applications, it is necessary that the latices have a special quality to meet their own application requirement.

Recently, it has been reported that the adsorption behavior of some water-soluble polymers (e.g., HPC) having a LCST depends significantly on the adsorption temperature. The maximum adsorption (A_s) of HPC at the LCST was 1.5 times as large as the adsorbance at room temperature [19]. Furthermore, the high

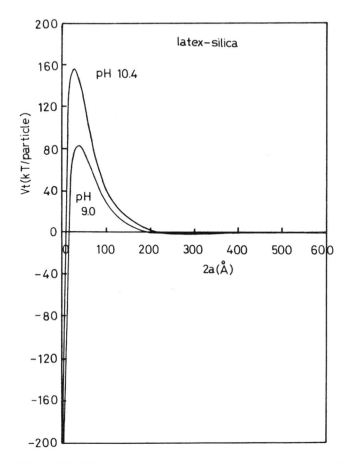

Figure 16 The total interaction potential curves between latex–silica particles at the various pH values in 5×10^{-3} KCl.

A_s value obtained at the LCST has been maintained for a long period at room temperature and the dense (or thick) adsorbed layer of HPC formed on the fine particles shows a strong protective action against flocculation of the particles [20].

In this subsection, we report that the dense (or thick) adsorbed layer of HPC formed at the LCST on the silica particles acts as a binder between the silica particles, and the growing polymer molecules in the emulsion polymerization of styrene in the silica particles act as seed. The dense HPC layer contributes to the synthesis of a new composite polymer latex with the silica particles in the latex particle cores. From these results, it is realized that this method indicates a new application of the adsorbed polymer layer and furnishes a new principle for the synthesis of a composite with core–shell structure.

1. Phase Diagram of HPC–Water System and Adsorption Behavior of HPC

The phase diagrams which were obtained for the aqueous solutions of three HPC samples (see Table 2) are shown in Fig. 17. All the systems show a LCST in the 47–52°C range; that is, the phase separation occurs at a definite temperature on heating and the systems become homogeneous again on cooling. It appears that the LCST is influenced both by the polymer concentration and by its molecular weight; that is, the LCST becomes lower both by an increase in the concentration and in the molecular weight of the polymer. These phenomena may be explained by the Flory–Schulz theory for the critical solution phenomenon [21].

Figure 18 shows adsorption isotherms of HPC-M on the silica particles at temperatures of 16°C and 48°C. It is evident that the isotherms are of the high-affinity types with a well-defined plateau showing temperature dependence. In Fig. 19, the temperature dependence of the saturated adsorption (A_s) of HPC-M is shown for the polystyrene latices and for the silica particles. After a constant value, there is a sharp increase in the A_s for both adsorbents. From comparison with the phase diagram shown in Fig. 17, it is evident that the trends in the A_s seen in Fig. 19 are based on the solvency of the medium (i.e., the reduction of solvency leads to an increased adsorption at the particle–water interface).

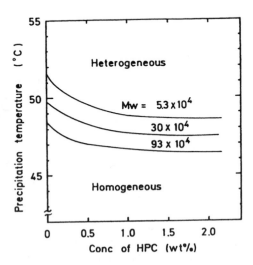

Figure 17   Phase diagram of HPC–water system.

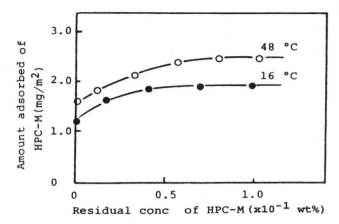

Figure 18 Adsorption isotherms of HPC-M on silica particles at the two medium temperatures.

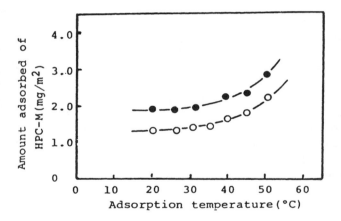

Figure 19 Temperature dependences of saturated adsorption on silica particles (●) and latex particles (○).

It is generally accepted that the time required for desorption of adsorbed polymer is very long and this process seems to be irreversible [22]. Accordingly, it is expected that the high adsorption value which appeared near the LCST may be maintained for a long time under the different temperature conditions. In Table 3, experimental results for irreversibility of adsorption in the HPC–latex systems are shown. After the HPC samples and the latex particles were mixed for 2 h at 48°C under otherwise the same conditions as in the case of the adsorption process, one

Table 3   Irreversibility of Adsorbed HPC Layer on Polystyrene Latices[a]

Sample	Cooling conditions	Separation temperature (°C)	Adsorbed amount of HPC (mg/m²)
HPC-M	—	48	1.66
HPC-M	48–20°C (gradually)	6	1.63
HPC-L	—	48	1.36
HPC-L	48–20°C (gradually)	6	1.38

[a]Adsorbed at 48°C in the HPC solution of 0.07 wt%.

portion of one of the samples was separated immediately by centrifugation at 48°C. The other half portion of the HPC-coated latex suspension was kept at room temperature for 24 h and then centrifuged at 6°C. As may be seen in Table 3, the amount adsorbed for both the samples is in close agreement. Furthermore, the same irreversibility of adsorption was also confirmed in the system of HPC–silica particles.

2.   Syntheses of Silica–Polystyrene Composite

Radical emulsion polymerization of styrene was carried out in the presence of bare silica particles and of the HPC-coated silica particles in water by using potassium persulfate as an initiator. Table 4 gives the typical ingredients used for these polymerizations. The HPC-coated silica particles were prepared under the same conditions as in the adsorption experiments. The polymerization temperature was kept at 50°C to protect the adsorption layer of HPC and the particles were polymerized for 24 h in the usual manner [1]. The degree of encapsulation of these particles was examined by comparing the electron micrograph and the electrophoretic mobility

Table 4   Typical Ingredients Used for Polymerization of Styrene in the Presence of Silica Particles

Sample	Bare silica particles (wt%)[a]	HPC-coated silica particles (wt%)[a]	Styrene (M)[a]	$K_2S_2O_8$ (M)[a]
SPL (−)	0.2	—	0.83	3.5×10^{-3}
SPL (HPC)	—	0.185	0.83	3.5×10^{-3}
SL	—	—	0.83	3.5×10^{-3}

Note. Polymerization temperature: 50°C; polymerization times: 24 h.
[a]Based on water.

of the produced particles, and by analyzing the molecular-weight distribution of the latex polymer using gel-permeation chromatography (GPC).

It is apparent that an important condition for controlling the synthesis of the silica–polystyrene composite is to reduce the residual concentration of HPC in the medium, because a slight amount of macromolecules dissolved in the solution brings about flocculation of the dispersion. To accomplish the above requirement, the dispersion of HPC-coated silica particles, which was treated at 48°C for 2 h, was heated again in an oil bath held at 70°C by rotating the bottles end-over-end for another 18 h to complete the adsorption. In Fig. 20, the residual amounts of HPC dissolved in the solution are plotted against the duration of the second adsorption process as the temperature was increased to 70°C and subsequently reduced to 25°C for several initial concentration of HPC-M. As seen from Fig. 20, as the temperature is raised to 70°C, the residual concentrations of the HPC gradually decrease with the lapse of time and attain final equilibrium values. Further, the final concentrations of HPC maintain the same values after the temperature is reduced to 25°C. These results indicate that raising the medium temperature contributes to the reduction in the concentration of HPC remaining in the solution and that the thick (or dense) adsorbed layer of HPC built up at the elevated temperature is maintained on the silica particles after reducing the temperature to 25°C.

Figures 21 and 22 show the electron micrographs of the final silica–polystyrene composites formed by the seeded polymerization using the bare silica particles and HPC-coated silica particles as the seed particles, respectively. Here, the HPC-coated silica particles were employed after treating the sample at 70°C. As may be seen from these electron micrographs, when the bare silica particles were

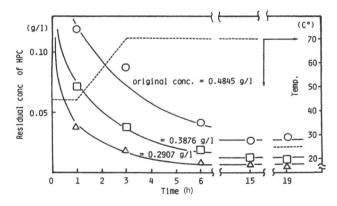

Figure 20 Residual amounts of HPC-M as a function of time elapsed during temperature raising and reducing. Original concentration of HPC: (○) 0.4845 g/L; (□) 0.3876 g/L; (△) 0.2907 g/L.

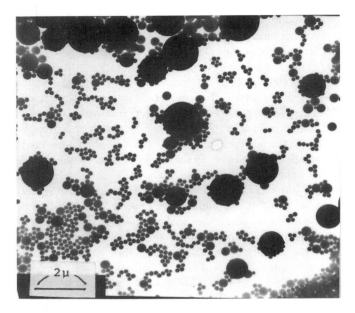

Figure 21   Electron micrograph of composite silica–polystyrene latex system, SPL(–),
prepared by using bare silica particles as the seed.

used in the polymerization, there was no tendency for encapsulation of silica particles and, indeed, new polymer particles were formed in the aqueous phase. On
the other hand, encapsulation of the seed particles proceeded preferentially when
the HPC-coated silica particles were used as the seed, and fairly monodispersed
composite latices with silica particles in the cores were generated. This indicates
that the dense adsorbed layer of HPC formed at the LCST acts as a binder between
the silica surface and the polystyrene molecules and contributes to the preparation
of composite polystyrene latices with silica particles in their cores.

3.   GPC Analysis for the Latex Polymer and Electrophoresis of the Composite Latices

The contribution of the adsorbed HPC layer to encapsulation was also illustrated
by GPC analysis for the latex polymer. The chromatogram of the latex polymer
separated from the silica–polystyrene composite is shown in Fig. 23. The chromatogram is composed of two separated broad peaks (peak A and peak B), which
indicates that the sample comprises two types of molecule synthesized under the

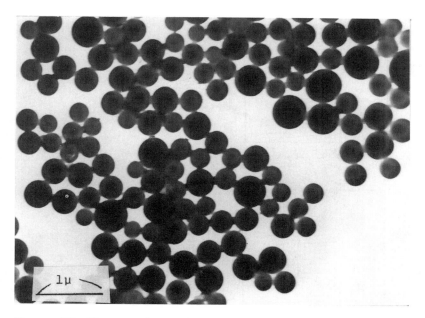

Figure 22 Electron micrograph of composite silica–polystyrene latex system, SPL(HPC), prepared by using HPC-coated silica particles as the seed.

different polymerization conditions. The polystyrene latices (SL in Table 4) prepared under similar polymerization conditions, but in the absence of silica particles, give a single molecular-weight peak situation in the range of peak B (see Fig. 23b). Thus, peak B in the silica–polystyrene composite appears to correspond to the isolated latex particles formed by initiation in the aqueous phase. On the other hand, the higher-molecular-weight peak A would correspond to encapsulating polymer molecules formed by polymerization in the layer surrounding the solid surface or, perhaps, the product of grafting polymerization with the adsorbed HPC [23].

For the composite generated, electrophoretic measurements were performed. Figure 24 shows the ζ-potential versus pH curves in 1×10^{-3} M KCl aqueous solution for the silica–polystyrene composite [SPL(HPC)], along with the data for the reference latices (SL), which were used after dialyzing exhaustively against 10^{-3} M KCl solution to remove the ionic impurities. As may be seen from Fig. 24, the ζ-potentials for both samples are almost same over the whole pH range examined. These results suggest that the surface of the composite is covered completely by the encapsulating polymer molecules and that its surface charges will be

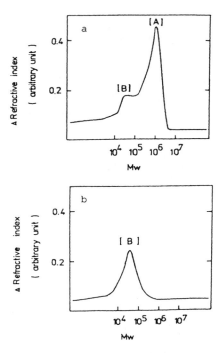

Figure 23 Gel-permeation chromatogram of latex polymer separated from composite silica–polystyrene latex system: SPL(HPC) (a) and polystyrene latex, SL, polymerized at 50°C in the absence of silica particles (b).

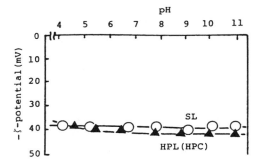

Figure 24 Zeta-potential versus pH curves in 1×10^{-3} M KCl aqueous solution: (▲) silica–polystyrene composite [SPL(HPC)]; (○) polystyrene latices (SL) polymerized at 50°C in the absence of silica particles.

–OSO$_3^-$ groups generated from K$_2$S$_2$O$_8$ used an initiator in the seeded polymerization.

4. Effect of Surfactant in the Syntheses of the Silica–Polystyrene Composite

To prepare a silica–polystyrene composite with a high yield, polymerization of the
composite was carried out in a system including a surfactant. Sodium dodecyl sulfate (SDS) was used as the surfactant under the same conditions given in Table 4.
The profile of the yield of composite versus the concentration of SDS is shown in
Fig. 25. It is known that the yield of composite increases with increasing concentration of SDS and approaches 100% near 0.008 M, which is the critical micelle
concentration (CMC) of SDS [25].

In the high-concentration runs of SDS, however, the generation of a new
type of composite was detected (i.e., the composite generated above the CMC of
SDS comprised a raspberry shape). Figure 26 shows a microphotograph of the
new type of composite. It is probable that the nucleation of new polymer particles
formed in the early stage of the emulsion polymerization would coagulate hetero-
geneously with the HPC-coated silica particles after some periods of seed polymerization. This new type of composite could be kept for a long time with excellent stability. It was also observed that no composite was generated in the runs
with a high concentration of SDS, when the bare silica particles have been used in
their polymerization as the seed.

All these results obtained in this work indicate that the dense (or thick)

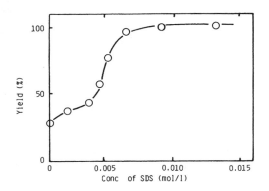

Figure 25 Relationship between the yield of silica–polystyrene composite and the concentration of SDS.

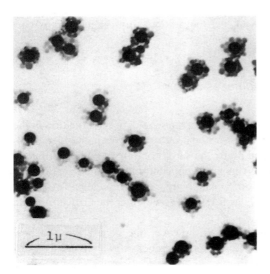

Figure 26   Electron micrograph of a new type of silica–polystyrene composite with a raspberry shape.

adsorbed layer of HPC formed on silica particles at the LCST is very effective in the encapsulation process with polystyrene via emulsion polymerization.

C.   Synthetic Process to Control the Total Size and Component Distribution of Multilayer Magnetic Composite Particles

In the synthesis of inorganic–organic composite particles, the establishment of a method to control the orientation of each element and their total sizes should be regard as an important factor because the characteristic of the composite particles largely depends on the local distribution of each component and on their total size. Aiming at the systematic resolution of these problems, the synthesis of composite particles with multilayers, comprised of usual anionic latices, magnetic ultrafine particles, and a polystyrene layer, has been investigated under various solution conditions. The magnetic ultrafine powder, $NiO \cdot ZnO \cdot Fe_2O_3$, offered by Sumitomo Chemical Co Ltd. Japan, was used without further purification. On the other hand, four different polystyrene latices were employed as organic species. The latex-L is a standard sample offered by Japan Synthetic Rubber Co. Ltd. The other three latex samples were synthesized by the surfactant-free emulsion polymerization in our laboratory [1]. All of these latex samples were used after sufficient dialysis in purified water and subsequent ion-exchange treatment with resin grains. The diameter and functional group of each sample are listed in Table 5.

Table 5 Diameter and Functional Groups of Samples

Sample	Diameter (nm)	Functional group
Latex-L	900	$-COO^-$
Latex-M	600	$-COO^-$
Latex-S	530	$-SO_3^-$
Latex-SS	180	$-SO_3^-$
$NiO \cdot ZnO \cdot Fe_2O_3$	20	—

1. Preparation of Magnetic Heterocoagulates

The dry powder of $NiO \cdot ZnO \cdot Fe_2O_3$ was compulsorily dispersed into ion-exchanged distilled water by intense ultrasonic emission for about 20 min. Nevertheless, the dynamic light-scattering measurement confirmed that the magnetic particles in the bulk were comprised of associated islands with the diameter of 250 nm, although the transmission electron micrographs showed the individual particle diameter to be 20 nm on average. It can be considered that this fact is due to their magnetism or to the peculiar hydrophobicity of dry powder.

The procedure for preparing the heterocoagulates from the latex-M and the magnetic particles is given here for illustration [10].

To begin with, 200 mL of pure water was added to a beaker containing 0.25 g of the magnetic power and sonicated for 20 min with a 600-W homogenizer (T-180; Ultrasonic Co. Ltd.) at the best tuning. Meanwhile, the original latex-M dispersion was diluted with purified water to control the particle number ratio to magnetic particles at 1/4500. These two suspensions were quickly blended and ultrasonic emission was used for another 3 min to promote complete mixing. Next, the pH of the blend suspension was adjusted with nitric acid to pH 2.5 to ensure that each particle acquired the opposite charge to one another. After the suspension was kept for 24 h at room temperature, the heterocoagulates, comprised of the latex-M (core) and the magnetic particles (shell), accumulated at the bottom of the vessel. The supernatant was replaced by purified water to remove the residual magnetic particles which did not adhere to the latex surface. The product was repeatedly washed in purified water and kept in a stock bottle. Approximately 0.5 g of the heterocoagulates can be obtained by this procedure.

Figure 27 are scanning electron micrographs showing four kinds of heterocoagulate provided from the suspensions of latices with different sizes and magnetic particles. All of these heterocoagulates were prepared by mixing the same $NiO \cdot ZnO \cdot Fe_2O_3$ dispersion and the respective latex dispersion at pH 2.5, where the ζ-potential of the magnetic particles has the sign opposite that of the latex particles, as shown in Fig. 28. The heterocoagulates generated from the latex-L, the latex-M, and the latex-S are composed of the uniform composite particles deposit-

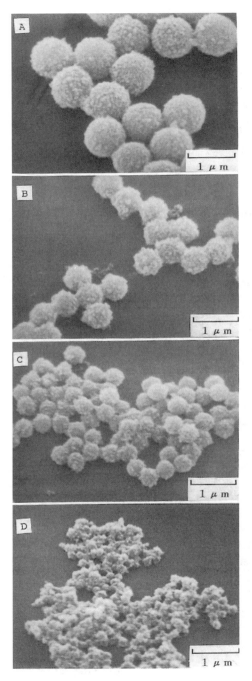

Figure 27 Electron micrographs showing the magnetic heterocoagulates prepared from different latex samples: (A) latex-L; (B) latex-M; (C) latex-S; (D) latex-SS.

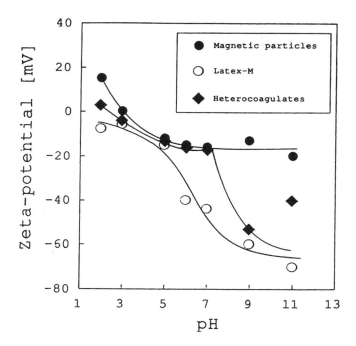

Figure 28 Zeta-potential of NiO·ZnO·Fe$_2$O$_3$ particles (●), latex-M (○), and heterocoagulates (◆) as a function of pH at 10^{-3} M KNO$_3$.

ing the small magnetic particles onto the large latex surfaces and all the heterocoagulates are separated as an isolated unit having an even surface. In contrast with these particles, the heterocoagulates generated from the latex-SS are made of large irregular aggregates and the regular heterocoagulates were difficult to recognize under any mixing conditions investigated. Furthermore, it should be noted that the heterocoagulates took over the uniformity as the original latex particles under the suitable mixing conditions. The composite particles were formed by the heterocoagulation of the magnetic particles onto the large latex particles with a constant layer, and they exactly maintain the uniformity of the latex particles after the heterocoagulation with the magnetic particles.

2. Characterization of Magnetic Heterocoagulates

In Fig. 28, the ζ-potential profile of the latex-M particles and the magnetic particles are presented as a function of the medium's pH, along with that of the heterocoagulates prepared by mixing their original suspensions at pH 2.5. The heterocoagulates display zero ζ-potential at pH 2.5 due to the charge neutralization between the two component particles. This suggests that the rapid sedimentation of the products after the heterocoagulation is attributed to the homoaggregation of the products rather than to the increase of their diameter or density. However, the dynamic light-scattering measurement and the transmission electron microscopy confirm that the aggregated products are redispersed into single particles by moderate ultrasonication without any desorption of the magnetic particles from the surfaces of the composite particles. Furthermore, Fig. 28 indicates that the ζ-potential of the heterocoagulates approaches that of the original magnetic particles in the alkaline medium. This behavior suggests that the magnetic particles, which cover the surface of the core latices in the acid and neutral pH systems, desorb from their surface in the alkaline system where the pH is above 9. This implied that the interaction between the surfaces of the two-component particles changes from lateral electrostatic attraction into electrostatic repulsion according to the increase of the medium's pH. From these results, it can be seen that the pH of the system largely dominates the following three aspects of heterocoagulates—the generation and sedimentation in the acidic system, the dispersion in the neutral pH region due to their high surface charge, and the decomposition in the alkaline system.

Figure 29 shows the thermograms of the magnetic particles, the latex-M particles, and the heterocoagulates prepared from these two kinds of particle. The rate of mass decrease for the magnetic particles was 2.0–2.3% during the temperature

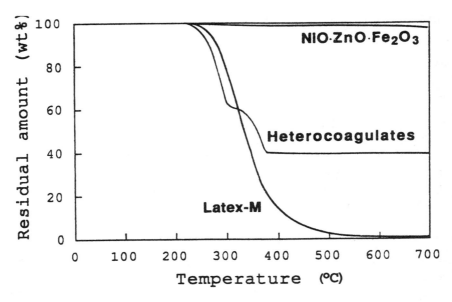

Figure 29   Thermograms of latex-M, NiO·ZnO·Fe$_2$O$_3$ particles, and heterocoagulates.

transition from 100 to 600°C, whereas the result of the latex-M was complete decomposition. The thermogram curve of the heterocoagulates exhibits a slow steady mass decrease below 200°C, a noticeable loss around 350°C, and a plateau above 400°C. The residual amount above 400°C is considered to be the weight of magnetic particles which did not decompose. As a result, the composition of the heterocoagulates is calculated to be 60.7% latex-M and 39.3% $NiO \cdot ZnO \cdot Fe_2O_3$ particles by weight percent. The application of the model proposed by Hansen and co-workers [4,25] (which assumes that the latex-M and $NiO \cdot ZnO \cdot Fe_2O_3$ are sphere particles with diameters of 600 and 20 nm, respectively) calculates the fractional coverage of the heterocoagulates (ϑ) to be ~1.0. This high coverage is attributed to the formation of the multiparticle layers of the magnetic particles on the latex surfaces, which resulted in the peculiar adsorption mechanism; the associated magnetic particles adhere on the surface of the core latices in the heterocoagulation process as if the islands were single particles.

3. Encapsulation of Magnetic Heterocoagulates

Concerning the heterocoagulates, both desorption of the magnetic particles and dissolution of ion species decrease their value as bioreactors because they inactivate biological functions. Hence, the optimal condition of encapsulation polymerization, in which the heterocoagulate surface is covered with a thin polystyrene layer, was surveyed. After 24 h on dialysis, a fixed amount of the heterocoagulates prepared from the latex-M was used to examine the eight polymerization conditions set up with regard to the amount of styrene monomer, the kind of initiator, and the presence of a surfactant. An initiator which provides a positive terminal group, azo-bis(isobutylamidine hydrochloric acid) was employed, and potassium persulfate was selected as a negative-terminal initiator. Moreover, a sodium oleate aqueous solution was adopted to promote hydrophobicity of the magnetic particles on the heterocoagulate surfaces. Approximately 0.2 g of the seed heterocoagulates was dispersed into 180 mL of water using the homogenizer, before it was moved into a special 200-mL vial with a screw cap. After that, distilled water containing a definite amount of $K_2S_2O_8$ and $C_{17}H_{33}COONa$ was carefully added to the vial. The pH of the sodium oleate aqueous solution was primarily adjusted to pH 6.0 with nitric acid, because the magnetic particles on the heterocoagulates desorb from the latex surface in the alkaline medium (see Fig. 28). This vial was set on the rotating disk of polymerization equipment and polymerized for 36 h at 60°C. The conditions and their results are shown in Table 6. As seen in the row labeled Product in Table 6, it is considered that the best condition in the list is the system No. 3 where ideal encapsulation was virtually accomplished. Under the conditions of samples 1, 2, 5, and 6 bridging flocculation among the seed particles or the generation of independent latex particles occurred in the course of the polymerization. Similarly, in samples system 4, 7, and 8, the flocculation and new par-

Table 6 Conditions for Encapsulation Polymerization of Heterocoagulates (Temp. 60°C, Scale 200 mL, Stirring Speed 30 rpm, Time of Polymerization 36 h)

	Sample number							
	1	2	3	4	5	6	7	8
Seed composite (g)	0.2	0.2	0.2	0.2	0.2	0.2	0.2	0.2
Styrene monomer (g)	0.2	1.0	0.2	1.0	0.2	1.0	0.2	1.0
KPS[a] (g)	10^{-3}	10^{-2}	10^{-3}	10^{-2}	—	—	—	—
AIBA·2HCl[a] (g)	—	—	—	—	10^{-3}	10^{-2}	10^{-3}	10^{-2}
Sodium oleate (g)	—	—	0.06	0.06	—	—	0.06	0.06
Product	×[b]	×	⊙[b]	×	×	×	△[b]	×

[a]KPS = potassium persulfate; AIBA·2HCl = azo-bis(isobutylamidine hydrochloric acid).
[b]× = flocculation or new particle generation, or both; ⊙ = good encapsulation; △ = encapsulation and new particle generation.

ticle generation were observed although a part of the seed particles were encapsulated. As a result, it can be considered that sodium oleate is indispensable for the encapsulation of the heterocoagulates and that potasium persulfate is a suitable initiator for this system.

Furthermore, as pointed out in Table 7, under similar conditions to sample 3 (the same amount of seed, potassium persulfate, and styrene monomer), the encapsulation polymerization was carried out at various sodium oleate concentration ranges from 0.003 to 0.60 g/200 mL. Below the concentration of 0.01 g/200 mL, the seed particles aggregated soon after the vials were equipped, whereas above 0.1 g/200 mL, desorption of the magnetic particles from the seed particle surface

Table 7 Conditions for Encapsulation Polymerization of Heterocoagulates (Temp. 60°C, Scale 200 mL, Stirring Speed 30 rpm, Time of Polymerization 36 h)

	Sample number						
	9	10	11	12	13	14	15
Seed composite (g)	0.2	0.2	0.2	0.2	0.2	0.2	0.2
Styrene monomer (g)	0.2	0.2	0.2	0.2	0.2	0.2	0.2
KPS (g)	10^{-3}	10^{-3}	10^{-3}	10^{-3}	10^{-3}	10^{-3}	10^{-3}
Sodium oleate (g)	0.003	0.006	0.03	0.06	0.12	0.30	0.60
Product	×[a]	×	⊙[a]	⊙	△[a]	×	×

[a]⊙ = good encapsulation; △ = encapsulation and partially no reaction; × = flocculation or decomposition.

occurred. Consequently, the concentration of sodium oleate has to be adjusted in the range from 0.01 to 0.1 g/200 mL for the encapsulation. This fact suggests that the adsorption state of oleate ion onto the seed particle surface largely affects the process of encapsulation polymerization.

Figure 30 is a transmission electron micrograph of the encapsulated composite particles whose surface was covered with a thin polymer layer, and it shows smoother surface than that of the seed particles (see Fig. 27).

4. Remarks of the Synthetic Process of Magnetic Composite Particles

Figure 31 shows the preparation process of the multilayer magnetic composite particles.

1. The heterocoagulates were produced from the magnetic ultrafine powder and the polymer latices with various sizes and functional groups. In this case, the particle size ratio and the ζ-potential of the two components should be noted in order to prepare the heterocoagulates whose fractional coverage is ~100%. The prepared heterocoagulates take over the uniformity of the latex particles adopted as the core. However, in the alkaline system above pH 9, the magnetic particles desorb from the surface of the core latices.

2. Using the heterocoagulates as the seeds, the optimum condition of encapsulation polymerization with styrene monomer was surveyed. In this case, it is indispensable to promote the hydrophobicity of the hete-

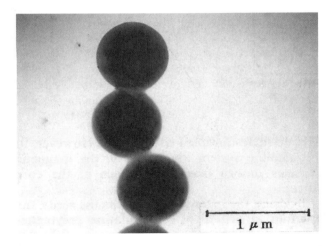

1 μm

Figure 30 Microphotograph showing the multilayer composite particles after the encapsulation polymerization.

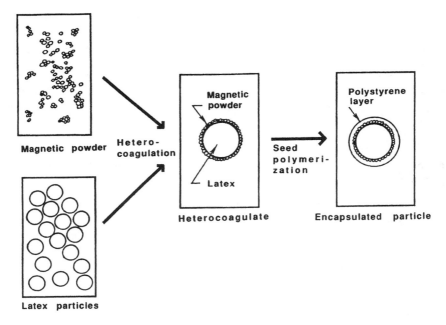

Figure 31 Schematic showing the process of synthesizing the multilayer composite particles.

rocoagulate surface with oleate ions. With the proper control of their adsorption state, the encapsulation is promoted efficiently.

3. As is seen in Fig. 30, the size of multilayer composite particles can be controlled by the diameter of the core latices which are applied primarily. This method realizes the localization of the magnetic particles on the outside of the composite particles covered with a thin polystyrene layer; therefore, they can give full function to the surrounding system.

D. Assembly of Latex Particles By Using Emulsion Droplets as Templates

For some applications, it might be of interest to obtain ordered aggregates of defined size and shape out of similar colloid particles. The simple coagulation of suspensions of colloid particles produces aggregates of dissimilar size and fractal appearance [26]. A few techniques allow the coagulation into clusters of approximately the same size, but these clusters could not be ordered and their shape is uneven and fractallike. Therefore, the fabrication of colloid particle aggregates that display ordering and have a symmetric and defined shape is still a challenge to the colloid investigator. In an attempt to tackle the above problems, we have applied a novel technique that allows the assembly of different types of colloid particle into ordered or multicomponent clusters [12,13] (these we sometimes denote as "supraparticles"). The size and the composition of the supraparticles during the assembly are controlled by gathering and confining their components in a restricted, colloid-size two-dimensional (2D) or three-dimensional (3D) space. Emulsion droplets are used to provide this restricted space. After the particles are assembled and fixed, the emulsion droplets are dissolved in the environment and the supraparticles are extracted as a colloid suspension.

1. General Description of the Emulsion Assembly Method

The different types of supraparticle suspension were assembled from monodisperse latex spheres of micrometer size. The colloid interactions between the latex particles (electrostatic, van der Waals, steric, hydration, hydrophobic, etc.) are principally known.

In this study, emulsion drops are used as 2D or 3D colloid "templates" whose size and shape control the overall size and shape of the obtained supraparticles. The basic principles of the "assembly-and-dissolution" scheme used in this study are illustrated in Fig. 32. The process starts with a suspension of latex microspheres in water. Emulsion drops of the oily phase are then introduced in the system. The particles could be adsorbed either on the droplet surfaces or in their bulk. The obtained ordered shells or balls of densely packed particles are fixed together by an appropriate agent. The carrier emulsion droplets are later dissolved by the addition of a mediator phase or solubilizing agent. Thus, the final product is supraparticle clusters resuspended in the water phase. The emulsion template scheme has the ability to assemble composite supraparticles comprising different types of colloid species, if a second type of particle is adsorbed on the droplets. In the study presented (Sec. II.D.2), we concentrate on the formation of spherical supraparticles from one type of latex bead only—the left branch in Fig. 32. Experimental data on the formation of ball-like and composite supraparticles (middle and right branches in Fig. 32) are presented in Secs. II.D.3 and II.D.4.

The initial investigation as to how the assembly and dissolution scheme could be experimentally implemented encountered a number of difficulties and even controversies in the required properties of the particles, the oil, and the dissolving agents. Being convinced that it will be close to impossible to find particles and phases that would a priori possess the proper surface properties, we made use of a number of agents that by physical or chemical means modify the colloid interactions within the particle/droplet system. Each of the steps of the multistage assembly process is carried out under specified and controllable conditions.

As the second phase for the oil–water emulsions, we chose octanol. This oil does not dissolve or cause swelling of the polystyrene microspheres. The mutual

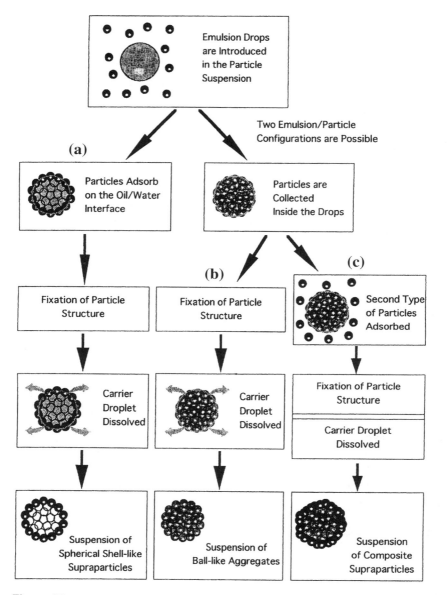

Figure 32 Possible modifications of the "emulsion template" method for supraparticle assembly: (a) spherical shell-like supraparticles; (b) ball-like assembly; (c) composite assembly. (From OD Velev, K Furusawa, K Nagayama. Langmuir 12:2374, 1996.)

solubility of octanol and water is small, and a clearly defined interface is formed. However, the mutual solubility drastically increases upon the addition of ethanol to the system, and we found that a 1 : 1 water–ethanol mixture can dissolve about 15 vol% octanol.

The practical assembly schemes and the experimental data for different types of latex are presented below.

2.   Ordered Hollow Assemblies from the Sulfate Latices

These supraparticles were assembled from the negatively charged "surfactant-free" sulfate–polystyrene latices. The stability of the original latex suspension is based only on the electrostatic repulsion between the microspheres [1,26]. The charge is provided by a mixture of sulfate and hydroxyl groups on the latex surface [1,27]. The largest part of the latex–water interface remains hydrophobic [28]. The success of the assembly scheme depends heavily on the surface properties of the used latex samples.

After experimenting with different substances and conditions, we formulated the assembly flowchart, presented in Fig. 33. The latex particles in their original state do not adsorb on the oil droplets. Therefore, the first step required is to modify their surface properties in such a way that the microspheres become adsorbable on the oil–water interface. This process is denoted hereafter as "sensitization". Note that the ideal modification of the latex surfaces should incite neither the flocculation of the beads in the suspension nor 2D coagulation after their adsorption on the interface, as both of these are disastrous for the 2D ordering. Out of the substances tried as sensitizers, the amino acid lysine worked best. Lysine was used in the form of its monohydrochloride salt at a concentration of 0.017 M. The pH of the thus prepared solution is neutral (5.5–6.0). The latex used, which was treated with lysine under these conditions, did not coagulate. It was, however, able to slowly adsorb on the octanol–water interface and to form 2D ordered structures. This process was greatly enhanced and quickened by the homogenizer.

The homogenization step of the assembly process was carried out by first introducing 0.3 vol% sulfate latex in aqueous solution of lysine hydrochloride. Five minutes after that, 1–1.5 vol% octanol was added. The system was then homogenized on the Ultra Turrax device. The adsorption of their latices onto the dispersed oil droplets during the homogenization process is noticeable by the change of the appearance of the white suspension from milky to opalescent and the buildup of oil–latex deposits on the wall of the test tube. We have tried both a continuous homogenization for 5–8 min and a cyclic scheme that combines the same time of treatment but split and prolonged into cycles of 60 s of homogenization and 30 s of rest. The adopted cyclic scheme worked better, possibly because the rest periods provide some time for rearrangement of the latex microspheres adsorbed on the droplets.

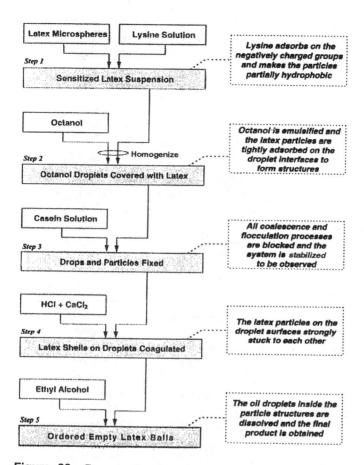

Figure 33 Concrete flowchart for the multistage, interaction-tailored assembly of ordered hollow supraparticles from the sulfate latex. The boxes on the right briefly describe the physical processes that take place at each step. (From OD Velev, K Furusawa, K Nagayama. Langmuir 12:2374, 1996.)

The outcome of the homogenization step, if the process is carried out thoroughly enough, is octanol emulsion droplets covered with a shell of 2D ordered microspheres. A panel of such latex–droplet assemblies is presented in Fig. 34. The Ultra Turrax homogenizer produced emulsion droplets with a wide polydispersity, which results in supraparticles of diverse sizes. To demonstrate this, in all panels we present pictures of widely different size droplets or particles obtained in one or more experiments. The mean size of the droplets depended on the input energy of the homogenizer. At 8000 rpm, the droplet diameter was within the

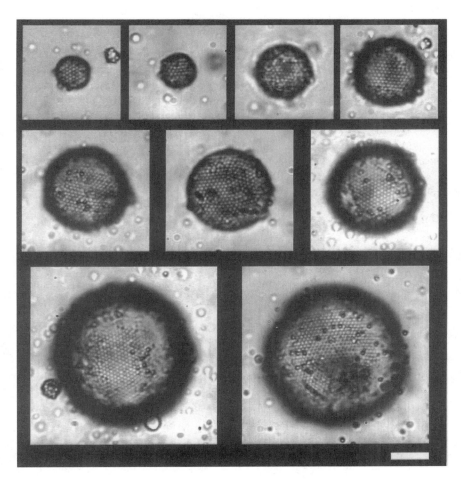

Figure 34 Digitized pictures of octanol droplets covered by ordered shells of the sulfate latex. These are obtained after Step 2 of the flowchart in Fig. 32. (From OD Velev, K Furusawa, K Nagayama. Langmuir 12:2374, 1996.)

10–100-μm range. At 14,000 rpm, the mean droplet size fell below 10 μm, but the coverage with particles become poor due to the very high interfacial area. All of the data reported hereafter were obtained in the range 8000–10,000 rpm. After the treatment with the homogenizer, it is very important to check by observation that the obtained droplets are fully covered by an ordered latex shell. A figure illustrating the appearance of the droplets in this check is included in Fig. 34. This observation should be done quickly, as the droplets covered with microstructured latex shells are highly unstable. If the test tubes are left at rest, the droplets will cream and coalesce in 5–10 min forming, a layer of oil and latices on top of the lysine solution.

To prevent breakdown the next immediate step undertaken in the overall assembly scheme was the stabilization of the latex–droplet assemblies. This was achieved by quick mixing of the emulsion with casein solution (Step 3 in Fig. 33). The concentration of casein in the system was adjusted at 0.5–1 mg/mL. A few minutes of vortexing was required to prevent creaming before the protein adsorption is completed. After stabilization with casein, the protected particle–droplet assemblies can be stored for tens of minutes and easily examined by microscope. The casein adsorption and steric stabilization step is irreversible.

We found that casein is able to gently bind the latex particles within the shells around the droplets. This possibly results from bridging of the adjacent particles by protein molecules adsorbed on their surfaces. The strength of this bridging is not sufficient to preserve the supraparticle structure during the dissolution of the carrier droplets. Therefore, the addition of a stronger binding agent (coagulant) is necessary. We used a mixture of HCl and $CaCl_2$, both of whose components are known to be strong coagulants of the used latex beads [29]. The final concentration of the substances in the system after Step 4 of the process (Fig. 33) amounted to 10^{-2} M HCl and 5×10^{-3} M $CaCl_2$.

It is important to note that in 2–3 min, the coagulant mixture was not able to induce aggregation of the latex–droplet assemblies, obviously due to their protective coverage with casein. However, it bound the compressed particles within the microstructured latex cores. The microstructured shells were thus prepared for the droplet extraction that was carried out by adding 50 vol% ethanol. We found empirically that the best yield is obtained when the ethanol is induced as quickly as possible—this was achieved by injection. After the carrier octanol droplets are dissolved, the supraparticle assemblies remain suspended in the water–ethanol environment.

The supraparticles obtained in this way possess the structure of empty ordered shells from the coagulated latices. Microscopic pictures of suspended assemblies are presented in Fig. 34. It is seen that the particles within the shell preserve the 2D ordered pattern that was formed during the adsorption on the droplet surface. As the focal depth of the objective is only a few microns, the larger supraparticles could be optically "cross sectioned" by focusing on their top, bottom, or equatorial planes. The "cross-sectioning" technique allows us to prove that most of the obtained spherical shells are ordered all around and that no oil droplet is present inside. Such pictures are arranged in Fig. 35.

3. Ball-Like Aggregates from the Amidine Latex

The experiments were carried out with latex microspheres of positive surface potential. The only functional group present on the particle surfaces is the amidine group $(C(NH_2){=}NH_2^+)$[27].

The overall scheme for obtaining ball-like aggregates is presented in Fig. 36.

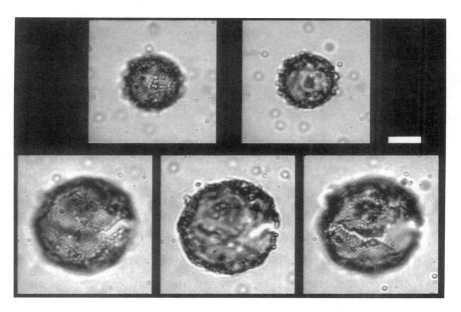

Figure 35 Optical "cross sectioning" of two of the obtained supraparticles. By moving the focal plane of the objective up and down, one is able to observe the outer shell structure or the inside of the assemblies. The crack in the large particle allows us to demonstrate that the oil droplet inside it has been dissolved. Bar = 10 μm. (From OD Velev, K Furusawa, K Nagayama. Langmuir 12:2374, 1996.)

The first step of the process is to modify the surface properties of the latex particles so that they adsorb and penetrate inside the octanol droplets. Of the substances tried, the addition of monobasic glutamic acid as a sensitizing agent could cause the slow penetration of the latex particles from the water phase into the bulk of the droplets. The monobasic glutamic acid, however, acts slowly and does not seem to affect all of the latex particles to the same extent, as some particles remain adsorbed on the interface or dispersed in the water. For this reason, we desorted the anionic surfactant SDS, which proved to be a very quick and efficient hydrophobizing agent for the amidine microspheres.

The required SDS concentration for the hydrophobization is very low. Concentrations of SDS on the order of 4×10^{-6} M were enough to induce latex penetration inside the droplets. Increasing the concentration further increased the speed and efficiency of the particle collection by the droplets until a "catastropic" event occurred at a concentration about 8×10^{-6} M: The particles quickly and irreversibly coagulated outside the droplets. For the experiments, we settled at a concentration of 7.5×10^{-6} M, at which SDS induced particle adsorption and penetration inside the oil droplets within a few tens of seconds. The octanol droplets were

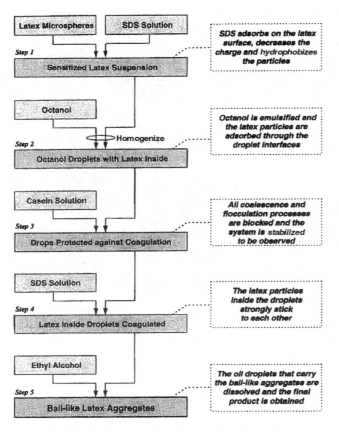

Figure 36 Flowchart of the multistage assembly of ball-like aggregates from the amidine latex. The boxes on the right briefly describe the physical processes that take place at each step. (From OD Velev, K Furusawa, K Nagayama. Langmuir 12:2385, 1996.)

introduced at a concentration of 1 vol%, the same as in the supraparticle experiment. The volume concentration of the latex was 0.5–0.6 vol%, so the droplet bulk was loosely filled with particles. The latex–octanol–SDS system was homogenized in five to seven cycles, each including 60 s of processing at 8000 rpm, followed by 30 s at rest. This resulted in the engulfment of the latex microspheres inside the octanol droplets (Fig. 37). The latex microspheres gathered in this way inside the droplets did not seem to be ordered in a 3D or 2D manner (even if there was some ordering, the conditions of microscopic observation did not permit its visualization). The overall shape of these aggregates, however, was precisely spherical, determined by the shape of the droplets.

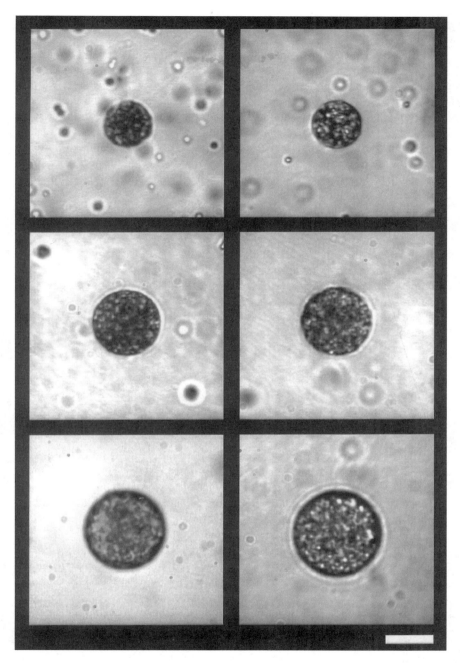

Figure 37 Digitized panel of emulsion droplets filled with amidine latex beads. SDS is used to induce the particle adsorption and penetration (Step 2 in Fig. 35). To outline the polydispersity of the samples, droplets of various sizes are presented. Bar = 10 μm. (From OD Velev, K Furusawa, K Nagayama. Langmuir 12:2385, 1996.)

Before extracting the spherical latex clusters, from the octanol droplets, it was necessary to introduce some steric stabilizer, protecting the oil droplets against coalescence and the latex balls from coagulation with each other. Similar to Sec. II.D.2, we used a casein solution at a concentration of 0.5 g/L. The droplet–particle complexes stabilized by casein could be stored for a prolonged period of time in a suspended state but could not survive the ethanol extraction step. As in the supraparticle experiments (Sect. II.D.2), the gathered particles required some further binding or coagulation in order to gain the necessary structural stability. This was achieved by adding a more concentrated SDS solution to raise the SDS concentration up to $1 \times 10^{-4} M$ (Step 4 in Fig. 35). When ethanol was injected at a 1 : 1 ratio to the system after 10 min (Step 5 in Fig. 35), the octanol droplets were dissolved and the ball-like aggregates remained intact and dispersed. As there is no cavity inside the spherical aggregates, their mechanical stability is higher than that of the shell-like supraparticles. After the octanol was removed from the surrounding environment (by centrifuging, washing, and resuspending the aggregates in a casein–SDS solution), latex balls could easily be deposited on the solid surface by drying. Micrographs of such deposited and dried balls are presented in Fig. 38. The amidine spheres are a basic component of the composite particles described in the following subsection.

4. Composite Assemblies

One interesting possibility, provided by the method, is the fabrication of supraparticles including more than one colloid component. We have assembled composite particles by using both the sulfate and amidine latices and by combining their assembly schemes.

These composite particles consist of a core of aggregated amidine microspheres surrounded by a shell of sulfate latices. The flowchart of their assembly is presented in Fig. 39. The process first included the formation of the ball-like aggregates from the amidine latex inside the octanol droplets. We then used the method for formation of sulfate latex shells on the droplet surfaces, reported in Sec. II.D.2, to surround the amidine latex with a sulfate shell.

The amidine latex was destabilized with the use of SDS, as described in the previous subsection. The system was homogenized in six cycles, each including 60 s of treatment and 30 s of rest between the cycles. In this process, the amidine beads penetrated and aggregated inside the octanol droplets (Step 2 in Fig. 39). Meanwhile, the sulfate beads were sensitized by a lysine hydrochloride solution. The dispersions of the octanol–amidine latex and the sensitized sulfate–latex suspension were mixed in a ratio of 1 : 1. The final concentration of the amidine and the sulfate latices after mixing was 0.13 vol%, and the oil amounted to 0.75 vol%. The mixture was briefly homogenized in three cycles of 30 s of processing followed by 30 s of rest (Step 4 in Fig. 39). This allowed the formation of

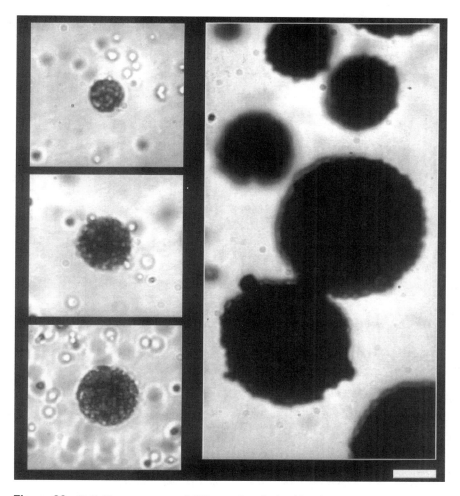

Figure 38 Ball-like aggregates of different size obtained by the scheme in Fig. 35. Left column; suspended aggregates; right column; aggregates dried over a glass surface. Bar = 10 μm. (From OD Velev, K Furusawa, K Nagayama. Langmuir 12:2385, 1996.)

sulfate–latex shells on the droplet surfaces. The adsorption of the sulfate–latex microspheres is possibly enhanced by the electrostatic attraction to the positively charged core. Some loss of latices and decrease of the final yield occurred during this step due to the buildup of heterocoagulate deposits on the walls of the test tubes.

After the core–shell structures were formed, the process followed the sul-

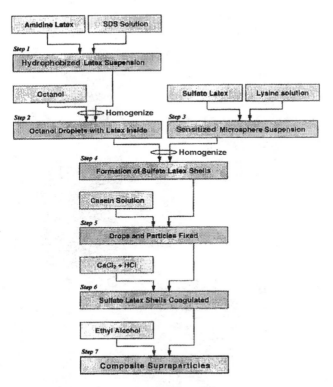

Figure 39 Flowchart for the assembly of composite supraparticles. The final product has a spherical core from aggregated amidine latices, over which a shell of sulfate latices is deposited. (From OD Velev, K Furusawa, K Nagayama. Langmuir 12:2385, 1996.)

fate–latex fixation and extraction scheme described in Sec. II.D.2. The composite particles were sterically stabilized by adsorption of casein (concentration = 1 g/L). The overlying sulfate shell was strongly coagulated in the next step by a mixture of 10^{-2} M HCl and 5×10^{-3} M CaCl$_2$ (Step 6 in Fig. 39). The octanol droplets inside the already fixed composite particles were dissolved by the addition of ethanol. We provide micrographs of the finally obtained composites in Fig. 40. The overall shape of the composite assemblies was usually close to spherical but not very smooth and regular. In the transmitted mode of illumination, one cannot distinguish which particles belong to the amidine type and which to the sulfate type, as their optical properties are identical. Still, a certain distinction between the core and the shell is noticed, possibly due to the difference in the densities of packing inside the aggregated latex phases. The yield of the process was below 10% due to losses by deposition and heterocoagulation. Once the composites are

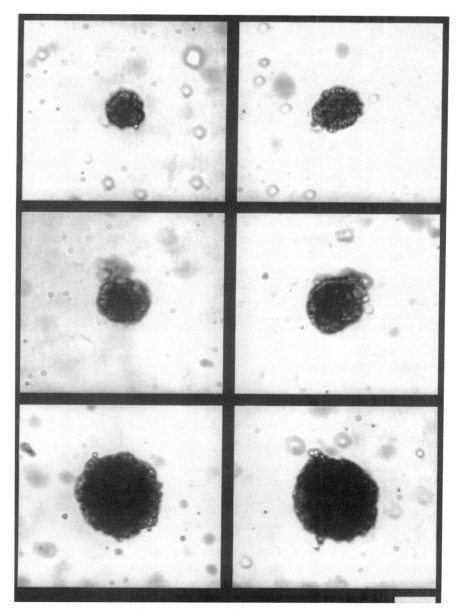

Figure 40 Amidine latex core–sulfate latex shell composite assemblies of various sizes in transmitted illumination. The core–shell structure is suggested by the different densities of the aggregated latices. Bar = 10 μm. (From OD Velev, K Furusawa, K Nagayama. Langmuir 12:2385, 1996.)

obtained, they can be concentrated and purified by centrifugation and stored in a suspended state or dried over a solid surface.

To prove and visualize the existence of core–shell structures, we carried out experiments in which the sulfate–latex was substituted with a sulfate–fluorescent latex. The fluorescent shell–amidine core supraparticles were observed both in transmitted and fluorescent illumination [13].

III. CONCLUDING REMARKS

This chapter concludes the description of shell-like supraparticle, ball-like, and composite aggregates obtained by using emulsion drops as templates for the assembly of latex beads. The data reported here do not encompass all of the possibilities of the assembly and dissolution emulsion schem. We believe that by choosing appropriate modifiers of the colloid interactions, composites and supraparticles comprising colloid species other than latex particles could be assembled.

The method is also potentially applicable for the production of more functionalized composites than the core–shell structures presented above. This could be achieved by assembling more than two types of colloid species or by introducing directionally specific particle–particle interactions.

REFERENCES

1. K Furusawa, W Norde, J Lyklema. Kolloid ZZ Polym 250:90, 1972.
2. E Matijevic. Chem Mater 5:412, 1993.
3. CM Cheng, FJ Micale, JW Vanderhoff, MS El-Aasser. J Colloid Interf Sci 150:549, 1992.
4. RD Harding. J Colloid Interf Sci 40:57, 1972.
5. N Kawahashi, E Matijevic. J Colloid Interf Sci 138:534, 1990.
6. K Furusawa, Y Kimura, T Tagawa. J Colloid Interf Sci 109:69, 1986.
7. H Honda, M Kimura, F Honda, T Matsuno, M Koishi. Colloids Surf A 82:117, 1994.
8. Y Otsubo, E Kazuya. J Colloid Interf Sci 168:230, 1994.
9. K Furusawa, C Anzai. Colloids Surf 63:103, 1992.
10. K Furusawa, K Nagashima, C Anzai. Colloid Polym Sci 272:1104, 1994.
11. A Rembaum, SPS Yen. J Macromol Sci Chem A 13:603, 1979; RS Molday, WJ Dreyer, A Rembaum, SPS Yen. J Cell Biol 64:74, 1975.
12. OD Velev, K Furusawa, K Nagayama. Langmuir 12:2374, 1996.
13. OD Velev, K Furusawa, K Nagayama. Langmuir 12:2385, 1996.
14. A Bleier, E Matijevic. J Chem Soc Faraday Trans 1 74:1346, 1978.
15. J Visser. Adv Colloid Interf Sci 3:331, 1972.
16. GR Wiese, TM Healy. Trans Faraday Soc 66:490, 1970.
17. RM Pashley. J Colloid Interf Sci 101:516, 1984.

18. H Fugita, K Ametani. Jpn J Appl Phys 18:753, 1979.
19. T Tagawa, S Yamashita, K Furusawa. Kobunshi Ronbunshu 40:273, 1983.
20. K Furusawa, T Tagawa. Colloid Polym Sci 263:353, 1985.
21. S Konno, S Saeki, N Kuwahara, M Nakata, M Kaneko. Macromolecules 8:799, 1975.
22. K Furusawa, K Yamamoto. Bull Chem Soc Japan 56:1956, 1983.
23. TI Min, A Klein, MS El-Aasser, JW Vanderhoff. Prep Org Coat Plast Chem 46:314, 1982.
24. PH Elworthy, AT Florence, CB Macfarlane. Solubilization by Surface Active Agents. London: Chapman & Hall, 1968.
25. FK Hansen, E Matijevic. J Chem Soc Faraday Trans–1, 76:1240, 1980.
26. Z Zukany, B Chu. Physica A 177:93, 1991.
27. GVF Seaman, JW Goodwin. Annu Clin Prod Rev 31, June 1986.
28. HJ Van den Hul, JW Vanderhoff. Electroanal Chem 37:161, 1972.
29. IDC Product Guide, Vol.7. IDC: Oregon, 1994.

7
Thermodynamic and Kinetic Aspects of Bridging Flocculation

H. Daniel Ou-Yang and Maria M. Santore
Lehigh University, Bethlehem, Pennsylvania

I. INTRODUCTION

Polymers added to colloidal dispersions maintain stability or induce flocculation through a variety of mechanisms [1]. Bridging flocculation, relevant to technologies such as wastewater treatment and papermaking, is classically defined in Fig. 1 to result from interparticle attractions occurring when polymer chains simultaneously adsorb onto two particles. A number of systems are known to undergo bridging flocculation, including polyethylene oxide adsorbing onto clay [2], hydrophobically modified cellulosics and polyethylene oxides adsorbing onto acrylic latex [3], and certain cationic polyacrylamides adsorbing onto polystyrene latex [4–7], clay [8–11], silica [12,13], cellulose fibers [14–16], and aluminum oxide [17]. Generally, bridging is expected when small amounts of high-molecular-weight polymers are added to dispersions where they adsorb to the particles but do not saturate the surface [18]. Without a careful series of experiments, however, bridging flocculation can be hard to distinguish from other means of colloidal destabilization. Furthermore, bridging is sometimes a transient process occurring prior to stabilization of the suspensions.

To understand the forces that drive bridging flocculation, a knowledge of polymer adsorption is needed. Homopolymers adsorbing at a single interface achieve configurations involving tails (sections of the backbone starting at the surface and extending into the solution), loops (sections of the backbone starting at the surface, looping through the solution, and returning to the surface), and trains (sections of the backbone completely on the surface), shown in Fig. 2a. In Fig. 2b, the tail region of the layer, extending beyond a distance R_g from the surface (where

Figure 1   Classical bridging flocculation.

R_g is the free coil radius of gyration), is extremely dilute with a volume fraction less than 1% [19]. Tails are, therefore, invisible to most techniques such as ellipsometry [20,21], neutron scattering [22], and reflectivity [23], but dominate the hydrodynamic properties of the layer [24] and long-range interparticle interactions. Loops and trains yield volume fractions exceeding 0.5 at the surface to about 0.1 further from the surface, at a distance near $R_g/2$ [25]. Loops and trains constitute most of a layer's mass and dominate the interparticle potential for encounters closer than R_g. The concentration profile of the loops decays exponentially away from the surface while the concentration profile of the tails reaches a broad maximum near a distance of R_g from the surface and then decays exponentially at greater distances [22].

Homopolymers of relatively high molecular weights, exceeding 1000 statistical segments, often have about 10–20% of their mass contributing to tails when the layer is saturated [25]. For lower surface loadings of high-molecular-weight polymers, chains tend to lie flat and there are fewer tails and loops, as shown in Fig. 2c. At both saturated and low surface loadings, low-molecular-weight polymers generally adsorb in flat configurations with very little mass of the layer contributing to tails [22]. Because tails are important in developing bridging forces, substantial molecular weight is needed if a polymer additive is to be used as a bridging flocculant. Indeed, the bridging flocculants used in the paper industry typically have a molecular weight of several million.

Prior to a discussion of the influence of adsorbed polymer on colloidal forces, one other feature of polymer adsorption should be mentioned: Adsorbed polymers generally maintain a dynamic equilibrium with chains free in solution [26]. Chains adsorb at an interface rapidly [27,28] and resist desorption into pure solvent [29], giving the appearance of irreversible binding. There is, however, a continued interchange between adsorbed and free chains [30] and between segments bound in trains and those in nearby loops and tails. The energy barrier for the segmental exchange process is a few kT, on the order of the segment adsorption energy. Therefore, significant segmental exchange occurs on experimental and process-related timescales.

(a)

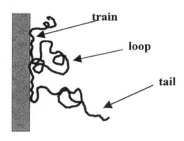

(b)

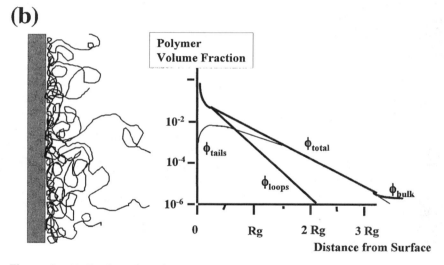

Figure 2 (a) Configuration of one adsorbed homopolymer as it would be in a saturated layer: tails, loops, and trains. (b) Concentration profile in a saturated homopolymer layer. Graph adapted from Ref. 24. (c) Configurations within a homopolymer layer of relatively low coverage.

II. MEASUREMENTS OF BRIDGING ATTRACTIONS

The surface forces apparatus (SFA) has been the primary instrument employed in precise measurements of colloid scale forces such as bridging attractions, steric repulsion, and electrostatic potentials [31,32,33]. Its crossed-cylinder geometry, in

(c)

Figure 2 Continued

Fig. 3, derives from two identical cylindrical lenses mounted perpendicular to one another, a configuration equivalent to a sphere interacting with a plate, or two interacting spheres. The molecularly smooth outer surface of mica sheets glued to the lenses traditionally comprises the surfaces of interest; although in recent years, the technique has been modified to accommodate self-assembled or adsorbed monolayers [34,35], silica, and alumina substrates [36,37]. The separation of the surfaces is measured via interferometry (the backs of the mica sheets are mirrored), and a cantilever spring allows determination of interparticle forces. The force–distance relationship between these macroscopic surfaces with a known radius of curvature can be rescaled so that the forces between small particles can be anticipated. The Derjaguin approximation yields [38]

Figure 3 The crossed-cylinder geometry of the surface forces apparatus.

$$F(D) = 2\pi RW(D) \tag{1}$$

where $F(D)$ is the force–distance relationship, $W(D)$ is the corresponding surface energy between flat surfaces, and R is the radius of curvature of the cylindrical lenses. D is the closest distance between the interacting surfaces. If Eq. (1) is applied to pairs of interacting spherical particles, the factor 2 is dropped from the equation and R is used as the spherical radius.

With the use of SFA, bridging attractions have been measured for partially saturated surfaces both above and below the theta temperature. For instance, Almog and Klein [39] studied the forces resulting from polystyrene adsorption on mica from cyclopentane at 23°C. This temperature exceeds the theta temperature of 19.6°C, yielding conditions of relatively good solvency. When the adsorption was conducted for 12 h with the crossed cylinders in relatively close proximity, at 30 μm, the diffusion of chains from the bulk solution to the adsorbing surface was limited, giving an adsorbed amount substantially less than the equilibrium coverage of 2.5 mg/m^2. When the surfaces were subsequently brought together, significant attractive forces, shown in Fig. 4a, were interpreted to result from bridging, with the tails from one layer penetrating the sparse tails and loops of the opposing layer to reach vacant surface sites. The onset of the attraction occurred just below a separation of $2R_g$, and the attractive force reached a maximum near a separation of $0.5R_g$. At closer separations, a repulsive core, resulting from the compression of the adsorbed chains, dominated the potential. Notably, in pure cyclopentane, no detectable interactions between the bare mica sheets were observed, except for the van der Waals attraction starting at separations of 10 nm. The bridging attraction was, therefore, substantially longer range than the van der Waals forces.

Similar bridging attractions, also shown in Fig. 4a, have been observed for polystyrene adsorbed on mica from cyclohexane at 26°C (poor solvent conditions, below the theta temperature of 34.5°C) [40]. For subsaturation coverages of 1.1 mg/m^2 (with saturation occurring at the higher level of 5.5 mg/m^2 due to the poor solvency), the onset of the attraction occurred at separations slightly smaller than $1.5R_g$. The attractive part of the force–distance curve took the shape of an exponential decay, with the decay constant very nearly equal to $R_g/4$. The potential minimum occurred at a separation near $R_g/4$ and was two to three times deeper than the corresponding minimum in good solvent conditions [40]. For bridging in a poor solvent, the interaction also contained a repulsive core, at separations depending on the amount of adsorbed polymer.

One of the distinguishing signatures of bridging attractions is that they diminish, if not disappear altogether, as more polymer adsorbs. In a good solvent— for instance, polystyrene adsorbed on mica from cyclopentane at 23°C in Fig. 4b—the interaction was found to become less attractive, and ultimately became completely repulsive as the surfaces were saturated [39]. As the polymer coverage was increased toward saturation and as the depth of the attractive minimum de-

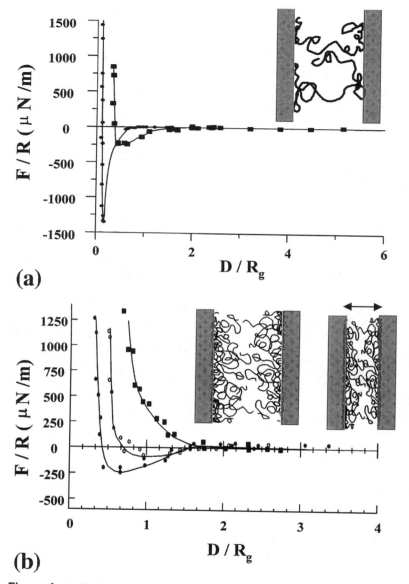

Figure 4 (a) Bridging attractions at low coverage, in a good (■) and poor (●) solvent. (Adapted from Refs. 38 and 39, respectively.) (b) Restabilization from bridging in a good solvent with increasing coverage, measured after exposure to polymer for 12 h at a separation of 0.1 mm (●), 2.5 more hours at a separation of 0.5 mm (○), and an additional 16 hours at 0.25 mm (■). (Adapted from Ref. 38.) Schematic emphasizes compression of layers. (c) Attractions between saturated surfaces in a poor solvent. (From Ref. 39.) Schematic emphasis attractions between tails. (d) Attractions between saturated surfaces in a modest solvent with significant local mobility of the train layer. (From Ref. 39.) See text for description of experiments.

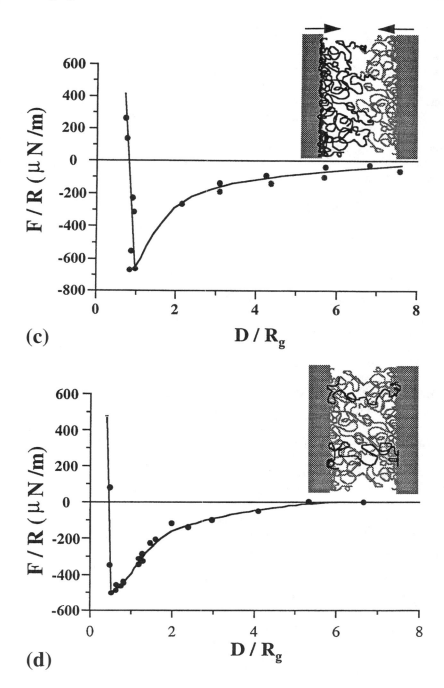

(c)

(d)

creased, both the attractive well and repulsive core increased in range, as a result of the larger amount of polymer in the gap contributing to an osmotic repulsion. When saturated surfaces were brought together in a good solvent, a repulsion began at a separation on the order of $2R_1$ (the free end–end distance). The repulsive force increased monotonically with decreasing separation [41,42]. It is thought that the approach of the surfaces causes a compression of the adsorbed layers, illustrated in Fig. 4b, rather then significant interpenetration [43]. Were the latter to occur, tail segments from one layer would reach the other surface and displace some of the trains on the second surface, lowering the energy and potentially causing a bridging attraction. The layer compression is related to the local osmotic pressure of the adsorbed layers and is the force responsible for steric stabilization.

The extent to which bridging attractions persist at more nearly saturated interfaces depends on the solvent quality (segment–segment versus segment–solvent interaction), strength of adsorption (segment–surface energy), and dynamics of segment–segment exchange in the layer nearest the surface. For instance, the potential between saturated polystyrene layers on mica in cyclohexane at 23°C and 26°C (below the theta temperature of 34.5°C) was found to exhibit an attractive minimum near R_g and a repulsive core; see Fig. 4c [40]. The range of the attraction, however, significantly exceeded that for the unsaturated surfaces in Fig. 4a, a clue that mechanisms other than bridging caused the attraction at saturation. With saturated layers (containing 5.5 mg/m^2 of polystyrene), the attraction became detectable at separations on the order of $5R_g$–$6R_g$, when the tails of the two layers just began to interact. This suggests that the effective segment–segment attractions resulting from the poor solvent conditions gave rise to colloid-scale attractions between the two surfaces! Because adsorption onto the opposing surface was thought not to occur, the phenomenon was not truly polymer bridging. The similar features of the two types of attractions, bridging versus poor solvency, mean that these two mechanisms for colloidal destabilization could be easily mistaken. Saturated layers in poor solvents have, in general, longer range and slightly weaker interactions than true bridging, and the former do not diminish with the further addition of polymer, as the surfaces are already saturated.

A second interesting situation resulting from competing surface, segment, and solvent interactions is exemplified by saturated polystyrene layers on mica in cyclohexane at 37°C in Fig. 4d. This temperature exceeds the theta temperature (T_θ), but apparently not enough to give adequate osmotic repulsion between approaching layers. Here, a weak attraction was observed prior to the repulsive core [40]. The onset of the attraction occurred for separations on the order of $3R_g$–$4R_g$, and the attractive minimum was slightly less than R_g, but a longer range than that observed with partially saturated interfaces. The remarkable feature of these data is that attractions are observed at all: Above the theta temperature, one expects steric stabilization, like that seen for polystyrene (PS) in cyclopentane at 23°C (3.4°C above T_θ) [39]. In cyclohexane at 37°C (2.5°C above T_θ), there is, apparently, some interpenetration of tails between the two layers. Furthermore because the polymer is rather weakly adsorbing, over the timescales of the experiment, tail segments from one layer can trade places with trains in the other layer to give bridging attractions. Clearly, if one hopes to avoid bridging, choosing the appropriate conditions may not be straightforward.

Although observations made with the SFA strongly suggested bridging in a number of cases, direct proof for bridging was not obtained because chain conformations were not measured. Strong arguments for the interpretation of the SFA data in the context of bridging come from calculations of polymer conformations between flat plates and the resulting potentials.

III. CALCULATIONS OF BRIDGING ATTRACTIONS

With theory in place to predict interfacial configurations and local concentration profiles of adsorbed polymers on a single surface [44–47], the calculation of concentration profiles, forces, and potentials between two surfaces bearing adsorbed polymer was the next advancement of the field. Both scaling and mean-field theories are now well developed for interactions between homopolymer and block copolymer layers. Most treatments take advantage of the fact that, frequently, the adsorbing polymers are much smaller than the particles onto which they adsorb. Then, it is appropriate to calculate interactions between flat plates bearing adsorbed polymer and later to apply the Derjaguin approximation to account for the particle curvature.

Treatments for flat-plate potentials calculate the free energy of a system (containing plates, adsorbed polymer, and polymer solution within and outside the gap between the plates) at a specified plate separation relative to the free energy at infinite separation. Calculations have recognized that with real colloidal collisions, several timescales must be distinguished. At short times, when surfaces approach, solvent will drain from the gap. At intermediate timescales, the chains reconfigure and the amount of segment–surface contacts evolves. Finally, at the longest timescales, chains may desorb and exit the gap if they are not permanently anchored. During the timescales of colloidal collisions, it is frequently presumed that solvent draining and chain reconfiguration including adjustment of segment–surface contacts occur, but there is insufficient time for the *adsorbed* chains to exit the gap. Free chains, however, might be removed with the solvent if they are not greatly entangled with those already adsorbed. Calculations following these assumptions present a "restricted-equilibrium" perspective on the interfacial potentials. True "full-equilibrium" calculations that allow chain desorption to maintain equilibrium (equal polymer chemical potentials) between the gap and a fluid reservoir at the bulk polymer concentration can give very different interaction potentials.

Scaling treatments provide broad perspective on interplate potentials in the presence of adsorbing homopolymer but are restricted in the chemical details and physical regimes that can be accommodated (close or far separation, good solvent, and weak adsorption such that there are significant number of vacant surface sites). DeGennes [48] predicted that at restricted equilibrium and full coverage in a good solvent, a purely repulsive potential between saturated layers scales as D^{-2} at large separations (on the order of R_g or greater), but as $D^{-5/4}$ at closer separations. Relaxing the constraint of restricted equilibrium leads to attractive potentials.

Mean-field theories for bridging attractions follow lattice calculations of the concentration profiles of adsorbed polymer and predict the potentials over a broader range of polymer architectures and chemistries than can be accommodated by a single scaling model. The lattice treatment of Scheutjens and Fleer [49–51] predicts that for surfaces sparsely loaded with adsorbing polymer (three equivalent monolayers, where a monolayer corresponds to the thickness of a statistical segment) in a theta solvent, at restricted equilibrium potentials have an attraction which begins near $2R_g$. The potential reaches an attractive minimum with depth $0.03 \, kT/a^2$ (where a is the statistical segment length) and a separation on the order of $0.5R_g$. The repulsive core is a direct result of the restricted equilibrium. Full-equilibrium calculations allow the polymer to leave the gap, giving a stronger attraction up to the point where the surfaces compress a single layer of polymer chains, the thickness of which corresponds to a statistical segment. Clearly, the restricted equilibrium assumption is a more realistic description for the behavior of polymer adsorbed in the SFA.

If more polymer is added to the surfaces, the attractive minimum diminishes and becomes longer range in the restricted equilibrium scenario [49–51]. With five equivalent monolayers of coverage, the interaction becomes purely repulsive. The mean-field predictions are in qualitative agreement with the SFA findings from Klein's group [39,41,42] except for their prediction of a repulsive potential for surfaces approaching saturation, even at theta conditions. The lattice treatment underestimates the influence of segment–segment attractions, which should give colloidal attractions near and below the theta conditions. Another interesting outcome from the mean-field theory was predictions for the number of bridges formed as the surfaces approached. For segment–surface attachment energies of 1 kT and separations below R_g, the number of bridges across the gap exceeds the number of tails. With increasing separation, the number of bridges diminishes and the number of tails increases, with the number of bridges and tails being roughly equal at $1.5R_g$. The number of bridges is negligible for separations exceeding $4R_g$.

Advances in mean-field treatments include continuum approaches [52,53] and the inclusion of different molar volumes for polymer segments and solvent molecules [54]. The predictions that follow for potentials between plates bearing various amounts of polymer [55] (with parameters chosen to match the experimental systems—for instance, polystyrene adsorbing on mica from cyclopentane) qualitatively agree with experimental data, but they overpredict bridging phenomena. Although there is clearly far to go, Ploehn was the first to attempt greater incorporation of physical–chemical detail into mean-field theories.

IV. DEPLETION VERSUS BRIDGING ATTRACTIONS

It is often difficult to distinguish bridging flocculation from a second mechanism, called depletion or volume restriction flocculation, which occurs for nonadsorbing polymers of substantial molecular weight. This is especially true if complex formulations make it difficult to determine the extent of polymer adsorption on suspended particles. When two particles approach in a solution of nonadsorbing polymer, chains leave the gap, generating a local region with almost pure solvent. The osmotic pressure in the fluid surrounding the particle pair exceeds that between the particles and forces the particles together, as illustrated in Fig. 5. Asakura and Oosawa [56,57] developed a model for the pair potential, $W(r)$, resulting from depletion attractions:

$$
W_d(r) \begin{cases} = \infty & \text{for } r < 2R \\ = -\dfrac{4\pi}{3} D_{\text{cross}}^3 \left(1 - \dfrac{3r}{4D_{\text{cross}}} + \dfrac{r^3}{16D_{\text{cross}}}\right) P_{\text{osm}} & \text{for } 2R \leq r \leq 2D_{\text{cross}} \\ = 0 & \text{for } 2D_{\text{cross}} < r \end{cases} \quad (2)
$$

Here, r is the center–center separation between two spheres of radius R: $r = D + 2R$. D_{cross} is the distance of closest approach between a particle and a polymer coil:

$$D_{\text{cross}} = R + R_g \quad \text{for dilute solutions} \tag{3a}$$

$$D_{\text{cross}} = R + \xi \quad \text{for semidilute solutions} \tag{3b}$$

Here, ξ is the screening length in the polymer solution.

Typical depletion attractions are long-range secondary minima on the order of 1–2 kT and cause reversible flocculation, which often yields a liquid–liquid or liquid–solid phase transition [58–60]. Because, in the dilute limit, the osmotic pressure scales as the bulk polymer concentration, the strength of the attraction grows with polymer concentration. Phase separation generally occurs at polymer concentrations significantly greater than those necessary for bridging, were the polymer to adsorb. Above c^*, however, intercoil interactions are screened, reducing the osmotic effects and allowing restabilization of the flocs, especially if the particles bear electrostatic charge preventing their close approach [59].

Insight into the differences between depletion and bridging flocculation can be gained by considering systematic increases in segment–surface attractions start-

Figure 5 Attractive forces from nonadsorbing polymer: depletion flocculation.

ing with the case of a nonadsorbing polymer. Here, we present the results from a statistical mechanical treatment of telechelic polymers, whose main backbones are nonadsorbing but which possess sticky end groups as a result of ionizable or hydrophobic chemistry [61]. This situation is relevant to several technologies, including rheological modifiers (such as hydrophobically modified water-soluble polymers which induce bridging flocculation in poor formulations,) and polyacrylamides with small amounts of cationic charge employed as flocculants for wastewater treatment. In the situation of interest, the end groups are attracted to the surface, but the treatment neglects micellization from the association between end groups, irrelevant to the general issue of bridging versus depletion attraction. The free energy between flat plates is calculated by taking into account the partition function for the polymer and the end-group interaction with the surface. The polymer coils are treated as Gaussian, and the influence of solvent quality derives from the second virial coefficient, v, which perturbs the Gaussian statistics. Three-body segment interactions (third virial coefficient) are neglected.

Other treatments for bridging attractions between the adsorbed telechelic polymer layers have recently appeared in the literature [62,63]. These employ a scaling treatment for polymer brushes where the end-group surface attraction is sufficiently strong to provide the energy needed to extend the polymer coil to several times its natural end-to-end distance. These treatments are not relevant to the case of weak adsorption, where one needs to explain the loss of depletion forces and the development of weak bridging.

Figure 6a illustrates that at theta conditions ($v = 0$), the osmotic pressure of the polymer solution with nonadsorbing end groups generates an attractive pair potential between the flat plates, starting at a separation near R_1 [61]. This full-equilibrium attraction would be even greater in a good solvent. As the end group–surface attraction increases, the depletion attraction diminishes. Finally, with about 5 kT sticking energy between the end group and the surface, the end-group adsorption overcomes the depletion, and a repulsive force appears due to compression of the adsorbed chains when the plates approach within R_1. With stronger end group–surface attractions, a bridging attraction appears, with the minimum occurring near separations of $1.5R_1$. The onset of this attraction is as

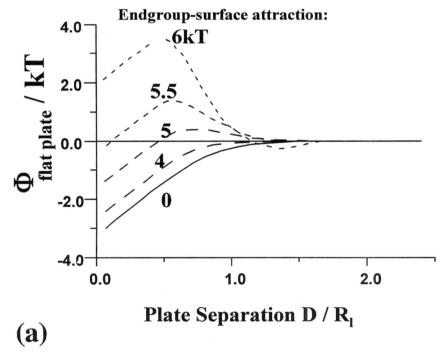

(a)

Figure 6 (a) Effect of end group–surface energy on flat-plate potential at full equilibrium for polymer chains of 1000 statistical segments and a concentration just below c^* and (b) at restricted equilibrium. (From Ref. 60.)

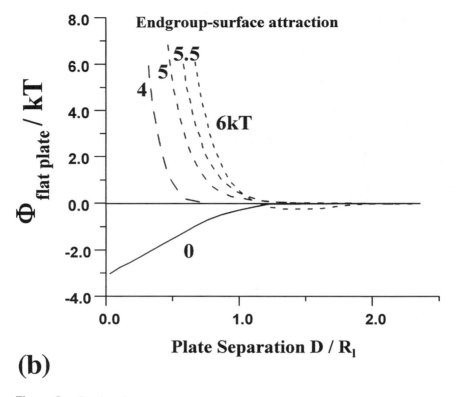

Figure 6   Continued

much as $2R_1$, suggesting chains stretch to bridge the gap. At full equilibrium, a local minimum in the potential occurs when chains leave the gap. For restricted equilibrium, in Fig. 6b, chains permanently caught in the gap produce a repulsive core, but loss of the depletion attraction and the growth of the bridging attraction occur, at restricted and full equilibrium, for similar values of the end group–surface energy. The bridging attractions for telechelic polymers at theta conditions are longer range than observations for homopolymer-induced bridging below the theta temperature [50,51,61], a result of the nonadsorbing backbone of the telechelic polymer.

Another factor, which affects the competition between depletion and bridging for telechelic polymers, is the placement of the surface-attracted groups relative to the chain ends. Figure 7 illustrates that as the sticky groups move away from the chain ends, the end group–surface interactions are minimized by the entropic loss of the backbone segments beyond the sticker [61]. If the stickers are hidden

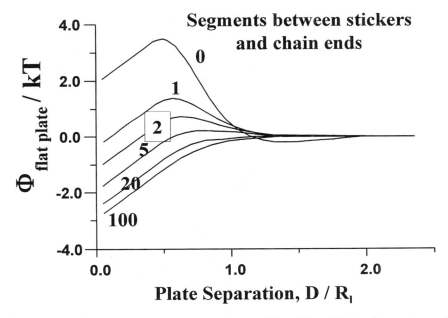

Figure 7 Effect of sticker placement on the potential at full equilibrium for a polymer of 1000 statistical segments and a sticker–surface energy of –6 kT.

inside the polymer coil, adsorption is diminished and depletion attractions occur for chemistries where the same stickers would have given bridging if they were placed on the chain ends. This effect is magnified in a good solvent. It is interesting, therefore, to note the importance of end-group chemistry in design polymers for bridging applications. They play crucial roles regardless of whether or not the main backbone adsorbs to the surface.

Several factors, similar to the ones affecting homopolymer bridging, diminish bridging attractions in telechelic systems. For end groups whose attraction is sufficient to cause bridging in a theta solvent, improvements in solvent quality cause the coils to swell and to increase the entropy loss for end-group adsorption. Hence, as shown in Fig. 8, improvements in solvent quality diminish bridging attractions [61]. These calculations were run for increases in solvent quality up to the point where the bridging attraction disappeared. The case of $v = 0.1$, however, is extreme and not expected to occur for most systems considered to be "good solvents." Another factor reducing bridging attractions is the amount of adsorbed polymer, which increases with the free polymer in solution. Figure 9 illustrates that in a good solvent, a weak bridging minimum becomes longer range and ulti-

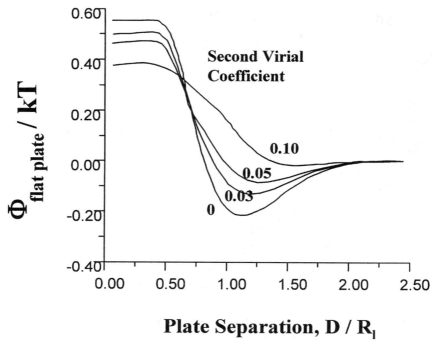

Figure 8 Effect of solvent quality on bridging attraction for a polymer of 500 statistical segments, in the dilute regime, two orders of magnitude below c^*. End group–surface attraction is −7 kT.

mately disappears with increased polymer concentration [61]. When these potentials were used in a statistical mechanical treatment of macroscopic phase separation, bridging flocculation was predicted at low polymer concentrations and restabilization occurred as the polymer concentration was increased [61]. The agreement with experiment was quantitatively acceptable for the specific chemical and physical parameters employed (particle size, polymer molecular weight, solvent quality, end-group sticking energy, etc.) [62].

V. BRIDGING OF AQUEOUS POLYELECTROLYTES AND OTHER COMPLEX SYSTEMS

Whereas most fundamental studies focus on forces generated by polymers adsorbing from organic solvent, or on flocculation behavior with nonionic adsorbing polymers, polyelectrolytes comprise a large market sector for flocculant additives.

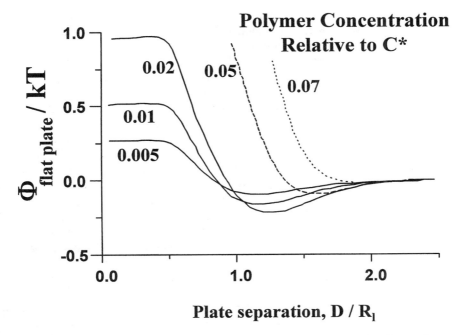

Figure 9 Effect of bulk polymer concentration on the bridging attraction for a polymer containing 1000 statistical segments, an end group–surface attraction of –7 kT, and a second virial coefficient of 0.02.

Cationic polyelectrolytes are popular flocculants because their adsorption reduces the electrostatic stabilization of negatively charged colloids by the addition of positive charge to the interface. For nonionic polymers, parameters such as molecular weight, polymer concentration, and solvent quality are important in controlling bridging flocculation. Polyelectrolytes bring the additional variable of charge density or the extent of backbone ionization, which can dominate colloidal interactions. To obtain bridging flocculation, cationic polyelectrolyte flocculants must be of relatively low charge density along their backbones [13,63]. This type of polymer adsorbs in tails, loops, and trains to negatively charged particles and appears to adsorb more slowly than analogs with higher charge density. Because the sparse backbone charges anchor the polymer to the negative colloids at distinct points, the adsorbed layers tend to be fluffy and extended [8,64]. Indeed, this extension is critical for generating the necessary long-range attractions needed to overcome the combination of viscous, diffusive, and internal forces between particles [65,66].

Several important features of bridging flocculation distinguish it from other flocculation mechanisms occurring with cationic polyelectrolytes. True bridging flocculation requires very high-molecular-weight polymers. For instance, bridging flocculation was found for a cationic polyacrylamide with a molecular weight of 2.7×10^6 g/mol, whereas a lower-molecular-weight version (8×10^5) of the same copolymer did not induce flocculation [65]. With the lower-molecular-weight analog, progressive adsorption canceled the negative charge on the primary particles; however, by the point neutralization was reached, sufficient polymer had adsorbed to sterically stabilize the dispersion. In the case of true bridging, the optimum flocculation concentration of added polyelectrolyte can be determined by the flocculation rate or clarity of the supernatant solution over the flocculated mass. The optimum flocculation concentration occurred when there was still a substantially negative zeta potential on the particles (including the addition of the positive charge through the polymer adsorption) [67].

Cationic polyelectrolytes of relatively high charge density are often effective flocculants for dispersions of negatively charged particles, through a mechanism of *charge neutralization,* an effect very different from bridging but difficult to distinguish. When polymer molecules of high cationic charge adsorb to a negatively charged surface they quickly assume a flat conformation, quite different from that in Fig. 2a, making bridging nearly impossible [68]. The positive charge from the polymer does, however, counteract the native negative surface charge, reducing the electrostatic stabilization of the particle. With densely charged cationic polymers, this charge neutralization occurs before enough polymer can be added to yield a steric stabilization. Indeed, the optimum flocculation concentration of polyelectrolyte for the charge-neutralization mechanism occurs when the zeta potential is near zero, in contrast with the negative zeta potentials associated with bridging flocculation. It has also been shown that the optimum charge density for adsorbing polyelectrolytes is one where the distance between cationic functionalities along the backbone is approximately equal to the negative charge spacing on the surface [68], such that adsorbing polymer does not bring extra charge to the interface. Also, charge neutralization works well over a broad range of molecular weights, whereas bridging flocculation is highly molecular-weight dependent and requires long chains. Indeed, we have observed charge-neutralization flocculation of anionic latex by the additive of an adsorbing rhodamine dye, with a molecular weight on the order of 500, far too low to induce bridging flocculation, but giving very distinct destabilization nonetheless [69].

It has been recognized recently that cationic polyelectyrolytes can destabilize suspensions by a third mechanism, the patchwise flocculation mechanisms in Fig. 10. This situation occurs with highly charged cationic polyelectrolytes, frequently where the spacing between cationic groups on the backbone is smaller than that between the negative charges on the surface [12]. Then, when a polymer coil adsorbs, it necessarily generates a region of local positive charge. At coverages below saturation, the positive regions of some particles adsorb to the bare negative regions of other particles. At higher coverages, the particle charge will be com-

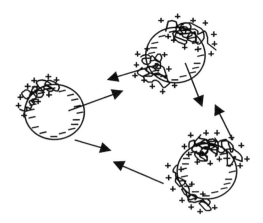

Figure 10 Patchwise flocculation mechanisms for cationic polyelectrolytes on negative particles. Primary charges are indicated. Counterions are not shown.

pletely reversed by the excess positive charge at the outer region of the adsorbed layer, beyond that needed to cancel the negative charges at the primary interface. Patchwise flocculation looks very much like bridging because in the final state, single polymer chains are in contact with two particles at once and because the addition of too much polymer will restabilize both bridging flocculation and patchwise flocculation. There are, however, some fundamental differences between the two mechanisms. Classical bridging flocculation, per the concept in Fig. 1, requires very high-molecular-weight chains, whereas patchwise flocculation can be achieved with lower molecular weights. Second, bridging requires a low charge density such that an extended layer is formed. Bridging may, however, occur as a transient during the adsorption of highly charged polymers, which later stabilize when they adsorb more completely. Patchwise flocculation depends less on adsorption history.

Finally, in more recent years there have been revelations about additional flocculation mechanisms occurring in complex systems involving several polymers or one polymer and several types of particles [70,71]. Some of these have been given names suggesting that they are related to the classical bridging flocculation in Fig. 1; however, this may not be the case. Of particular note, van de Ven has discovered a heterobridging flocculation process that appears to be distinctly different than the classical bridging described here. In asymmetric bridging, two types of particles are flocculated by a single polymer; for instance, polyethylene oxide has been shown to flocculate a combination of clay and wood pulp fibers [72]. Interestingly poly(ethylene oxide) (PEO) flocculation of clay by bridging is already established [2]; however, PEO is known not to adsorb onto this particular wood pulp [72]. PEO, does, however, flocculate mixtures of the two particles. The hypothesized explanation is that after PEO adsorbs onto the clay, its entropic loss makes the subsequent adsorption of the PEO-covered clay onto the fibers easier [72]. Alternate explanations include that the larger size of the PEO-covered clay particles increase their collisions with the fibers, but this does not explain the ultimate (thermodynamic) observation that fibers wind up in the flocs. Although the molecular-weight dependence of this process has not been examined, it seems intuitive that this process may have some common features with the patchwise bridging mechanism in the previous paragraph. It is through modification of the surface of the some of the particles that they are able to stick to other particles. It appears that the molecular weights needed to accomplish this asymmetric bridging are, therefore, lower than those needed for classical bridging and that this phenomenon will be the study of future work.

The complexity involved with bridging flocculation occurs because our current understanding of adsorbed polymer conformations and their influence on interparticle forces is based on a static equilibrium. When processes require flocculation, the appropriate polymer is injected into a dispersion and adsorption occurs. Although the ultimate state of the adsorbed polymer may well be described as tails loops and trains, the layer must evolve before this can be achieved. Depending on the relative rates of polymer adsorption, interfacial relaxation, and colloidal collisions, bridged conformations may be achieved prior to layer equilibrium, and if the bridging pulls the particles into the van der Waals minimum, flocculation may be permanent, with additional polymer adsorbing on the outside of the flocs. On the other hand, if the bridging forces generate only a secondary minimum beyond the electrostatic barrier, additional polymer adsorption may disrupt flocs and bridging may only occur as a transient. This brings us to considerations of the kinetic aspect of bridging flocculation.

VI. KINETIC ASPECT OF BRIDGING FLOCCULATION

When polymeric flocculant is injected into a stable particle suspension, several steps are involved in destabilizing the suspension. The schematic diagram, Fig. 11 by Gregory [73], depicts the process: Mixing of the polymer molecules among the particles, adsorption of polymer chains on the particles, and, finally, flocculation and/or floc breakup. The kinetics of these steps involve four basic timescales: (1) the mixing time, (2) the time for the polymers in the bulk solution to adsorb onto the colloidal surface, τ_A, (3) the time for adsorbed chains to conform/relax at the surface, τ_R, and (4) the collision time between colloidal particles or between colloidal clusters, τ_C.

The time of mixing depends on the process by which the polymer solution is added to the suspension of particles. In practice, a small amount of concentrated polymer flocculant is added to a large volume of uniformly suspended particles. It

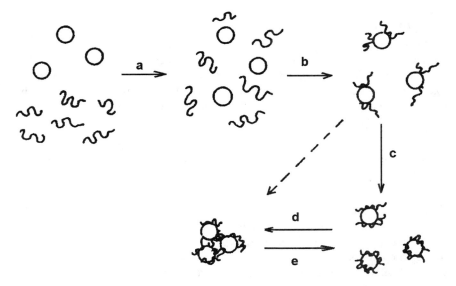

Figure 11 (a) Mixing of the polymer molecules with the particles, (b) adsorption of polymer chains onto the particle, (c) polymer molecule reconformation, (d) flocculation, and (e) floc breakup. (From Ref. 73.)

takes time for the polymer molecules to disperse uniformly before localized bridging flocculation occurs. The effect is thought to be the origin of residual haze in some flocculated suspensions [73,74]. Although it is difficult to quantitatively investigate mixing kinetics, transient polymeric-bridging flocculation can sometimes be used to study the kinetics of floc breakup [75]. We will return to a more detailed review of the kinetics of floc breakup in a later section.

Both the polymer adsorption time τ_A and the particle flocculation time τ_C involve considerations of the collision rate and the effectiveness of a collision, or the reactivity. On the other hand, the chain relaxation/reconformation time τ_R depends on the chain flexibility and mobility at the solvent–solid interface. It was found in a study by Yu and Somasundaran that the adsorbed chain conformation can affect the reactivity of the particle collisions, thus affecting bridging flocculation kinetics [76–78]. This motivates a discussion of the theory by von Smoluchowski [79] for binary collisions in colloidal suspensions, which applies to both polymer molecule–particle and the particle–particle collisions.

According to the theory of von Smoluchowski [79], the number of encounters between particles i and j per unit time at unit concentration is

$$k_{ij} = \frac{4}{3} G(a_i + a_j)^3 + \frac{2kT}{3\eta}(a_i + a_j)(a_i^{-1} + a_j^{-1}) \tag{4}$$

where G is the shear rate, a_i and a_j are radii of the particle species i and j, respectively, k is the Boltzmann constant, T is the absolute temperature, and η is the solvent viscosity. The first term is caused by the shear-induced collision (i.e., the orthokinetic effects). The second term is the collisions caused by Brownian motion (i.e., the perikenitic effects). At low shear rates, the Brownian diffusion effect dominates, whereas at higher shear rates, the orthokinetic effect dominates. The crossover shear rate depends on temperature and the particle radii. The Peclet number, P_e, defined as the ratio of the shear effect to the Brownian diffusion effect, can be expressed as

$$P_e = \frac{2G\eta a^3_i}{kT} r(1 + r) \tag{5}$$

where a_i is the radius of the larger particle and r is defined to be the ratio of particle radii a_j/a_i. For example, taking the radius of the particle $a_i = 0.5$ μm and the polymer molecule radius $a_j = 0.1$ μm, $r = 0.2$, $G = 100$ s^{-1}, $\eta = 0.01$ P, and $T = 300$ K, one gets $P_e = 1.45$. The shear effect and the diffusion effect are about equal. At the same conditions for collisions between monodisperse particles with $a_i = 0.5$ μm, the Peclet number $P_e = 12$; the shear effect dominates.

A. Adsorption Time τ_A

Assume that particle number density N_0 does change significantly before all the polymers are consumed by the particles and that the added polymer molecules coat only partially the particle surface. The time derivative of the free-polymer molecule concentration can be written as [78]

$$\frac{dc}{dt} = -\alpha_{pm}k_{pm}N_0c \tag{6}$$

where c is the number density (concentration) of the polymer molecules, k_{pm}, obtainable from Eq. (4), is the rate coefficients for particle–polymer molecule collision, and α_{pm} is the reactivity of the collision. In what follows, the reactivity $\alpha_{pm} = 1$ is assumed. Equation (6) has a simple solution:

$$c(t) = c_0 \exp(-\alpha_{pm}k_{pm}N_0t) \tag{7}$$

where c_0 is the initial polymer concentration. The adsorption process has a well-defined characteristic time:

$$\tau_A = \frac{1}{\alpha_{pm}k_{pm}N_0} \tag{8}$$

The assumption that N_0 is constant throughout the process is, of course, a simplification. In reality, the rate constant k_{pm} decreases as more polymer molecules are adsorbed, and the particle number density N_0 also decreases when flocculation starts to occur. Therefore, the adsorption time τ_A given here serves as a lower bound.

B. Collision Time Between Colloidal Particles τ_C

The consideration of particle–particle collisions leading to flocculation is similar to that of polymer–particle collisions leading to polymer adsorption. By considering only binary collisions between singlet particles in a monodisperse particle suspension, the time derivative of the number of singlet particles can be written as

$$\frac{dN_1}{dt} = -\alpha_{pp}k_{pp}N_1^{\,2} \tag{9}$$

where α_{pp} is the reactivity of the particle collision, k_{pp} is the number of binary particle collisions per unit time, and N_1 is the singlet particle number density. Solving Eq. (9) gives

$$N_1 = \frac{N_0}{1 + 2\alpha_{pp}k_{pp}N_0t} \tag{10}$$

The characteristic collision time τ_C, the time in which the number density of the singlet particles reduces to half of its initial value, can be shown to be

$$\tau_C = \frac{1}{2\alpha_{pp}\mathrm{k}_{pp}\mathrm{N}_0} \tag{11}$$

C. Reactivity α_{pp}

According La Mer, the half-surface-coverage criterion for effective polymer bridging flocculation [18], the reactivity is proportional to the term $\theta(1-\theta)$, where θ is the fraction of the particle surface covered by polymer molecules. The La Mer criterion implies that the maximum reactivity occurs at $\theta = 0.5$. Others have modified the criterion by considering particle reorientation after collision [80]. This criterion works well for systems where the adsorbed polymers remain active for bridging at all times, a situation referred to equilibrium flocculation (EF).

A study by Pelssers et al. [78] illustrated a situation where the adsorbed polymer molecules could change their bridging reactivity by reconformation after they had landed on the particle surface. Their theoretical approach combines the von Smoluchowski formulation and a time-dependent bridging reactivity to give a quantitative analysis of bridging kinetics. We first discuss their theoretical model and then, briefly, their experiments.

VII. TREATMENT BY PELSSERS ET AL.

Nonequilibrium flocculation (NEF) is a process in which the adsorbed polymers can undergo significant conformational relaxation [78]. The adsorbed polymer chains, shortly after landing on the particle, assume an extended configuration and, over time, the chains relax and assume a flattened configuration, as shown in Fig. 12. In the model by Pelssers et al. [78], the transition between active chains to inactive chains depends on the ionic screening length λ of the solution. The number of active polymer molecules per particle can be written in terms of a simple differential equation [78]:

$$\frac{dn}{dt} = k_{pm}c - k_i n \tag{12}$$

where k_i is the rate coefficient for inactivation. The free-polymer concentration, c, in the bulk solution starts at $c(t = 0) = c_0$ and is depleted according to Eq. (7). The inactivation coefficient k_i is empirical and can be obtained by fitting to experimental data. In most practical cases, k_i^{-1} is on the order of a few seconds [81,82]. (When k_i is zero, the equation describes normal equilibrium flocculation.) With the initial condition $n(t = 0) = 0$, Eq. (12) can be solved to give [78]

$$n(t) = \frac{k_{pm}c_0}{k_{pm}N_0 - k_i} [\exp(-k_i t) - \exp(-k_{pm}N_0 t)] \tag{13}$$

In the limit when k_i approaches zero, the number of active polymer molecules reach the asymptotic value c_0/N_0 at $t = \infty$.

The quantity $k_{pm}N_0$ is estimated using Eq. (4). For example, if the radius of

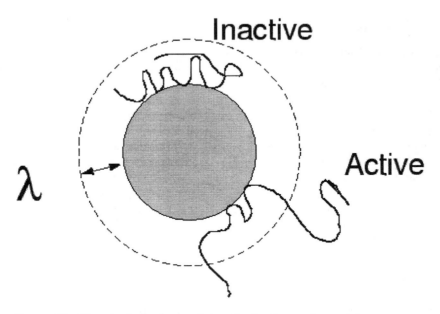

Figure 12 The adsorbed polymer chain, after landing on the particle, assumes an extended configuration and, over time, the chain relaxes to a flattened configuration (inactivated). The dashed line indicates the range of electrostatic repulsion defined by Debye length λ.

the particle $a_i = 0.5$ μm, the polymer molecule radius $a_j = 0.1$ μm, $r = 0.2$, $G = 100$ s^{-1}, $\eta = 0.01$ P, $T = 300$ K, and $N_0 = 1.0 \times 10^{10}$ cm^{-3}, one obtains a value of $k_{pm}N_0 = 0.5$, comparable to k_i.

Following the La Mer criterion that the maximum reactivity occurs at $\theta = 0.5$, and supposing the ratio of the radii of the polymer molecule to that of the particle to be $r = 0.2$, a rough estimate for the value of $n_{max} = c_0/N_0$ follows:

$$n_{max} = \frac{1}{2} \frac{4\pi a_p^2}{\pi a_m^2} = \frac{2}{r^2} = 50 \tag{14}$$

The estimated value obtained by Eq. (14) is not very different from experimental values obtained by Pelssers et al. for PEO adsorption on polystyrene particles [78].

A. Critical n_c Below Which Particle Reactivity α_{pp} is Zero:

One of the important assumptions made in the NEF model is the existence of a critical number of active polymer molecules n_c for a pair of particles to bind upon collision. The assumption is that the reactivity $\alpha_{pp} = 1$ during the period of time that $n(t) \geq n_c$; otherwise, $\alpha_{pp} = 0$. In this model, the value of n_c ($\#n_{max}$) depends on the type of polymer flocculant and on the screening length of the charged particles.

Figure 13a shows a plot of $n(t)$ for $N_0 = 2.0 \times 10^{10}$ cm^{-3} and $k_i = 0.3$ s^{-1}. The initial polymer concentration c_0 is $n_{max}N_0$ and a value of $n_{max} = 50$ is chosen. Between the period $t1$ and $t2$, $n(t) \geq n_c = 25$, and, as shown in Fig. 13b, the reactivity $\alpha_{pp} = 1$ between $t1$ and $t2$, and $\alpha_{pp} = 0$ everywhere else. Figure 13c shows the calculated normalized singlet concentration $N_1(t)\backslash N_0$ according to Eq. (10).

Figure 14 shows a plot of the number of active polymer molecules per particle $n(t)$ for systems with the radius of the particle $a_i = 0.5$ μm, polymer molecule radius $a_j = 0.1$ μm, $r = 0.2$, $G = 100$ s^{-1}, $\eta = 0.01$ P, $T = 300$ K, and for $N_0 = 5.0 \times 10^9$ (a), 1.0×10^{10} (b), 2.0×10^{10} (c), 5.0×10^{10} (d), and 1.0×10^{11} cm^{-3} (e).

Figures 13 and 14 illustrate for this model that the number of singlets at long times depends critically on the time interval $\Delta t = t2 - t1$, which, in turn, depends on the choice of the cutoff number n_c. In Fig. 15, the time interval Δt is plotted as a function of N_0 for n_c equals (a) 10, (b) 20, and (c) 25. It can be seen that for each n_c, there is a cutoff concentration for the initial primary particle concentration N_0 below which no flocculation can occur. It is interesting to note that for initial primary particle concentrations above the cutoff N_0, the model predicts a constant time interval during which flocculation occurs with unit reactivity i.e., the long time singlets N_1/N_0 ratio is insensitive to N_0.

B. Shear-Rate Dependence

The NEF model also provides a means to inspect shear-rate-dependent flocculation. The shear effect comes into the model in Eq. (4), where the collision rates k_{pm}

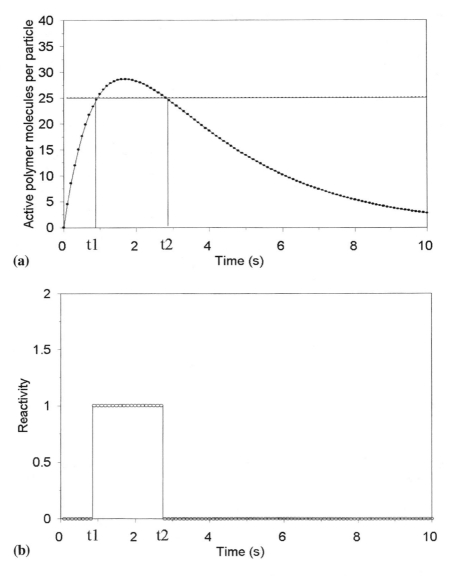

Figure 13   (a) A plot of $n(t)$ for $N_0 = 2.0 \times 10^{10}$ cm^{-3} and $k_i = 0.3$ s^{-1}. The initial polymer concentration c_0 is $n_{max}N_0$ and we use $n_{max} = 50$. Between the period $t1$ and $t2$, $n(t) \geq n_c = 25$. (b) The reactivity $\alpha_{pp} = 1$ between $t1$ and $t2$; $\alpha_{pp} = 0$ everywhere else. (c) The calculated normalized, singlet concentration $N_1(t)/N_0$ according to Eq. (10).

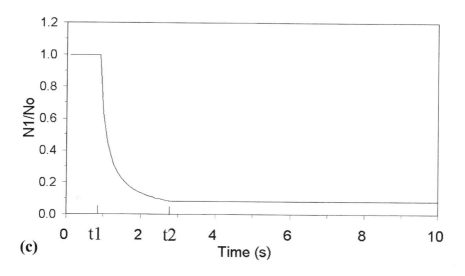

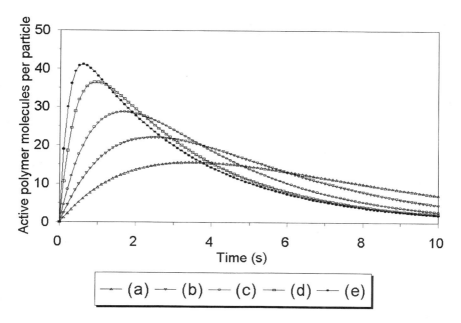

Figure 14 A plot of the number of active polymer molecules per particle $n(t)$ for systems with the radius of the particle $a_i = 0.5$ μm, polymer molecule radius $a_j = 0.1$ μm, $r = 0.2$, $G = 100$ s^{-1}, $\eta = 0.01$ P, $T = 300$ K, and for $N_0 = 5.0 \times 10^9$ (a), 1.0×10^{10} (b), 2.0×10^{10} (c), 5.0×10^{10} (d), and 1.0×10^{11} cm^{-3} (e).

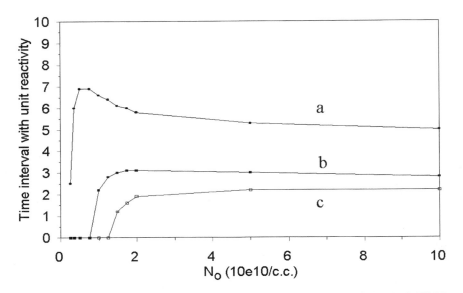

Figure 15 The time interval Δt is plotted as a function of N_0 for n_c equals (a) 10, (b) 20, and (c) 25. For each n_c, there is a cutoff concentration for the initial primary particle concentration N_0 below which no flocculation can occur. For initial primary particle concentration above the cutoff N_0, the model predicts a constant time interval during which flocculation occurs with unit reactivity i.e., the long time singlets N_1/N_0 ratio is insensitive to N_0.

and k_{pp} are calculated. Figure 16A shows $n(t)$, the number of active polymer molecules per particle, for shear rates $G = 0$, 20, 40, 60, 80, and 100 s^{-1}. The particle and polymer concentrations are 2.0×10^{10} and 1.0×10^{12} cm^{-3}, respectively. The particle radius is $a_i = 0.5$ μm; the polymer molecule radius $a_j = 0.1$ μm. Figure 16B shows the N_e/N_0, the normalized singlet particle number density at long times, as a function of shear rate for (a) $n_c = 20$ and (b) $n_c = 25$. Note that for sufficiently large shear rate G, k_{pp} is linear in G. At long times, the left-hand side of Eq. (10) behaves like G^{-1}, resulting in $N_e/N_0 \propto G^{-1}$ for large G, as is shown in Fig. 16b. The fact that N_e/N_0 is similar to G^{-1} was observed experimentally by Pelssers et al. [78].

C. The Limit When k_i Approaches Zero: Equilibrium Flocculation

The model by Pelssers et al. can describe equilibrium flocculation (EF) in the limiting case when the deactivation rate constant k_i approaches zero. The number of

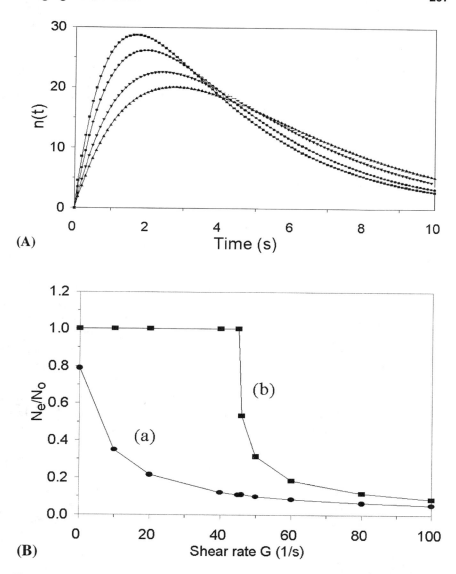

Figure 16 (A) The active polymer molecules per particle as a function of time, $n(t)$, are plotted for shear rates $G = 0, 20, 40, 60, 80,$ and 100 s^{-1}. The particle and polymer concentrations are 2.0×10^{10} and 1.0×10^{12} cm^{-3}, respectively. The particle radius is $a_i = 0.5$ μm; the polymer molecule radius $a_j = 0.1$ μm. (B) The N_e/N_0, the normalized singlet particle number density at long times, as a function of shear rate for (a) $n_c = 20$ and (b) $n_c = 25$. For a sufficiently large shear rate G, k_{pp} is linear in G, resulting in $N_e/N_0 \propto G^{-1}$ for large G.

active polymer molecules increases steadily with time according to the following expression:

$$n(t) = n_{max}[1 - \exp(-k_{pm}N_0t)] \tag{15}$$

When $n(t)$ reaches n_c, every binary particle collision becomes effective, and the number of primary (singlet) particles, N_1, decays according to Eq. (10) until all of the primary particles are consumed.

VIII. FLOCCULATION MONITORING

Light-scattering experimental techniques are used in various configurations to monitor the dynamic aspects of flocculation [83,84]. Pelssers et al. [85] developed a single-particle optical sizer (SPOS). In the SPOS, a narrow stream of dispersion is hydrodynamically focused and then passed through a focused elliptical laser beam. The dispersion is dilute enough that the laser beam hits at most one cluster at a time. Because the scattering intensity is proportional to the cluster size, the technique yields excellent resolution of the time-dependent aggregate size distribution. Using SPOS, Pelssers et al. obtained results for PEO with molecular weights of 3×10^6 and 7.7×10^7 g/mol as flocculant for 696-nm-diameter polystyrene latex particles [78]. The experimental work was carried out to determine the optimum polymer dose, particle concentration dependence, time dependence, and shear-rate dependence of the bridging flocculation efficiency.

To summarize, the NEF model was found to be quite successful in its ability to quantitatively explain the data. The key features of the model are (1) the inclusion of the concept of inactivation rate coefficient k_i in the calculation of the time-dependent number of active adsorbed chains and (2) the concept of the cutoff number n_c for the active polymer molecules on the particle, below which the binary particle collisions are not effective.

It should be noted that several assumptions are made in the model: (1) The rate coefficient k_i is assumed to be a constant, whereas it may, in some cases, depend on n, the number of active polymer molecules. (2) The model also assumes that for $n \geq n_c$, the reactivity $\alpha_{pp} = 1$. A more realistic assumption is a smoother function of $\alpha_{pp}(n)$. (3) The collision events between polymer molecules and the particles are assumed to happen before the particles start to flocculate. In other words, the timescale of the adsorption process is assumed to be much less than that of particle collision. This is consistent with the model in that effective particle collisions take place only when $n \geq n_c$. A study has shown, however, that this assumption could lead to significant error [86]. (4) Only binary particle collisions between the singlet particles were considered; singlet–doublet collisions and all the higher-order terms were ignored.

IX. FLOC BREAKUP

In irreversible flocculation, polymer-bridged particle clusters can break up only by the application of shear. The size of the clusters depends on the shear rate and the floc strength [87,88]. Because a precise determination of shear rate and floc strength cannot be made accurately, quantitative understanding is difficult to accomplish for irreversible systems. However, one might learn from investigating the reversible bridging flocculation where locally formed droplets of transient flocs can break up without an applied shear.

What makes studying the breakup of a reversible floc possible is the creation of a transient floc by mixing a small amount of concentrated polymer solution in a large volume of dilute and uniform suspension of particles. During mixing, small droplets composed of many particles and polymer molecules are formed locally. If the total amount of added polymer molecules are low enough that they cannot reach the minimum number of molecules per particle to form bridges, then the system will eventually (at long times) reach a state in which the particles are restabilized. Because these droplets take time to disperse, experiments can be designed to study the breakup of the small droplets, as shown in Fig. 17.

The experiments by Dewalt et al. [75] were carried out by adding a small amount of high-concentration (0.5% by weight) telechelic PEO polymer to a uniform, dilute suspension (6×10^{-3} by volume) of polystyrene colloidal particles. This mixture yields 10 polymer molecules per sphere on average, a state where the system was known to be stable. A slightly opaque sample was produced initially, and after a few minutes, the sample turned transparent.

Telechelic PEO molecules in aqueous solution are linear chains terminated with one sticky hydrophobe on each end. Each of these polymer molecules can form a bridge between two colloidal polystyrene (PS) particles by attaching one hydrophobe onto each particle. The polymer molecules were C20-116 PEO, molecules with 1×10^6 g/mol PEO with both ends capped by $C_{20}H_{41}$ hydrophobes. The molecular structure is $C_{20}H_{41}$–O–(DI–PEO)$_7$–DI–O–$C_{20}H_{41}$ (Union Carbide, Cary, NC), where DI is the isophorone di-isocyanate linkage. The particles were 28-nm-diameter PS (Duke Scientific, Palo Alto, CA).

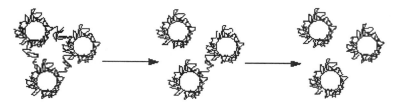

Figure 17 Transient bridged state breakup from a triplet to singlets.

The transient polymer-bridged colloidal clusters and the relaxation of the cluster decay were monitored by a dynamic light-scattering technique and by measuring the light-scattering intensity. Light-scattering (DLS) techniques that measure both the polarized scattering (VV: vertically polarized light in and vertically polarized light out) and depolarized scattering (VH: vertically polarized light in and horizontally polarized light out) intensities were employed simultaneously. Two photomultiplier detectors were used, each positioned at symmetric scattering angles of ±45° from the incident laser beam (Spectra Physics 2025, Ar ion at 488 nm). The VV and VH intensity data were accumulated independently by two Brookhaven Instruments BI-2030 autocorrelators and were averaged over periods of 30–60 s. This time was chosen to provide adequate time resolution for the kinetics of relaxation. The experimental setup is shown in Fig. 18, with AVH being the depolarized analyzer and AVV being the polarized analyzer.

The scattered intensity from N primary particles of mass m can be written [89]

$$I_0 \propto N_m^2 \tag{16}$$

When aggregates are formed by flocculation, the polarized scattered intensity will increase linearly with the number of primary particles per aggregate because the

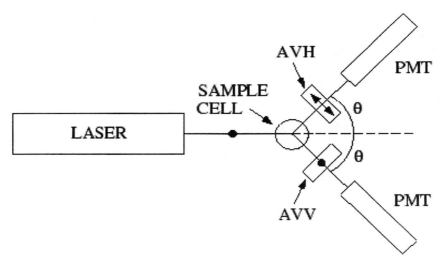

Figure 18 Two photomultiplier detectors were used, each positioned at symmetric scattering angles of ±45° from the incident laser beam (Spectra Physics 2025, Ar ion at 488 nm). The VV and VH intensity data were accumulated independently by two Brookhaven Instruments BI-2030 autocorrelators. AVH is the depolarized analyzer and AVV is the polarized analyzer.

total particle number is conserved. The scattering intensity $I_{cluster}$, measured from a polydisperse system of particles with a different aggregation number, provides a measure of the average cluster size:

$$\frac{I_{cluster}}{I_0} = n_{ave} \tag{17}$$

The scattered intensity was measured as a function of time while the system decayed from the clustered state to the single particle state that one expects at long times.

Figure 19 shows the normalized scattered intensity as a function of time. Time $t = 0$ is the point at which the data collection began (i.e., several seconds after addition and subsequent mixing of the polymer with the uniform colloidal suspension). The initial low scattering intensity shown in Fig. 19 is believed to be caused by multiple scattering, because the sample appeared hazy at this stage. As the clusters began to disintegrate, the scattering intensity at first increased and then started to decrease.

The data from Fig. 19 are plotted on a semilog scale in Fig. 20, which indicates two relaxation times except at the very short time regime (less than 300 s) where multiple scattering could be seen visually. The fast decay has a characteris-

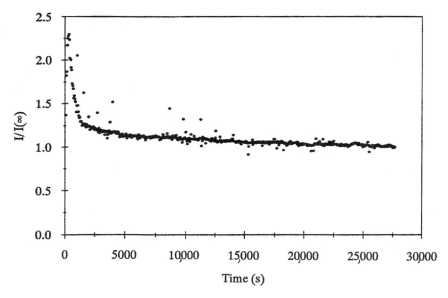

Figure 19 Normalized VV scattering intensity for a mixture of 10 C20-116 chains per 28-nm PS particle at 25 nm.

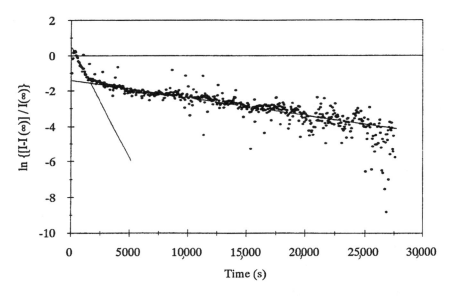

Figure 20   Semilog plot of the VV scattering for the sample in Fig. 19.

tic decay time of 800 s, whereas the slow decay has a characteristic decay time of 10,000 s.

Depolarized scattering comes from two effects: (1) multiple scattering in the highly flocculated system at early times or (2) in the single scattering regime, from clusters with high anisotropic shape (e.g., the doublets). VH scattering is negligible from a system of primary particles. Figure 21 shows the VH scattering data, taken simultaneously with those of Figs. 19 and 20, in a semilog plot. The VV and VH scattering data acquisitions were synchronized within an error of about 5 s, short compared to the observed relaxation time. Again, two clear relaxation times are apparent with characteristic decay times of 350 and 10,000 s.

The VV scattering intensity–intensity autocorrelation functions were also calculated. These autocorrelation functions were averaged over the collection period of about 60 s. The average hydrodynamic radius of the scattering clusters was then calculated from a cumulant expansion of the autocorrelation function [90]. Figure 22 shows the cluster size obtained from dynamic light-scattering analysis of the VV scattering data for the same flocculation event recorded in Figs. 19–21. Each point in Fig. 22 represents an individual correlation function at the corresponding time with an average over 60 s.

Temperature-dependent measurements of the VV scattering were also made to investigate the energy involved in polymer bridging. In Fig. 23, the intermedi-

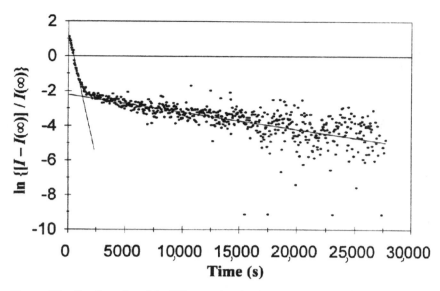

Figure 21   Semilog plot of the VH scattering for the same sample in Figs. 19 and 20.

ate characteristic decay times obtained from VV scattering are plotted versus $1/T$. The Arrhenius behavior indicates an activation energy of 30 kT.

The initial observation that the sample increased in opacity when the polymer was introduced suggests that significant flocculation was occurring. However, because this flocculation was only transient, large aggregates break up into smaller ones and the sample returned to being transparent within several minutes. The fact that, at long times, the suspension became and remained transparent, and that at long times, the DLS measured particle sizes similar to the bare particle size (with a very small amount of adsorbed polymers) confirm the expectation that individual singlet particles dominate the suspension. Thus, the experimental observations suggest that mixing polymer solution with the suspended 28-nm PS particles led to a transient nonequilibrium state where large flocculated clusters could form initially, and with time, these large aggregates relax to an equilibrium state composed of single particles with polymer adsorbed to the surface, as shown in Fig. 17. By analyzing the decay in the light-scattering intensity, we can follow the relaxation process of the cluster breakup. From the relaxation data shown in Figs. 19–22, three time regimes are apparent: The short time covers about the first 300 s, the intermediate time regime from about 300 to 3000 s, and the long time regime from 3000 to 30,000 s.

Just after mixing the concentrated polymer solution into the particle suspen-

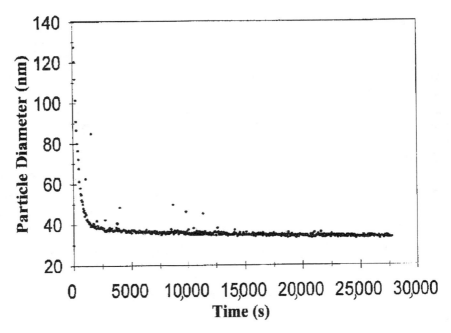

Figure 22   Particle size versus time plot for the same sample used in Figs. 19–21.

sion, a population of transient polymer-bridged colloidal clusters was formed. In the short time regime, it is difficult to quantitatively interpret the data because it is difficult to reproduce the initial conditions from run to run. The samples appear milky due to multiple scattering, which is difficult to analyze by conventional light-scattering techniques. However, the sample is transparent in the intermediate and long time regimes.

Figures 20 and 21 show two distinct regimes with well-defined relaxation times. In the intermediate regime, the average number of particles in a cluster is on the order of 2 to 5. There exist several relaxation channels for the clusters to break up. For example, a cluster of five particles can relax by breaking up into clusters of one, two, three, or four particles. The long-time relaxation occurs in the time domain where the average cluster size is relaxing from about 1.2 to 1 particle. It is reasonable to assume that the population of clusters in this time range was dominated by doublets and singlets, in which case, there exists only one decay channel (i.e., from doublets to singlets). The possibility of recombination of singlets to yield doublets is low because at polymer coverage as low as 10 molecules per particle, both hydrophobes tend to bind on the particle strongly and are, thus, not available for rebridging.

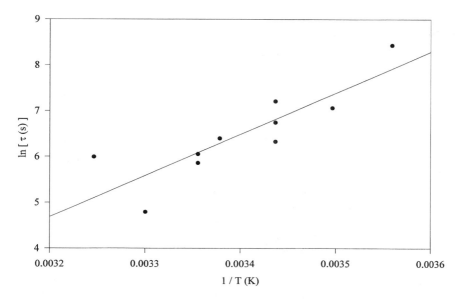

Figure 23 Semilog plot of the relaxation time versus $1/T$ for result from VV static scattering intensity. The solid line is an Arrhenius function fit to data with an activation energy of 30 kT.

At very long times, the average particle diameter approaches 32 nm with a very low polydisersity [75]. This implies that the 10 polymer chains form an adsorbed layer of 2 nm thickness, not an unreasonable value based on what is known of similar polymer molecules adsorbed on PS latex particles [91]. This further supports the point that, at long times, singlet particles dominate.

The temperature-dependent measurements in the intermediate time regimes show that the aggregates break up through an activation process. The activation energy of 30 kT suggests that the particles were bridged, on the average, by more than one chain because the adsorption energy per end group is expected to be on the order of 10 kT for the $C_{20}H_{41}$ hydrophobe.

X. CONCLUSIONS

The time-dependent configurations of adsorbed polymers are crucial in determining the extent and kinetics of bridging flocculation. Bridging requires polymers of high molecular weight with strong affinity for the colloidal surface, at least at the ends of the polymer chains. If this criterion is not met, flocculation may still occur, but through other mechanisms. In addition to chain architecture, processing conditions must also be carefully manipulated to achieve maximum bridging. This includes choosing an amount of polymer less than that needed to saturate the particle surface and good solvent conditions to maximize chain extension from the surface.

Bridging flocculation kinetics involves several time scales: mixing time, polymer adsorption time, particle collision time, and polymer reconformation time. The classic theory of von Smoluchovski is introduced for the calculation of particle–particle and polymer–particle collision rates. The reconformation time of the adsorbed polymer molecules is considered in the phenomenological level in a theoretical approach by Pelssers et al. for treating nonequilibrium flocculations. The approach by Pelssers et al. provides a quantitative analysis of kinetics of flocculation by taking into account the inactivation of polymer molecules, caused by adsorbed polymer reconformation, to bridging and its effects to flocculent efficiency. The second part ends with an experimental study of transient bridging and floc breakup by light scattering.

Inhomogeneous mixing of polymer and particle can produce transient-state flocs that eventually break up into stabilized particle suspension. We demonstrated an example where a nonequilibrium bridged state produced in a reversible bridging system by inhomogeneous mixing. The transient bridged flocs were produced so that they break up into stabilized individual particles. The relaxation appears to proceed in three stages: a fast first stage lasts less than 300 s, the intermediate stage relaxes in about 3000 s, and the third stage lasts about 30,000 s before the system reaches stabilized singlet particles. During the intermediate stage, the floc breakup was found to follow Arrhenius behavior with an activation energy of about 30 kT, an energy corresponding to multiple bridges. In the future, further temperature-dependence studies of the slow relaxation could help us understand the doublet–singlet transitions better.

ACKNOWLEDGMENTS

MMS gratefully acknowledges the support of the National Science Foundation (CTS-9209290). HDO acknowledges the support of the National Science Foundation (CTS-9805887). Both of the authors appreciate the support of the NSF–IUCRC for Polymer Interfaces at Lehigh University.

REFERENCES

1. DH Napper, Polymeric Stabilization of Colloidal Dispersions. London: Academic Press, 1983, pp. 412–413.

2. L Lapcik, B Alince, TGM van de Ven. J Pulp Paper Sci 21:19, 1995.
3. PR Sperry, JC Thibeault, EC Kostansek. Adv Org Coat Sci Technol Ser 9:1–11, 1985.
4. BC Bonekamp, J Lyklema. J Colloid Interf Sci 113:67, 1986.
5. J Gregory. J Colloid Interf Sci 42:448, 1973.
6. J Gregory. J Colloid Interf Sci 55:35, 1976.
7. T Cosgrove, TM Obey, B Vincent. J Colloid Interf Sci 111:409, 1986.
8. G Durand-Piana, F Laufma, R Audebert. J Colloid Interf Sci 119:474, 1987.
9. R Denoyel, G Durand, F Laufma, R. Audebert. J Colloid Interf Sci 139:281, 1990.
10. RIS Gill, TM Herrington. Colloids Surf 22:51, 1987.
11. RIS Gill, TM Herrington. Colloids Surf 25:297, 1987.
12. F Mabire, R Audebert, C Quivoron. J Colloid Interf Sci 97:120, 1984.
13. GM Lindquist, RA Stratton. J Colloid Interf Sci 55:45, 1976.
14. H Tanaka, L Odberg, L Wagberg, T Lindstrom. J Colloid Interf Sci 134:219, 1990.
15. T Lindstrom, C Soremark, L Eklund. Pulp Paper Mag Trans Tech Sect 3:114, 1977.
16. M Falk, L Odberg, L Wagberg, G Risinger. Colloids Surf 40:115, 1989.
17. Y Xiang, P Somasundaran J Colloid Interf Sci 177:283–287, 1996.
18. VK La Mer, TW Healy. Rev Pure Appl Chem 13:112, 1963.
19. JMHM Scheutjens, GJ Fleer, MA Cohen Stuart. Colloids Surf 21:285–306, 1986.
20. A Takahashi, M Kawaguchi. Adv Polym Sci 46:1–65, 1982.
21. M Kawaguchi, K Hawakawa, A Takahashi, Macromolecules 16:631–635, 1983.
22. T Cosgrove, TL Crowley, K Ryan, JRP Webster, Colloids Surf 51:255, 1990.
23. EM Lee, RK Thomas, AR Rennie. Eurphys. Lett. 13:135, 1990.
24. MA Cohen Stuart, FHWH Waajen, T Cosgrove, B Vincent, TL Crowley. Macromolecules 17:1825, 1984.
25. GJ Fleer, MA Cohen Stuart, JMHM Scheutjens, T Cosgrove, B Vincent. Polymers at Inteferfaces. London: Chapman Hall, 1992, pp. 234–236.
26. A Pefferkorn, Carroy, R Varoqui. J Polym Sci Polym Phys Ed 23:1997, 1985.
27. JC Dijt, MA Cohen Stuart, GJ Fleer. Macromolecules 27:3219, 1994.
28. Z Fu, MM Santore. Colloids Surf A: Physiochem Eng Aspects 135:63–75, 1998.
29. JC Dijt, MA Cohen Stuart, GJ Fleer. Macromolecules 25:5416–5423, 1992.
30. P Frantz, S Granick. Phys Rev Lett 66:899, 1991.
31. JN Israelachvili, D Tabor. Proc Roy Soc London Ser A 331:19–38, 1972.
32. JN Israelachvili GE Adams. J Chem Soc Faraday Trans I 74:975–1001, 1978.
33. JN Israelachvili. Acc Chem Res 20:415–421, 1987.
34. CS Lee, G Belfort. Proc Natl Acad Sci USA 86:8392–8396, 1989.
35. S Yamada, JN Israelachvili. J Phys Chem B 102:234, 1998.
36. RG Horn, DR Clarke, MT Clarkson. J Mater Res 3:413–416, 1988.
37. RG Horn, DT Smith, W Haller. Chem Phys Lett 162:404–408, 1989.
38. BV Derjaguin. Kolloid Z 69:155–164, 1934.
39. Y Almog, J Klein. J Colloid Interf Sci 106:33–44, 1985.
40. JN Israelachvili, M Tirrell, J Klein, Y Almog. Macromolecules 17:204–209, 1984.
41. J Klein, PF Luckham. Macromolecules 17:1041–1048, 1984.
42. PF Luckham, J Klein. Macromolecules 18:721–728, 1985.
43. WB Russel, DA Saville, WR Schowalter. Colloidal Dispersions. Cambridge: Cambridge University Press, 1989, p. 201.
44. JMHM Scheutjens, GJ Fleer. J Phys Chem 83:1619–1635, 1979.

45. JMHM Scheutjens, GJ Fleer. J Phys Chem 84:178–190, 1980.
46. PG deGennes. Macromolecules 14:1637, 1981.
47. J des Cloizeaux. J Phys 36:281–291, 1975.
48. PG deGennes. Macromolecules 15:492–500, 1982.
49. JMHM Scheutjens, GJ Fleer. Adv Colloid Interf Sci 16:361–380, 1982.
50. JMHM Scheutjens, GJ Fleer. Macromolecules 18:1882, 1985.
51. JMHM Scheutjens, GJ Fleer. J Colloid Interf Sci 111:504, 1986.
52. HJ Ploehn, WB Russel, CK Hall. Macromolecules 21:1075, 1988.
53. HJ Ploehn, WB Russel. Macromolecules 22:266, 1989.
54. HJ Ploehn. Macromolecules 27:1617–1626, 1994.
55. HJ Ploehn. Macromolecules 27:1627–1636, 1994.
56. S Asakura, F Oosawa. J Chem Phys 22:1255, 1954.
57. S Asakura, F Oosaawa. J Polym Sci 33:183, 1958.
58. AP Gast, CK Hall, WB Russel. J Colloid Interf Sci 96:251–267, 1983.
59. AP Gast, CK Hall, WB Russel. Faraday Discuss Chem Soc 76:189–201, 1983.
60. MM Santore, WB Russel, RK Prud'homme. Macromolecules 22:1317–1325, 1989.
61. MM Santore, WB Russel, RK Prud'homme. Macromolecules 23:3821–3832, 1990.
62. MM Santore, RK Prud'homme, WB Russel. Faraday Discuss Chem Soc 90:323, 1990.
63. G Durand, F Laufma, R Audebert. Prog Colloid Polym Sci 266:278, 1988.
64. T Wang, R Audebert. J Colloid Interf Sci 121:32, 1988.
65. K Takamura, HL Goldsmith, SG Mason. J Colloid Interf Sci 82:175, 1981.
66. K Takamura, HL Goldsmith, SG Mason. J Colloid Interf Sci 82:190, 1981.
67. T Wang, R Audebert. J Colloid Interf Sci 119:459–465, 1987.
68. L Eriksson, B Alm, P Stenius. Colloids Surf A: Physchem Eng Aspects 70:47–60, 1993.
69. E Mubarekyan, MM Santore. Langmuir 14:1597–1603, 1998.
70. E Bohmer. In: JP Casey, ed. Pulp and Paper Chemistry and Chemical Technology. New York: Wiley, 1981, p. 1515.
71. TGM van de Ven, Nord Pulp Paper Res J 1:130, 1993.
72. TGM van de Ven, B Alince, J Colloid Interf Sci 181:73–78, 1996.
73. J Gregory. Colloids Surf 31:231–253, 1988.
74. RW Slater, JA Kitchener. Discuss Faraday Soc 42:267, 1966.
75. LE Dewalt, Z Gao, HD Ou-Yang. In: JE Glass, ed. Hydrophilic Polymers; Performance with Environmental Acceptability. Washington, DC: American Chemical Society, 1996, pp. 395–405.
76. X Yu, P Somasundaran. J Colloids Interf Sci 177:283–287, 1996.
77. X Yu, P Somasundaran. J Colloids Interf Sci 178:770–774, 1996.
78. EGM Pelssers, MA Cohen Stuart, GJ Fleer, J Chem Soc Faraday Trans 86:1355–1361, 1990.
79. M von Smoluchowski, Phys Z 17:539, 1916; Z Phys Chem 92:129, 1917.
80. A Molsli, R Hogg. J Colloid Interf Sci 102:2320, 1984.
81. MA Cohen Stuart, M Tamai, Macromolecules 21:1863, 1988.
82. MA Cohen Stuart, M Tamai, Langmuir 4:1184, 1988.
83. J Gregory, DW Nelson, In: J Gregory, ed. Solid-Liquid Separation Chichester: Ellis Horwood, 1984, pp. 172–182.

84. J Eisenlauer, D Horn. Colloids Surf 14:121, 1985.
85. EGM Pelssers, MA Cohen Stuart, GJ Fleer, J Colloid Interf Sci 137:350–361, 362–372.
86. Hsu and Lin, Colloids Polym 270, 1992.
87. AF Horn, EW Merrill, Nature, 312:140, 1984.
88. W Ditter, J Eisenlauer, D Horn. In: Tadros, ed. The Effect of Polymers on Dispersion Process. London: Academic Press, 1982, pp. 323–341.
89. BJ Berne, R Pecora. Dynamics Light Scattering. New York: Wiley, 1976.
90. B Chu. Laser Light Scattering, New York: Academic Press, 1991, p. 247.
91. HD Ou-Yang, ZH Gao. J Phys II (France) 1:1375, 1991.

8
Metal Complexation in Polymer Systems

Tohru Miyajima
Saga University, Saga, Japan

I. INTRODUCTION

Whether naturally occurring or synthetic in origin, polymer ligands with metal-coordination atoms, such as oxygen and nitrogen atoms, sometimes show strong and variable metal-ion binding ability, which is closely related to their delicate functions of these polymers. Their coordination behavior is much more sensitive to solution conditions, such as pH and added salt concentration levels, than the simple ligand molecules, whose structures mimic the polymer functional groups. Despite the essential role of metal complexation of these polymer ligands in biological and environmental systems as well as in separation science, very little has so far been disclosed on the equilibrium, kinetic, and structural aspects of the macromolecular complexes. Information gained by systematic studies about simple or oligomeric ligand molecules have not always fully been applied to the chemistry of polymer ligand complexation. In most cases, their experimental results have been interpreted insufficiently and/or improperly because of the neglect of the "polymer effects." Several difficulties encountered in polymer ligand complexation still remain unsolved at the present time, even though the information gained by overall or macroscopic complexation should finally be related to individual or microscopic complexation at a particular reaction site, after correcting for the polymer effects.

Because most of the polymer ligands are charged negatively or positively, depending on the solution property, it is of primary importance to take into account their polyelectrolytic nature. The strong electric field formed at the polymer surface varies mainly with pH and the added salt concentration levels. In this chapter, the complexation equilibria of polymer ligands with linear structure will be explored, because their behavior is believed to give fundamental aspects on metal complexation in polymer systems. Polyacrylic acid (PAA) and poly(*N*-vinylimidazole) (PVIm), are used as representatives of weak acidic polyelectrolytes [1–3] and weak basic polyelectrolytes [3,4], respectively. The chemical structures as well as abbreviations of the linear polymers discussed in this chapter are illustrated in Fig. 1. When a PAA molecule is neutralized with a base, carboxylic acids repeated on the linear backbone are dissociated to endow negative charges to the polymer molecules. Accumulation of the negative charges induces countercation attraction in the vicinity of the polymer skeleton and, hence, protonation as well as metal complexation with the polymer will be enhanced. On the other hand, when a PVIm molecule is treated with acid, the fixed neutral imidazole groups are associated with protons to endow the polymer molecules with positive charges. In this instance, the crowded positive charges at the polymer surface repel metal cations from the polymer, which suppresses metal complexation with imidazole groups on the polymer ligands. The free metal ions at the surface of the charged polymers are extremely concentrated or diluted compared with the bulk-solution phase, and the free metal ion concentration in the bulk-solution phase experimentally accessible cannot be used directly for the equilibrium calculation of "intrinsic constants." The metal ions that are located in the vicinity of the polymer molecules but are essentially free are designated as "territorially bound" [5,6], and the metal ions bound directly to the functional groups of the polymer ligands accompanied by desolvation are designated "site bound" [5,6], which include contact ion-pairs and chelate complexes.

Another important feature of polymer complexation is the multidentate coordination property due to the immobilization of ligating groups on the polymer molecules. This sometimes assists simultaneous metal coordination with several functionalities belonging to identical and/or different polymer segments. This aspect is discussed in detail in this chapter by exemplifying the complexation of both PAA and PVIm ligands. Monodentate as well as multidentate complex formation of these polymers have been revealed [1,2,4] after correcting for the polyelectrolytic effect of the polymers, whose behavior is apparently controlled by the average spacing of the free-ligand functionalities arrayed on the linear backbone. Spectroscopic evidence for the multidentate coordination of the polymer ligand has also been provided, which validates the present thermodynamic approach.

Compared with these weakly acidic or basic polyelectrolytes where some portion of metal ions are bound in the manner of "site bound", the nature of metal ion binding to strong acidic polyelectrolytes (Fig. 1), such as polystyrene sulfonate (PSS) and poly(vinyl sulfonate), (PVS), is considered to be purely electrostatic (i.e., the bound metal ions are considered completely in "territorially bound" fashion). However, when mobile hydrophobic ligands are added to the metal ion–polymer system, the metal complexation with the ligand molecules is sometimes remarkably enhanced due to preferential binding of hydrophobic complexes formed to polymer molecules. This indicates the importance of the hydrophobicity of the

a) weak acidic polyelectrolyte

PAA
Polyacrylic acid

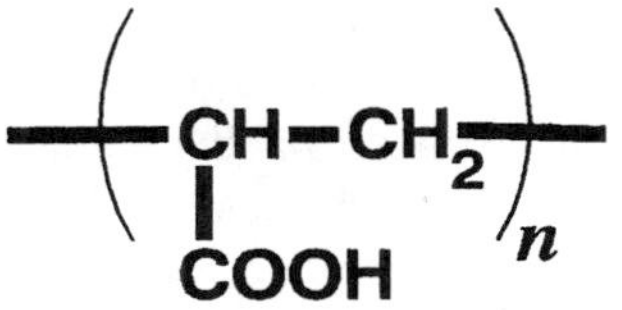

b) weak basic polyelectrolyte

PVIm
Poly(N-vinylimidazole)

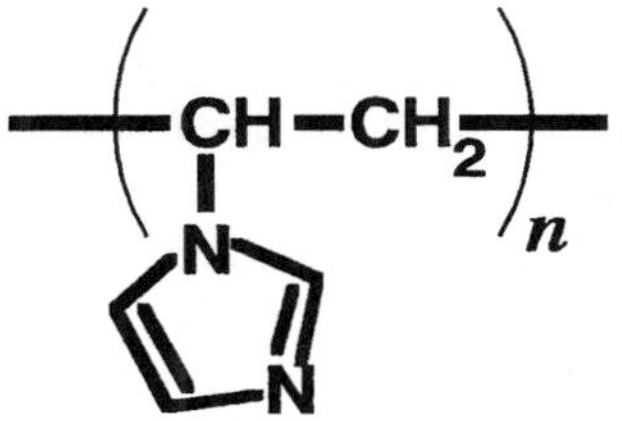

c) strong acidic polyelectrolyte

PSS
polystyrenesulfonate ion

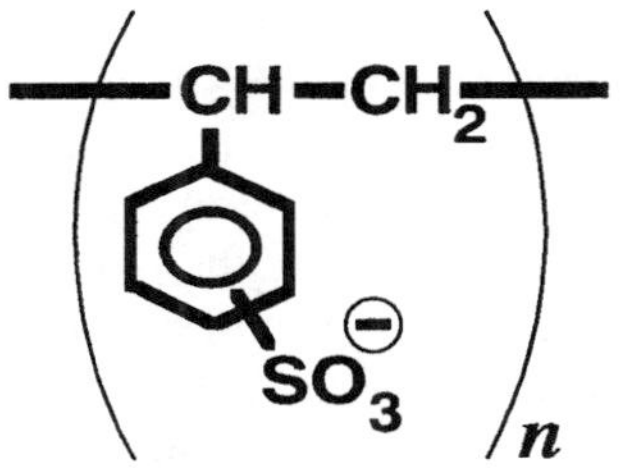

PVS
polyvinylsulfonate ion

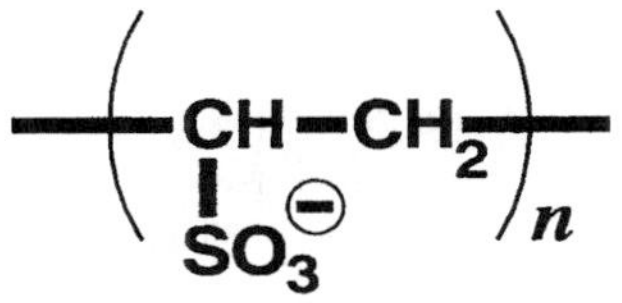

d) strong basic polyelectrolyte

PVBTMA⁺
polyvinylbenzyltrimethylammonium ion

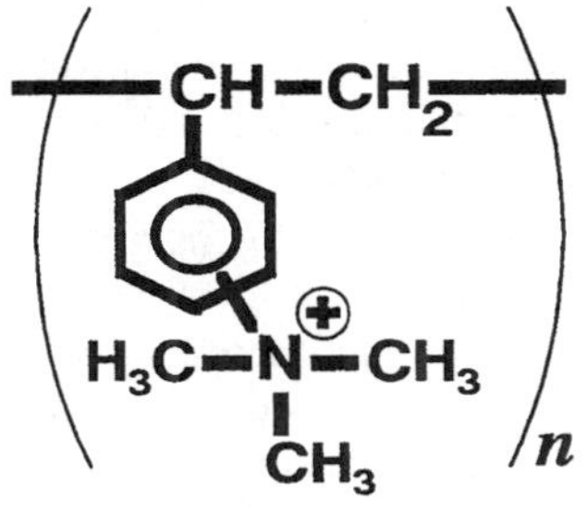

Figure 1 The chemical structures and the abbreviations of the polymers discussed in this chapter.

polymer molecules as well as the metal complexes involved. This aspect is most pronounced in the combination of anionic complexes and a strong basic polymer (Fig. 1), and a spectroscopic study on Co^{2+}/SCN^- complexation in the presence of poly(vinyl benzyltrimethylammonium) ions (PVBTMA$^+$), is performed in order to investigate the remarkable hydrophilicity/hydrophobicity effect of supporting electrolytes on the complexation [7]. Because it has been revealed that water uptake by positively charged polymers is particularly dependent on the chemical nature of the polymer molecules compared with negatively charged polymers, it is stressed in this chapter that special attention should be paid to the complexation with anionic mobile ligands in the presence of positively charged polymers. This important aspect is directly related to anion selectivity of anion exchangers as well. The hydrophilicity/hydrophobicity nature of monovalent cations and anions present in excess of the complexation equilibrium is discussed precisely.

All these ionic reactions involved in polymer systems have been analyzed in a unified manner based on a Gibbs–Donnan concept, and fundamental aspects are described in the following section. The equilibrium analyses of the acid-dissociation reaction of both PAA and the conjugate acids of PVIm, PVImH$^+$, have been discussed in Section III. A rationale for the straightforward assessment of the polyelectrolytic nature of a polymer molecule at a particular solution condition, such as the degree of the polyacid, α, and the inert salt concentration levels, C_S (mol/dm^3) by the use of acid-dissociation equilibrium as a probe, has been presented in this section. Metal complexation equilibria of polyelectrolytes with weakly acidic or basic functionalities are analyzed in a unified manner (Section IV). The equilibrium analyses based on the Gibbs–Donnan concept have been verified by exemplifying PAA and PVIm molecules as polymer ligands. After correcting for the electrostatic effect inherent in the apparent or macroscopic metal complexation equilibria, intrinsic binding constants, independent of C_S, are expressed as a function of α. The electrostatic nonideality term estimated by the acid-dissociation equilibrium analyses has successfully been applied to electrostatic correction for the overall metal complexation equilibria. Multicoordination properties of these linear polymer molecules are extracted and compared with each other in this section. In section V, the abnormally high complexability of Co^{2+}/SCN^- system in the presence of PVBTMA$^+$ is illustrated; the complexation equilibria have also been analyzed according to the Gibbs–Donnan logic. All these equilibrium data presented and discussed in this chapter were compiled at 298 K unless otherwise stated.

Even though the polymer ligands examined in this chapter are restricted to one-dimensional or linear polymer molecules, it is believed that the fundamental concept presented here can be extended directly to other polymer molecules of higher dimensionalities [i.e., two-dimensional (surface) polymers such as ionic micelles, ionic latexes, and metal oxides] in which the present readers are particularly interested.

II. POLYELECTROLYTIC NATURE—A MODELING

Theoretical approaches to modeling of the polyelectrolytic nature of linear polymers proposed up to the present time can be classified into two categories: (1) approaches based on the Poisson–Boltzmann (P-B) equation [8,9] and (2) Manning's counterion–condensation (C-C) theory [10]. According to the P-B equation, the electrostatic surface potential of polyions can be calculated directly, where most of the linear polyacids are approximated as rodlike molecules [8,9]. The acid-dissociation behaviors of polyacids have been explained by computing the electrostatic nonideality terms with two arbitrary parameters: the closest approach of the counterions to the polyion surface and the linear charge separation of the polyions. Compared with the P-B equation, only one adjustable parameter, ξ, is needed for the C-C theory, ξ being relatable to the structure of the polyion and defined as $\xi = 7.14/b$, where b (Å) is the averaged linear charge separation of a rigid and infinitely long linear polyion whose skeletal volume is neglected [10]. It is anticipated for all the linear polyions with $\xi > 1$ that univalent counterions are "condensed" or "accumulated" in the vicinity of the polyion molecules until the ξ value is reduced to unity. Accordingly, discontinuity is predicted by this model to appear in the titration curve corresponding to polycarboxylic acid-dissociation equilibria at $\xi = 1$; however, such a peculiar phenomenon has not clearly been observed in their thermodynamic properties [10], which has been a matter of debate [11]. Ambiguities in the geometrical parameters such as the "closest approach" and the "linear charge separation" needed for the computations by the theoretical approaches become serious when they are applied to predict the equilibria encountered in the polyions with heterogeneous functionalities on flexible backbones, such as naturally occurring polymer molecules.

With regard to the state of counterions "bound" to polyions, it is generally accepted that some fraction of counterions "bound" in a thermodynamic sense can move freely in a "bound region" formed in the vicinity of the polyion skeleton [12,13]. This "naive" picture of the "purely electrostatic binding" in polyions has been conceptualized as "counterion–condensation" by Manning [12,13], and such a binding is termed "territorial." In the C-C theory, the "bound region" is defined in the vicinity of polymer skeleton as a "polyelectrolyte phase," being separated from the bulk-solution phase [12,13]. Due to this simplification, the model does not take into consideration the concentration gradients of mobile small ions (i.e., counterions and co-ions) from the polyion surface to the bulk solution, but gives the averaged concentrations of these ions in both polyelectrolyte and bulk solution phases. It has been proven in most cases, however, that the averaged concentrations calculated by the C-C theory are in good agreement with those calculated by the P-B equation [14,15].

Because of the essential equality of the "two-phase" approach in the treatment of ion-binding equilibria of linear polyions to water-insoluble cross-linked linear polyion systems (i.e., ion exchangers), Marinsky has analyzed the ion-binding equilibria of water-soluble linear polyions based on the Gibbs–Donnan concept in a manner similar to ion exchangers, even though no phase boundary is observable between the two phases [16,17]. The ion-exchange equilibria of counterions between the two phases is quantified according to the Gibbs–Donnan equilibrium. No geometrical parameter is needed for this approach; it only uses the logic of "two phase" in a polyelectrolyte/excess inert salt system. A "polyelectrolyte phase volume," V_p, is defined and the distribution of counterions and co-ions between the polyelectrolyte phase and the bulk solution phase is clearly described by the Gibbs–Donnan equation. Even though this concept has originally been indicated by Alexandrowicz and Katchalsky [18] and Oosawa [19], its applicability to the analyses of various categories of ionic reactions in polyion systems has recently been verified by several research groups, including the present author [5,6]. The theoretical computations by the P-B equation or C-C theory require hypothetical molecular geometries of polyions as well as ambiguous structural parameters; however, the Gibbs–Donnan approach determines the electrostatic nonideality terms directly by the use of ion-binding equilibria as a probe [5,6], which is experimentally available.

III. ACID-DISSOCIATION EQUILIBRIUM ANALYSES OF LINEAR POLYELECTROLYTES

Because of the primary importance in characterization of ion binding in polyion systems, thermodynamic analyses of acid-dissociation equilibria of linear polyelectrolytes with weakly acidic or basic functionalities have extensively been studied in various disciplines of chemistry. Acid-dissociation equilibria of the monomeric functionalities of a weak acidic polyelectrolyte, $(HA)_n$, and the conjugate acid of a weak basic polyelectrolyte, $(BH^+)_n$, can generally be expressed respectively as follows:

$$HA \rightleftarrows H^+ + A^-$$

$$BH^+ \rightleftarrows H^+ + B$$

Both equilibria may be quantified in a unified manner by defining an apparent acid-dissociation constant of the polyelectrolytes (pK_{app}), which is calculable as follows by the use of the values of pH and α at equilibrium:

$$pK_{app} = pH - \log\left(\frac{\alpha}{1 - \alpha}\right) \tag{1}$$

pK_{app} indicates a macroscopic or overall equilibrium constant, which is greatly influenced by the degree of neutralization of the polymer (i.e., the electric potential at the surface of the polyion), as well as by the concentration level of added salt, through a Debye–Hückel-type ion-screening effect. We can define a hypothetical equilibrium constant (i.e., an intrinsic acid-dissociation constant) pK_0, which corresponds to a microscopic equilibrium constant for respective functionalities. By its definition, pK_0 is independent of electrostatic effect and is constant irrespective of the change in α or C_S.

Based on the Donnan's logic, pK_0 can be defined as follows by the use of the averaged pH value in the polyelectrolyte phase, $(pH)_p$:

$$pK_0 = (pH)_p - \log \left\{ \frac{\alpha}{1 - \alpha} \right\} \tag{2}$$

where the subscript p indicates the polyelectrolyte phase. The pK_0 value must be equal or close to the pK_a value of the monomeric acid analog of the polymer functionality. Combining Eqs. (1) and (2), pK_{app} at a defined α value can be related to pK_0 as

$$pK_{app} = pK_0 + \log \left\{ \frac{(a_H)_p}{a_H} \right\} \tag{3}$$

In the presence of an excess inert salt (e.g., NaCl in the present case), the following relationship is anticipated [2] by equating the activities of NaCl in both phases:

$$(a_{Na})_p (a_{Cl})_p = a_{Na} \, a_{Cl} \tag{4}$$

Similarly, by equating the activities of HCl in both phases, the following can be obtained:

$$(a_H)_p (a_{Cl})_p = a_H \, a_{Cl} \tag{5}$$

Because of the strong electric potential formed at the polyion surface, counterions (i.e., Na^+ ions for PAA anions and Cl^- ions for PVIm H^+ cations) are highly accumulated in the polyion domain. The activity ratios of the counterions between the two phases $(a_{Na})_p/a_{Na}$ and $(a_{Cl})_p/a_{Cl}$ are related to the singly charged H^+ ion activity ratio by the use of Eqs. (4) and (5), respectively:

$$\frac{(a_{Na})_p}{a_{Na}} = \frac{(a_H)_p}{a_H} \tag{6}$$

and

$$\frac{(a_{Cl})_p}{a_{Cl}} = \frac{a_H}{(a_H)_p} \tag{7}$$

Because the $(a_H)_p/a_H$ term in Eq. (3) can be replaced by the counterion activity ratios expressed by Eqs. (6) and (7), the effect of C_S on the acid-dissociation equilibria of the polyacid, $(HA)_n$, and the conjugate acid of the polybase, $(BH^+)_n$, can be expressed respectively as

$$pK_{app} = pK_0 + \log(a_{Na})_p - \log a_{Na} \tag{8}$$

and

$$pK_{app} = pK_0 - \log(a_{Cl})_p + \log a_{Cl} \tag{9}$$

It has already been indicated in our previous works [20,21] that the activities of counterions in the polyelectrolyte phase can be calculated by the use of the specific polyelectrolyte-phase volume V_p/n_p, where n_p indicates the amount of the functional groups fixed on the polyelectrolyte. For the $(HA)_n$ system, for example, $(a_{Na})_p$ can be expressed as the product of the activity coefficient of the Na^+ ion in the polyelectrolyte phase, $(y_{Na})_p$, and the free Na^+ ion concentration in the polyelectrolyte phase $[Na]_p$ as $(a_{Na})_p = (y_{Na})_p[Na]_p$. $[Na]_p$ can be expressed as the sum of two concentration terms: (1) Na^+ ions present in the polyelectrolyte phase to neutralize the free A^- groups and (2) Na^+ ions imbibed in the polyion domain in the form of Na^+Cl^-. Escape of Na^+ ions from the polyion domain due to their thermal motion produces a site vacancy of the polyion which should also be taken into account. This fraction of site vacancy is experimentally available as a "practical osmotic coefficient," $\phi_{p,Na}$, which can simply be related to the linear charge separation of the polyion, α/b. It has been revealed experimentally that $\phi_{p,Na}$ is not affected by the change in the polymer concentration, C_p, expressed in monomol/dm^3 or C_S; this is known as an additivity rule [22,23]. Because the $\phi_{p,Na}$ values determined in the salt-free system can be substituted for those for the excess-salt system, $(a_{Na})_p$ in Eq. (8) can be expressed as

$$(a_{Na})_p = (y_{Na})_p \left\{ \frac{(1 - \phi_{p,Na}) \, \alpha n_p}{V_p} + (C_{NaCl})_p \right\} \tag{10}$$

The $\phi_{p,Na}$ values for PAA, for example, have been summarized by Katchalsky and the relationship between $\phi_{p,Na}$ and α/b has been shown in the literature [24].

At sufficiently low concentrations of the added salt, the $(C_{NaCl})_p$ term in Eq. (10) can be neglected and pK_{app} for $(HA)_n$ is expressed as

$$pK_{app} = pK_0 + \log \left\{ \frac{(y_{Na})_p(1 - \phi_{p,Na}) \, \alpha n_p}{V_p} \right\} - \log a_{Na} \tag{11}$$

Because the volume term in Eq. (11), (n_p/V_p), can be regarded constant irrespective of the change in C_S, at a specified α value when α is close to unity [6], the second term in Eq. (11) remains unvaried at sufficiently low C_S, which results in a straight line whose slope is exactly -1 when pK_{app} is plotted against $\log a_{Na}$. With an increase in C_S, the salt imbibement effect on the $[Na]_p$ term cannot be neglected, and at the higher salt concentration levels, the $(C_{NaCl})_p$ term becomes predominant over the $(1 - \phi_{p,Na})\alpha n_p/V_p$ term. At this extreme condition, where $(a_{Na})_p$ can be approximated to a_{Na}, it is anticipated from Eq. (11) that pK_{app} becomes equal to pK_0. This condition can also be fulfilled by lowering the charge density of the polyions (i.e., at an α value close to zero, where the polyelecrolytic nature of $(HA)_n$ diminishes).

For $(BH^+)_n$, on the other hand, the counteranion activity in the polyelectrolyte phase, $(a_{Cl})_p$, can be expressed in a manner similar to Eq. (10) as:

$$(a_{Cl})_p = (y_{Cl})_p \left\{ \frac{(1 - \phi_{p,Cl})(1 - \alpha)n_p}{V_p} + (C_{NaCl})_p \right\} \tag{12}$$

where $(y_{Cl})_p$ represents the activity coefficient of the Cl^- ion in the polyelectrolyte phase and $\phi_{p,Cl}$ is the practical osmotic coefficient of the chloride solution of $(BH^+)_n$. In a manner similar to that for $(HA)_n$, it is anticipated that the $\phi_{p,Cl}$ value is just dependent on the linear charge separation of the polyions (i.e., at a specified α value of nearly zero, the $\phi_{p,Cl}$ value is not influenced by C_S). The combination of Eqs. (9) and (12) gives the following equation:

$$pK_{app} = pK_0 - \log \left\{ \frac{(y_{Cl})_p(1 - \phi_{p,Cl})(1 - \alpha)\, n_p}{V_p} \right\} + \log a_{Cl} \tag{13}$$

Based on a discussion similar to that for for $(HA)_n$, the following relationship is anticipated for $(BH^+)_n$: At sufficiently low C_S, pK_{app} at a low degree of dissociation can simply be expressed as a linear function of $\log a_{Cl}$, the slope being exactly equal to 1. With an increase in C_S, the slope is expected to decrease to approach 0. At an extremely high C_S, where the $(a_{Cl})_p$ term becomes equal to a_{Cl}, pK_{app} approaches pK_0. In a sufficiently high α region, where the linear charge density of the positively charged polyions is fully decreased, the polyelectrolyte nature of $(BH^+)_n$ disappears and pK_{app} values obtained at various C_S converge to a constant value of pK_0.

The plots of pK_{app} versus α for PAA and $PVImH^+$ are shown in Figs. 2 and 3, respectively, where the average charge per functional group, Z, is shown in the upper abscissa (i.e., $Z = -\alpha$ for the PAA system and $Z = 1 - \alpha$ for the $PVImH^+$ system). pK_{app} increases with α in both curves, whereas the effect of C_S is entirely opposite to each other. Depression of pK_{app} with increasing C_S is remarkable in the PAA system, whereas the substantial increase in pK_{app} with the addition of salt is observed for the $PVImH^+$ system. It should be pointed out that the pK_{app} versus α curves determined under various C_S converge to specific values of pK_0 for both systems at $Z = 0$, where the surface charge is fully diminished (i.e., at the complete neutral condition). The pK_0 values thus estimated for PAA and $PVImH^+$ are ~4.3 and ~7.0, respectively, being quite close to the pK_a values of the monomer analogs of the polymer functionalities {i.e., acetic acid ($pK_a = 4.56$) [25] for PAA and imidazole ($pK_a = 6.99$) [26] for $PVImH^+$}.

The pK_{app} change upon neutralization of the polyacids as well as the effect

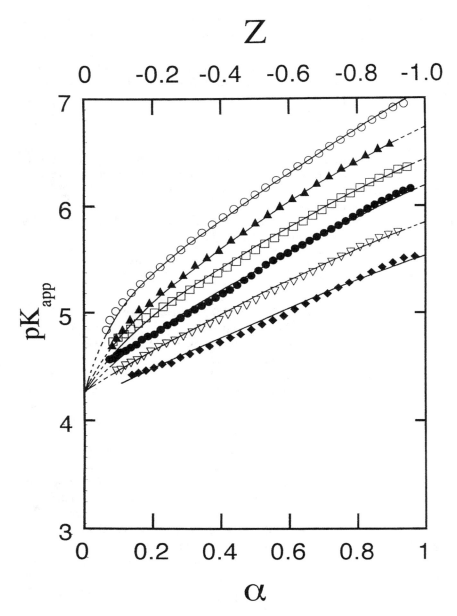

Figure 2 Salt (NaCl) concentration effect on pK_{app} versus α plots of PAA: (○) $C_S = 0.01$; (▲) $C_S = 0.02$; (□) $C_S = 0.05$; (●) $C_S = 0.10$; (▽) $C_S = 0.20$; (◆) $C_S = 0.50$ mol/dm^3. (From Ref. 3.)

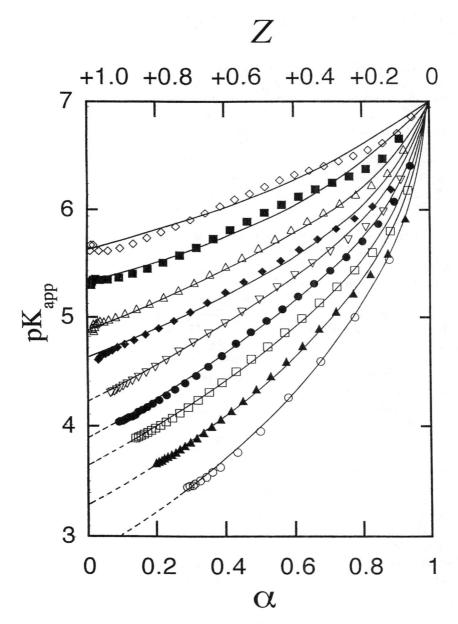

Figure 3 Salt (NaCl) concentration effect on pK_{app} versus α plots of PVImH$^+$: (○) $C_S = 0.01$; (▲) $C_S = 0.02$; (□) $C_S = 0.05$; (●) $C_S = 0.10$; (▽) $C_S = 0.20$; (◆) $C_S = 0.50$; (△) $C_S = 1.00$; (■) $C_S = 2.00$; (◇) $C_S = 3.00$ mol/dm^3. (From Ref. 3.)

of C_S can be rationalized qualitatively by taking into account the attraction or repulsion between H$^+$ ions and the charged polyions. Due to the crowded negative charges formed at the PAA polymer surface, the concentration of the free H$^+$ ions in the vicinity of the polymer molecules is higher than the bulk solution, which results in the greater increase in pK_{app} than pK_0 (= 4.3). This electrostatic attraction between the negatively charged polyion surface and H$^+$ ions is reduced by the presence of excess salt as shown in Fig. 3 through the Debye–Hückel-type screening effect, which is most pronounced at the highest C_S (i.e., at C_S = 0.5 mol/dm^3 examined in the present study). The PAA sample used in this study formed precipitation at a high salt concentration level of 1.0 mol/dm^3, which prevented precise pH determination at a higher C_S region than 0.5 mol/dm^3. With decreasing C_S, the repulsion becomes serious and the discrepancy between pK_{app} and pK_0 is pronounced. Contrary to the PAA system, H$^+$ ions are strongly repelled from the surface of the positively charged PVImH$^+$ molecules, and the decreased concentration of the mobile H$^+$ ions in the vicinity of the fixed imidazole groups of PVImH$^+$ molecules leads to a greater decrease in pK_{app} than pK_0 (= 7.0). The ion-screening effect due to the added inert salt is most effective at the highest C_S level as well (i.e., at C_S = 3.00 mol/dm^3, as shown in Fig. 3).

The pK_{app} values determined at specified α values are plotted in Fig. 4 against log a_{Na}, as suggested by Eq. (11). Good linearities are obtained at a relatively low C_S region for any combinations of pK_{app} determined at specified α values. It is notable that the slope of the straight line of the linear portion is quite close to −1 at α = 1, where all the functionalities are completely charged, which is, indeed, expected by the Gibbs–Donnan concept. Decreasing α to 0, the slope decreases to 0, where no salt concentration dependence is observed in accordance with the prediction of Eq. (11). These relationships found in the pK_{app} versus log a_{Na} plots clearly indicate the phase-separation property of PAA molecules dissolved in aqueous solution, even though the polymer solution appears homogeneous and no boundary between the two phases is observable. The fixed carboxylate groups on the linear polymers were endowed with negative charges by H$^+$ ion dissociation, and Na$^+$ ions must be accumulated in the vicinity of the polymer skeleton because of the strong attraction between the negatively charged polyion surface and Na$^+$ ions. Even though the mobile H$^+$ ion concentration in the vicinity of the negatively charged polyion molecules must be much higher than the bulk solution phase as well, the concentration of free H$^+$ ions is reduced by the Na$^+$ ion binding to the polyion surface, which is controlled directly by the concentration of Na$^+$ ions in the bulk solution phase. The correlation between H$^+$ and Na$^+$ ion activities between the polyelectrolyte domain (or polyelectrolyte phase) and the bulk solution phase is characterized by Donnan's law, which is expressed by Eq. (6).

The validity of the Gibbs–Donnan approach must be confirmed by analyzing the acid-dissociation equilibria of the conjugate acids of weak basic polymers based on the same concept. The acid-dissociation equilibria of PVImH$^+$ were ex-

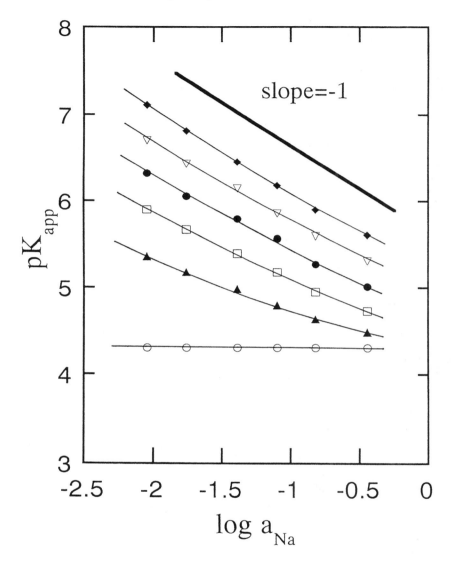

Figure 4 pK_{app} of PAA as a function of log a_{Na}: (○) $\alpha = 0$; (▲) $\alpha = 0.2$; (□) $\alpha = 0.4$; (●) $\alpha = 0.6$; (▽) $\alpha = 0.8$; (◆) $\alpha = 1.0$. (From Ref. 3.)

amined for this purpose at various salt concentration levels (i.e., $C_S = 0.01–3.00$ mol/dm^3. The pK_{app} values determined at specified values of α are plotted in Fig. 5 against log a_{Cl^-}. It is notable that all the plots form approximately straight lines at any α values when C_S is sufficiently low; in particular, at $\alpha = 0$ (i.e., $Z = 1$, the slope of the linear plots is quite close to 1, as expected from Eq. (13). On the contrary, at $\alpha = 1$ ($Z = 0$), the slope becomes equal to 0, resulting from the invasion effect of the inert salt to the polyelectrolyte phase.

It is apparent from Eq. (3) that the difference between pK_{app} and pK_0, ΔpK, can be related directly to the free H$^+$ ion activity ratio between the two phases. By approximating the activity coefficient ratio of the free H$^+$ ion between the two phases to be unity, the ΔpK term can be expressed as the free H$^+$ ion concentration ratio:

$$\Delta pK = \log \left(\frac{[H]_p}{[H]} \right) \tag{14}$$

ΔpK reflects the magnitude of the concentration or dilution of free H$^+$ ions from the bulk solution phase to the polyelectrolyte phase across a hypothetical Donnan membrane. The ΔpK values for PAA are therefore always positive, whereas the ΔpK values for PVImH$^+$ are always negative. In order to compare the magnitude of the nonideality terms observed in both PAA and PVIm systems directly, the absolute value of ΔpK, $|\Delta pK|$, is plotted in Figs. 6 and 7 against Z respectively for PAA and PVIm, because both linear polymer molecules are composed of the same backbone structure. A discrepancy is noted between the $|\Delta pK|$ versus Z plots obtained for both systems; however, much higher $|\Delta pK|$ values are given for the PVIm system, even though it has already been revealed that the ΔpK term can be related to the electrostatic effect straightforwardly in the case of PAA [2,9]. This discrepancy observed between the two polymer systems is of particular interest and can be attributed to a preferential binding of Cl$^-$ ions to the positively charged imidazolium groups, even though the binding of Na$^+$ ions to the carboxylate groups of PAA molecules is estimated to be purely electrostatic in nature. This additional interaction to the electrostatic one operating between positively charged polymer surface and supporting anions may be categorized as a hydrophobic interaction originating from the low degree of hydration of imidazolium groups, as well as Cl$^-$ ions. Further systematic investigation should be carried out on the effect of the nature of the counteranions on the potentiometric titration of positively charged polymers [27–29].

IV. METAL COMPLEXATION EQUILIBRIUM ANALYSES OF LINEAR POLYELECTROLYTES

In the preceding section, it was revealed that acid-dissociation equilibria of linear polymers can be analyzed in a unified manner based on the two-phase model by

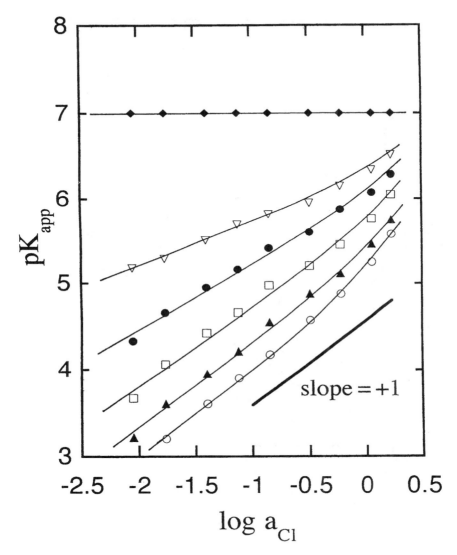

Figure 5 pK_{app} of PVImH$^+$ as a function of log a_{Cl}: (○) α = 0; (▲) α = 0.2; (□) α = 0.4; (●) α = 0.6; (▽) α = 0.8; (◆) α = 1.0. (From Ref. 3.)

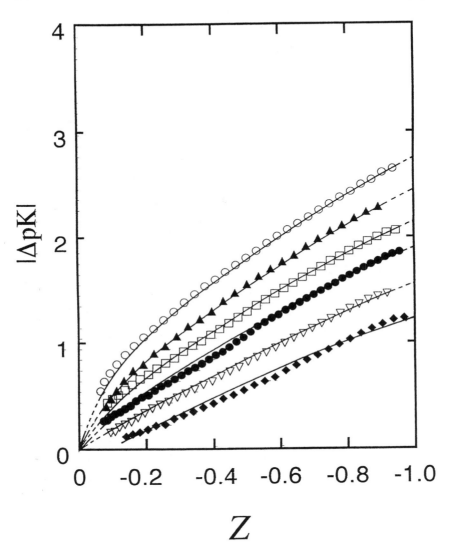

Figure 6 $|\Delta pK|$ as a function of Z for PAA: ($\bigcirc$) $C_S = 0.01$; ($\blacktriangle$) $C_S = 0.02$; ($\square$) $C_S = 0.05$; ($\bullet$) $C_S = 0.10$; ($\triangledown$) $C_S = 0.20$; ($\blacklozenge$) $C_S = 0.50$ mol/dm^3. (From Ref. 3.)

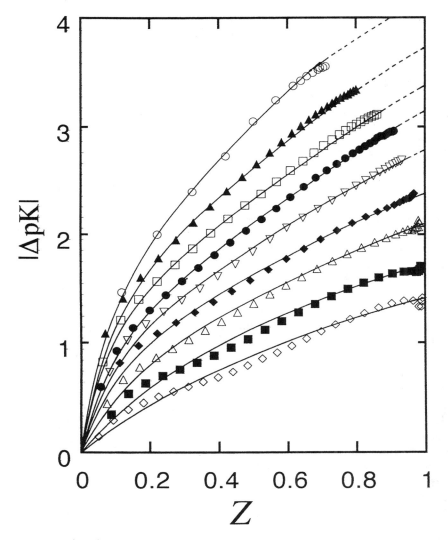

Figure 7 $|\Delta pK|$ as a function of Z for PVImH$^+$: ($\bigcirc$) $C_S = 0.01$; ($\blacktriangle$) $C_S = 0.02$; ($\square$) $C_S = 0.05$; ($\bullet$) $C_S = 0.10$; ($\triangledown$) $C_S = 0.20$; ($\blacklozenge$) $C_S = 0.50$; ($\triangle$) $C_S = 1.00$; ($\blacksquare$) $C_S = 2.00$; ($\lozenge$) $C_S = 3.00$ mol/dm^3. (From Ref. 3.)

which total solution volume is divided into two portions: a polyelectrolyte-phase volume and a bulk solution phase volume. Ion distribution equilibria between the two phases across the hypothetical membrane can be rationalized by a Gibbs–Donnan logic. In this section, the same feature is tested for the equilibrium analyses of metal complexation. In Sec. IV.A, the equilibrium analysis of PAA complexation with monovalent and divalent metal ions is presented. It is shown that the equilibria of a weak acidic or basic polyelectrolyte can equally analyzed.

A. Polyacrylic Acid

Complexation equilibria of polycarboxylic acids, such as PAA and poly-(methacrylic acid) (PMA), as well as their cross-linked gel analogs have been studied extensively because of their practical significance and theoretical interest. Despite a large body of studies carried out up to the present time; however, the complexation equilibria as well as the structures of the complexes have not yet been described fully. For example, the coordination number of carboxylate groups bound to a metal ion and the "equilibrium constant" of the complexes of monovalent and divalent metal ions with PAA and PMA determined by the use of several different thermodynamic techniques are not always in accordance with each other. Gregor et al. [30–32] and Mandel and Leyte [33] have analyzed the complexation equilibria based on a modified Bjerrum method [30,33] and have reported that divalent metal ions, such as Mg^{2+}, Ca^{2+}, Mn^{2+}, and Zn^{2+} form complexes with PAA molecules whose maximum coordination number is 2 [32,33]. Later, Marinsky and co-workers pointed out the inappropriate assumptions inherent in the modified Bjerrum method [34–36] and proposed an alternative experimental approach to analyze the complexation equilibria based on the two-phase model [35,37–39]. They concluded that divalent metal ions, such as Ca^{2+}, Co^{2+}, and Zn^{2+} ions, with PAA and PMA form only contact ion-pairs [34,35,38] (i.e., the coordination number is unity), whereas Cu^{2+} ions form 1 : 2 complexes with PAA [37]. Comparison of the results obtained by both thermodynamic and spectroscopic measurements has been expected to give the definitive conclusion on this debate, and Cu^{2+}–PAA and Cu^{2+}–PMA systems have frequently been examined for this purpose. Ultraviolet (UV) and visible (VIS) spectroscopies together with potentiometry have often been applied to such studies. However, because of inconsistency in the interpretation of the spectra, unified interpretation on the coordination structures of Cu^{2+} ions to these polyacids has not yet been established.

The complexation equilibria of these polycarboxylate ions are strongly dependent on the experimental conditions (i.e., the total concentrations of metal ions and the polyions, as well as the ratio of the two). Also, they are greatly influenced by solution properties, such as pH and ionic strength. For the sake of quantitative analyses of the equilibria, the following two complicating factors should be taken into account: (1) the electrostatic effect due to the polyelectrolyte nature of the negatively charged polyions and (2) the multidentate property of the polymer ligand molecules. Because two adjacent carboxylate groups fixed on the polycarboxylate molecules are quite close to each other, formation of bidentate ligand complexes are anticipated in some cases, in addition to monodentate ligand complexes.

In this section, complexation equilibrium analyses of PAA with metal ions ($M^{Z+} = Ag^+$, Ca^{2+}, Cu^{2+}, Cd^{2+}, and Pb^{2+}) examined in the presence of an excess of Na^+ ions are presented. A concurrent potentiometric titration measurement of pH and pM of the equilibrium mixture solutions was performed by the use of a commercially available glass electrode and metal ion-selective electrodes. In addition, a ^{113}Cd nuclear magnetic resonance (NMR) study on Cd^{2+} ion complexation was undertaken in order to verify the results obtained by the thermodynamic approach. The ^{113}Cd NMR chemical shift measurement is one of the most straightforward methods for determining the structures of Cd^{2+}–polymer ligand complexes in aqueous solution because the chemical shift is highly sensitive to the donor atom type, coordination number, and geometry.

When M^{Z+} ions are added to the polyion system, they are distributed between a polyelectrolyte phase and a bulk solution phase, as with H^+ ions. By the Donnan relation, the activity ratio of the free M^{Z+} ions between the two phases, $(a_M)_p/a_M$, can be related to the activity ratio of monovalent ions, such as H^+ ions, by the following equation:

$$\left\{ \frac{(a_M)_p}{a_M} \right\} = \left\{ \frac{(a_H)_p}{a_H} \right\}^Z \tag{15}$$

By the use of concentrations instead of activities, Eq. (15) can be written as

$$\left\{ \frac{[M]_p}{[M]} \right\} = G \left\{ \frac{[H]_p}{[H]} \right\}^Z \tag{16}$$

where G represents the activity coefficient quotient $[G = y_M(y_M)_p^{-1}(y_H)_p^{Z}(y_H)^{-Z}]$. Assuming that the G value is constant and equal to unity [6,40], Eq. (16) can be written as

$$\left\{ \frac{[M]_p}{[M]} \right\} = \left(\frac{[H]_p}{[H]} \right)^Z = 10^{Z\Delta pK} \tag{17}$$

It can be seen that the concentration of free M^{Z+} ions in the polyelectrolyte phase can be related to the concentration of free M^{Z+} ions in the bulk solution phase by the ΔpK term, which is determined by the use of acid-dissociation equilibria of the polymer as a probe. The Donnan potential term (i.e., the ΔpK term) for PAA is estimated to be as large as ~2.8 at $C_S = 0.01$ mol/dm^3 and $\alpha = 1.0$ (Fig. 6). Under this condition, the concentration ratio of divalent metal ions in the polyion domain to the bulk solution phase can be estimated to be as high as $\sim 4 \times 10^5$. To make possible the evaluation of the electrostatic effect at defined α and C_S by the use of the ΔpK terms, it is essential to keep the concentration level of M^{Z+} ions as low as possible so that perturbation due to complexation to acid-dissociation equilibria can be neglected.

According to the two-phase model [13,19], the overall or apparent complexing of M^{Z+} ions with polyions can be divided into the following two processes. The first is the distribution of free M^{Z+} ions between the bulk solution phase and the polyelectrolyte phase [Eq. (18)]; the free M^{Z+} ions in the polyelectrolyte phase are bound "territorially." The second is the direct binding of the free M^{Z+} ions in the polyelectrolyte phase to carboxylate group(s) fixed on the polyion skeletons to form monodentate ligand complexes, MA [Eq. (19)], and bidentate ligand complexes MA_2 [Eq. (20)]; these complexed M^{Z+} ions are categorized as "site bound," and participate in the following equilibria:

$$M^{Z+} \rightleftarrows (M^{Z+})_p \tag{18}$$

$$(M^{Z+})_p + (A^-)_p \rightleftarrows (MA)_p^{(Z-1)+} \tag{19}$$

$$(MA)_p^{(Z-1)+} + (A^-)_p \rightleftarrows (MA_2)_p^{(Z-2)+} \tag{20}$$

Note that the first complexation reaction [Eq. (19)] leading to monodentate ligand complex formation is a bimolecular reaction (i.e., an intermolecular reaction), whereas the second reaction [Eq. (20)] is a unimolecular reaction (i.e., an intramolecular reaction) because the second A^- groups to react with the monodentate ligand complexes, $MA^{(Z-1)+}$, are fixed on the same polyion backbone. In order to express these successive complexation equilibria, an equilibrium constant, k_{intra}, can be defined by the concentration ratio of the two complexed species as

$$k_{intra} = \frac{[(MA_2)_p^{(Z-2)+}]}{[(MA)_p^{(Z-1)+}]} \tag{21}$$

The overall or macroscopic binding equilibrium of a metal ion to a polyion can generally be expressed as [19,40]

$$M^{Z+}(\text{free}) \rightleftarrows M^{Z+}(\text{bound}) \tag{22}$$

and an apparent binding constant, $(K_M^0)_{app}$, defined in the presence of negligible amount of metal ions, can be calculated by the following equation [13,40]:

$$(K_M^0)_{app} = \frac{[M]_{bound}}{[M]\alpha C_p} = \frac{C_M - [M]}{[M]\alpha C_P} \tag{23}$$

where C_M and C_p indicate the total concentrations of M^{Z+} ions and carboxylate groups of the polyacids at each titration point, respectively. C_p is expressed in monomole/dm^3. The αC_p term corresponding to the concentration of free carboxylate groups of PAA molecules is needed for normalization of the distribution quotient.

In order to study the successive complexation equilibria (i.e., monodentate ligand complexation followed by intramolecular complexation), it is important to keep the concentration of the polyion sufficiently low in order to prevent the formation of bidentate ligand complexes by which a metal ion bridges the ligand sites belonging to the different polyion segments. When a linear polyion with n coordination sites (A$^-$) is expressed as P (i.e., P = nA$^-$), the first step complexation to form a 1 : 1 complex (one M^{Z+} ion to one ligand site) can be expressed as follows:

$$M + P \rightleftarrows MP \tag{24}$$

The intrinsic equilibrium constant, $(K_1)_p$, for this bimolecular complexation reaction can be written as

$$(K_1)_p = \frac{[MP(1)]_p}{[M]_p[P]_p} \tag{25}$$

The i value in the parentheses of MP(i) in Eq. (25) indicates the number of sites on the macromolecules that participate in complexation. If this bimolecular reaction is followed by a bidentate ligand complexation (i.e., unimolecular complexation reaction), then the following equilibrium reaction should be taken into account:

$$MP(1) \rightleftarrows MP(2) \tag{26}$$

and the concentration ratio of the two complexed species can be expressed by the use of an equilibrium constant for the intramolecular reaction, k_{intra}, as

$$\frac{[MP(2)]}{[MP(1)]} = k_{intra} \tag{27}$$

Because the $[M]_{bound}$ term in Eq. (23) can be displaced by the sum of the complexed species formed in the polyelectrolyte phase and the αC_p term can be substituted by the free ligand concentration expressed in monomole in the polyelectrolyte phase, $[A]_p$, $(K_M^0)_{app}$ can be written as

$$(K_M^0)_{app} = \frac{[MP(1)]_p + [MP(2)]_p}{[M][A]_p} \tag{28}$$

Note that the concentrations of complexes as well as free ligands are expressed based on the total solution volume in Eq. (23), whereas the corresponding concentrations in Eq. (28) are based on the polyelectrolyte phase volume. $[A]_p$ can be related to the polyion concentration expressed on the polyion basis, $[P]_p$, as

$$[A]_p = n[P]_p \tag{29}$$

By combining Eqs. (25)–(29), $(K_M{}^0)_{app}$ can be rewritten as

$$(K_M{}^0)_{app} = \frac{[MP(1)]_p(1 + k_{intra})}{[M]n[P]_p} = \left(\frac{[M]_p}{[M]}\right)\left\{\frac{(K_1)_p}{n}\right\}(1 + k_{intra}) \tag{30}$$

Taking the logarithmic form of Eq. (30) by knowing that the $[M]_p/[M]$ term can be approximated by $Z\Delta pK$, Eq. (20) can be rearranged into the following form:

$$\log(K_M{}^0)_{app} = Z\Delta pK + \log\left(\left\{\frac{(K_1)_p}{n}\right\}(1 + k_{intra})\right) \tag{31}$$

By plotting $\log(K_M{}^0)_{app}$ versus ΔpK, a straight line of slope Z, the charge of the metal ion, is anticipated, provided that the second logarithmic term in Eq. (31) is constant irrespective of the change in α. In this case, the intercept of the linear plots at the ordinate corresponds to an intrinsic complex formation function, $\log(K_M{}^0)_{int}$, which can be written

$$\log(K_M{}^0)_{int} = \log\left(\left\{\frac{(K_1)_p}{n}\right\}(1 + k_{intra})\right) \tag{32}$$

It is evident that the $(K_1)_p/n$ term on the right-hand side of Eq. (32) corresponds to the equilibrium constant of the 1 : 1 complex, $(\beta_1)_p$, which is defined by the use of the polyion concentration expressed in monomole/dm³; that is, $(\beta_1)_p = [MA]_p/[M]_p[A]_p$; this value is comparable to the stability constant of the one-to-one complexes with a ligand of a monomer analog of the polyion ligand, A⁻. The $(1 + k_{intra})$ term, on the other hand, reflects the bidentate ligand complex formation ability of the metal ion. We can see that the product $(\beta_1)_p(1 + k_{intra})$ is comparable to the β_1 value of the complexes of the corresponding bidentate ligands (dicarboxylate anions), which resemble in the molecular structures of the macromolecular ligands.

Log$(K_{Ag}{}^0)_{app}$ and $\log(K_{Ca}{}^0)_{app}$ determined under various medium concentration levels are plotted in Figs. 8 and 9, respectively, versus α. Apparently, the macroscopic binding equilibrium quotients are strongly dependent on α and C_S because of the polyelectrolytic nature of PAA. They increase with α and decrease with C_S, as with the pK_{app} versus α plots (Figs. 2 and 3). It should be noted that this α and C_S dependence of $\log(K_M{}^0)_{app}$ is more pronounced for the doubly charged Ca^{2+} ion than the singly charged Ag^+ ion, as expected by the difference in the nonideality terms affected by the exposure of metal ions to the same electric potential. By plotting these $\log(K_M{}^0)_{app}$ values versus the corresponding ΔpK values calculated by the use of the pK_{app} values determined by the concurrent pH measurement, we can see that all the plots converge to form straight

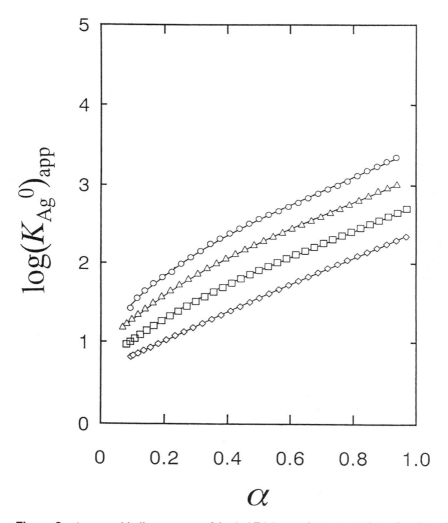

Figure 8 Apparent binding constant of the Ag$^+$/PAA complex expressed as a function of α and C_S: ($\bigcirc$) $C_S = 0.01$; ($\triangle$) $C_S = 0.02$; ($\square$) $C_S = 0.05$; ($\diamond$) $C_S = 0.10$ mol/dm^3. (From Ref. 2.)

lines, as shown in Figs. 10 and 11, whose slopes are +1 for the Ag$^+$–PAA system and 2 for the Ca^{2+}–PAA system, respectively.

It should be stressed again that the straight lines were obtained for these two binding systems in the whole pH and C_S regions examined by this study. This clearly indicates that only one kind of complexed species are formed in both

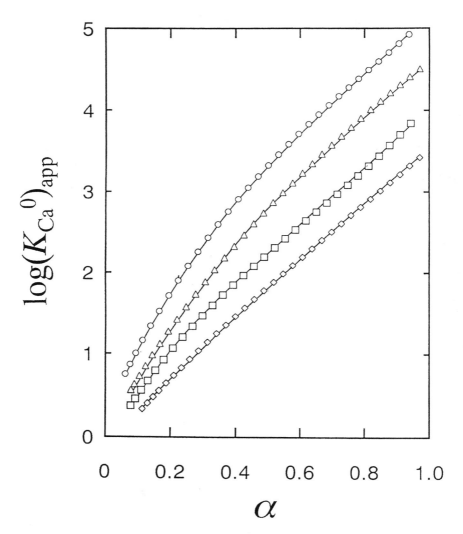

Figure 9　Apparent binding constant of the Ca^{2+}/PAA complex expressed as a function of α and C_S: (◯) $C_S = 0.01$; (△) $C_S = 0.02$; (□) $C_S = 0.05$; (◇) $C_S = 0.10$ mol/dm^3. (From Ref. 2.)

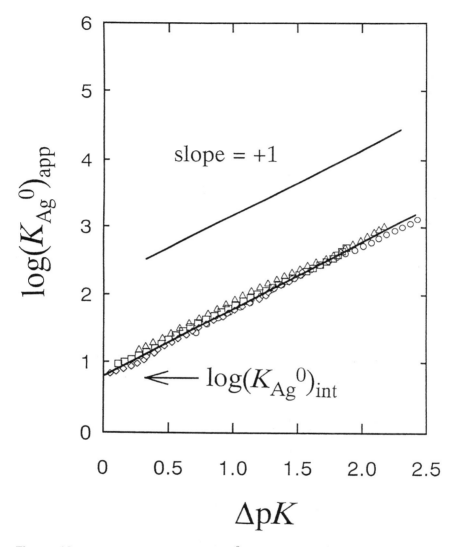

Figure 10 Relationship between $\log(K_{Ag}^{0})_{app}$ and ΔpK for PAA: ($\bigcirc$) $C_S = 0.01$; ($\triangle$) $C_S = 0.02$; ($\square$) $C_S = 0.05$; ($\diamond$) $C_S = 0.10$ mol/dm^3. (From Ref. 2.)

Ag$^+$–PAA and Ca^{2+}–PAA systems. The intercepts at the $\Delta pK = 0$ axis correspond to $\log(K_M^{0})_{int}$, as has been suggested by Eq. (31). By comparing the $\log(K_{Ag}^{0})_{int}$ value (= 0.7) with the $\log \beta_1$ value of the 1 : 1 complex of Ag$^+$ (acetate) (= 0.73) [41] and the $\log(K_{Ca}^{0})_{int}$ value (= 0.2) with the $\log \beta_1$ value of the 1 : 1 complex of Ca^{2+} (acetate) (= 0.45) [25], it can be concluded that monodentate carboxylate

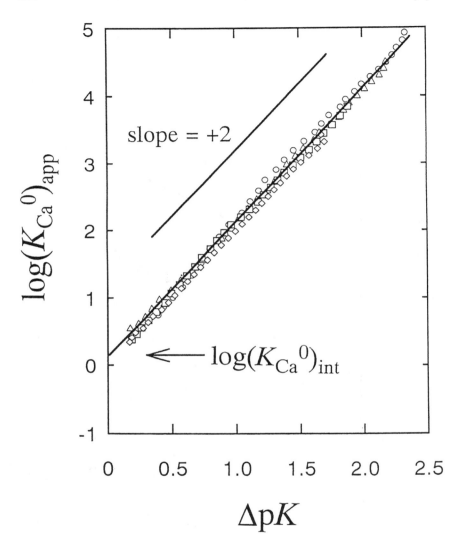

Figure 11 Relationship between $\log(K_{Ca}^{0})_{app}$ and ΔpK for PAA: (○) $C_S = 0.01$; (△) $C_S = 0.02$; (□) $C_S = 0.05$; (◇) $C_S = 0.10$ mol/dm^3. (From Ref. 2.)

complexes are predominantly formed in the Ag$^+$–PAA and Ca^{2+}–PAA systems in all the experimental conditions examined in this study.

In contrast to these monotonous changes observed for Ag$^+$ and Ca^{2+} systems, the plots of $\log(K_M^{0})_{app}$ versus ΔpK for the Cu^{2+}, Cd^{2+}, and Pb^{2+} systems show sigmoidal curves, which are composed of two straight lines as in Figs. 12,

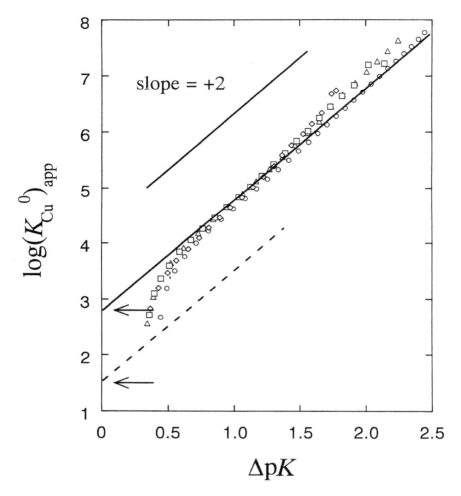

Figure 12 Relationship between $\log(K_{Cu}^0)_{app}$ and ΔpK for PAA. The solid line refers to the 1 : 2 complex formation. The upper arrow indicates the $\log[(\beta_1)_p(1 + k_{intra})]$ value (= 2.8). The broken line refers to the 1 : 1 complex formation. The lower arrow indicates the $\log(\beta_1)_p$ value (=1.6). (○) $C_S = 0.01$; (△) $C_S = 0.02$; (□) $C_S = 0.05$; (◇) $C_S = 0.10$ mol/dm^3. (From Ref. 2.)

13, and 14. Note that the slopes of the respective straight lines are equal to 2, the charge of the metal ions investigated. The $\log(K_M^0)_{int}$ values calculated by extrapolating the lower lines (broken lines) to $\Delta pK = 0$ for the Cu^{2+}, Cd^{2+}, and Pb^{2+} systems are 1.6, 1.5, and 2.2, respectively. By comparison with the $\log \beta_1$ values of the corresponding 1 : 1 complexes of Cu^{2+} (acetate), Cd^{2+} (acetate), and Pb^{2+} (ac-

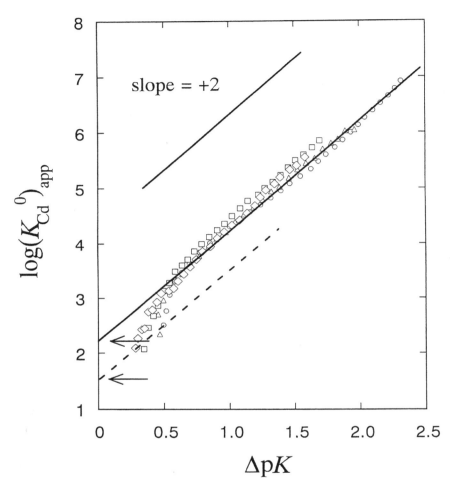

Figure 13 Relationship between $\log(K_{Cd}^{0})_{app}$ and ΔpK for PAA. The solid line refers to the 1 : 2 complex formation. The upper arrow indicates the $\log[(\beta_1)_p(1 + k_{intra})]$ value (= 2.2). The broken line refers to the 1 : 1 complex formation. The lower arrow indicates the $\log(\beta_1)_p$ value (= 1.5). ($\bigcirc$) $C_S = 0.01$; ($\triangle$) $C_S = 0.02$; ($\square$) $C_S = 0.05$; ($\diamondsuit$) $C_S = 0.10$ mol/dm^3. (From Ref. 1.)

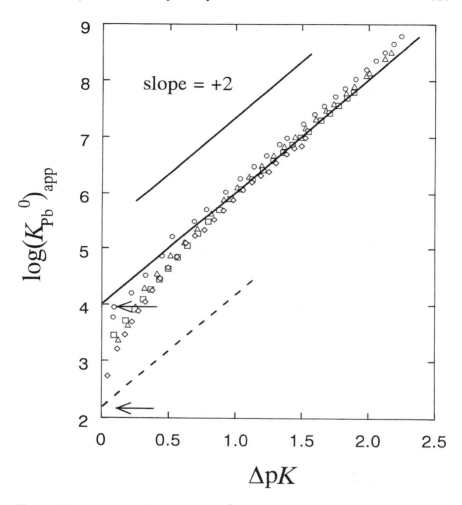

Figure 14 Relationship between $\log(K_{Pb}{}^0)_{app}$ and ΔpK for PAA. The solid line refers to the 1 : 2 complex formation. The upper arrow indicates the $\log[(\beta_1)_p(1 + k_{intra})]$ value ($= 4.0$). The broken line refers to the 1 : 1 complex formation. The lower arrow indicates the $\log(\beta_1)_p$ value ($= 2.2$). ($\bigcirc$) $C_S = 0.01$; ($\triangle$) $C_S = 0.02$; ($\square$) $C_S = 0.05$; ($\diamond$) $C_S = 0.10$ mol/dm^3. (From Ref. 2.)

etate), 1.8 [42], 1.6 [42], and 2.2 [42], respectively, it is apparent that each lower straight line indicates monodentate carboxylate complex formation, which is predominant in the lower α region.

On the other hand, the intercepts determined by extrapolation of the upper straight lines (solid lines) to $\Delta pK = 0$ are much larger compared with the stability constants of respective monodentate ligand complexes and can be regarded as the stability constants of bidentate carboxylate complexes where two carboxylate groups fixed on the same polyion backbone are involved in simultaneous complexation. Indeed, the $\log[(\beta_1)_p(1 + k_{intra})]$ values thus evaluated by extrapolation (i.e., 2.8 for Cu^{2+}, 2.2 for Cd^{2+}, and 4.0 for Pb^{2+} systems) are comparable with the $\log \beta_2$ values of $Cu(acetate)_2$, $Cd(acetate)_2$, and $Pb(acetate)_2$, 3.1 [43], 1.9 [43] and 3.5 [43], respectively. From this comparison, it is reasonable to conclude that at $\alpha > \sim 0.3$, the formation of bidentate carboxylate complexes are predominant in Cu^{2+}, Cd^{2+}, and Pb^{2+} complexation systems.

It is of significance to verify this estimate of bidentate ligand complex formation of divalent transition metal ions with PAA molecules with a more straightforward experimental technique. For that purpose, the ^{113}Cd NMR chemical shift change upon complexation has been measured as a function of α of the PAA polymer in order to study the coordination states of Cd^{2+} ions to fixed carboxylate groups. The $^{113}Cd^{2+}$ ion chemical shift observed, $(\delta_{Cd})_{obsd}$, can be expressed as the weighed average of the complexed and free Cd^{2+} ions, depending on the free Cd^{2+} ion fraction of the total Cd^{2+} ion, f_{Cd}:

$$(\delta_{Cd})_{obsd} = f_{Cd}\delta_{Cd} + (1 - f_{Cd})(\delta_{Cd})_{corr} \qquad (33)$$

where δ_{Cd} indicates the eigenvalue for free Cd^{2+} ions and is equal to zero, as free Cd^{2+} ions are usually taken as a reference for the shift measurement. The f_{Cd} value can be determined potentiometrically by using a Cd^{2+} ion-selective electrode. Because $\delta_{Cd} = 0$, the corrected δ_{Cd} value, $(\delta_{Cd})_{corr}$, which corresponds to the shift value of complexed Cd^{2+} ions, can be expressed by the use of f_{Cd} and $(\delta_{Cd})_{obsd}$ as

$$(\delta_{Cd})_{corr} = \frac{(\delta_{Cd})_{obsd}}{1 - f_{Cd}} \qquad (34)$$

The $(\delta_{Cd})_{corr}$ values thus calculated are plotted in Fig. 15. against α. It is apparent that the $(\delta_{Cd})_{corr}$ values determined at $0.2 < \alpha < 1.0$ are almost unvaried (i.e., approximately -35 to -38 ppm, being quite close to the intrinsic $^{113}Cd^{2+}$ shift values determined for the 1 : 2 complex (-40.0 ppm) of Cd^{2+}–acetate [44,45] and Cd^{2+}–glutarate ($^-OOC(CH_2)_3COO^-$) [44,45], which are the analogs of bidentate complexes of PAA molecules. At low α values ($0 < \alpha < 0.2$) however, $(\delta_{Cd})_{corr}$ shifts upfield upon decreasing α, showing that the limiting value at $\alpha = 0$ corresponds to approximately -20 ppm. This value, indeed, corresponds to the intrinsic shift value of the 1 : 1 complex (-22.1 ppm) of Cd^{2+}–acetate [44,45], which is the monomer analog of PAA. By the comparison of the chemical shift values, it is apparent that monodentate ligand complex formation is predominant at the lower α

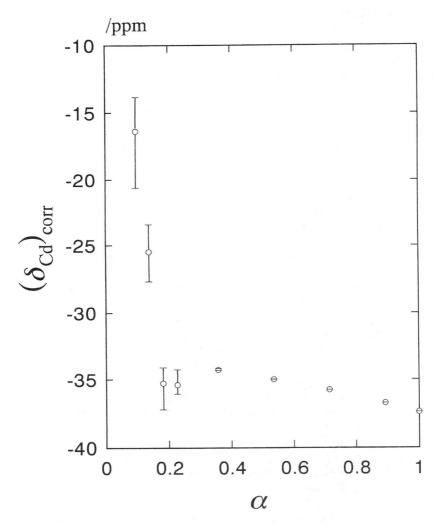

Figure 15 $(\delta_{Cd})_{corr}$ versus α plots for PAA. Error bars correspond to calculated f_{Cd} variation due to the potentiometric measurement of pCd in the sample solutions with an error of ± 0.1 mV. (From Ref. 1.)

region $(0 < \alpha < 0.2)$ and the bidentate ligand complex formation is predominant at the higher α region $(0.2 < \alpha < 1.0)$. Even though the experimental conditions for the ^{113}Cd NMR measurement and the thermodynamic study were not identical (i.e., the total concentrations of Cd^{2+} ions and PAA ions examined by the NMR method were almost 20 times higher than those applied to the potentiometric titration experiments due to the low sensitivity of the ^{113}Cd nuclei [1]), this chemical

Table 1 $\mathrm{Log}(\beta_1)_p$ and $\mathrm{Log}(1 + k_{\mathrm{intra}})$ Values of
PAA Complexes

System	$\mathrm{Log}(\beta_1)_p$	$\mathrm{Log}(1 + k_{\mathrm{intra}})$ at $\alpha > 0.3$
Ag^+	0.7	—
Ca^{2+}	0.2	—
Cu^{2+}	1.6	1.2
Cd^{2+}	1.5	0.7
Pb^{2+}	2.2	1.8

shift change of $^{113}Cd^{2+}$ ions upon α increase is quite consistent with this estimate by the thermodynamic approach. This consistency validates the sound interpretation of the thermodynamic results obtained by the Gibbs–Donnan-based concept.

The logarithmic values of $(\beta_1)_p$ and $(1 + k_{\mathrm{intra}})$ determined for Ag^+, Ca^{2+}, Cu^{2+}, Cd^{2+}, and Pb^{2+} ion bindings are listed in Table 1 for comparison to the multidentate complexation properties of these metal ions. Also, in order to compare the transition behavior from 1 : 1 complexes to 1 : 2 complexes, $\log(K_M^0)_{\mathrm{int}}$ $[= \log(K_M^0)_{\mathrm{app}} - Z\Delta pK)]$ is plotted in Fig. 16 against α for all the systems examined. No or little salt concentration effect has been observed in these plots, which indicates that the distribution of monodentate and bidentate complexes can be expressed purely as a function of α. It is apparent that for metal ions with low $(\beta_1)_p$ values (i.e., with weak coordination ability), only monodentate ligand complex formation is observed in the whole α region; that is, no further successive reaction to form bidentate ligand complexation can be observed in Ag^+ and Ca^{2+} binding systems. On the other hand, for metal ions with a strong coordination ability, the formation of bidentate ligand complex becomes appreciable as α increases to reach a constant $\log(K_M^0)_{\mathrm{int}}$ value. Because glutarate ions, $^-OOC(CH_2)_3COO^-$, can be regarded as a monomer analog ligand resembling the bidentate ligand site of PAA, it is of particular interest to compare the $\log\{(\beta_1)_p(1 + k_{\mathrm{intra}})\}$ values with the $\log\beta_1$ values of complexes with the dicarboxylate ion. It is necessary to correct for the difference in the ligand concentration definitions in the equilibrium constant of $\{(\beta_1)_p(1 + k_{\mathrm{intra}})\}$ for polyions and β_1 for bidentate ligand complexes of a simple monomer analog; that is, as the denominator of $(K_M^0)_{\mathrm{int}}$, the ligand concentration is expressed in monomole/dm^3, whereas the ligand concentration of the dicarboxylic acid ligand is defined based on molecules composed of two carboxylate groups. Keeping this difference in the ligand concentration definition in mind and assuming that only bidentate carboxylate complexation is predominant in the complexes with glutarate ions, the $\log\{(\beta_1)_p(1 + k_{\mathrm{intra}})\}$ values estimated for the systems of Cu^{2+}–PAA, Cd^{2+}–PAA, and Pb^{2+}–PAA of 2.8, 2.2, and 4.0, respectively, are compared with the $\log\beta_1$ values of 1 : 1 complexes of Cu^{2+} (glutarate), Cd^{2+} (glutarate), and Pb^{2+} (glutarate) of 5.1 [42], 2.6 [42], and 3.1 [42] respec-

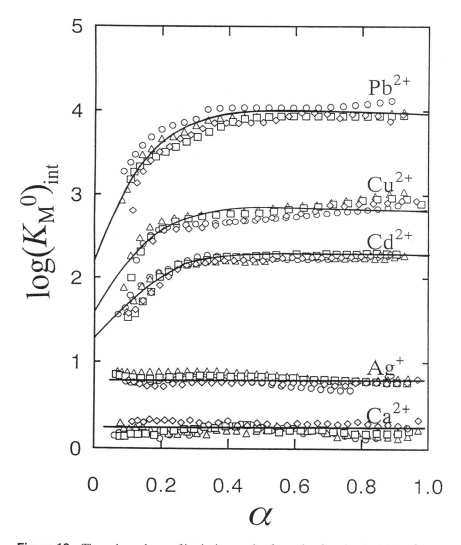

Figure 16 The α dependence of intrinsic complex formation function for PAA. ($\bigcirc$) $C_S =$ 0.01; ($\triangle$) $C_S = 0.02$; ($\square$) $C_S = 0.05$; ($\diamondsuit$) $C_S = 0.10$ mol/dm^3. (From Ref. 2.)

tively. It should be stressed that the stability constants of this dicarboxylate ligand complexes are in the order $Cu^{2+} > Pb^{2+} > Cd^{2+}$, whereas the order found in the bidentate ligand complexes formed in the PAA system is $Pb^{2+} > Cu^{2+} > Cd^{2+}$; in the PAA complexation system, Pb^{2+} ions have much stronger coordination ability than Cu^{2+} ions, even though Cu^{2+} ions bind much stronger than Pb^{2+} ions for the glutarate ion. Such a delicate difference observed in the order of the stability constants of the complexes with bidentate ligands formed on polycarboxylate ions can be attributed to the coordination geometry of respective metal ions as well as the availability of the second coordinating carboxylate groups fixed on the polymer backbone. This information is of special importance in designing the chemical structures of these carboxylate polymer molecules.

It is notable in Fig. 16 that the transition from monodentate to bidentate complexes is observed in a similar α range of 0.2–0.3, in spite of the large difference in the $\log(k_{intra})$ values among Cu^{2+}, Cd^{2+}, and Pb^{2+} systems. In spite of the difference in the coordination nature of these divalent metal ions, similarity observed in the transition of PAA complexation must reflect a common binding property of PAA molecules. We can see that the transition from monodentate ligand complex formation to bidentate ligand complex formation can be observed at a specified ligand spacing, where the averaged distance between the neighboring free carboxylate sites is estimated to be ~10Å.

B. Poly(N-vinylimidazole)

It has been proven by the above discussion that nonideality originating from electrostatic effect in metal complexation equilibria of negatively charged polyions can successfully be estimated by the use of the corresponding acid-dissociation equilibria as a probe. Insights into microscopic complexation at the polyion sites gained by this analysis can be related directly to the complexation reaction with monomeric ligands which resemble the functionalities of the polyion molecules. It has also been revealed in Section III that the nonideality term due to electrostatic repulsion between H^+ ions and positively charged polyions observed in acid-dissociation equilibria of PVIm can be quantified based on the "separate phase model" as well. In order to examine the validity of this approach to the analyses of the ion-binding equilibria of weakly basic polyelectrolytes, the systems PVIm–Ag^+ and PVIm–Cu^{2+} have been investigated by a potentiometric titration method with a procedure similar to the PAA complexation study [4].

The maximum coordination number of imidazole groups on the PVIm molecule to a metal ion has been reported by Gold and Gregor to be 2 for Ag^+ ion complexes [46] and 4 for Cu^{2+} ion complexes [47]. The presence of 1 : 4 complexes for Cu^{2+}–PVIm complexation has also been verified by an electronic spectroscopic measurement [47]. These measurements were carried out at specified high salt concentration level of 1 mol/dm^3 of $NaNO_3$ or KNO_3 used as the supporting electrolyte, where the free imidazole concentration monomer basis needed for the complexation analyses has been calculated by the empirical "modified Henderson–Hasselbalch equation" [46,47]. It is anticipated that positive charges produced by protonated imidazole groups endow the polymer molecule a positively charged electric field, which suppresses metal complexation because of electrostatic repulsion between the positively charged PVIm molecules and metal ions. The magnitude of the electrostatic repulsion is expected to be highly sensitive to C_S as well as α.

Both pH and pM (M^{Z+} = Ag^+ or Cu^{2+}) of the equilibrium solutions at each titration point have been determined simultaneously in the presence of $NaNO_3$ at 25°C. Even though it has been revealed that the overall complexation equilibria are strongly affected by both C_S and α, the electrostatic nonideality term has been estimated by the acid-dissociation property of the conjugate acid polymers (i.e., $PVImH^+$). Whether polymer molecules are charged negatively or positively (i.e., the polymer functionalities are acidic or basic), the magnitude of the polyelectrolytic effect inherent in the ion–polyion binding equilibria can be assessed by studying the acid-dissociation equilibria of the charged polymers as a probe. An apparent acid-dissociation equilibrium constant, pK_{app}, of the $PVImH^+$ polymer defined by Eq. (1) is calculated for this assessment.

In Fig. 17, pK_{app} calculated by the use of pH and α determined under various C_S is plotted against α. By comparing the pK_{app} versus α plots determined in the presence of NaCl and $NaNO_3$, a small difference can be observed between the two series of the plots, which may indicate participation of supporting anions on the acid-dissociation equilibria of $PVImH^+$. For any C_S, pK_{app} increases drastically with α to reach a constant value of ~7.0 at $\alpha = 1$, where all the imidazole groups of PVIm molecules are completely dissociated. Even though these curves are highly sensitive to C_S at lower α, it is apparent that they converge to the specified value at $\alpha = 1$, being quite close to the pK_a values of the conjugate acids of imidazole (pK_a = 7.00) [26] or methylimidazole (pK_a) = 7.06) [48]; both acids are the monomer analogs of $PVImH^+$. This indicates that intrinsic acid-dissociation equilibria of the imidazole sites of $PVImH^+$ molecules mimic those of the corresponding monomer analogs. The difference between the intrinsic equilibrium constant, pK_0 (= 7.0), and pK_{app} at a specified α and C_S, which is designated as ΔpK, may reflect the polyelectrolytic nature of the positively charged polymers. It is notable that the deviation becomes serious at lower α and at lower C_S. This strong C_S dependence of pK_{app} may primarily be rationalized by the Debye–Hückel-type ion-screening effect on the electrostatic repulsion between H^+ ions and the positively charged polymer surface. Indeed, as shown in Fig. 2, the pK_{app} versus α plots of PAA is remarkably dependent on C_S; however, the direction of the shift of the pK_{app} versus α plots owing to the C_S change is just the opposite of the plots observed in the $PVImH^+$ system.

As a result of electrostatic repulsion, free H^+ ions in the vicinity of $PVImH^+$ molecules is highly diluted compared with the bulk solution phase, which results in a substantial decrease in pK_{app} compared with pK_0. By definition, pK_0 must be free from the polyelectrolytic effect and remain constant at any α or C_S. By combining Eqs. (2) and (3), it is apparent that ΔpK can be related directly to the free H^+

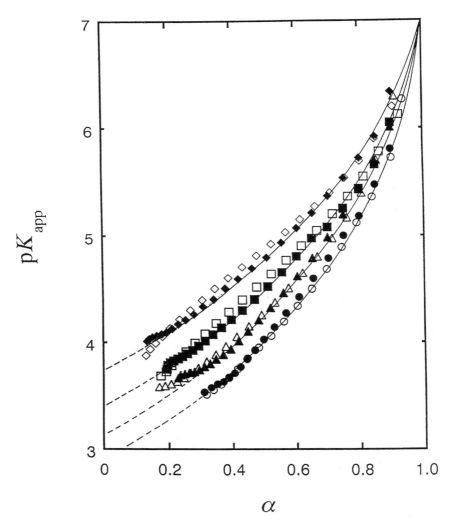

Figure 17 Salt (NaNO$_3$) concentration effect on pK_{app} versus α plots of PVImH$^+$. ($\bigcirc$) C_S = 0.005; ($\triangle$) C_S = 0.01; ($\square$) C_S = 0.02; ($\diamond$) C_S = 0.05 mol/dm^3. Closed, PVI mH$^+$/NaNO$_3$; Open Ag$^+$/PVImH$^+$/NaNO$_3$. (From Ref. 4.)

ion concentration ratio between two phases as shown by Eq. (14). ΔpK reflects the magnitude of dilution of free H$^+$ ions in the polyelectrolyte phase compared with the bulk solution phase across a hypothetical Donnan membrane; for example the $\{[\text{H}]_p/[\text{H}]\}$ term is roughly estimated to be 10^{-4} at $\alpha = 0$ and at $C_S = 0.01$ mol/dm^3, as indicated in Fig. 17.

Because the logic to analyze the complexation equilibria of basic polymer ligands is identical to that for acidic polymer ligands, the analysis procedure is duplicated here briefly to recall the definition of respective equilibrium quotients. M^{Z+} ion complexation to basic polymer ligands can be described by M^{Z+} ion distribution between the polyion phase (bound) and the bulk solution phase (free). A macroscopic binding quotient, $(K_M)_{app}$ can be defined in an identical manner using Eq. (22) as the distribution of M^{Z+} ions between the two phases. Here, the αC_p term corresponding to the concentration of free imidazole groups is used again in order to normalize the distribution quotient. In the presence of quite a small amount of M^{Z+} ions, $(K_M)_{app}$ is not influenced by the degree of M^{Z+} ion loading to the polymer, and this quotient can be denoted by $(K_M{}^0)_{app}$.

Even though $(K_M{}^0)_{app}$ is still a function of α and C_S, an intrinsic binding quotient, $(K_M{}^0)_{int}$, free from the electrostatic effect, can be defined as follows by the use of a hypothetical free M^{Z+} ion concentration in the polyelectrolyte phase, $[M]_p$, instead of $[M]$:

$$(K_M{}^0)_{int} = \frac{[M]_{bound}}{[M]_p \alpha C_p} = (K_M{}^0)_{app} \left\{ \frac{[M]}{[M]_p} \right\} \tag{35}$$

$(K_M{}^0)_{int}$ corresponds to the microscopic equilibrium quotient. The $\{[M]_p/[M]\}$ term needed to relate $(K_M{}^0)_{app}$ to $(K_M{}^0)_{int}$ can be evaluated as follows. Even though the concentrations of cationic species in the positively charged polyion domain are much smaller than those of the bulk solution phase due to electrostatic repulsion, we can relate the activities of the coions in the two phases in the following Donnan's equation:

$$\left\{ \frac{(a_M)_p}{a_M} \right\} = \left\{ \frac{(a^H)_p}{a_H} \right\}^Z \tag{36}$$

This equation is identical to Eq. (15) derived for counterion distribution equilibria in negatively charged polyion systems. The concentration ratio of co-ions (i.e., M^{Z+} and H^+ ions) between the two phases can be expressed as follows instead of the activities:

$$\left\{ \frac{[M]_p}{[M]} \right\} = G \left\{ \frac{[H]_p}{[H]} \right\}^Z \tag{37}$$

where G stands for the activity coefficient quotient $[G = y_M(y_M)_p{}^{-1})(y_H)_p{}^Z(y_H)^{-Z}]$. Assuming again that the G value is constant and equal to unity, and combining Eqs. (14) and (37), $\{[M]_p/[M]\}$ can be expressed as

$$\left\{ \frac{[M]_p}{[M]} \right\} = \left\{ \frac{[H]_p}{[H]} \right\}^Z = 10^{Z\Delta pK} \tag{38}$$

It is apparent that the free metal ion concentration ratio between the two phases is available from the ΔpK term. For example, the ratio $\{[Ag]_p/[Ag]\}$ at $C_S = 0.01$ mol/dm^3 and at $\alpha = 0$ is estimated as low as $\sim 10^{-4}$. By combining Eqs. (35) and (38), $\log(K_M^0)_{app}$ can be expressed by the ΔpK term as

$$\log(K_M^0)_{app} = Z\Delta pK + \log(K_M^0)_{int} \qquad (39)$$

It is notable that the above equation derived for the equilibrium analysis of basic polymer ligands is exactly the same as derived for acidic polymer ligands [i.e., Eq. (31)].

As shown in Fig. 18, $(K_{Ag}^0)_{app}$ increases drastically with an α increase primarily due to the decreased electrostatic repulsion between PVImH$^+$ molecules and Ag$^+$ ions. $(K_{Ag}^0)_{app}$ is also affected by the concentration of inert 1 : 1 salt. This C_S dependence of the apparent binding quotient is consistent with the acid-dissociation properties of PVImH$^+$ and can be rationalized by the ion-screening effect as well. By comparing the two curves shown in Figs. 17 and 18, it is obvious that the $\log(K_{Ag}^0)_{app}$ versus α plots are not monotonous, as the pK_{app} versus α plots are, showing the tendency of lowering of the complexation ability at α lower than ~ 0.3, probably due to the change in availability of free imidazole groups.

When $\log(K_{Ag}^0)_{app}$ is plotted against ΔpK instead of α (Fig. 19), more straightforward insights into the microscopic equilibria can be gained. Note that the ΔpK values for PVIm complexation are always negative, which is just the opposite trend compared with PAA complexation (Fig. 10). The most enlightening is the convergency of the four separate curves corresponding to different salt concentration levels into a unique curve, which appears in Fig. 19. This, indeed, validates the Donnan-based correction procedure for the electrostatic effect operating in the Ag$^+$–PVIm complexation. As is predicted by Eq. (39), the slope of the linear portion of the plots is exactly 1, 1 being the charge of the Ag$^+$ ion. The intercept at ΔpK = 0 corresponds to $\log(K_{Ag}^0)_{int} = 7.22$. This value, however, is much higher than the stability constant of the 1 : 1 complex of the Ag$^+$ ion with imidazole (i.e., log $\beta_1 = 3.17$ [49]), indicating the formation of multidentate ligand complexes in the case of the Ag$^+$–PVIm system. When nitrogen atoms are involved in complexation with transition metal ions, the binding is expected to be much stronger than the case of carboxylate oxygen atoms. This enables metal ions to multicoordinate with nitrogen atoms attached to one polymer ligand molecule. Due to the lack of the reported equilibrium constants of Ag$^+$ complexes with bifunctional imidazole ligands (i.e., α, ω di-N-imidazolylalkane), the formation constant of the 1 : 2 complex of Ag$^+$–imidazole, log $\beta_2 = 6.94$ [49] was compared to $\log(K_{Ag}^0)_{int}$ (= 7.22) instead. Consistency between the two values indicates a bidentate ligand complex (1 : 2 complex) chelate formation in Ag$^+$–PVIm complexes. It should be pointed out that $\log(K_{Ag}^0)_{app}$ versus ΔpK plots deviates drastically downward from the straight line at $\alpha < 0.3$, indicating the remarkable suppression of bidentate ligand complex formation with these highly protonated PVIm molecules. Availability of free imidazole groups needed for the formation of bidentate ligand complexes is considerably

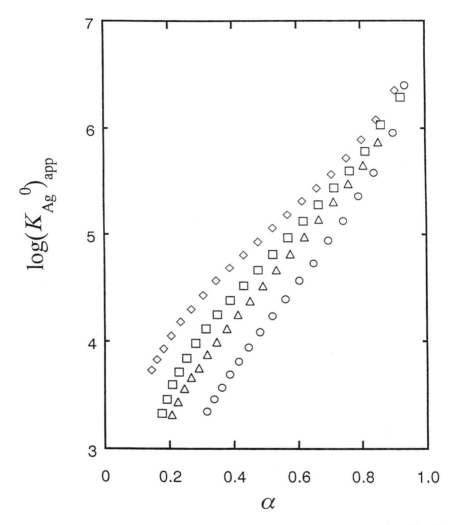

Figure 18 Apparent binding constant of the Ag^+/PVIm complex expressed as a function of α and C_S: (○) $C_S = 0.005$; (△) $C_S = 0.01$; (□) $C_S = 0.02$; (◇) $C_S = 0.05$ mol/dm³. (From Ref. 4.)

reduced under these acidic conditions, and simultaneous coordination of two neighboring imidazole groups to one Ag^+ ion may be prohibited. In these circumstances, monodentate complexation becomes predominant; its stability constant becomes quite close to that of the 1 : 1 complex of the Ag^+ ion with imidazole (i.e., $\log \beta_1 = 3.17$ [49]). This value is too small to draw the corresponding line in the same figure.

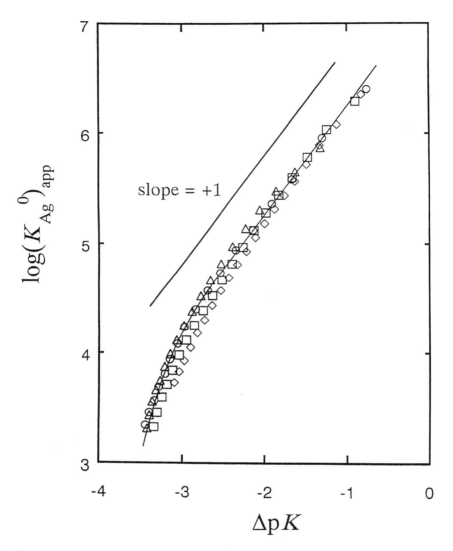

Figure 19 Relationship between $\log(K_{Ag}^{0})_{app}$ and ΔpK for PVIm: ($\bigcirc$) $C_S = 0.005$; ($\triangle$) $C_S = 0.01$; ($\square$) $C_S = 0.02$; ($\diamond$) $C_S = 0.05$ mol/dm^3. (From Ref. 4.)

The complexation of Cu^{2+} ions to PVIm molecules is much stronger than to Ag^+ ions. Based on the same data analysis procedure applied to Ag^+–PVIm complexation, microscopic insights into Cu^{2+} ion complexation have been gained (Fig. 20). It should be noted that although separate lines corresponding to C_S are obtained in the plots of $\log(K_{Cu}^0)_{app}$ against α, they converge to the identical line when $\log(K_{Cu}^0)_{app}$ is plotted against ΔpK (Fig. 21). It is also obvious that the slope of the linear portion of the line is approximately 2, the charge of Cu^{2+} ions. These results indicate that the change in the electrostatic effect due to the C_S change can be corrected fully for the Cu^{2+}–PVIm system as well as the Ag^+–PVIm system. The intercept of the line at $\Delta pK = 0$ corresponds to 14.7. Because the stability constants of successive Cu^{2+}–Im complexes are reported to be $\log \beta_1 = 4.18$, $\log \beta_2 = 7.66$, $\log \beta_3 = 10.51$, and $\log \beta_4 = 12.6$ [50], it is estimated that Cu^{2+} ions form 1 : 4 complexes at the fully dissociated state ($\alpha = 1$) by comparing the value of $\log (K_{Cu}^0)_{int} = 14.7$ with $\log \beta_4$ ($= 12.6$). It can be estimated that adjacent two imidazole groups fixed onto the same linear polymer form one chelate ligand site and one Cu^{2+} ion binds the two chelate sites to form a 1 : 4 complex. The $\log(K_{Cu}^0)_{app}$ versus ΔpK plots, indeed, reflect the stepwise complexation process with an α increase. In order to examine this trend in detail, the $\log(K_{Cu}^0)_{int}$ values calculated according to the following equation are plotted in Fig. 22 against α:

$$\log(K_{Cu}^0)_{int} = \log(K_{Cu}^0)_{app} - 2\Delta pK \tag{40}$$

Because we can see two transitions at $\alpha = \sim 0.3$ and ~ 0.8 in this Fig. 22, it can be estimated that the multidentate complexation property of PVIm is highly dependent on α, which controls the available distance between the neighboring ligand groups. Also, as can be seen in Fig. 22b, the transition observed in the Ag^+–PVIm system occurs at $\alpha = \sim 0.3$; the calculated distance between the neighboring free imidazole groups is ~ 10 Å. In the case of the Cu^{2+}–PVIm system, stepwise complexation has also been observed clearly at $\alpha = \sim 0.3$ and ~ 0.8. Taking into account the average distance of the neighboring groups, it is apparent that the transition occurs at a particular distance independent of the nature of metal ions. It can be estimated that a specified geometry of PVIm molecules in aqueous solution may be responsible for this transition.

V. ABNORMAL COMPLEXATION BEHAVIOR OBSERVED IN THE PRESENCE OF HYDROPHOBIC LINEAR POLYELECTROLYTES

Apparent binding constants for metal ions to strongly acidic polyions, such as PSS and PVS (Fig. 1) are usually much smaller in magnitude than corresponding

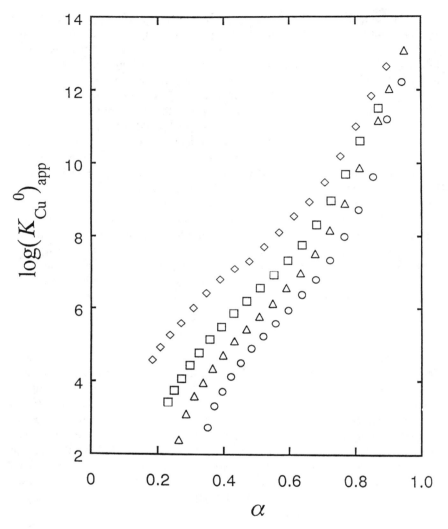

Figure 20 Apparent binding constant of Cu^{2+}–PVIm complex expressed as a function of α and C_S: (○) $C_S = 0.005$; (△) $C_S = 0.01$; (□) $C_S = 0.02$; (◇) $C_S = 0.05$ mol/dm^3. (From Ref. 4.)

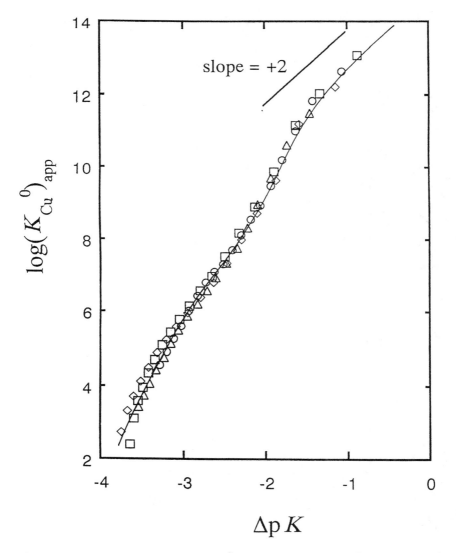

Figure 21 Relationship between $\log(K_{Cu}^{0})_{app}$ and ΔpK for PVIm: (○) $C_S = 0.005$; (△) $C_S = 0.01$; (□) $C_S = 0.02$; (◇) $C_S = 0.05$ mol/dm³. (From Ref. 4.)

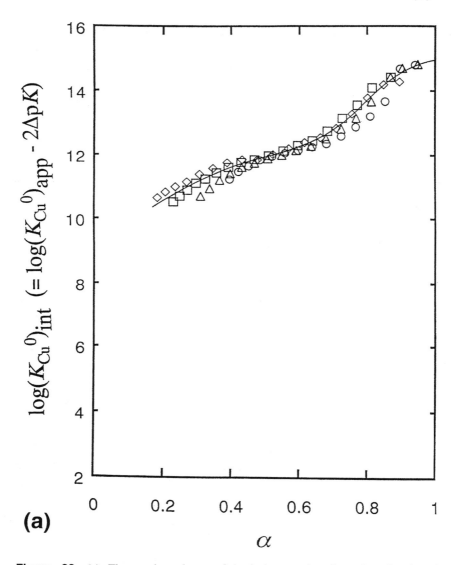

Figure 22 (a) The α dependence of intrinsic complex formation function for Cu^{2+}–PVIm: (O) $C_S = 0.005$; ($\triangle$) $C_S = 0.01$; ($\square$) $C_S = 0.02$; ($\diamond$) $C_S = 0.05$ mol/dm^3. (b) The α dependence of intrinsic complex formation function for Ag$^+$–PVIm: (O) $C_S = 0.005$; ($\triangle$) $C_S = 0.01$; ($\square$) $C_S = 0.02$; ($\diamond$) $C_S = 0.05$ mol/dm^3. (From Ref. 4.)

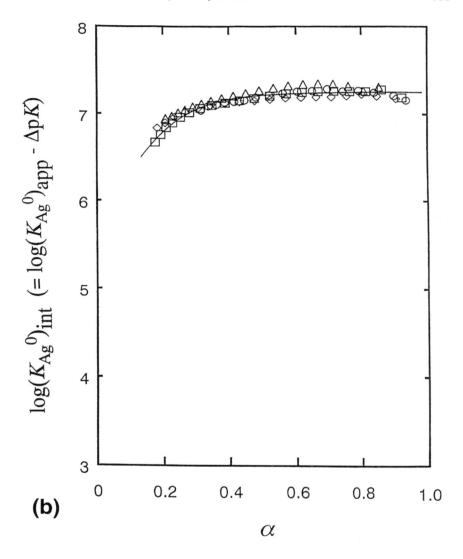

(b)

weak acidic polyions with a similar linear charge density, as binding to strong acidic polyions is primarily purely electrostatic in nature. High sensitivity of the binding equilibria to added salt concentration levels is duplicated in the strong acidic polyion systems as well. Because the detailed interpretations based on the Gibbs–Donnan concept of these purely electrostatic binding have already been given elsewhere [5,6,51], no further discussion will be provided in this chapter;

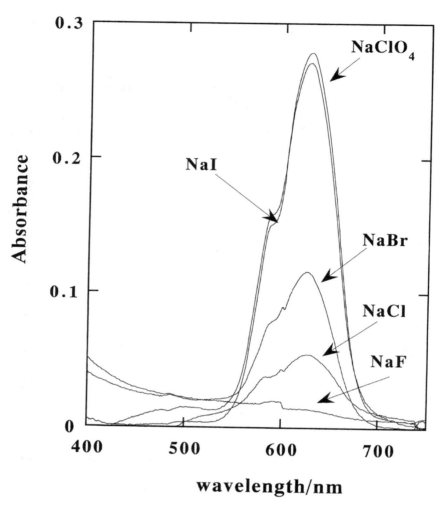

Figure 23 Electronic spectra of a resin phase equilibrated (Dowex 1×2 Cl⁻ form, 100–200 mesh, 0.6 mEq) with mixture solutions (12 cm³) of $CoCl_2(2.5 \times 10^{-1}$ mol/dm³) + NaSCN (0.1 mol/dm³) + NaX (1 mol/dm³; X = F⁻, Cl⁻, Br⁻, I⁻, and ClO_4^-).

however, it is stressed that the binding is just due to the electrostatic attraction of metal ions to the polyion surface from the bulk solution, not to direct "complexation" such as contact ion-pair formation and chelation [5,6]; that is, the hydration state of "bound" metal ions is quite similar to that of free metal ions in the bulk solution.

Another important aspect of the effect of charged polymers on metal complexation will be discussed in this section. When a polyion is added to a complexation system of metal cation–anionic small ligand, the degree of complexation is sometimes remarkably enhanced. One of the well-known and important examples of this phenomenon is abnormal complexability observed in a strongly basic anion-exchange resin [52–54]. A typical commercially available anion-exchange resin, Dowex 1 (Dow Chemical Co., Midland, MI, U.S.A.), is widely used as a separation material for various chemical species, including anionic metal complexes and oxoanions. This polyion gel is a cross-linked linear polyion [poly(vinyl benzyltrimethylammonium) ion; PVBTMA$^+$]. An example of a peculiar complexation in Dowex 1 resin phase was given by Kraus and Nelson [55,56], who expressed the distribution ratio of various metal ions to Dowex 1 resin as a function of hydrochloric acid concentration levels. The high magnitude of the distribution ratio of a metal ion to the resin phase has been ascribed to preferential formation of anionic chloro complexes in the resin phase; that is, the higher degree of metal complexation with Cl$^-$ ions in the resin phase than in the corresponding bulk solution phase may be attributed to the concentrated Cl$^-$ ions in the resin phase. However, quantitative analyses that followed have revealed that the degree of complexation is much higher than originally calculated by the use of effective Cl$^-$ ion concentration levels inside the resin phase [52–54]. Also, it has been pointed out that the distribution behavior is strongly dependent on the kind of univalent cations present in excess [55]. These phenomena have also been observed in other anionic ligand complexation systems, including the Co^{2+}–SCN$^-$ system [7,52], whose complexation is enhanced even in a cation-exchange resin. Some workers have attributed this exceptionally high complexability in these ion-exchanger resins to a high internal pressure, a low dielectric constant, and so forth in an ion-exchange resin phase [52–54]. These observations may indicate that complexation at a surface of linear polymer is essentially quite different from the bulk solution phase because the physicochemical environment inside the polyion gels must mimic the corresponding linear polymer surface. It seems appropriate to give a rationale for these unexpectedly high complexability of the anionic ligands inside the resin phase. It should be notified, however, that in these complexation systems, the ligands involved are not fixed on the polymer skeletons, whereas the ligating groups with polymer molecules are fixed on the polymer skeletons whose complexation equilibria have been analyzed in the previous sections.

A novel interpretation on the abnormally high complexability in anion-exchange resin phase is shown in this section by exemplifying the complexation of Co^{2+}–SCN$^-$ in the presence of PVBTMA$^+$ ions, the linear polymer analog of Dowex 1 resin. The magnitude of complexability of the Co^{2+}–SCN$^-$–PVBTMA$^+$–A$^+$X$^-$ system has been expressed as a function of the concentrations of the polymer, C_p, the supporting electrolyte A$^+$X$^-$, C_S, as well as the concentration of the SCN$^-$ ion, C_{SCN}, by measuring the electronic spectra of these mixture solutions. The peak maxima of octahedral $[Co(H_2O)_6]^{2+}$ ions and tetrahedral $[Co(NCS)_4]^{2-}$ ions are 510 nm and 625 nm, respectively, which allowed quantitative determination of the complexes. Representative spectra of the Co^{2+}–SCN^- system equilibrated with Dowex 1×2 resin in the presence of excess Na^+X^- (X^- = F^-, Cl^-, Br^-, I^-, and ClO_4^-) salts are shown in Fig. 23. By comparing the reported spectrum of $[Co(NCS)_4]^{2-}$ obtained in DMF [57], it was concluded that $[Co(NCS)_4]^{2-}$ complexes are formed predominantly in the resin phase.

Two characteristics must be notified for the complexation behavior of anionic ligands in the anion-exchange resin phase (i.e., (1) anionic complexes have the highest number of coordinating ligands and (2) the apparent complexability is strongly influenced by the nature of anions and cations of added salt [58]. Figure 23 apparently indicates that the complexation is remarkably enhanced by the addition of hydrophobic anions, such as ClO_4^- and I^- ions, and no or quite a small degree of complexation takes place in the presence of hydrophilic F^- ions. An identical phenomenon has been observed with the linear polymer analog of Dowex 1 resin, PVBTMA$^+$ ions, where an apparent molar extinction coefficient of Co^{2+} species measured at 625 nm, ε_{app}, is expressed in Fig. 24 as a function of the concentration of X^- anions, C_X. ε_{app} is defined by the following equation:

$$\varepsilon_{app} = \frac{A}{C_{Co}} \tag{41}$$

where the cell length is maintained constant at 1 cm and A and C_{Co} correspond to the overall absorbance and total Co^{2+} ion concentration of the mixture solution, respectively. It is striking that enhancement of the formation of $[Co(NCS)_4]^{2-}$ due to hydrophobic anions is so pronounced, because it is usually anticipated that the increase in C_X suppresses the binding of anionic complexes to the polymer.

As shown in the plots of ε_{app} versus C_A (Fig. 25), cation effect is not as pronounced as the anion effect, although the hydrophilicity/hydrophobicity nature of univalent cations added in excess also influences the magnitude of the formation of tetrahedral complexes. The effectiveness sequence of alkali metal ions is $Li^+ > Na^+ > K^+ > Rb^+ > Cs^+$, just the reverse hydrophobicity/hydrophilicity order of the anion effect order (i.e., $ClO_4^- > I^- > Br^- > Cl^- > F^-$). One of the possible explanations for these anion and cation effects is that the hydration degree of hydrophobic PVBTMA$^+$ polymer surface is suppressed by electrostatic attraction of hydrophobic anions and/or electrostatic repulsion of hydrophilic cations, which stabilizes the hydrophobic $[Co(NCS)_4]^{2-}$ complexes in the vicinity of the polymer skeleton.

As has been demonstrated, $[Co(NCS)_4]^{2-}$ is predominantly formed in the presence of PVBTMA$^+$ ions or Dowex 1, even though the overall stability constant of the tetrahedral complex is quite small in aqueous solution (i.e., $\log \beta_4 = -0.08$) [59]. Preferential formation of $[Co(NCS)_4]^{2-}$ can, therefore, be attributed to the specific nature of the positively charged polymer domain itself (i.e., $[Co(NCS)_4]^{2-}$

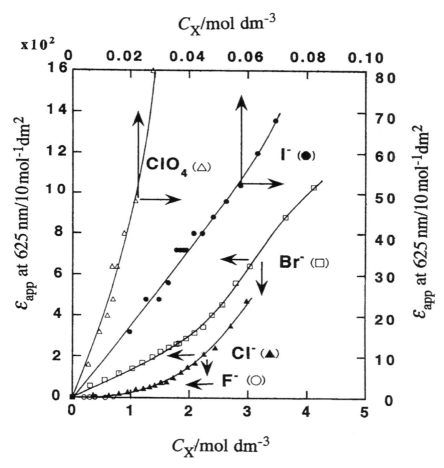

Figure 24 ε_{app} plotted against C_X. $CoCl_2$ (2.5×10^{-3} mol/dm^3) + NaSCN (2×10^{-2} mol/dm^3) + PVBTMA $^+$Cl$^-$ (0.10 mol/dm^3).

must be stabilized in the polymer domain). The present author attributes this stabilization to the formation of an "ion-pair" between the tetrahedral anionic complex and the cationic site, trimethylbenzyl group, R$^+$ in the polymer domain. In order to express the macroscopic complexation equilibria of the Co^{2+}– SCN$^-$–PVBTMA$^+$ system, it is assumed that Co^{2+} and SCN$^-$ ions are distributed between a polyelectrolyte phase (volume: V_p) and a bulk solution phase (volume: V_s) as shown schematically in Fig. 26. In this diagram, the formation of the [Co(NCS)$_4$]$^{2-}$ complex in the polyelectrolyte phase is taken into consideration, as

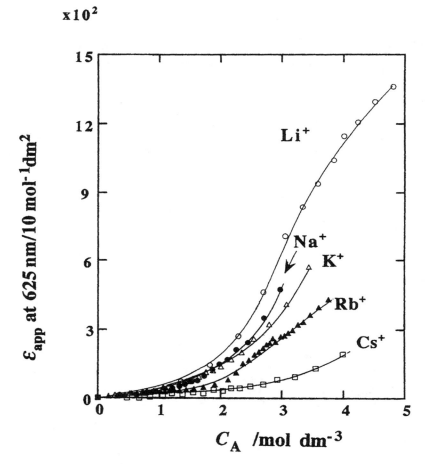

Figure 25 ε_{app} plotted against C_A. CoCl$_2$ (2.5×10^{-3} mol/dm^3) + NaSCN (2×10^{-2} mol/dm^3) + PVBTMA$^+$Cl$^-$ (0.10 mol/dm^3).

its concentration term is necessary for the equilibrium calculation of the "ion-pair" {[Co(NCS)$_4$]$^{2-}$R$^+$} in the polymer domain.

The magnitude of the formation of the complex is monitored as ε_{app}. Based on the two-phase model, both absorbance terms due to the bulk solution phase, A_s, and the polyelectrolyte phase, A_p, contribute to A. A can be expressed as follows by taking into account the volume fractions of both phases, V_s/V_{tot} and V_p/V_{tot}, where V_{tot} indicates the total solution volume (i.e., $V_{tot} = V_s + V_p$):

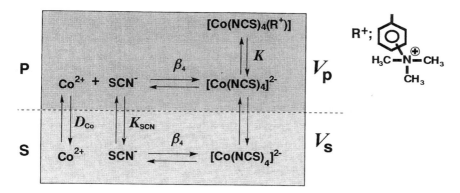

Figure 26 Schematic representation of the complexation equilibrium analysis based on the two-phase model. All the equilibrium constants are defined in the text.

$$A = \frac{A_s V_s}{V_{tot}} + \frac{A_p V_p}{V_{tot}} \tag{42}$$

Because the contribution of the free Co^{2+} ion absorbance at the wavelength of the peak maximum of $[Co(NCS)_4]^{2-}$ (625 nm), can be neglected, A can be related directly to the concentration of the contact ion-pair in the polyelectrolyte phase, $[Co(NCS)_4R]_p$, with the molar extinction coefficient of the tetraisothiocyanato complex, $\varepsilon_{[Co(NCS)_4]}$:

$$A = \frac{\varepsilon_{[Co(NCS)_4]} [Co(NCS)_4R]_p V_p}{V_{tot}} \tag{43}$$

Also, C_{Co} can be expressed as

$$C_{Co} = \frac{[Co]_s V_s + [Co(NCS)_4R]_p V_p}{V_{tot}} \tag{44}$$

Combining Eqs. (43) and (44), we obtain

$$\varepsilon_{app} = \frac{\varepsilon_{[Co(NCS)_4]}[Co(NCS)_4R]_p V_p}{[Co]_s V_{tot} + [Co(NCS)_4R]_p V_p} \tag{45}$$

where V_s can be assumed quite close to V_{tot} in value.

In order to express all the equilibria presented in Fig. 26, the following four equilibrium constants must be defined:

1. The Co^{2+}–SCN^- complexation equilibrium in the polyelectrolyte phase:

$$Co^{2+} + 4SCN^- \rightleftarrows [Co(NCS)_4]^{2-}, \qquad \beta_4 = \frac{[Co(NCS)_4]_p}{[Co]_p[SCN]_p^4} \qquad (46)$$

where the overall stability constant, β_4, in the polyelectrolyte phase is estimated to be consistent with the reported value (log $\beta_4 = -0.08$ [59]).

2. The ion-pair formation equilibrium of $[Co(NCS)_4]^{2-}$ with R^+ in the polyelectrolyte phase:

$$[Co(NCS)_4]^{2-} + R^+ \rightleftarrows [Co(NCS)_4R], \qquad K = \frac{[Co(NCS)_4R]_p}{[Co(NCS)_4]_p[R]_p} \qquad (47)$$

3. The distribution equilibrium of Co^{2+} ions between the two phases.

$$(Co^{2+})_s \rightleftarrows (Co^{2+})_p, \qquad D_{Co} = \frac{[Co]_p}{[Co]_s} \qquad (48)$$

Because Co^{2+} ions are repelled from the positively charged polymer region, the distribution equilibrium can totally be controlled by a Donnan relation.

4. The equilibrium of SCN^- ion binding to $PVBTMA^+$ polymers:

$$(SCN^-)_{free} \rightleftarrows (SCN^-)_{bound}, \qquad K_{SCN} = \frac{[SCN]_{bound}}{[SCN]_s C_p} \qquad (49)$$

where C_p denotes the polyion concentration expressed in monomole/dm^3. The bound SCN^- concentration based on V_{tot}, $[SCN]_{bound}$, can be related to $[SCN]_p$ based on V_p by the following equation. Note that the C_p term is introduced in Eq. (49) in order to normalize the equilibrium quotient, $[SCN]_{bound}/[SCN]_s$:

$$[SCN]_{bound} = \frac{[SCN]_p V_p}{V_{tot}}$$

Because $[SCN]_p$ can be related to C_{SCN} through Eq. (50) as

$$[SCN]_p = \frac{[K_{SCN} C_{SCN} C_p/(1 + K_{SCN} C_p)]V_{tot}}{V_p} \qquad (50)$$

and the V_{tot}/V_p term in the above expression can be replaced by C_p, $V_{tot}/V_p = n_p/V_p C_p^{-1}$, the $[Co(NCS)_4R]_p$ term appearing in Eq. (47) can finally be expressed as

$$[Co(NCS)_4R]_p = K[Co(NCS)_4]_p[R]_p$$

$$= K[Co(NCS)_4]_p\left(\frac{n_p}{V_p}\right) = \frac{K\beta_4 D_{Co}K_{SCN}^4(n_p/V_p)^4 C_{SCN}^4 C_p}{(1 + K_{SCN}C_p)^4} \tag{51}$$

where n_p indicates the total amount of benzyltrimethylammonium units (in monomoles) fixed on the polymer molecules. Combining Eqs. (45) and (51), ε_{app} can be written as

$$\varepsilon_{app} = \left(\frac{\varepsilon_{[Co(NCS)_4]} GC_{SCN}^4 C_p}{(1 + K_{SCN} C_p)^4}\right)\left(1 + \frac{GC_{SCN}^4 C_p}{(1 + K_{SCN} C_p)^4}\right)^{-1} \tag{52}$$

where G represents the product, $K\beta_4 D_{Co} K_{SCN}^4 (n_p/V_p)^4$.

It is obvious that ε_{app} is as a function of C_{SCN} as well as C_p if the volume term, V_p/n_p, remains constant throughout all the experimental conditions. At constant C_p, Eq. (52) can be simplified as

$$\varepsilon_{app} = \frac{\varepsilon_{[Co(NCS)_4]} GkC_{SCN}^4}{(1 + GkC_{SCN}^4)} \tag{53}$$

where k corresponds to $C_p/(1 + K_{SCN} C_p)^4$.

At constant C_{SCN} on the other hand, ε_{app} can be expressed as

$$\varepsilon_{app} = \left(\frac{\varepsilon_{[Co(NCS)_4]} aC_p}{(1 + K_{SCN} C_p)^4}\right)\left(1 + \frac{aC_p}{(1 + K_{SCN} C_p)^4}\right)^{-1} \tag{54}$$

where a represents GC_{SCN}^4. Taking the logarithmic form of Eq. (54) and differentiating $\log \varepsilon_{app}$ with respect to $\log C_p$, it is apparent that linearities are expected in the $\log \varepsilon_{app}$ versus $\log C_p$ plots in sufficiently low and high C_p regions; the slopes of 1 and −3 are obtained in the lower and higher C_p regions, respectively. Qualitatively, suppression of the formation of $[Co(NCS)_4]^{2-}$ observed in the high C_p region can be attributed to dilution of the effective concentrations of Co^{2+} and SCN^- ions in the polyion domain, whose volume is not affected by the C_p change. Some of the representative plots are shown in Fig. 27. The $\log \varepsilon_{app}$ versus $\log C_p$ plots obtained for various concentrations of C_S and C_{SCN} have been analyzed by the least-squares curve-fitting method, and the adjustable parameters of K_{SCN} and G thus obtained are summarized in Table 2. The $\varepsilon_{[Co(NCS)_4]}$ value estimated by the present study is ~1600, being consistent with the reported values of ~1900 [56]

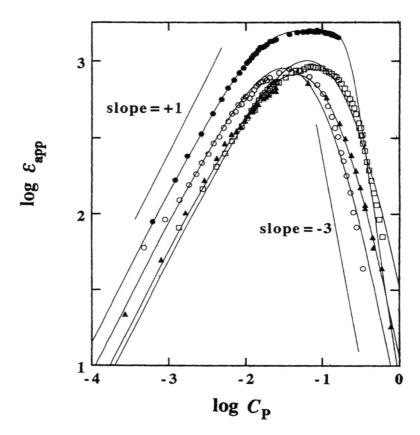

Figure 27 Representative log ε_{app} versus log C_p plots. $CoCl_2$ (5×10^{-4} mol/dm³) + NaSCN (5×10^{-2} mol/dm³) + NaCl (C_S mol/dm³). C_S in mol/dm³: (○): 0.20; (▲): 0.50; (□): 1.00; (●): 3.00. Solid lines refer to the curves calculated according to Eq. (54) by the use of the parameters listed in Table 2.

and 1750 [57]. The present quantitative analysis of the dilution effect based on the two-phase model has been proven to be valid under all the experimental conditions, which supports the assumption of the presence of a volume element in the vicinity of the polymer skeleton. The increase in log ε_{app} at the lower C_p region corresponds to the binding of SCN⁻ ions to the polymer, and the maximum plateau corresponds to the sufficient formation of $[Co(NCS)_4]^{2-}$ complexes in the polymer domain. The C_p term in the numerator of Eq. (52) is due to the ion-pair formation between $[Co(NCS)_4]^{2-}$ and the fixed R⁺ group, and the term in the denominator corresponds to the complex formation of $[Co(NCS)_4]^{2-}$. The observed slopes in

Table 2 Calculated Values for K_{SCN} and G

C_S (mol/dm^3)	K_{SCN} (mol/dm^3)	G $(=KD_{Co}\beta_4 K_{SCN}{}^4(n_p/V_p)^4)$
0.20	11	1.3×10^7
0.50	8.3	7.9×10^6
1.00	5.7	8.3×10^6
2.00	3.2	1.1×10^6

Note: $C_{Co} = 5.0 \times 10^{-4}$ mol/dm^3, $C_{SCN} = 0.05$ mol/dm^3, $\varepsilon_{[Co(NCS)_4]}$ = 1580/10 mol/dm^2.

both C_p regions are exactly 1 and -3, as shown in Fig. 27, indicating the validity of the assumed equilibria in the polyion domain. Significant evidence gained by the present plots is the fact that the V_p/n_p term appearing in Eqs. (51)–(54) remains essentially unchanged irrespective of the change in C_p.

As has been indicated by Eq. (53), ε_{app} should also be a function of C_{SCN}. In order to examine the validity of this formulization and to verify preferential formation of $[Co(NCS)_4]^{2-}$ complexes in the polyelectrolyte phase, ε_{app} determined under constant C_p is plotted in Fig. 28 against C_{SCN}. When C_{SCN} is close to C_p in value, the best fit between the experimental plots and the calculated curve has been obtained, whereas at extreme cases where $C_{SCN} \gg C_p$ or $C_{SCN} \ll C_p$, the discrepancy between the calculated and experimental values is appreciable, which is attributable to the change in K_{SCN} in Eq. (53) with C_{SCN}.

An abnormally high degree of complexability of a metal ion with hydrophobic anion ligands in an anion-exchanger resin phase can thus be interpreted as hydrophobic adsorption of anionic complexes formed onto positively charged polymer surface by exemplifying a linear cationic polymer as the resin analog. Reasonings presented earlier for the higher complexability in ion-exchanger phases, such as high internal pressure and low dielectric constant [47–49], are not always valid, because the same phenomenon has also been observed in the non-cross-linked cationic polymer system, and the effect of the kind of cations and anions is so pronounced. Despite the difference in appearance of a linear polymer (PVBTMA$^+$) and a cross-linked linear polymer (anion-exchange resin), the Co^{2+}–SCN$^-$ complexation behavior in both cases could be treated in an identical manner to that based on the two-phase model. The magnitude of the formation of $[Co(NCS)_4]^{2-}$ has fully been reproduced by the use of the polymer concentration, which validates the introduction of the polyelectrolyte phase volume term for the interpretation of ionic reactions in linear polymer systems. Finally, it should be stressed that the interaction between hydrophobic ions plays quite an important role in the ionic reactions related to charged polymer molecules.

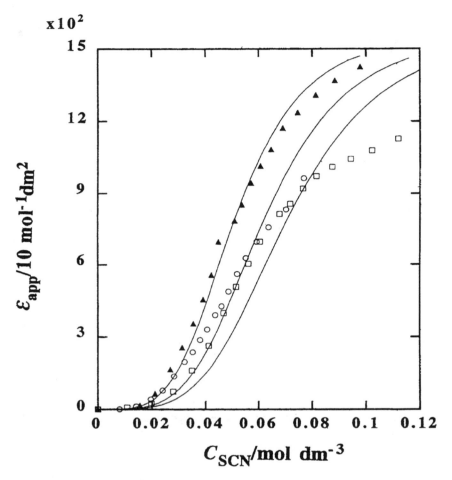

Figure 28 Representative ε_{app} plots against C_{SCN}. $CoCl_2$ (2.5×10^{-4} mol/dm³) + NaCl (1.0 mol/dm³) + PVBTMA $^+$Cl$^-$. C_p in mol/dm³: (◯): 10^{-2}; (▲): 5×10^{-2}; (☐): 2×10^{-1}. Solid lines refer to the curves calculated according to Eq. (53).

VI. CONCLUSIONS

Complicating problems inherent in metal complexation equilibria with polymer molecules (i.e., polyelectrolytic property, multidentate coordinating property, and hydrophobic property) have been examined in detail in this chapter by exemplifying charged linear (one-dimensional) polymers. It has been proved valid the promise that nonideality terms originating from these essential problems can be evaluated based on a "two-phase property" of a polymer–small ion mixture in an aqueous solution. For any polymer molecules examined, a volume element (a polyelectrolyte phase) can be assumed around the polymer skeleton, which is separated by a hypothetical Donnan membrane from a bulk solution. This thermodynamic simplification facilitates the calculation of the ion distribution equilibrium between the two phases [i.e., the polymer domain (a polyelectrolyte phase) and the bulk solution], and this procedure for averaging electrostatic potential in the domain is essentially important for the equilibrium analyses of all the ionic reactions involved in polymer solutions.

For the analysis of ion-binding equilibria of charged polymers, the "polyelectrolytic property" should be evaluated prior to estimation of the magnitudes of the secondary properties. The magnitude of "polyelectrolytic effect" due to a strong and variable electrostatic field formed at charged polymer molecules has proven to be accessible experimentally at any particular solution conditions. For the equilibrium analyses of metal complexation of weak acidic polyelectrolyte (e.g., PAA) and weak basic polyelectrolytes (e.g., PVIm), averaged concentrations of metal cations in the vicinity of polymer ligands can be related to the bulk solution phase by the use of acid-dissociation equilibria of PAA and PVImH$^+$ as a probe. The magnitude of concentration of metal cations in the polymer domain from the bulk solution for negatively charged PAA polymer and the magnitude of metal cation dilution for positively charged PVImH$^+$ polymer can be calculated by the use of the concentration ratio of H$^+$ ions between the two phases through a Donnan's relation. The H$^+$ ion concentration ratio is accessible by the nonideality term for acid-dissociation equilibria of PAA and PVImH$^+$ molecules, by assuming that the intrinsic acid-dissociation constants of the repeating monomeric functional groups fixed on these homologous polymer molecules are unvaried throughout any solution conditions. The fact that the intrinsic stability constants evaluated by this "two-phase" approach for both PAA and PVIm polymer ligands determined at various degrees of dissociation of the polyacids and under different salt concentration levels are consistent with the stability constants of metal complexes with corresponding monomeric ligands validates the present approach. The solution property of the polyelectrolyte phase must correspond to the concentrated simple electrolyte solution, although the activity coefficients of metal ions, H$^+$ ions, as well as supporting cations and anions in the polyelectrolyte phase must be greatly influenced by the "ionic atmosphere" of the ion-concentrated region and are much different from those in the bulk solution phase. However, the activity coefficient quotients appearing in Eqs. (16) and (37) for PAA and PVIm, respectively, must remain constant throughout the experimental conditions, which has been validated by the present studies as well.

After correcting for the polyelectrolytic effect on the complexation equilibria of charged polymers, information on multidentate coordination property of linear polymers has been extracted. Polymer complexes with weak complexing ability predominantly from monodentate complexes. Ag^+–PAA and Ca^{2+}–PAA systems are examples of this kind. With a much stronger complexing ability, polymer ligands form an appreciable amount of multidentate complexes. PAA complexes with Cu^{2+}, Cd^{2+}, and Pb^{2+} ions are examples of this kind coordination. They form bidentate complexes together with monodentate complexes depending on solution conditions, such as pH and the added salt concentration levels. The distribution of the two types of complexes has been revealed to be dependent on the average spacing of free carboxylate groups fixed on the linear polymer molecules (i.e., only on the degree of dissociation of PAA). At a lower degree of dissociation, monodentate coordination is preferential, whereas with an increase in the degree of dissociation of the polyacid, bidentate complexation becomes appreciable. Similarity and dissimilarity of the intrinsic stability constants of bidentate complexes for PAA ligand to the corresponding β_2 values of the complexes with monomeric acetate ligands and with the β_1 value of glutarate ligands have revealed that the multidentate coordination behavior in linear polymer ligand complexation is highly dependent on metal coordinating orientation property.

It has been revealed that PVIm ligands with nitrogen atoms show much stronger complexing ability than PAA with oxygen atoms to transition metal ions, such as Ag^+ and Cu^{2+} ions, due to nitrogen atom as coordinating metal ions. At complete dissociation of $PVImH^+$, Ag^+, and Cu^{2+} ions show maximum coordination numbers of 2 and 4 with PVIm ligand, respectively. In the similar manner with the functionality spacing dependence of PAA molecules, the number of coordinating ligating groups (i.e., imidazole groups) decreases with decreasing the degree of neutralization. It has been revealed that this transition takes place stepwisely as a function of the degree of dissociation of linear polymer ligands. Interestingly, the linear charge spacings that correspond to the transition of coordination for PAA and PVIm ligands for different metal cations are consistent with each other, which may be due to the identical skeletal structure of these two linear polymers.

Abnormal enhancement complexability of metal cation–small ligand anion system in the presence of hydrophobic cationic linear polymers has been quantified based on the two-phase approach. For the Co^{2+}–SCN^- complexation system, preferential formation of the complexes with the maximum coordination number, $[Co(NCS)_4]^{2-}$, has been revealed, although the overall stability constant of the complex in aqueous solution is quite small. The nature and the concentrations of anions and cations remarkably affect the extent of the formation of tetrahedral complex; however, the order of the effectiveness of these supporting ions is just the opposite of each other; the more hydrophobic the anion is and the more hydrophilic the cation is, the complexation is enhanced. The complexation equilibrium has successfully been analyzed by assuming an ion-pair formation between hydrophilic anionic complexes, $[Co(NCS)_4]^{2-}$ with a hydrophobic cationic functional groups, and trimethylbenzylammonium group in the polyelectrolyte phase.

This assumption enables quantitative expression of the PVBTMA+ polymer concentration dependence of the magnitude of complexation, expressed by an apparent molar extinction coefficient of Co^{2+} ions. The intrinsic ion-pair formation constant thus determined by the present procedure indicates the possibility of a quantitative expression for the hydrophobic interaction in polymer systems. It should be stressed that the acid-dissociation equilibria of PVImH+ polymers are strongly affected by the nature of supporting anions. The slight difference observed in the pK_{app} versus α plots for NaCl (Fig. 3) and $NaNO_3$ (Fig. 17) systems corresponds to the direct hydrophobic interaction between these anions, with the cationic sites fixed at the linear polymer surface. It can be estimated that the potentiometric titration behavior of weak basic polymers is much more strongly influenced by the nature of supporting anions than that of weak acidic polymers by the nature of supporting cations.

"Polymer effects" characteristic for ionic reactions related to "two-dimensional" (i.e., surface) polymers should be interpreted equally [60] by evaluating the same properties for "one-dimensional" polymers, although the polyelecrolytic and hydrophobic properties for surface polymers must be much more pronounced. Due to the extreme restriction of the magnitude of functional group freedom for ionic latex, the multidentate coordination behavior is expected to be "frozen" in the case for the corresponding flexible polymers. This interesting property is closely related to the origin of "molecular and ionic recognition with imprinted polymers," where multipoint interactions play an essential role [61] and are worth further detailed study.

REFERENCES

1. T Miyajima, M Mori, S Ishiguro, KH Chung, C-H Moon. J Colloid Interf Sci 184:279–288, 1996.
2. T Miyajima, M Mori, S Ishiguro. J Colloid Interf Sci 187:259–266, 1997.
3. H Kodama, T Miyajima, M Mori, M Takahashi, H Nishimura, S Ishiguro. Colloid Polym Sci 275:938–945, 1997.
4. T Miyajima, H Nishimura, H Kodama, S Ishiguro. React Funct Polym, 38:183–195, 1998.
5. JA Marinsky. In: JA Marinsky, Y Marcus, eds. Ion Exchange and Solvent Extraction— A Series of Advances. New York: Marcel Dekker, 1993, Vol 11, Chap. 5.
6. T Miyajima. In: JA Marinsky, Y Marcus, eds. Ion Exchange and Solvent Extraction— A Series of Advances. New York: Marcel Dekker, 1995, Vol 12, Chap. 7.
7. T Miyajima, H. Kodama, K Tajiri, S Ishiguro. Abnormal complexability of Co^{2+}/SCN⁻ system in the presence of polyvinylbenzyltrimethylammonium ions. Proceedings of "Polyelectrolyte 98 (Inuyama)," 1998.
8. M Nagasawa, T Murase, K Kondo. J Phys Chem 69:4005–4012, 1965.
9. M Nagasawa. Pure Appl Chem 26:519–536, 1971.
10. GS Manning, A Holtzer. J Phys Chem 77:2206–2212, 1973.

11. GS Manning. J Chem Phys 89:3772–3777, 1988.
12. GS Manning. Acc Chem Res 12:443–449, 1979.
13. GS Manning. Quart Rev Biophys 2:179–246, 1978.
14. M Fixman. J Chem Phys 70:4995–5005, 1979.
15. CF Anderson, MT Record Jr. Ann Rev Phys Chem 33:191, 1982.
16. JA Marinsky. J Phys Chem 96:6484–6487, 1992.
17. JA Marinsky. J Phys Chem 100:1858–1866, 1996.
18. Z Alexandrowicz, A. Katchalsky. J Polym Sci Pt A 1:3231–3260, 1963.
19. F Oosawa. Polyelectrolytes. New York: Marcel Dekker, 1971.
20. JA Marinsky, T Miyajima, E Högfeldt, M Muhammed. React Polym 11:279–289, 1989.
21. JA Marinsky, T Miyajima, E Högfeldt, M Muhammed. React Polym 11:291–300, 1989.
22. A Katchalsky, Z Alexandrowitcz. J Polym Sci A1:2093–2099, 1963.
23. L Kotin, M Nagasawa. J Chem Phys 36:873–879, 1962.
24. A Katchalsky, Z Alexandrowicz, O Kedem. In: BE Conway, RG Barradas, eds. Chemical Physics of Ionic Solution. New York: Wiley, 1996, p 295.
25. R Portnanova, P DiBernade, A Cassol, E Tondello, L Magon. Inorg Chim Acta 8:233–240, 1974.
26. SP Datta, AK Grybowski. J Chem Soc (B) 136–140, 1966.
27. DH Gold, HP Gregor. Z Phys Chem (Frankfurt) 15:93–102, 1958.
28. DH Gold, HP Gregor. Z Phys Chem 64:1461, 1968.
29. K Lin, HP Gregor. J Phys Chem 69:1252–1259, 1965.
30. HP Gregor, LB Luttinger, EM Loebl. J Phys Chem 59:34–39, 1955.
31. HP Gregor, LB Luttinger, EM Loebl. J Phys Chem 59:559–560, 1955.
32. HP Gregor, LB Luttinger, EM Loebl. J Phys Chem 59:990–991, 1955.
33. M Mandel, JC Leyte. J Polym Sci A2:2883–2899, 1964.
34. JA Marinsky. In: JA Marinsky, Y Marcus, eds. Ion Exchange and Solvent Extraction. New York: Marcel Dekker, 1973, Vol 4, Chap. 5.
35. WM Anspach, JA Marinsky. J Phys Chem 79:433–439, 1975.
36. JA Marinsky. Coord Chem Rev 19:125–171, 1976.
37. JA Marinsky, N Imai, MC Lim. Isr J Chem 11:601–622, 1973.
38. C Travers, JA Marinsky. J Polym Sci Sympos 47:285–297, 1974.
39. JA Marinsky, WM Anspach. J Phys Chem 79:439–444, 1975.
40. T Miyajima, K Yoshida, Y Kanegae, H Tohfuku, JA Marinsky. React Polym 15:55–62, 1991.
41. FH MacDougall, LE Topel. J Phys Chem 56:1090–1093, 1952.
42. M Yasuda, K Yamasaki, H Ohtaki. Bull Chem Soc Jpn 33:1067–1070, 1960.
43. P Gerding. Acta Chem Scand 21:2015–2027, 1967.
44. KH Chung, E Hong, Y Do, C-H Moon. J Chem Soc Chem Commun 2333, 1995.
45. KH Chung, C-H Moon. J Chem Soc Dalton Trans 75, 1996.
46. DH Gold, HP Gregor. J Phys Chem 64:1461, 1960.
47. DH Gold, HP Gregor. J Phys Chem 64:1462, 1960.
48. NC Li, JM White, E Doody. J Am Chem Soc 76:6219–6223, 1954.
49. SP Datta, AK Grzybowski. J Chem Soc (A) 1059, 1966.
50. WL Koltum, RN Dexter, RE Clark, FRN Gurd. J Am Chem Soc 80:4188–4194, 1958.

51. JA Marinsky. J Phys Chem 89:5294–5302, 1985.
52. H Waki, Y Miyazaki. Polyhedron 8:859–864, 1989.
53. H Waki. In: JA Marinsky, Y Marcus eds. Ion Exchange and Solvent Extraction. New York: Marcel Dekker, New York, 1995, Vol 12, Chap. 5.
54. Y Miyazaki, H Waki. Polyhedron 11:3031–3036, 1992.
55. KA Kraus, F Nelson, F Clough, RC Carlton. J Am Chem Soc 77:1391, 1955.
56. KA Kraus, DC Michelson, F Nelson. J Am Chem Soc 81:3204–3207, 1959.
57. S Ishiguro, K Ozutsumi. Inorg Chem 29:1117–1123, 1990.
58. F Helfferich. Ion-Exchange. New York, McGraw Hill, 1962, Chap. 5.
59. HA Asilber, MA Marguia. Inorg Chem 24:3794–3802, 1985.
60. JA Marinsky. J Phys Chem 100:1858–1866, 1996.
61. T Miyajima, K Sohma, S Ishiguro, M Ando, S Nakamura, M Takagi. In: R Bartsch, M Maeda, eds. Molecular and Ionic Recognition with Imprinted Polymers, ACS Symposium Series 703, Oxford: Oxford University Press, 1998, Chap. 20.

9

Adsorption of Quaternary Ammonium Compounds at Polymer Surfaces

Sharad G. Dixit and Ajay K. Vanjara
University of Mumbai, Mumbai, India

I. INTRODUCTION

Adsorption at the solid–liquid interface has been an area of great interest to the researchers. This interest basically stems from the role adsorption plays in influencing and affecting a large number of processes commonly used in laboratory and industrial practice. The important applications where solid–liquid interface plays a major role include froth flotation of minerals, flocculation and selective flocculation, dispersion, sedimentation, processing of fertilizers, and production of paints. A large number of investigations have been carried out on the adsorption of surfactants on the surfaces of inorganic substances such as ores, minerals, inorganic oxides, metals, and fertilizers.

The polymer substrates have become important only in recent years due to their increasing applications in industry, biomedical and biochemical fields. Primarily, the adsorption of surfactants from aqueous solution has been investigated. Studies on adsorption from nonaqueous solution have been rather scanty. The investigations have been basically carried out with two goals in mind: (1) to find process conditions for a specific application and (2) to investigate the factors underlying the adsorption process with a view to develop better understanding. This understanding, in turn, leads to development of an appropriate model that describes the process. It is necessary to have a system which is well defined and simple to investigate basic aspects. A vast body of literature is available with regard to the adsorption on the inorganic substances of concern. All types of surfactant (cationic [1–10], anionic [11–20], nonionic [21–25], and amphoteric [5]) have been studied. The understanding of the adsorption of surfactant at the inorganic substrate–liquid interface is of great help in understanding the adsorption of the surfactant on the polymer substrate. However the polymer substrates have their own peculiarities. In this chapter, it is proposed to overview the work done by various investigators on the adsorption of quaternary ammonium compounds on the polymer substrate.

II. AMINO SURFACTANTS

The amino surfactants belong to the category of cationic surfactants. There are several types of amino surfactants, such as primary amine, secondary amine, tertiary amine, quaternary ammonium salt, diamine, triamine, and pyridinium salt. Their typical structures are given in Table 1. It is to be noted from Table 1 that whereas primary, secondary, and tertiary amines do not have a charge on the nitrogen atom, the quaternary ammonium compound, diamine, triamine, and pyridinium salts have a net positive charge on their nitrogen atom. Furthermore, it has been shown that during the formation of the N–H bond, there is some electron transfer from hydrogen to the nitrogen which makes nitrogen more electronegative as compared to hydrogen [26] and there is a partial positive charge in the vicinity of nitrogen atom.

The following characteristics of the amino surfactants influence the adsorption:

1. Number and length of the hydrocarbon chain
2. Branching and double bonds present in the hydrocarbon chain
3. Presence of an ionic group
4. Cross-sectional area of groups
5. Orientation of the chains with respect to one another and the surface

The solution properties of the surfactants such as solubility, base strength, micellization characteristics, kraft point, and the surface tension are, therefore, also dependent on the surfactant characteristics. Some of the solution characteristic of primary, secondary, tertiary and quaternary ammonium surfactants are given in Table 2.

A. The Strength of Amines as Bases

Because of the presence of a lone pair of electrons on the nitrogen, all amines, whether primary, secondary or tertiary, can act as potential bases and can accept protons from the Brønsted acid. In case of normal acid, it results in the formation of substituted ammonium ions with the formation of a hydrogen bond:

$$AH + NH_2R \rightleftharpoons A^{-}\cdots\cdots HN^{+}H_2R$$

Table 1 Structures of Various Cationic Surfactants

Cationic surfactant	Structure[a]
Primary amine	$R-N\langle^H_H$
Secondary amine	$^R_{R'}\rangle N-H$
Tertiary amine	$^R_{R'}\rangle N-R''$
Quaternary ammonium Salt	$^R_{R'}\rangle \overset{+}{N} X^- \langle^{R''}_{R'''}$
Diamine	$R-\overset{+}{\underset{H}{N}}-(CH_2)_X\overset{+}{N}\langle^H_H$
Triamine	$R-\overset{+}{\underset{H}{N}}-(CH_2)_X-\underset{H}{\overset{H}{C}}-(CH_2)_X\overset{+}{N}\langle^H_H \quad (\overset{+}{N}H_2 \text{ above})$
Pyridinium salt	pyridinium ring $-\overset{+}{N}\langle^{X^-}_R$

[a]R is long hydrocarbon chain (C_8 or more); R', R'', and R''' are short hydrocarbon chains, X^- is a halide ion.

However, with weak proton donors such as alcohol, the interaction leads to the formation of a hydrogen-bonded complex:

$$AH + NH_2R \rightleftharpoons AH\cdots\cdots NH_2R$$

In aqueous solutions, the strength of the base can be defined in terms of the equilibrium constant K_b of the reaction:

$$B + H_2O \rightleftharpoons BH^+ + OH^-$$

$$K_b = \frac{C_{BH^+} C_{OH^-}}{C_B} \frac{f_{BH^+} f_{OH^-}}{f_B}$$

At low concentration, the activity coefficient term f, can be taken as unity; hence,

$$K_b = \frac{C_{BH^+} C_{OH^-}}{C_B} \quad \text{and} \quad pK_b = -\log K_b$$

The substituted ammonium ion RN^+H_3 is the conjugated acid of the base RNH_2; similarly, the base RNH_2 is the conjugated base of the acid RNH_3, which ionizes as

$$RN^+H_3 \rightleftharpoons RNH_2 + H^+$$

The equilibrium constant of acid is K_a and is defined as

$$K_a = \frac{C_{R\ddot{N}H_2} C_{H^+}}{C_{RN^+H_3}} \quad \text{and} \quad pK_a = -\log K_a$$

As $K_w = [H^+][OH^-] = 10^{-14}$, the general relationship $pK_a = 14 - pK_b$ holds for equilibria involving amine and their conjugate acid. General practice is to report the pK_a values for all types of amine. These values are given in Table 2. It will be seen from Table 2 that pK_a values for primary amines of interest are more or less constant at 10.62 ± 0.01. Similarly for the secondary amines of interest, pK_a values remain more or less constant at 11. There is greater variation of pK_a values with chain length [33] in case of tertiary amines. Table 2 also gives the solubility values and the limited data presented indicates that solubility decreases with increasing chain length in primary amines. Similarly, the critical micelle concentration (CMC) values decrease with increasing chain length in the case of primary amines. Table 3 gives CMC values of amines and pyridinium salts. In case of salts also, CMC decreases with increasing chain length in primary, secondary, tertiary as well as quaternary ammonium surfactants and alkyl pyridinium surfactants. Thus, it is quite obvious that although the basic characteristics of amines are not affected by chain length significantly, aggregation characteristics as reflected in the CMC are greatly affected in the case of all types of amine and their salts as well

Table 2 Ionization Constants, Solubility, and CMC of Amine Surfactants

Amine	$pK_a(25°C)$ ionization constant	Solubility of molecular species (mol/dm^3)	CMC (mol/dm^3)	Ref.
Primary amine				
n-Decyl amine	10.64	5.0×10^{-4}	3.2×10^{-2}	27–29
n-Dodecyl amine	10.63	2.0×10^{-5}	1.3×10^{-2}	27–29
n-Tetradecyl amine	10.62	1.0×10^{-6}	4.1×10^{-3}	27–29
n-Hexadecyl amine	10.61	—	8.3×10^{-4}	27–29
Secondary amine				
N-Methyl-n-dodecyl amine	11.00	1.2×10^{-5}		30, 31
Di-n-pentyl amine	11.18	—		27
Di-n-hexyl amine	11.01	—		27
Di-n-octyl amine	11.01	—		27
Tertiary amine				
N,N-Dimethyl dodecyl amine	9.74	7.2×10^{-6}		30, 31
N,N-Dimethyl-n-propyl amine	9.99	—		32

as quaternary ammonium compounds. It is well known that the decrease in the surface tension due to the addition of these surfactants is greater for a surfactant with a longer chain length.

III. POLYMER SURFACES

The surfaces of polymers are unique in that they have a very low surface energy (Table 4). Chemically, the polymer surfaces are regarded as rather inert and, therefore, the chemical interaction with the surfactants such as quaternary ammonium compounds is expected to be rather weak. In some cases, the polymer surface can act as both a hydrophobic and oleophobic surface. However, the method of preparation of polymers by the emulsion polymerization process [48,49] involves the use of an initiator as well as an emulsifying agent. These processes follow the free-radical mechanism, and chemical grouping from the initiator and emulsifiers may be either incorporated on the surface or remain adsorbed on the surface. The chemical groupings impart some hydrophilic characteristics to the polymer surfaces. Thus, adsorption of a surfactant on the polymer surface is interesting both from academic as well application points of view. The initial studies on adsorption of surface-active agents on polymer were carried out on the latex particle surfaces pre-

Table 3 Critical Micelle Concentration of Amine and Pyridinium Salts

Amines/pyridinium salts	CMC (mol/dm³)	Ref.
Primary amine salt		
n-Decyl ammonium chloride	4.0×10^{-2} (60°C)	29, 34
n-Dodecyl ammonium chloride	1.3×10^{-2} (60°C)	29, 34
n-Tetradecyl ammonium chloride	4.0×10^{-3} (60°C)	29, 34
n-Hexadecyl ammonium chloride	3.0×10^{-4} (60°C)	29, 34
Secondary amine salt		
N,N-Dimethyl-n-octyl ammonium chloride	2.7×10^{-2} (30°C)	35, 36
N,N-Dimethyl-n-decyl ammonium chloride	2.1×10^{-2} (30°C)	35, 36
N,N-Dimethyl-n-dodecyl ammonium chloride	1.8×10^{-4} (30°C)	35, 36
Tertiary amine salt		
N,N,N-Trimethyl-decyl ammonium chloride	7.5×10^{-2} (60°C)	35, 37
N,N,N-Trimethyl-dodecyl ammonium chloride	1.75×10^{-2} (60°C)	35, 37
N,N,N-Trimethyl-tetradecyl ammonium chloride	4.2×10^{-3} (60°C)	35, 37
N,N,N-Trimethyl-hexadecyl ammonium chloride	1.0×10^{-3} (60°C)	35, 37
Quaternary ammonium salt		
Decyl trimethyl ammonium chloride	6.1×10^{-2} (25°C)	38
Dodecyl trimethyl ammonium chloride	2.0×10^{-2} (25°C)	39
Tetradecyl trimethyl ammonium chloride	4.5×10^{-3} (25°C)	38
Hexadecyl trimethyl ammonium chloride	1.3×10^{-3} (25°C)	40
Alkyl pyridinium salt		
Dodecyl pyridinium chloride	1.5×10^{-2} (25°C)	41
Hexadecyl pyridinium chloride	9.0×10^{-4} (25°C)	42
Octadecyl pyridinium chloride	2.4×10^{-4} (25°C)	43

Table 4 Surface Energy of Various Polymer Materials

Polymer	Surface energy (mJ/m²)	Ref.
Polytetrafluoroethylene	18	44
Polyethylene	34.3	45
Polystyrene	33	46
Polyvinylchloride	34	47

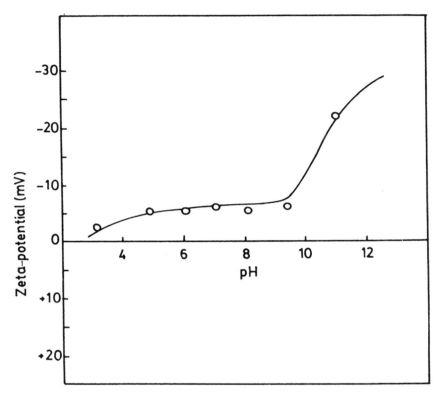

Figure 1 Zeta-potential of PTFE in 0.001 M KCl at various pH values. (From Ref. 51.)

pared by dispersion polymerization. The particles are extensively dialysed with distilled water, which generally removes the physically adsorbed emulsifiers to a great extent, but the chemical groups incorporated in the surface of polymer cannot be removed completely with distilled water. Hilary et al. [50] have reported a surface charge density of 0.68 ± 0.02 $\mu C/cm^2$ as determined by conductometric titration. The electrophoretic mobility and ζ-potential measurement have also been made. Typical behavior with respect to poly(tetrafluoroethylene) (PTFE) and polystyrene are shown in Figs. 1 and 2, respectively [51,52]. As seen from the figures, the ζ-potential for both the PTFE powder and polystyrene in 10^{-3} M KCl gives a negative charge value over the entire pH range from pH 2 to 10. The numerical values first increase slightly and then remain more or less constant over a wide pH range of 5–9.5 at 6 ± 1 mV and appear to increase in a higher alkaline range beyond pH 10

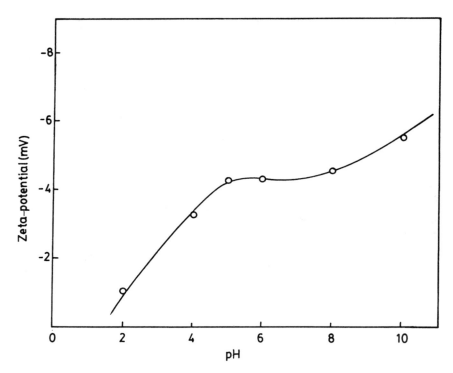

Figure 2 Zeta-potential of polystyrene in 10^{-3} M KCl solution at various pH values. (From Ref. 52.)

in the case of PTFE. Polystyrene also shows a similar behavior with the ζ-potential remaining constant over a pH range of 4–10 at a value of 5 ± 1 mV. Thus, it is evident that these polymer surfaces are negatively charged. The charge is clearly due to the fact that there are ionic centers present on the polymer surfaces. Conductometric titration have been carried out to measure the percentage of surface occupied by acquired ionic groups. These values are given in Table 5. Various workers have attributed the negative charge to the presence of carboxylic groups which ionize to form the carboxylate ion $-COO^-$ in aqueous solution [50–56].

A. Infrared Spectral Study

Fourier transform infrared (FTIR) studies have been carried out on PTFE powder and polystyrene powder by Vanjara and Dixit [51] and Desai and Dixit [52], respectively. The aim of the study was to look for any acquired functional groups such

Table 5 Percentage of Area of Surface Occupied by Ionic Groups

Polymer material	% Area occupied by carboxyl group	Ref.
Polytetrafluoroethylene latex particle	2	50
Polytetrafluoroethylene powder	29	51
Polystyrene powder	20.83	53

as the carboxylic (–COOH) functional group. The spectra were recorded after the powder was brought into contact with water having different pH values over 24 h. The spectra for PTFE for various pH values are shown in Fig. 3. The spectrum corresponding to pH 4.1 shows a well-defined peak at 1710 and 1745 cm^{-1} (Fig. 3a). The peak at 1710 cm^{-1} may be ascribed to the carboxylic (–COOH) groups in the dimer, whereas the peak at 1745 cm^{-1} may be ascribed to monomeric –COOH groups. As the pH increases to 6.4, peaks at 1724 and 1743 cm^{-1} can still be seen (Fig. 3b). These peaks are also due to antisymmetric stretching vibration of —C= O in the –COOH groups. The spectra corresponding to pH 9.1 show strong peaks at 1710 and 1529 cm^{-1} (Fig. 3c), whereas the spectrum around 1725 cm^{-1} appears to have merged with the 1710-cm^{-1} peak to give a broad peak. The appearance of a new peak at 1529 cm^{-1} clearly indicates the reaction of the alkali with –COOH groups, because the peak at 1529 cm^{-1} is characteristic of–COO$^-$ groups. The FTIR study clearly establishes the presence of –COOH groups on the surface of PTFE. This was further confirmed by conductometric titration of PTFE powder suspended in a water–ethanol mixture against 0.001 N NaOH solution. A clear neutralization point can be seen in Fig. 4. A similar infrared (IR) study was conducted on polystyrene powder [52]. Figure 5 shows the FTIR spectra of the polystyrene powder used in the investigation as well as the polystyrene powder after contacting with water having different pH values. The relevant portions between 1800 and 1500 cm^{-1} only are shown. The experimental polystyrene exhibits bands at 1773 and 1705 cm^{-1}, which may be assigned to carbonyl stretching vibrations. Normally, these disappear upon salt formation in alkaline pH and is replaced by asymmetric (1600–1550 cm^{-1}) or symmetric (1420–1330 cm^{-1}) modes associated with carboxylate ions [57]. In fact, the band at 1705 cm^{-1} did not disappear in the spectra of samples contacted at pH 8 and 10.5, suggesting that all the acid groups did not ionize. However, a new peak appeared at 1531 cm^{-1} in the spectrum at pH 10.5. This clearly indicates the reaction of alkali with carboxylic (–COOH) groups, as this is the band characteristic of the asymmetric stretching of carboxylate ions. Thus, FTIR study clearly indicates the presence of ionizable carboxylic acid groups on the surface of polymer.

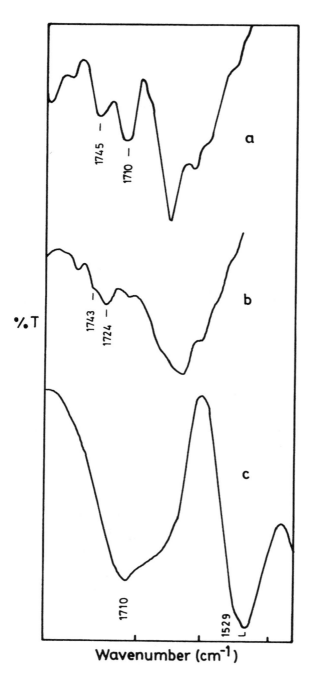

Figure 3 Infrared spectra of PTFE at various pH values: (a) pH 4.1; (b) pH 6.4; (c) pH 3.1. (From Ref. 51.)

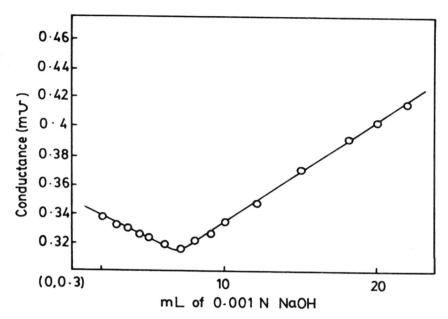

Figure 4 Conductometric titration of PTFE with 0.001 M NaOH. (From Ref. 51.)

Scanning electron microscopic studies of Teflon and Dacron were carried out by Yao and Strauss [58,59]. These studies reveal that polymer surfaces have threadlike structures with channels formed in between. These channels play an important role in the adsorption of surfactants on these polymer surfaces in that they may promote multilayer formation.

IV. ADSORPTION OF QUATERNARY AMMONIUM COMPOUNDS

A. Polystyrene Surface

Adsorption of surfactants on polymer surfaces from their aqueous solution has not received much attention until recently. The earlier work in this area was initiated by Alexander and co-workers [60,61] and Ottewill and co-workers [50,62]. However, the detailed work on the adsorption of alkyl trimethyl ammonium ions on the polystyrene–latex particles has been reported by Connor and Ottewill [62]. The surface area of the latex particle was determined by using particle size distribution. Typical adsorption isotherms for cetyltrimethyl ammonium ion (CTA$^+$), dodecyltrimethyl ammonium ion (DTA$^+$), decyltrimethyl ammonium ion (DeTA$^+$), and

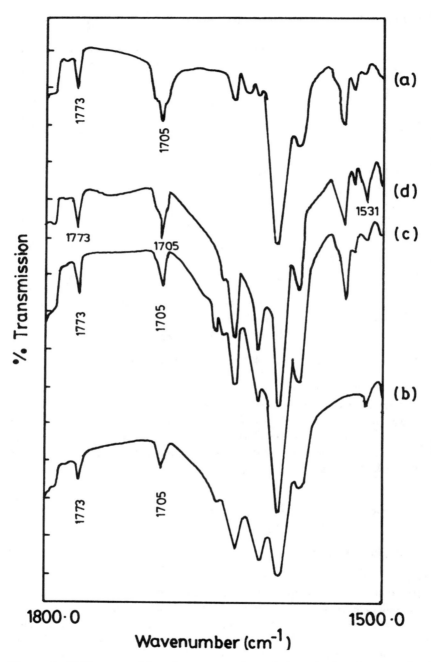

Figure 5 FTIR spectra of the polystyrene powder under investigation after contacting for 24 h with water at various pH values: (a) polystyrene powder; (b) pH 4.4; (c) pH 8; (d) pH 10.5. (From Ref. 52.)

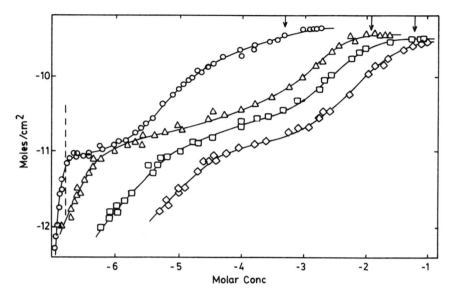

Figure 6 Adsorption isotherms on particles of latex-G at pH 8.0 in 10^{-3} M KBr solution: (○) hexadecyltrimethyl ammonium; (△) dodecyltrimethyl ammonium; (□) decyltrimethyl ammonium; (◇) octyltrimethyl ammonium; (↑) CMC values; (¦) reversal of charge concentration for hexadecyltrimethyl ammonium ion. (From Ref. 62.)

octyltrimethyl ammonium ion (OTA⁺) on polystyrene latex-G (surface area of 26.9 m²/g) in the presence of 10^{-3} M KBr at pH 8 are shown in Fig. 6. All the isotherms are characterized by a well-defined knee. For all the compounds, the knee occurs at almost identical values of the amount adsorbed but at different values of the equilibrium concentrations as the chain length of the compound decreased. Table 6 gives parameters obtained from the adsorption isotherms. After the knee, a second upward rise in the adsorption occurs at a concentration which was lower than the CMC value which is shown by an arrow in Fig. 6. The adsorption increases with increase in the chain length. Thus, the area occupied by the surfactants at the saturation increases with decrease in the chain length, and in all the cases, it was higher than the value calculated by assuming a close-packed trimethyl ammonium ion adsorption (35 Å²).

The effect of pH was also studied for CTA⁺ and DTA⁺ adsorption. Typical results for adsorption of CTA⁺ ions are given in Fig. 7. It will be seen that a decrease in pH from 8 to 4.6 results in the decrease in the amount adsorbed at the knee of the isotherms. Thus, the area occupied per ion becomes 6310 Å² at pH 4.6 as against 2625 Å² at pH 8. Similar results were obtained for the adsorption of

Table 6 CMC and Number of Parameters Obtained from the Adsorption of n-Alkyl Trimethyl Ammonium Halides on Polystyrene–Latex-G at pH 8.0 in 10^{-3} M KBr

n-Alkyl trimethyl ammonium halide	CMC (mol/dm³)	Adsorbate[a]	Area occupied at "knee" (Å²)	Equilibrium conc. at "knee" (M)	$\dfrac{d \log \Gamma}{d \log C}$	Area occupied at saturation (Å²)
n-Hexadecyl trimethyl ammonium bromide (CTAB)	4.8×10^{-4} 8.8×10^{-5} (5×10^{-2} M KBr)	CTA⁺	2625 6310 (pH 4.6)	1.26×10^{-7} 1.59×10^{-7} (pH 4.6)	7.28	47 59 (pH 4.6) 36 (5×10^{-2} M KBr)
n-Dodecyl trimethyl ammonium bromide (DTAB)	1.2×10^{-2}	DTA⁺	2345 5010 (pH 4.6)	5.62×10^{-7} 2.82×10^{-7} (pH 4.6)	1.78	56 68 (pH 4.6)
n-Decyl trimethyl ammonium bromide (DeTAB)	6.8×10^{-2}	DeTA⁺	2600	5.62×10^{-6}	1.00	65
n-Octyl trimethyl ammonium iodide (OTAI)	9.8×10^{-2}	OTA⁺	2580	3.16×10^{-5}	1.18	71

[a]CTA⁺: n-hexadecyl trimethyl ammonium; DTA⁺: n-dodecyl trimethyl ammonium; DeTA⁺: n-decyl trimethyl ammonium; OTA⁺: n-octyl trimethyl ammonium ions.

Note: Some parameters at different pH and KBr concentration are also included with specific mention of the same.

Source: Ref. 62.

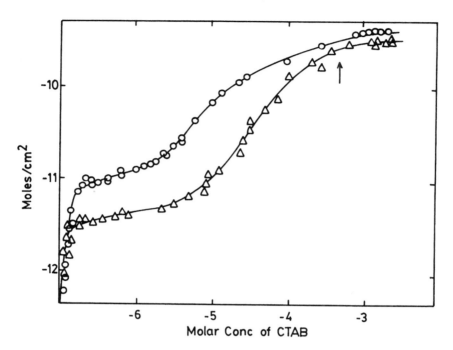

Figure 7 Effect of pH on the adsorption of hexadecyltrimethyl ammonium ions on particles of latex-G in 10^{-3} M KBr solution: (○) pH 8.0; (△) pH 4.6; (↑) CMC values. (From Ref. 62.)

DTA$^+$. Increase in the ionic strength of KBr results in an overall increase of adsorption over the entire range of equilibrium concentration studied, as can be seen from Fig. 8. Thus, the saturation area per cation was found to be 36 Å^2 at 5×10^{-3} M KBr as compared to 47 Å^2 in 10^{-3} M KBr. The higher salt concentration decreases the electrical repulsion between ionic heads of the surfactants and allow them to adsorb in a close-packed manner. The electrophoretic mobility and ζ-potential measurements made by Ottewill and co-authors [62] are shown in Fig. 9 for CTAB adsorption. Charge reversal takes place at a CTAB concentration of 1.59×10^{-7} mol, which is the same as the equilibrium concentration at the knee of the adsorption isotherm, thus indicating that neutralization of negative charge on latex surface occurs at that point. The above authors suggest that latex particles have two types of sites for adsorption:

1. Charge sites which interact with cationic head groups of quaternary ammonium ions
2. Hydrophobic sites on which alkyl chains are adsorbed

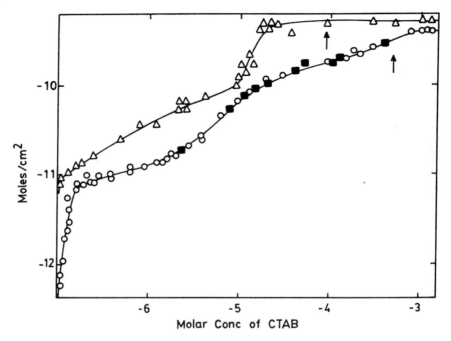

Figure 8 Effect of ionic strength on adsorption of hexadecyl trimethyl ammonium ion on particles of latex-G at pH 8.0: ($\bigcirc$) results in 10^{-3} M KBr by radiotracer measurements; ($\blacksquare$) by colorimetric measurements; ($\triangle$) results in 5×10^{-2} M KBr. ($\uparrow$) CMC values (From Ref. 62.)

The adsorption up to the knee involves the negatively charged surface carboxylate groups, as indicated by the effect of pH. At pH 4.6, the area per adsorbed cation is twice that at pH 8, as expected from pK_a values of the carboxylic group. The variation of adsorption with the chain length indicates that Traube's rule is obeyed. This is further confirmed by plotting log (C/C_0) versus adsorption density, as shown in Fig. 10. Except for CTA$^+$ isotherms, the data for other surfactants fall on a single reduced isotherm. The reason for deviation in the case of CTA$^+$ is not fully understood. It may be due to the dimerization or premicelleziation likely to occur in the case of a long-chain surfactant. The fact that Traube's rule is obeyed indicates that in this region, the alkyl chains are lying flat on the surface in hydrophobic association. Above the knee, the adsorption takes place on a surface which has a net positive charge and hence the electrostatic contribution is very low, if any. The adsorption process in this region is motivated by the decrease in free energy occurring upon removal of the alkyl chain from the solution to hydrophobic sites

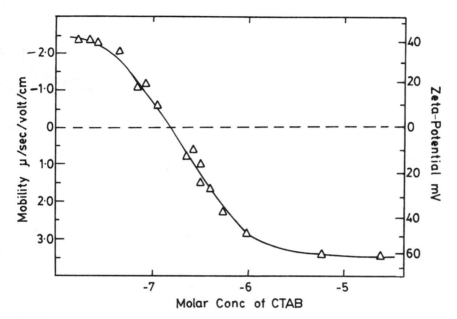

Figure 9 Mobility versus log molar concentration of CTAB for particles of latex-G at pH 8.0 in 10^{-3} M KBr. (From Ref. 62.)

on the latex surface. The association of alkyl chains on the surface takes place at a concentration much below the CMC of the solution. A small area of occupation per cation at saturation indicates that a high proportion of chains may be vertically oriented with their head groups toward the solution side.

Zeta-potential studies of the above system were carried out by Kayes [63] using polystyrene latex. Typical results are given in Fig. 11. It will be seen that the curve has an extended S shape as reported by Ottewill et al. [64,65] In the initial stage, the adsorption occurs primarily by an ion exchange between the surfactant cation and the inorganic cations such as Na^+ ions, which are present in the Stern layer. The orientation of the surface-active ions on the surface is reported to be flat, as the surface is not covered fully. The positive ζ-potential in all the cases increases linearly with concentrations until a maximum value is reached and then virtually remains constant with the further increase in the concentration. In this region, the surfactant cation is adsorbed in a reverse orientation with their head groups toward solution phase. Clearly, the main driving force for adsorption comes from the hydrophobic effect.

Zhao and Brown [66] have studied the adsorption of alkyl trimethyl ammonium bromides on polystyrene–latex particles by dynamic light-scattering mea-

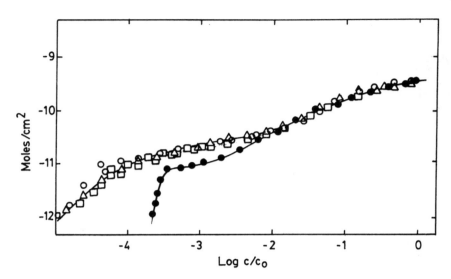

Figure 10   Reduced isotherms for adsorption on particles of latex-G at pH 8.0 in 10^{-3} M KBr: ($\bullet$) hexadecyl trimethyl ammonium; ($\bigcirc$) dodecyl trimethyl ammonium; ($\triangle$) decyl trimethyl ammonium; ($\square$) octyl trimethyl ammonium. (From Ref. 62.)

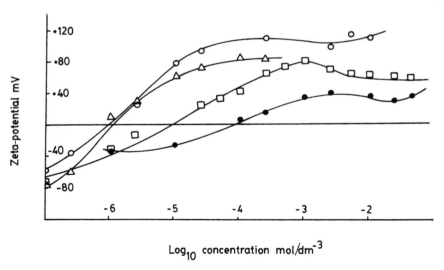

Figure 11   Zeta-potential versus concentration: C_{10}TAB ($\bullet$); C_{12}TAB ($\square$); C_{14}TAB ($\bigcirc$); C_{16}TAB ($\triangle$). (From Ref. 63.)

surement. According to the authors, the latex surface comprised of flexible polymer chain terminated in polar entities such as sulfate groups. The conformation of their chains is a key factor in determining adsorption patterns of the surfactant. The adsorption isotherm shows a significant two-step characteristic. A first plateau occurs at 4×10^{-5} mol/g and corresponds to the maximum value of the hydrodynamic radius R_h. Initially, DTA^+ ions adsorbed due to the charge attraction from a negatively charged sulfate ion resulting in charge neutralization. Further adsorption of DTA^+ ions takes place around initially adsorbed DTA^+ ions. The alkyl tails have contact with polymer chain and also with each other, whereas the cationic head extends in the solution. Thus, aggregates are formed. This results in a steep increase in adsorption. The addition of salts such as NaBr significantly affects the adsorption, as can be seen from Fig. 12. The presence of the bromide ion effectively screened the electrostatic repulsion between head groups of the surfactant and, therefore, more DTAB can be adsorbed because of the hydrophobic interaction of alkyl chains of the surfactant. Above the CMC, a further strong adsorption takes place because of the enhanced growth of the surface aggregates and eventually a closed-packed adsorbed layer is formed.

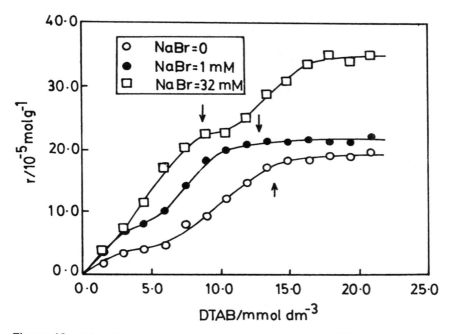

Figure 12   Adsorption isotherms for a latex suspension versus DTAB concentration at different ionic strengths of added NaBr at 25°C. Vertical arrows indicate the CMC at each ionic strength. (From Ref. 66.)

Kawabata et al. [67] studied the adsorption of cationic surfactants such as dodecyl trimethyl ammonium chloride (DTACl), hexadecyl trimethyl ammonium bromide (HDABr), and hexadecyl pyridinium chloride (HDC) on cross-linked poly(p-hydroxystyrene) (PHS resin). The PHS resin was found to be an excellent selective adsorbent for cationic surfactants in aqueous solution. The capacity of resin for adsorption is dependent on the content of p-vinylphenol in the resin. The breakthrough capacities of cationic surfactants DTAC, HDABr, and HDC was reported to be 226, 391, and 312 mg/g, respectively, on PHS-7 containing 25 mol% p-vinylphenol. An acid–base interaction between the cationic surfactant and phenolic hydroxyl group on PHS resin has been suggested.

B.   Polycarbonate Surface

Keesom et al. [68] have investigated the adsorption of cationic surfactants such as dodecyl trimethyl ammonium bromide (DTAB) and hexadecyl trimethyl ammonium bromide (HTAB) on a hydrophobic surface of a track-etched polycarbonate membrane using the electrokinetic method. Polycarbonate membranes are extensively used as filters. The pores of the membrane are weakly charged due to the presence of ionizable carboxylic surface groups. It was found that the negatively charged polycarbonate surface attracts cationic surfactants. The hydrophobic tails of these surfactants interact due to hydrophobic forces. The polycarbonate shows a negative ζ-potential over a wide range of pH from 2 to 10. The ζ-potential of the polycarbonate membrane was measured as a function of DTAB concentration at pH 7 and is shown in Fig. 13. Keesom et al. suggest that surfactant adsorption on hydrophobic surface is due to the attraction of the surfactant hydrocarbon chain to the hydrophobic surface. This attraction is due to hydrophobic bonding interactions. A monolayer is formed. The authors have given surface adsorption theory to quantify the adsorption. They postulated a limited number of specific adsorption sites on a surface in addition to ionogenic substrate acid groups. The adsorption takes place as follows:

$$P - A^{\pm} \underset{}{\overset{K_a}{\rightleftarrows}} P + A^{\pm}$$

where P denotes an adsorption site, A^+ denotes a cationic surfactant, A^- denotes an anionic surfactant, and K_a is the equilibrium constant. The equilibrium relation corresponding to above equation is

$$K_a = \frac{(P)[A^{\pm}] \exp(\mp F\zeta/RT)}{(P - A)}$$

which includes the Boltzmann electrostatic correction term involving the ζ-potential. The above equation combined with the condition of conservation of adsorption site gives the expression for the adsorbed charge from which the ζ-potential

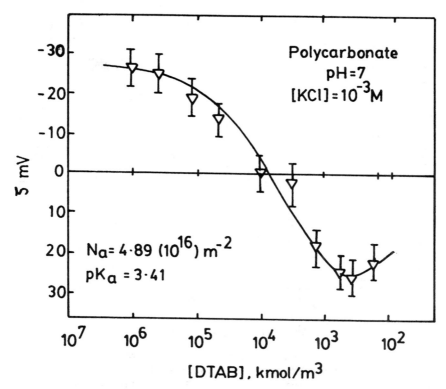

Figure 13 Zeta-potential of polycarbonate membranes as a function of DTAB concentration. The solid line is fit by the adsorption model. (From Ref. 68.)

can be calculated. For DTAB adsorption, $\Delta G°$ has been calculated to be $\Delta G° = -7.9RT$ and the maximum adsorption site density $N_a = 4.89 \times 10^{16}$ sites/m². These are in good agreement with the earlier work of Osseo-Asare et al. [69] and Ottewill et al. [64]. The above model also predicts the effect of pH on the ζ-potential

C. PTFE and Dacron Surface

Yao and Strauss [58] have used chromatographic Teflon beads having a surface area of 8.96 m²/g for their adsorption studies. The binding levels and mobilities of quartenary ammonium surfactants hexadecyl trimethyl ammonium chloride (CTAC), didodecyl dimethyl ammonium bromide (DDAB), and tridodecyl methyl ammonium chloride (TD-MAC) containing one, two, and three alkyl chains, respectively, were studied when they were adsorbed on PTFE surface. Adsorption

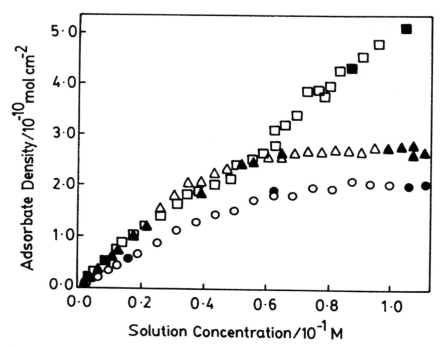

Figure 14 Adsorption isotherms of quaternary ammonium surfactants on PTFE particle surface adsorbed from methanol. Open points; surfactants alone; filled points; surfactants with 12-DOXYLME spin probe co-adsorbed. (□, ■) TDMAC; (△, ▲), DDAB; (○, ●) CTAC. (From Ref. 58.)

isotherms were determined and these are shown in Fig. 14. Isotherms for single- and double-chained surfactants showed saturation, indicating a monolayer formation. However, the triple-chained surfactant TD-MAC did not show any saturation, thus indicating formation of a multilayer. The limiting areas per surfactant molecule were 83 Å^2 for CTAC and 61 Å^2 for DDAB, as calculated from the amount adsorbed at saturation and on a specific surface area of the polymer. Both of these values have a magnitude which indicates that the adsorbents are vertically oriented monolayers formed on the polymer surface. The sudden upturn in the case of TD-MAC after reaching a plateau is considered to be due to the formation of a multi-layer. An electron-spin resonance (ESR) study using spin probes showed that all the surfactants were lying flat on the surface at low adsorption levels. With increasing surface density, they become vertically oriented as monolayers in case the of CTAC and DDAB and multilayered in the case of TD-MAC. The above work was further extended [59] to study the effect of surfactants on vascular grafts made of PTFE in the form of expanded Teflon and polyethylene phthalate in the form of knitted Dacron, which are used in surgical practice. These materials have a highly structured surface topography. The following aspects were studied:

1. Hydrophobic and oleophobic interactions of adsorbate with the polymer substrate and their effect on orientation and mobility of the adsorbate film
2. The effect of narrow pores and channels in the polymer surface
3. The ability of the adsorbed surfactant layer to bind a secondary adsorbate

As in the previous case, a two-chain surfactant DDAB showed a saturation adsorption limit on both Teflon and Dacron (Fig. 15), whereas TD-MAC, with three alkyl chains, was found to adsorb without a saturation limit on both of these polymers (Fig. 16). This indicated that DDAB was adsorbed as a monolayer, whereas TD-MAC was adsorbed as a multilayer. These differences appear to arise from the presence of a third alkyl chain in TD-MAC. When TD-MAC is in a close-packed state one of the alkyl chains may extend upward away from substrate surface, thus creating an additional site for surfactant adsorption by hydrophobic interaction.

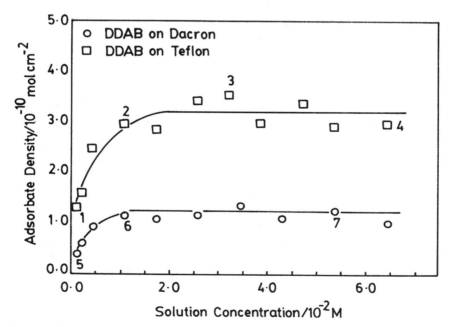

Figure 15 Adsorption isotherms of DDAB on Teflon and Dacron grafts. The numbers refer to the data points listed in Table 7. (From Ref. 59.)

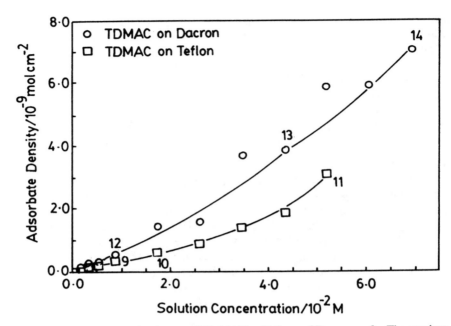

Figure 16 Adsorption isotherms of TD-MAC on Teflon and Dacron grafts. The numbers refer to the data points listed in Table 7. (From Ref. 59.)

This may occur even before all of the surface is covered, resulting in irregular clusters. Contrary to the work reported by Hillary et al. [50], no knee in the adsorption isotherm was found in the case of adsorption on PTFE or Dacron.

1. Mobility Studies

The mobility of the adsorbate was characterized by ESR spectra using (spin probe methyl 12-doxylstearate [12-DOXYL-ME]) coadsorbed with the surfactant. The air-dried surfactant-coated polymer sample gave spectra indicating almost complete immobilization at all binding levels. Only the hydrated samples showed characteristic changes in mobility depending on the binding levels of surfactant. On each polymer, two limiting states could be recognized from the maximum hyperfine splitting $2A_{max}$. From these A_{max} values, the order parameter, S, was found to have a value of 0.11 for the mobile limit and a value of 0.78 and 0.96 for immobile extremes for Teflon and Dacron, respectively. The fraction of adsorbate in the mobile state, f_m, was calculated. The data are given in Table 7. It may be noted that the adsorbate existed in two distinct states, indicating two-phase equilibrium, rather than being in a uniform state of some intermediate degree of mobility. The

Table 7 Mobile and Immobile Adsorption Densities

System	Data point (1–7, Fig. 15) (8–14, Fig. 16)	f_m	Adsorbate density (10^{-9} mol/cm^2)		
			Total	Mobile	Immobile
DTAB/Teflon	1	0.25	0.125	0.031	0.094
	2	0.75	0.292	0.219	0.073
	3	~1.0	0.323	0.323	~0
	4	~1.0	0.323	0.323	~0
DTAB/Dacron	5	0	0.034	0	0.034
	6	0	0.118	0	0.118
	7	0	0.126	0	0.126
TD-MAC/Teflon	8	0.35	0.170	0.060	0.110
	9	0.70	0.353	0.247	0.106
	10	0.85	0.608	0.517	0.091
	11	~0.97	3.16	~3.06	~0.1
TD-MAC/Dacron	12	0.40	0.547	0.219	0.328
	13	~0.9	3.89	~3.5	~0.39
	14	~0.95	7.10	~6.75	~0.35

Source: Ref. 59.

DDAB on Teflon showed a gradual increase in f_m with the increase in binding level of the surfactant until all of the surfactant was in the mobile phase ($f_m = 1$). This point coincides with the binding level at which surface saturation was reached. The cross-sectional area of 52 Å^2 per molecule was calculated. This cross section is consistent with the value for vertically oriented amphiphiles with two saturated alkyl chains and an –N(CH$_3$)$_2$ head group. The mobile fraction f_m therefore can be regarded as a measure of the fractional surface covered by a vertically oriented, closely packed, adsorbate layer.

Adsorption of DDAB on Dacron showed a completely different behavior, indicating an immobile state ($f_m = 0$) at all binding levels up to saturation. The limiting saturation surface density of 1.2×10^{-10} mol/cm^2 (Fig. 14) was much lower than Teflon. The density corresponds to a monolayer cross section of 138 Å^2, which suggested that DDAB forms a flat-lying monolayer at saturation coverage. It is further suggested that only the hydrocarbon chain lies on the surface with the head group curving upward. Table 7 also lists the adsorbent densities as well as the mobile fractions of the adsorption of TD-MAC. The immobilities for TD-MAC remains constant with varying total adsorbate densities. Immobile densities were 0.1 $\times 10^{-9}$ and 0.35×10^{-9} mol/cm^2 on Teflon and Dacron, respectively. The immobile fraction of constant density arises due to the topography of Teflon and Dacron surfaces. The scanning electron micrograph show threads on the surface in close proximity. The resulting channels are filled with multilayers of TD-MAC which are immobile due to their attachment to the opposing surfaces of the channels. This effect is not observed in DDAB because it does not form multilayers. It has been shown that the secondary adsorbate binds only to the mobile primary adsorbate which is vertically oriented. Thus, the above studies bring out the effect of (1) the number of alkyl chains of the surfactant, (2) the nature of substrate surface, and (3) its topography. The orientation at the surface determines the mobility of the adsorbate molecules. A closely packed vertical layer on the surface has a liquidlike mobility. The surfactants with three alkyl chains like TD-MAC form close-packed vertical layers with one of the chains pointing away from the substrate. This chain enables the formation of subsequent layers resulting in multilayer adsorption.

Vanjara and Dixit [51] have studied the adsorption of series of alkyl trimethyl ammonium bromides CTAB, TTAB, and DTAB and alkyl pyridinium chlorides cetyl pyridinium chloride (CPC) and dodecyl pyridinium chloride (DPC) on the PTFE surface. The adsorption isotherms of alkyl trimethyl ammonium bromides and alkyl pyridinium chlorides on PTFE surface are given in Figs. 17 and 18, respectively. Adsorption isotherms having characteristics similar to

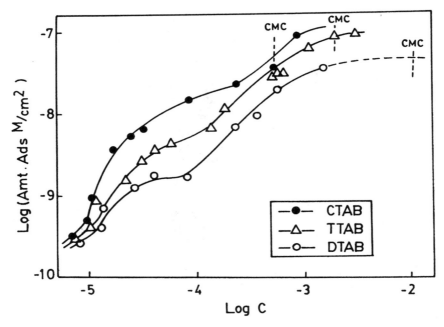

Figure 17 Adsorption isotherms of alkyl trimethyl ammonium bromide surfactants. (From Ref. 51.)

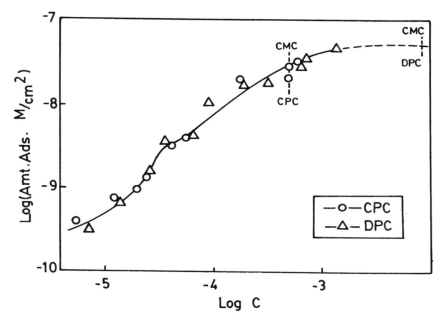

Figure 18 Adsorption isotherms of alkyl pyridinium chloride surfactants. (From Ref. 51.)

those of two plateau regions were observed for all these surfactants. In case of alkyl trimethyl ammonium bromides, there is a considerable effect of chain length which could be attributed to the arrangement of surfactant molecules being tilted away from the perpendicular. In the case of alkyl pyridinium chlorides an insignificant effect of chain length is due to a predominantly perpendicular arrangement. The calculated values of various parameters from adsorption isotherms such as mean aggregation number of hemimicelle (n_{hm}), free-energy change for hemicellization ($\Delta G°$), and free-energy change per mole of hemimicelle ($\Delta G°/n_{hm}$) are given in Table 8.

V. COADSORPTION OF SURFACTANTS

In many applications, the use of surfactant mixtures significantly improves performance over single-surfactant systems. Therefore, an understanding of adsorption of surfactants from their mixture is important from academic as well as industrial points of view.

Dixit and co-workers [52,70] have studied adsorption from the aqueous so-

Table 8 Mean Aggregation Number of Hemimicelles (n_{hm}), Equilibrium Constant (K), and Free Energy of Hemimicellization of Surfactants on PTFE Surface

Surfactant in 0.001 M KCl medium	CMC (M)	n_{hm}	K	$-\Delta G°$ for 1 mol of hemimicelle (kJ)	$-\Delta G°/n_{hm}$ for 1 mol of surfactant (kJ)
CTAB (16)	5×10^{-4}	7	$1.6 \times 10^{+22}$	128.8	18.4
TTAB (14)	2×10^{-3}	6	$1.82 \times 10^{+18}$	103.07	17.1
DTAB (12)	1.2×10^{-2}	5	$2.99 \times 10^{+10}$	76.47	15.1
CPC (16)	4.8×10^{-4}	4	$2.73 \times 10^{+10}$	59.22	14.80
DPC (12)	9.6×10^{-3}	4	$1.48 \times 10^{+10}$	57.71	14.43

Source: Ref. 51.

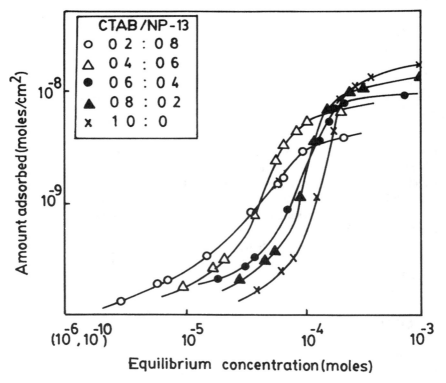

Figure 19 Adsorption isotherms of CTAB form CTAB–NP-13 mixtures on PTFE. Ionic strength = 10^{-3} M KCl. (From Ref. 70.)

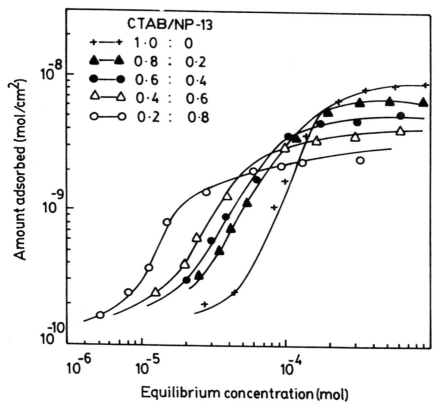

Figure 20 Adsorption isotherms of CTAB form CTAB–NP-13 mixtures on polystyrene. Ionic strength = 10^{-3} M KCl. (From Ref. 52.)

lution of mixtures of surfactant CTAB and nonyl phenyl ethoxylates (NP-n) (n = 13, 20, and 30) on PTFE as well as polystyrene surfaces in a wide range of concentrations.

A. Adsorption of CTAB

The adsorption isotherms for CTAB from the CTAB–NP-13 mixture at various mole fractions are given in Fig. 19 for PTFE and in Fig. 20 for polystyrene. The adsorption isotherms can be divided into two regions: (1) the preplateau region below the CMC and (2) the plateau region above the CMC.

1. Preplateau Region

It is seen from Figs. 19 and 20 that the affinity of CTAB to both PTFE and the polystyrene surface is enhanced by the presence of a nonionic surfactant. This is clearly indicated by the continuous shift of the CTAB isotherms toward the left with the increase in the mole fractions of nonionic surfactant NP-13. This behavior may be attributed to the synergistic coadsorption of CTAB and NP-13. In the initial stages, the adsorption of CTAB takes place because of electrostatic interactions with negatively charged polymer surfaces. As the concentration of CTAB increases, there is a chain–chain interaction between the CTAB and nonionic surfactant chains that are adsorbed on adjacent sites. The free-energy change from such a chain–chain interaction is responsible for the reduction in residual concentration of CTAB required to achieve a certain level of adsorption, thus resulting in the shift of isotherms toward the left.

2. Plateau Region

The adsorption isotherms indicate that there is a reverse trend in the plateau region (i.e., as the mole fraction of NP-13 increases, the adsorption of CTAB decreases). It has been reported by Rubingh [71] that because of the mixed micellization of cationic–nonionic surfactants, there is a decrease in the CMC which reduces the monomer concentration in the solution. Also, there is a decrease in the mole fraction of CTAB itself. Both of these lower the saturation limit as the mole fraction of NP-13 is increased.

B. Adsorption of NP-13

The adsorption isotherms of NP-13 from the mixture of CTAB and NP-13 are given in Fig. 21 for adsorption on PTFE and in Fig. 22 for adsorption of polystyrene. In the presence of CTAB, there is an overall increase in the adsorption of nonionic surfactants. This becomes more evident in the plateau region. In the preplateau region, the individual molecules are adsorbed on the surface initially. However, as the concentration increases, interactions between the chains become predominant and this results in an increase in adsorption. A similar behavior was noted for CTAB–NP-20 and CTAB–NP-30 mixed surfactant systems [52,70]. It was also noted that whereas NP-13 adsorbs at all mole fractions from 0.2 to 0.8, adsorption of NP-20 occurs at mole fraction from 0.4 to 0.8 of NP-20, and adsorption of NP-30 occurs at 0.4 and 0.8 mole fractions of NP-30 [70]. The nuclear magnetic resonance measurements for the above systems [72] have shown that the cationic–nonionic interaction increases with the increase in oxyethylene chain length (i.e., the interaction increases in the order CTAB–NP-30 > CTAB–NP-20 > CTAB–NP-13. Surfactants prefer to remain in the mixed micellar state in the so-

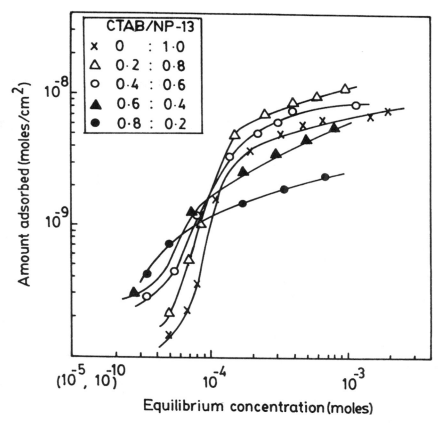

Figure 21 Adsorption isotherms of NP-13 form CTAB–NP-13 mixtures on PTFE. Ionic strength = 10^{-3} M KCl. (From Ref. 70.)

lution if the interaction is greater. Thus, in the case of NP-30, no adsorption was obtained at the lower mole fractions of NP-30. The NP-13 has the least interaction and, therefore, adsorbs at all mole fractions.

VI. SUMMARY

Studies of the adsorption of quaternary ammonium compounds on polymer surfaces are rather limited. Therefore, it is premature to make sweeping generalizations. The available data indicate that the polymer surfaces are charged due to the

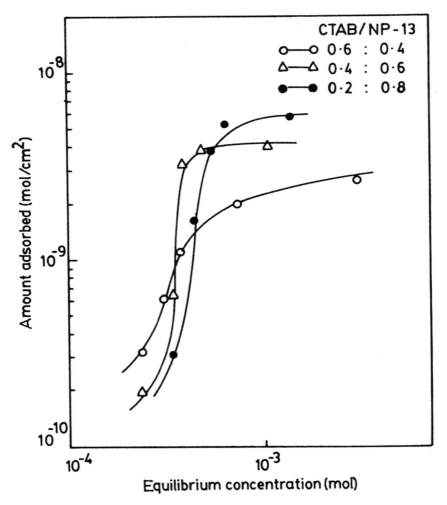

Figure 22   Adsorption isotherms of NP-13 form CTAB–NP-13 mixtures on polystyrene. Ionic strength = 10^{-3} M KCl. (From Ref. 52.)

presence of acquired anionic groups such as $-COO^-$ and $-SO_4^{2-}$. These act as centers for electrostatic attraction of the cationic polar head of the surfactant. As a result, at lower concentrations the adsorption takes place due to Coulombic attraction between the individual surfactant ions and the negatively charged sites on the polymer surface. As the concentration increases, the chain–chain interaction due to hydrophobic bonding results in a steep increase in adsorption density. Some isotherms show a two-step adsorption. The first knee of such an isotherm is due to the neutralization of charges. The saturation limit in most of the adsorption isotherms is reached beyond the CMC at which the orientation of surfactant molecule is generally vertical. The characteristics of the surfactants, such as chain length and the number of side chains, affect the adsorption significantly. The topography of the polymer surface which has a threadlike appearance also influences the adsorption. There is also an interaction between the surfactant and polymer chains resulting in a flat orientation. Coadsorption from cationic–nonionic mixtures shows a synergistic effect for both cationic as well as nonionic adsorption in the preplateau region. The adsorption is basically affected by the formation of mixed micelles in solution and mixed clusters on the surface.

REFERENCES

1. P Somasundaran, W Healy, DW Fuerstenau. J Phys Chem 68:3562–3566, 1964.
2. ME Ginn. In: E Jungermann, ed. Cationic Surfactants. New York: Marcel Dekker, Inc. 1970, pp 341–367.
3. B Ball, DW Fuerstenau. Discuss Faraday Soc 52:361–371, 1971.
4. MJ Schwuger, W Von Rybinsk, P Krings. In: RH Ottewill, CH Rochester, AL Smith, eds. Adsorption from Solution. London: Academic Press, 1983, pp 185–196.
5. RW Smith. In: P Somasundaran, BM Moudgil, eds. Reagents in Mineral Technology. New York: Marcel Dekker, Inc., 1987, pp 219–256.
6. EA Simister, RK Thomas. Langmuir 6:103, 1990.
7. JCJ Van der Donk, GEJ Vaessen, HN Stein. Langmuir 9:3553–3557, 1993.
8. AK Vanjara, SG Dixit. Adsorption Sci Technol 13:397–407, 1996.
9. K Esumi, S Uda, M Goino, K Ishiduk, T Suhara, N Fukui, Y Koide. Langmuir 13:2803–2807, 1997.
10. P Jingjun, Y Guanying, H Buxing, V Haike. J Colloid Interf Sci 194:276–280, 1997.
11. P Somasundaran, DW Fuerstenau. J Phys Chem 70:90–96, 1966.
12. SG Dick, DW Fuerstenau. J Colloid Interf Sci 37:595–602, 1971.
13. G Goujon, JM Cases, B Mutaftschiev. J Colloid Interf Sci 56:587–595, 1976.
14. MJ Rosen, Y Nakamura. J Phys Chem 81:873–897, 1977.
15. JF Scamehorn, RS Schechter, WH Wade. J Colloid Interf Sci 85:463–478, 1982.
16. S Chander, DW Fuerstenau, D Stigter. In: RH Ottewill, CH Rochster, AL Smith, eds. Adsorption from Solution. London: Academic Press, 1983, pp 197–210.
17. P Somasundaran, R Middleton, KV Viswanathan. In: MJ Rosen, ed. Structure/Performance Relationship in Surfactant. ACS Symposium Series 253. Washington, DC: American Chemical Society, 1984, pp 269–290.
18. BL Robert, JF Scamehorn, JH Harwell. In: JF Scamehorn, ed. Phenomena in Mixed Surfactants Systems. Washington, DC: American Chemical Society, 1986, pp 200.
19. A Couzis, E Gulari. Langmuir 9:3414–3421, 1993.
20. K Esumi, H Toyoda, M Goino, T Suhara, H Fukui. Langmuir 14:199–203, 1998.

21. RH Ottewill. In: MJ Schick, ed. Nonionic Surfactant. New York: Marcel Dekker, Inc., chap 19, 1967.
22. P Levitz, H van Damme, D Keravis. J Phys Chem 88:2228–2235, 1984.
23. M Lindheimer, E Keh, S Zaini, S Partyka. J Colloid Interf Sci 138:83–91, 1990.
24. MW Rutland, TJ Senden. Langmuir 9:412–418, 1993.
25. F Portel, PL Desbene, C Treiner. J Colloid Interf Sci 194:379–391, 1997.
26. R Dauden. In: S Patai, ed. Chemistry of the Amino Group, London: Wiley, 1968, pp 1–35.
27. CW Hoerr, MR McCorkle, AW Ralston. J Am Chem Soc 65:328–329, 1943.
28. RW Aplan, DW Fuerstenau. In: DW Fuerstenau, ed. Principles of Nonmetallic Mineral Flotation. Froth Flotation, New York: AIME, 1962, pp 170–214.
29. AW Ralston. Fatty Acids and Their Derivatives. New York: Wiley, 1948.
30. RW Smith, S Akhtar. In: MC Fuerstenau, ed. Flotation, AM Gaudin Symposium Volume. New York: American Institute of Mining Engineers, 1976, pp 87–116.
31. RW Smith. Trans Soc Min Eng AIME 254:353–357, 1973.
32. J Hansen. Svensk Kem Tidskr 67:256, 1955.
33. JW Smith. In: S Patai, ed. Chemistry of the Amino Group. London: Wiley, 1968, pp 161–204.
34. AW Ralston, CW Hoerr. J Am Chem Soc 64:772–776, 1942.
35. H Gerrens, G Hirsch. In: J Brandrup, EH Immergul, eds. Polymer Handbook Vol. II. (Quoted in "Reagents in mineral processing" ed. by P. Somasundaran et al. Chap. 8, Ref. no. 33.) New York: Wiley, 199, p 448.
36. AW Ralston, DN Eggenberger, PL Du Brow. J Am Chem Soc 70:977, 1948.
37. HB Klevens. J Phys Colloid Chem 52:130, 1948.
38. CW Hoyer, A Marmo. J Phys Chem 65:1807, 1961.
39. J Osugi, M Sato, N Ifuku. Rev Phys Chem Japan 35:32, 1965.
40. AW Ralston, DN Eggeneberger, HJ Harwood. J Am Chem Soc 69:2095, 1947.
41. W Ford, RH Ottewill, HC Parreira. J Colloid Interf Sci 21:522–533, 1966.
42. GS Hartley. J Chem Soc 1968, 1938.
43. EC Evers, CA Kraus. J Am Chem Soc 70:3049–3054, 1948.
44. MK Bernett, WA Zisman. J Phys Chem 63:1241–1249, 1959.
45. RH Deltre, RE Johnson. J Colloid Interf Sci 21:367–377, 1966.
46. AH Ellison, WA Zisman. J Phys Chem 58:503–506, 1954.
47. AH Ellison, WA Zisman. J Phys Chem 58:260–265, 1954.
48. CA Sperati, HW Starkweather. Fortschr Hochpolym Forsch 2:465, 1961.
49. S Sherratt. Kirk-Othmer Encyclopedia of Chemical Technology. 2nd ed. New York: Interscience, 1966, vol. 9, pp 805.
50. EB Hilary, RH Ottewill, D Rance. In: RH Ottewill, CH Rochechter, eds. Adsorption from Solutions. New York: Academic Press, 1982, pp 155–171.
51. AK Vanjara, SG Dixit. J Colloid Interf Sci 177:359–362, 1996.
52. TR Desai, SG Dixit. Adsorption Sci Technol 15:391–405, 1997.
53. TR Desai. Studies in Surface Phenomena of Coadsorption of Cationic–Nonionic Mixed Surfactants on Polymer Surfaces and Their Solution Properties. PhD dissertation, Bombay University, Mumbai, 1995.
54. JN Shaw, MC Marshall. J Polym Sci 6(A1):449, 1968.
55. RH Ottewill, JN Shaw. Kolloid ZZ Polym 215:161, 1967.

56. RH Ottewill, JN Shaw. Kolloid Z Z Polym 218:34, 1967.
57. LJ Bellamy. Infrared Spectra of Complex Molecules. London: Methuen, 1958.
58. J Yao, G Strauss. Langmuir 7:2353–2357, 1991.
59. J Yao, G Strauss. Langmuir 8:2274–2278, 1992.
60. AE Alexander, DH Napper. Chem Ind (London) 1936, 1967.
61. DJ Robb, AE Alexander. Soc Chem Ind. Sympo Wetting 25:292, 1967.
62. P Connor, RH Ottewill. J Colloid Interf Sci 37:642–651, 1971.
63. JB Kayes. J Colloid Interf Sci 56:426–442, 1976.
64. RH Ottewill, MC Rastogi, A Watanabe. Trans Faraday Soc 56:854–865, 1960.
65. RH Ottewill, MC Rastogi, Trans Faraday Soc 56:880–892, 1960.
66. J Zhao, W Brown. Langmuir 11:2944–2950, 1995.
67. N Kawabata, K Sumiyoshi, M Tanaka. Ind Eng Chem Res 29:1889–1893, 1990.
68. WH Keesom, RL Zeleka, CJ Radke, J Colloid Interf Sci 125:575–585, 1988.
69. K Osseo-Asare, DW Fuerstenau, RJ Otewill. ACS Sympo Ser 8:63, 1975.
70. TR Desai, SG Dixit. J Colloid Interf Sci 179:544–551, 1996.
71. DN Rubingh. In: KL Mittal, ed. Solution Chemistry of Surfactants. New York: Plenum Press, 1978, pp 3383.
72. TR Desai, SG Dixit. J Colloid Interf Sci 177:471–477, 1996.

10

Adsorption onto Poly(tetrafluoroethylene) from Aqueous Solutions

Watson Loh, Josias R. Lopes, and Antonio C. S. Ramos
Universidade Estadual de Campinas, Campinas, São Paulo, Brazil

I. INTRODUCTION

The substitution of hydrogen atoms by fluorine in the chemical structure of some polymers causes many changes in the physical and chemical properties of these materials. Typically, this substitution turns the fluoropolymers both hydrophobic and incompatible with apolar compounds, as well as conferring them high chemical stability and thermal resistance.

Poly(tetrafluoroethylene) (PTFE) is the most successful example of such an approach. This polymer is prepared by polymerization of tetrafluoroethylene, producing a high-molecular-weight and mostly straight-chain polymer of formula $-(CF_2CF_2)_n-$. This polymer was discovered in 1938 and some of its current commercial names are Teflon (DuPont), Halon (Allied Chemical), Fluon (ICI), and Hostaflon (Hoechst), among others. Other fluorinated polymers are commonly used as, for instance, hexafluoropropylene, fluoroethylpropyl-tetrafluoroethylene, commonly known as FEP-Teflon, or poly(vinylidene fluoride) $(-(CF_2CH_2)_n-)$.

Poly(tetrafluoroethylene) usually appears as a white-to-translucid solid, with a high crystalline degree. Virgin PTFE is found to possess crystallinity in the range of 92–98% [1], confirming its linear structure. Because of its low solubility in normal solvents, molecular-weight determinations are difficult to perform, but some estimates are in the range 10^6–10^7 g/mol. An empirical relationship between PTFE specific gravity (SG) and its molecular weight (M_n) has also been proposed [1]:

$$SG = 2.612 - 0.058 \log M_n \tag{1}$$

Poly(tetrafluoroethylene) chemical properties arise from the high stability of the C—F and C—C chemical bonds, imparting a remarkable chemical inertness and stability, which has been the basis for most of its applications. It is interesting to note that other fluorinated compounds with, for instance, Si—F or P—F chemical bonds are very reactive toward hydrolysis. PTFE resistance to solvents arises from the incompatibility of perfluorinated compounds with both polar and apolar solvents, referred to as both hydrophobic and oleophobic. Many examples of such a unique behavior are found in mixtures of liquids [2] or in assemblies of surfactant molecules [3]. Usually, this behavior has been interpreted as a consequence of volumetric mismatch between the perfluorinated chains and the hydrocarbon chains or water [4].

II. PTFE SURFACE PROPERTIES

The chemical structure of PTFE provides many interesting surface properties, especially a very low coefficient of friction with high surface lubricity. Also, due to its chemical inertness, PTFE is highly biocompatible, sometimes assumed as inert toward adsorption. All these properties, allied to its mechanical features, made this material one of the most used polymers for biomedical applications. Among them, one may cite heart valves and other vascular prostheses [5] using porous PTFE (commercially known as Gore-Tex), which is inert toward blood, but because of its porosity, it allows the anchoring and growth of endothelial cells.

As PTFE is prepared by dispersion polymerization, which is free-radical-initiated, some polar groups may remain in the polymer structure. From the mechanism of radical propagation, the carboxyl groups

$$-(CF_2CF_2)_nCH_2CH_2COOH \quad \text{and} \quad -(CF_2CF_2)_nCOOH$$

may occur in the latex particle. As the pK_a values of these groups are smaller than 4, they should be ionized under the conditions of normal experiments carried out in aqueous solutions. Bee and co-workers [4] have analyzed this hypothesis determining, by conductometric titration, the existence of ionic groups on the surface of PTFE latices. Their experiments lead to surface charge density of 0.68 $\mu C/cm^2$. Using an average area per perfluoroalkyl carboxylate group, these authors reached a final conclusion that 2% of the surface would be occupied by carboxyl groups. As these polar groups would decrease the polymer surface energy, it is possible that this fraction will be larger at the latex surface than in the bulk polymer. These results point out that although pristine PTFE samples are unique examples of inert surfaces, because of their polymerization process some ionic groups may remain in the polymer structure and, due to the reduction they cause in the surface energy, they might be concentrated on the polymer surface. The effects of these groups on adsorption experiments will be discussed in more detail in the next section.

Another property derived from the PTFE chemical structure is its low surface energy, ~18.6 mJ/m^2 [6] and low wettability. These properties are positive for some applications where adsorption is to be reduced, but restrict many others that require good bondability, such as in adhesives or painting and printing processes. Therefore, many PTFE surface modification procedures have been envisaged in order to improve its adhesive features.

A selective chemical modification of PTFE surface by aluminum deposition was performed by McKeown et al. [7]. The polymer surface was modified by sequential deposition of thin metal layers from the vapor phase, followed by aqueous removal after each step. This technique was shown to be very effective and selective, producing controlled surface modification. The PTFE surface became more hydrophilic, with its water contact angle being reduced from 114° to 70° after the third Al deposition.

Brace and co-workers [8] have alternatively used an electrochemical treatment of PTFE surface, using magnesium in liquid ammonia, resulting in the formation of double C—C bonds and polar functional groups such as carboxyl, hydroxyl, and carbonyl groups. This treatment provided more hydrophilic surfaces, as confirmed by contact-angle measurements with water and organic solvents [9], and is associated with increases in the polymer surface energy. Both effects are ascribed to partial defluorination of the PTFE surface due to its exposition to solvated electrons.

In another approach, Badey et al. [10] used microwave plasma to modify PTFE surfaces, following the changes in surface properties by water contact angles and ESCA measurements. Different types of plasma and experimental setups have been tested and, in the most effective case (using NH$_3$ as the gas plasma), the contact angle was reduced from 115° (pristine sample) to 70°. These changes are ascribed to the chemical modification of the polymer surface, introducing hydrocarbon groups, as well as functional groups containing nitrogen and oxygen.

The gas plasma was also used by Youxian et al. [11] to increase the surface energy of PTFE. Atmospheres of argon, oxygen, water, and air were tested and the water contact angles were found to decrease from 112° to 74°–91°, depending on the gas used. In all cases, surface defluorination was observed, along with the production of surface cross-links. Hydrogen plasma was used by Yamada and co-workers [6] in order to introduce hydrogen atoms into the PTFE structure. Water contact-angle measurements revealed an increase surface hydrophilicity with values of 67°–89° degrees, confirming the success of this approach.

Chemical modification of PTFE surfaces was also obtained by ultraviolet (UV) irradiation in vacuum and in ammonia atmosphere, as reported by Heitz et al [12]. Polymer films were irradiated by Vacuum UV using excimer krypton and xenon lamps, at 146 and 172 nm, respectively, under continuous ammonia flow or in vacuum. Surface modification was followed by contact-angle measurements, using both water and ethanol. In all cases, extreme changes were observed, with contact angle decreases from 125° (advancing) and 100° (receding) degrees with water and 35° with ethanol, to 20° and 0° in water and 0° degrees in ethanol, after only 20 s of irradiation. These drastic changes were attributed to removal of fluorine atoms as well as to insertion of oxygen, nitrogen, and hydrogen in PTFE structure.

All these approaches demonstrate different procedures for increasing the PTFE surface hydrophilicity and, therefore, adhesive properties. However, an ubiquitous concern is the depth of chemical modification, as more aggressive treatments would not only change polymer surface compositions but also its advantageous bulk properties.

III. ADSORPTION ONTO PTFE FROM SOLUTION

A. Sorption/Absorption by PTFE

As previously stressed, PTFE is regarded as an essentially inert polymer, with little or no interaction with common solvents. Therefore, sorption or absorption of solvents by this polymer are only slight, even at high temperatures. Similarly, permeation through PTFE films is very low, enabling its use as a barrier, except for fluorocarbon oils. However, as the polymer structure contains some voids, these might be filled with small amounts of solvents; this swelling leading to a polymer weight increase. These weight increases were found to be negligible for common acids (hydrochloric or nitric acids at 10%) and bases as 59% sodium hydroxide, even at temperatures as high as 100°C and after 12 months of exposure [1]. For organic solvents like acetone, ethanol, and toluene, the amount of swelling was found to be less than 1%, at the same conditions. The only slightly significant weight increase was observed with carbon tetrachloride, being 2.5% at 100°C and 3.7% at 200°C, after 8 h of exposure.

This behavior can be explained by using Hildebrand's solubility parameter approach. The solubility parameter, δ, is proposed as a measure of the cohesive energy in liquids or solids and has been extensively used in predicting solubilities [13]. According to this approach, the closer the δ values of the two compounds, the higher their miscibility. Perfluorinated compounds present very low δ values, between 12 and 13 $MPa^{1/2}$, whereas solvents like water or carbon tetrachloride, for instance, possess δ values of 32–36 and 18, respectively [13]. Starkweather has shown that the sorption of some chemicals by PTFE is closely related to their solubility parameters [14].

Allying this approach to the use of a very low-surface-tension liquid (iron pentacarbonyl), with an estimated δ value of 8, Galembeck was able to achieve a high degree of sorption by PTFE [15]. The amount of $Fe(CO)_5$ incorporated into the polymer was ~4%, after long equilibration times (up to 200 h). The long equilibration times confirmed the occurrence of the $Fe(CO)_5$ deep penetration within the bulk polymer. In addition, this sorption was found to differ significantly with the polymer crystallinity, being almost halved when changing from PTFE with 44% to a sample with 64% crystallinity. This process was interpreted as a permeation of iron carbonyl through the amorphous regions of the polymer, leading to changes in its crystalline regions, as confirmed by x-ray crystallography of the treated polymer samples. This procedure allows the generation of iron oxides at the polymer surfaces, which has been shown to decrease the polymer surface energy, as revealed by contact-angle measurements [16]. Other solvents were also tested, but the only extent of sorption above 1% was displayed by chloroform.

One of the compounds whose sorption was determined to be low was acetic acid (less than 0.5%). However, even small quantities of acids when adsorbed by PTFE may cause notable problems, as described by Jardim et al. [17]. They have used PTFE beakers to prepare ^{51}Cr species with high activity and purity, having observed, occasionally, unacceptably low yields. These observations prompted careful analysis of the materials used, including the polymer containers. As one of the steps involved chromium oxidation in a slightly basic medium, following an acid cleaning of the PTFE beakers, there was suspicion that the cause of experimental problems might be acid absorbed by the polymer. Further experiments showed that 10.8 µmol of acid was released from the acid-treated beaker, a considerable amount when compared to the 10 µmol of base used in some of their synthetic procedures. This quantity was found to increase when pressure was applied to the system, as well as with the polymer specific surface, becoming quite significant for small PTFE spheres (Chromosorb T). On the other hand, the absorption of base by PTFE was found to be much smaller. These findings seem to be in line with the previously discussed occurrence of carboxyl groups on the polymer surface.

B. Adsorption of Solutes onto PTFE

Another example of unexpected problems arising from adsorption onto PTFE when conducting diffusion measurements in liquids was reported by Loh and co-workers [18]. These experiments were performed using the Taylor dispersion technique, which involves the injection of a small amount of solute into solvent pumped through a long tubing. During this experiment, the solute suffers a combination of dispersion processes, leading to a Gaussian-like concentration profile, which allows the determination of its diffusion coefficient. The solute dispersion should occur undisturbed; therefore, a series of requirements referring to laminar flow and to the geometric arrangement of the apparatus should apply. Measurements of the diffusion coefficients of a homologous series of alkyl p-hydroxybenzoates in water, NaOH solution, and ethanol have been performed [18] and the results are summarized in Fig. 1.

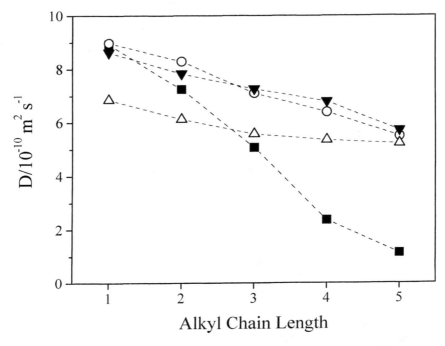

Figure 1 Diffusion coefficients for the homologous series of alkyl *p*-hydroxybenzoates in (■) water, using PTFE tubing; (○) water, using stainless-steel tubing; (△) ethanol; (▼) 10% NaOH. (Data from Refs. 18 and 19.)

Diffusion coefficients for a family of compounds in a specific liquid should follow the Stokes–Einstein equation, which predicts a linear decrease of D with the increase in solute radius. However, as shown in Fig. 1, the diffusion coefficients obtained in water using a PTFE tubing present a steeper decrease than the one predicted by the solute size trend or as observed in ethanol or NaOH solutions. These changes were observed to follow increases in the solute retention times, as well as the appearance of asymmetries ("tails") in the dispersion curves. In such experiments, the solute retention time is determined only by the pump rate and, therefore, its increase would suggest the occurrence of solute interaction with the PTFE tubing walls. In this case, the dispersion would be disturbed in a way similar to separation through capillary chromatography. In fact, a correction of the diffusion coefficients was attempted through Golay's equation for capillary chromatography by using the solute retention times. All the observed changes and the retention factors derived from Golay's approach were found to increase with the solute alkyl chain length, revealing a process of hydrophobic nature. Further experiments, repeated in a similar apparatus, using a stainless-steel tubing produced diffusion coefficients which follow more closely the Stokes–Einstein behavior [19], as can be observed in Fig. 1, pointing out the importance of careful selection of tubing material. In addition, no change in retention times or anomalous diffusion coefficient variation along the series was verified in ethanol or NaOH solutions. In both solvents, the solutes are much more soluble than in water—in the former due to their apolar nature, and in the latter due to formation of the conjugate anion.

The hypothesis that these deviations were caused by solute adsorption onto the tubing was finally confirmed by direct determination of their adsorption isotherms [20]. Using the same PTFE tubing as in the previous diffusion measurements, the amounts of removed solute were determined. The polymer surface was analyzed by attenuated total reflectance (ATR)–FTIR spectroscopy, revealing no indication of $C=O$, $O-H$, or $C-H$ absorption bands, ruling out contributions from polar groups or hydrocarbon contributions to the solute–polymer interaction. A kinetic analysis of the solute removal process was first performed, and the results for pentyl p-hydroxybenzoate are shown in Fig. 2. These results suggest the

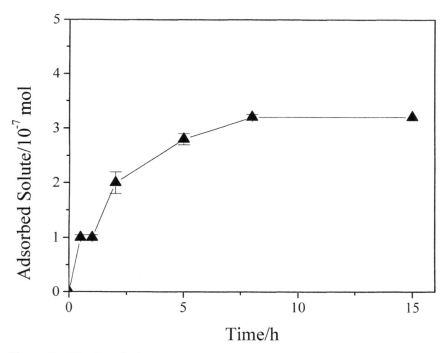

Figure 2 Kinetics of solute removal onto the surface of a PTFE tubing. Pentyl p-hydroxybenzoate at initial concentration of 7×10^{-4} mol/dm^3. (Adapted from Ref. 20.)

occurrence of two adsorption processes, the last one taking ~5 h to reach equilibrium. Estimates of the solute size by its van der Waals radius [18] lead to an area of cross section of 41 Å. By using this value and the amount of removed solute at the first plateau, the obtained area is close to the internal surface provided by the tubing. Therefore, this leads to the assumption that the tubing surface may be saturated with solute molecules lying flat. At the second plateau, the average area per solute molecule is reduced to a third, what might suggest solute rearrangement at the polymer surface, or even some slight extent of solute penetration into the polymer. The mean solute displacement after 5 h may be estimated as smaller than 0.2 mm by using its diffusion coefficient [18] and the Einstein relation; this displacement being too small to indicate a significant extent of sorption, especially considering than its diffusion within the polymer would be much slower.

Adsorption isotherms were determined after equilibrium time, as shown in Fig. 3. The extents of removal for the methyl and ethyl homologs were too small to be determined; this is in line with the previous observations of negligible adsorption effect on their diffusion [18]. For the other homologs, propyl, butyl, and

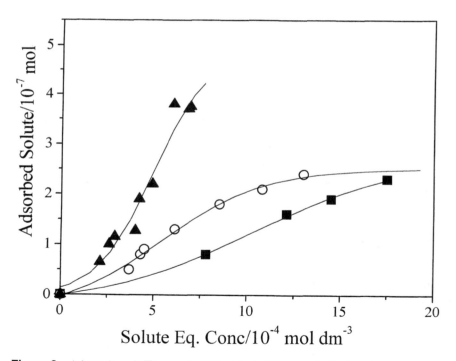

Figure 3   Adsorption of (■) propyl, (○) butyl, and (▲) pentyl p-hydroxybenzoates onto PTFE tubing. (Adapted from Ref. 20.)

pentyl *p*-hydroxybenzoates, the shape of the adsorption isotherms resembles the ones for the low-affinity interaction [21] and may indicate some degree of cooperativity for the adsorption process. The maximum amounts of removed solute, though, are small: roughly 10%, 5%, and 3%, respectively, for pentyl, butyl and, propyl *p*-hydroxybenzoates.

During the attempts to maximize solute removal, some tests were performed using PTFE latices, which were not considered due to poor reproducibility of the data caused by flotation of polymer particles. However, for the higher homologs, some particles were observed to sink into the aqueous solution, indicating an increased PTFE wettability. This observation prompted the determination of PTFE contact angles with these aqueous solutions [22]. First, a kinetic analysis of the required equilibration times was performed, producing the results in Fig. 4. Again, the process of changing the polymer wettability is slow, attaining equilibrium after ~4–5 h, the same time interval observed during the adsorption measurements. This agreement suggests a common process as the cause of both effects: solute removal and change in contact angle. Equilibrium contact angles were determined for so-

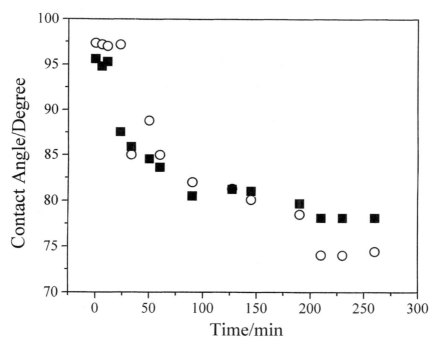

Figure 4 Kinetic analysis of changes in PTFE wettability with (■) methyl and (○) heptyl *p*-hydroxybenzoates at concentrations, respectively 1.4×10^{-2} and 3.3×10^{-5} mol/dm^3. (Data from Ref. 22.)

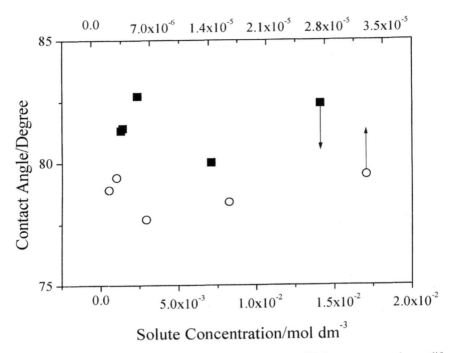

Figure 5 Changes in PTFE wettability, represented by equilibrium contact angles, at different (■) methyl and (○) heptyl p-hydroxybenzoates concentrations. (Data from Ref. 22.)

lutions of methyl and heptyl p-hydroxybenzoates with PTFE, and the results are shown in Fig. 5. The contact angles are reduced in the presence of these solutes, but, within the studied concentration range (the upper limit is their saturated solutions), no significant concentration dependence is observed. This concentration range is the same as that used in the adsorption studies, and these results confirm an earlier hypothesis [20] that the more hydrophobic solutes cause larger wettability increases, in parallel to their higher extent of adsorption. The differences among the homologs, however statistically significant, are small.

However, one has to bear in mind that contact angles are the consequence of a mechanical balance at the polymer surface involving solid–liquid, solid–vapor, and liquid–vapor surface tensions, as represented in Fig. 6. This balance is summarized by Young's equation, as follows:

$$\gamma_{SL} = \gamma_{SV} - \gamma_{LV} \cos \theta \tag{2}$$

where γ represents the different interfacial tensions and θ represents the contact angle between the liquid and solid.

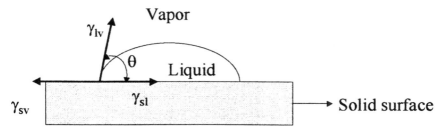

Figure 6 Schematic representation of the force balance affecting contact angles.

Usually, changes in contact angle are ascribed to adsorption onto the solid–liquid interface. However, this balance may be also changed by variations in the liquid–vapor interfacial tension, leading to contact-angle changes which are not the consequence of adsorption onto the polymer. Decreases in γ_{LV}, for instance, would produce smaller contact angles. To verify this possibility, surface tensions of aqueous solutions of methyl and heptyl p-hydroxybenzoates were measured in the studied concentration range. The results, shown in Fig. 7, demonstrate that in this concentration range, the smaller homolog is the most effective in reducing the surface tension, as opposed to the trend shown for reduction of contact angles. These findings confirm the role of solute adsorption in the wettability changes and demonstrate the importance of contact-angle measurements in assessing processes related to adsorption from solution.

The general picture that arises from this series of experiments is that this homologous series of compounds adsorb onto PTFE in a process of a hydrophobic nature. The solute removal is slow, but not enough to be ascribed to a sorption process. The average area occupied per solute at the polymer–solution interface, the hydrophobic nature of the process, and the consequent increase in PTFE wettability suggest that the solute molecules adsorb oriented with their polar moieties toward the aqueous solution and the alkyl chain in contact with the polymer. Therefore, the driving forces for solute adsorption will be the removal of its apolar moiety from the aqueous environment allied to a decrease in the interfacial energy between polymer and solution. This evidence demonstrates that although PTFE may be regarded as relatively inert toward adsorption, care should be taken when dealing with aqueous solutions of small, although hydrophobic, solutes.

C. Adsorption of Surfactants and Polymers onto PTFE

There are a series of reports on the adsorption of surfactants of different head groups and apolar chains onto PTFE. Bee et al. [4]. have investigated the adsorption of nonionic, anionic, and cationic hydrocarbon surfactants and of ammonium

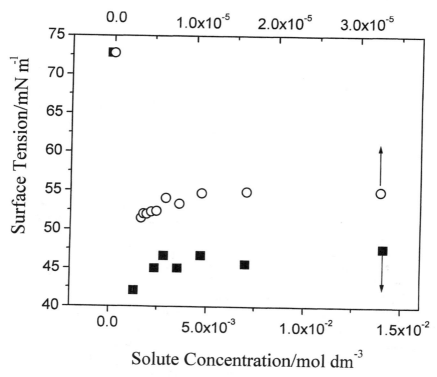

Figure 7 Changes of air–solution interfacial tension in (■) methyl and (○) heptyl *p*-hydroxybenzoates solutions. (Data from Ref. 22.)

perfluorooctanoate (APFO) onto PTFE latices. As discussed in the previous subsection, these authors have verified the presence of carboxylic groups on the polymer surfaces (~2% of the surface). The adsorption isotherms for APFO and sodium dodecanoate onto these latices are shown in Fig. 8. The perfluorinated surfactant is found to interact much more strongly with the polymer surface, reaching a plateau well below its CMC (2.3×10^{-6} mol/dm^3). At this plateau, the average area per surfactant molecule is ~54 Å^2, which is very close to the value estimated at the solution–air interface (47 Å^2), suggesting a vertical orientation for the molecule. The hydrocarbon surfactant, on the other hand, does not reach a saturation point even close to its solubility limit, and its average area at maximum occupancy is 120 Å^2, much larger than the value obtained at polystyrene surfaces, ~30 Å^2. These findings confirm the lack of affinity between the surfactant hydrocarbon chain and the perfluorinated polymer.

Similar measurements were made with a nonionic surfactant, C$_{12}$EO$_6$, and another sample of PTFE latex. For this surfactant, the adsorption plateau was

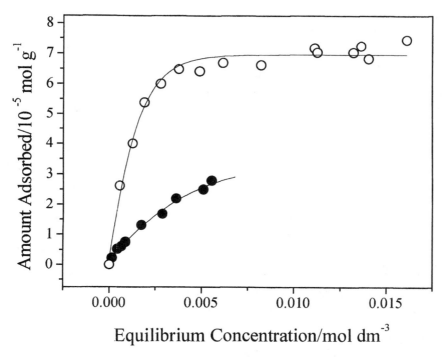

Figure 8 Adsorption of (○) APFO and (●) sodium dodecanoate onto PTFE latex. (Adapted from Ref. 4.)

reached only at its CMC, producing average molecule areas at the polymer surface of ~65 Å², which is larger than the value of 40 Å² derived from experiments with polystyrene lattices. In addition, the effect of the addition of magnesium sulfate on the latex stability was determined in the presence of this surfactant, revealing a dual behavior: The surfactant decreases the latex stability at low concentrations, increasing it more than 30 times at higher concentrations. The nonionic surfactant is thought to adsorb initially via its ethylene oxide chain, turning the latex hydrophobic and less stable. At higher concentrations, further surfactant molecules may interact with the exposed hydrocarbon chains, leading to steric stabilization of the latices.

Studies with cationic decyl trimethyl ammonium bromide (C_{10}TAB) also revealed a weak interaction with PTFE latex, the plateau being reached only at its CMC, the average area per molecule (118 Å²) being much larger than the value obtained for negatively charged polystyrene lattices (65 Å²). At a lower surfactant concentration, a plateau region is observed, concurrent with latex coagulation due to the neutralization of its surface charges. Critical coagulation concentrations were determined for three homolog surfactants, and the results are shown in Table 1, along with those determined for negatively charged polystyrene latices.

It is interesting to note that for PTFE latices, little effect is presented by the increase in surfactant hydrophobicity, whereas this change is a leading factor for the surfactant interaction with polystyrene latex. This finding, once again, supports the lack of interaction between the surfactant hydrocarbon chain and the PTFE surface, with the electrostatic interaction the dominant contribution.

Despite their weak interaction with perfluorinated surfaces, Yao and Strauss have studied PTFE interaction with cationic surfactants with one [cetyl trimethyl ammonium chloride CTAC], two [didodecyl ammonium bromide (DDAB)], and three [tri dodecyl ammonium chloride (TDAC) alkyl chains] [23,24]. Two distinct behaviors were observed depending on the number of surfactant alkyl chains, as shown by the adsorption isotherms represented in Fig. 9. The curves for single- and double-chain surfactants reached a saturation point, producing, respectively, average molecular areas per adsorbed surfactant of 83 and 61 Å^2, values close to that expected for vertical orientation. For TDAC, however, no plateau was reached, and the adsorption indicated the formation of surfactant multilayers. In addition, PTFE particles covered by TDAC are wetted by water, whereas bare PTFE is not, which suggests that the surfactant is oriented at the polymer surface with its polar head group toward the aqueous phase. This change in PTFE wettability was also reported to occur during the adsorption of alkyl p-hydroxybenzoates [20], as discussed in the previous subsection.

Further studies [24] were performed on the interaction of DDAB and TDAC with the biomedical materials–expanded PTFE and knitted Dacron [poly(ethyleneterephtalate)]. In both cases, the adsorption isotherms confirmed the earlier observations; plateau regions for DDAB and the indication of multilayer formation for TDAC. When comparing both materials, the interaction of both surfactants was more intense with Dacron than with Teflon, and the limiting areas per DDAB molecule suggest a vertical surfactant orientation on PTFE surface, in contrast with a flat-lying DDAB layer on Dacron. Both observations are consistent with the reported low affinity of hydrocarbon chains and polar head groups by PTFE surface.

Table 1   Studies on the Coagulation of PTFE Latex by Cationic Surfactants

Agent	Critical coagulation concentrations (mol/dm^3)	
	PTFE latex	Polystyrene latex
C_8TAB	4.5×10^{-5}	1.6×10^{-4}
C_{10}TAB	3.1×10^{-5}	2.0×10^{-5}
C_{12}TAB	2.9×10^{-5}	1.3×10^{-6}

Source: From Ref. 4.

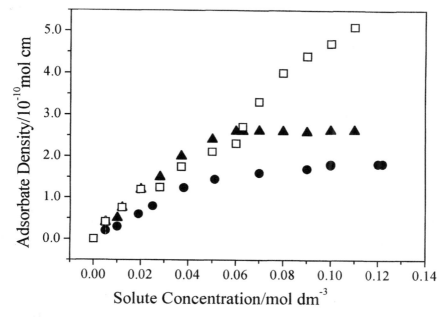

Figure 9 Adsorption of cationic surfactants (●) CTAC, (▲) DDAB, and (□) TDAC onto PTFE particles. (Adapted from Ref. 23.)

An important consequence of these series of investigations is the verified possibility of modifying these polymer surfaces by controlled deposition of surfactant layers, leading to the prospect of secondary adsorption of lipid-soluble substances like drugs or antibiotics.

Another study in connection with potential biomedical applications is the one of adsorption of ethylene oxide–propylene oxide–ethylene oxide (EO–PO–EO) block copolymers onto PTFE [25]. These copolymers possess surface-active features, enabling their use in a variety of industrial applications, including cosmetic and pharmaceutical formulations, as well as in emulsions or as dispersing agents. An interesting property they have shown is increasing in vivo lifetimes of solid dispersions [26], which has been related to their mode of adsorption onto the solid–liquid interface, affecting their interaction with proteins and antibodies [27]. Their interaction with PTFE surfaces was investigated through contact-angle measurements, leading to the results summarized in Fig. 10. A kinetic analysis of the changes in contact angles was also performed, and typical results are represented in Fig. 11, revealing equilibrium times similar to those observed for smaller solutes (as shown in the previous subsection), and consistent with an adsorption process. These copolymers were chosen due to the proximity of their poly(propy-

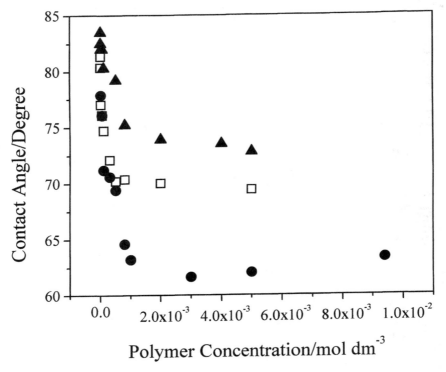

Figure 10   Changes in PTFE wettability in contact with block copolymer aqueous solu-

Table 2   Nominal Composition and CMC of the
Studied Copolymers

Polymer	NEO	NPO	$\langle M \rangle$ (g/mol)	CMC[a] (mol/dm^3)
P 103	2×17	62	4950	7.4×10^{-4}
P 105	2×37	58	6500	8.0×10^{-4}
F 108	2×127	64	12000	5.1×10^{-4}

[a]Data from Ref. 28.

lene oxide) content and variable ethoxylated chain size, as summarized in Table 2. Examination of the equilibrium contact angles reveals that the polymer with the smallest EO/PO ratios, (i.e., the most hydrophobic) produces the largest increase in wettability. The variations observed for the other two polymers are similar. However, as briefly discussed in the previous subsection, the contact angle is part

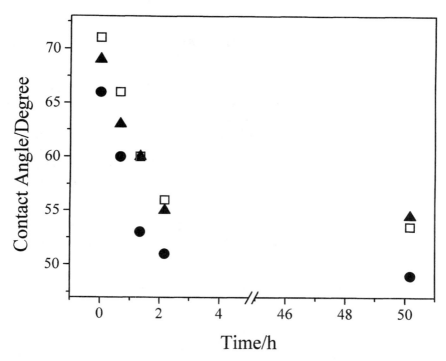

Figure 11 Kinetic analysis of wettability changes with (□) F 108, (●) P 103, and (▲) P 105 solutions. (Data from Ref. 25.)

of a mechanical balance represented by Young's equation [Eq. (2)]. Therefore, changes in contact angle cannot be ascribed only to adsorption at the solid–liquid interface.

Gu and Rosen [29] have proposed an approach to the discrimination between the different simultaneous processes that might affect contact angles. The adsorption onto solid–liquid interfaces may be described by the Gibbs equation as follows:

$$\left(\frac{d\gamma_{SL}}{d\ln C}\right) = -RT\Gamma_{SL} \qquad (3)$$

where Γ_{SL} represents the solute excess concentration at the solid surface and C is its concentration in the liquid phase. By differentiating Young's equation [Eq. (2)] and considering γ_{SV} constant, as PTFE is a low-surface-energy solid, one reaches

$$d\gamma_{SL} = -d(\gamma_{LV}\cos\theta) \qquad (4)$$

which can be substituted into Eq. (3), producing

$$d(\gamma_{LV} \cos \theta) = RT\Gamma_{SL} \, d(\ln C) \tag{5}$$

By using a previous determination of the F 108 effect on the water–air interfacial tension [28], the data of Fig. 10 was plotted according to Eq. (5), as shown in Fig. 12. This treatment confirms the polymer adsorption onto the PTFE surface and enables estimates of its average molecular area at the solid–liquid interface as 583 Å^2. This value is smaller than the one expected for full coverage of PO units (~860 Å^2) and much smaller than the area which would be occupied by the polymer EO units (~2900 Å^2), suggesting that the copolymer interacts with PTFE via its hydrophobic PO units. A similar observation was derived from data on the adsorption of two other copolymers of this family onto polystyrene latex [30], which produced average adsorbed areas consistent with total PO adsorption. Again, the surfactant interaction with PTFE seems to be weaker than with hydrocarbon polymers.

Another interesting observation is that the break displayed by the curves of Fig. 10, which occurs at the copolymer CMC, is not related to its adsorption onto

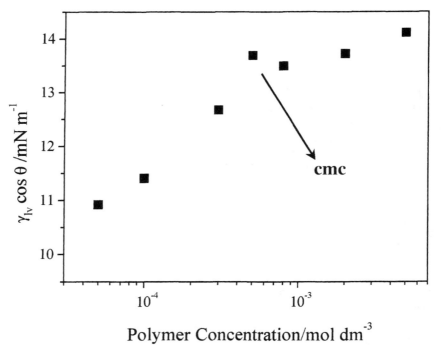

Figure 12 Analysis of F 108 interaction with PTFE surface derived from contact-angle data according to Eq. (5).

PTFE (as no discontinuity is revealed by the treatment of Fig. 12) but to the copolymer effect on the water–air interfacial tension [28]. This behavior suggests that the copolymer adsorption mode remains the same after its CMC [different from the one reported for some binary surfactant mixtures [29]], where a break is observed. Once again, this point reinforces the importance of the approach summarized in Eq. (5) in examining contact-angle data as a means of investigating adsorption.

IV. CONCLUSION

Poly(tetrafluoroethylene) is certainly a unique polymer and among its outstanding properties one may cite its low surface energy and consequent reduced adsorption capacity. Some procedures are available for surface modification aimed at increased adhesion, but the studies described in this chapter reveal that even pristine PTFE surfaces may interact with some molecules in aqueous solution. This interaction is favored by the solute low Hildebrand solubility parameter and by its reduced aqueous solubility. The issue of solute orientation at the polymer surface reveals somewhat unexpected features when compared to observations related to other polymer surfaces, and certainly deserves further investigations. In addition, the use of contact-angle measurements as a tool to assess adsorption onto PTFE is discussed and the presented data reinforce the importance of a more comprehensive approach, considering the contribution of other process than adsorption at the solid surface.

REFERENCES

1. SV Gangal. In: M Graynen, ed. Kirk–Othmer Encyclopedia of Chemical Technology, Vol. II. 3rd ed. New York: Wiley, 1980, pp 1–24.
2. J Aracil, RG Rubio, MD Peña, JAR Renuncio. J Chem Soc Faraday Trans I 84:539–550, 1988.
3. P Lo Nostro, S-H Chen. J Phys Chem 97:6535–6540, 1993.
4. HE Bee, RH Ottwewil, DG Rance, RA Richardson. In: RH Ottwewil, CH Rochester and AL Smith, eds. Adsorption from Solution. London: Academic Press, 1983, pp 155–171.
5. HR Allock, FW Lampe. Contemporary Polymer Chemistry. 2nd ed. Englewood Cliffs, NJ: Prentice-Hall, 1990, pp 503, 564, 578–579.
6. Y Yamada, T Yamada, S Tasaka, N Inagaki. Macromolecules 29:4331–4339, 1996.
7. NB McKeown, PG Kalman, R Sodhi, AD Romaschin, M Thompson. Langmuir 7:2146–2152, 1991.
8. K Brace, C Combellas, M Delamar, A Fritsch, F Kanoufi, MER Shanahan, A Thiébault. Chem Commun 403, 1996.

9. K Brace, C Combellas, E Dujardin, A Thiébault, M Delamar, F Kanoufi, MER Shanahan. Polymer 38:3295–3305, 1997.

10. JP Badey, E Urbaczewski-Espuche, Y Jugnet, D Sage, TM Duc, B Chabert. Polymer 35:2472–2479, 1994.

11. D Youxian, HJ Griesser, AW-H Mau, R Schmidt, J Liesegang. Polymer 32:1126–1130, 1991.

12. J Heitz, H Niino, A Yabe. Appl Phys Lett 68:2648–2650, 1996.

13. AFM Barton. Handbook of Solubility. Boca Raton, FL: CRC Press, 1985.

14. HW Starkweather. Macromolecules 10:1161–1163, 1977.

15. F Galembeck. J Polym Sci, Polym Chem Ed 16:3015–3017, 1978.

16. F Galembeck. J Polym Sci, Polymer Lett 15:107–109, 1977.

17. ICSF Jardim, M Sartoratto, PR Saliba, C Archundia, KE Collins. Appl Radiat Isotopes 40:643, 1989.

18. W Loh, CA Tonegutti, PLO Volpe. J Chem Soc Faraday Trans. 89:113–118, 1993.

19. W Loh, AE Beezer, JC Mitchell. Langmuir 9:3431–3435, 1994.

20. W Loh, PLO Volpe. J Colloid Interf Sci 179:322–323, 1996.

21. SJ Gregg, KSW Sing. Adsorption, Surface Area and Porosity. London: Academic Press, 1982, Chap 5.

22. AC Ramos, W Loh, unpublished results.

23. J Yao, G Strauss. Langmuir 7:2353–2357, 1991.

24. J Yao, G Strauss. Langmuir 8:2274–2278, 1992.

25. JR Lopes, W Loh. Contact angle measurements to investigate solute and polymer adsorption onto poly(tetrafluoroethylene), International Conference on Colloids, Brazil, 1998.

26. S Stonilk, NC Felumb, CR Heald, MC Garnett, L Illum, SS Davis. Colloids Surf A 122:151–159, 1997.

27. J-T Li, KD Caldwell. Colloids Surf. B 7:9–22, 1996.

28. JR Lopes, W Loh. Langmuir 14:750–756, 1998.

29. B Gu, MJ Rosen. J Colloid Interf Sci 129:537–553, 1989.

30. S Gaisford, W Loh, AE Beezer, JC Mitchell. A calorimetric study of polymer binding to polystyrene microspheres. 13th IUPAC Conference Chemical Thermodynamics, Clermont-Ferrand, 1994.

11

Adsorption from Polymer Mixtures at the Interface with Solids

Yuri S. Lipatov, Tamara T. Todosijchuk, and Valentina N. Chornaya
Institute of Macromolecular Chemistry, National Academy of Sciences of Ukraine, Kiev, Ukraine

I. INTRODUCTION

Adsorption phenomena play an important role in many fields of the application of polymers (stabilization and flocculation of dispersion, formation of dispersion, production of coatings and composite materials, etc.). The adsorption phenomena determine the structure of the border layers at the polymer–solid interface in the most important practical cases such as surface layers at the interface with a solid [1], adhesion joining [2], properties of the thin films at the interface and in pores [3], and stabilization of colloidal systems [4].

Presently, there is a great deal of theoretical and experimental work dedicated to the adsorption of polymers from solutions and to adsorption layers and their role in a wide class of colloidal materials. The present state of the theory of adsorption of polymers from solutions is given in Refs. 4–8. In all the work dedicated to adsorption, a large number investigate the structure of adsorption layers, which determines the properties of materials and colloidal dispersions—the forces acting between interfaces in solutions.

At the same time, almost all theoretical and experimental work are restricted by the analysis of adsorption from dilute solutions. Meanwhile, adsorption from semidiluted and concentrated solutions has specific features that were analyzed in detail in our earlier work [1,9,10]. The specifics of adsorption from semidiluted and concentrated solutions are determined by the formation in such a solution of molecular aggregates or associates that is in dynamic equilibrium with free polymer chains. By increasing the solution concentration, statistical coils begin to overlap, and as a result, the compression and decrease in coil size occurs [11]. In this region of concentration, according to de Gennes two regimes exist [12]: semi-diluted and concentrated solutions. In semidilute solutions, overlapping of macro-molecular coils occurs at the critical concentration of overlapping, $C^* = 1/[\eta]$, where $[\eta]$ is the intrinsic viscosity corresponding to the limit molecular mass.

The processes of the formation of molecular aggregates proceed both below and above C^* and are connected with the dependence of the thermodynamic parameter of the polymer–solvent interaction χ_{12} on concentration. The types of aggregate and their number depend on the intermolecular interactions among macromolecules, concentration of solution, and thermodynamic quality of the solvent that is characterized by the parameter χ_{12}. If $\chi_{12} < 0.5$, a solvent is considered to be a good one; if $\chi_{12} = 0.5$, a solvent is considered ideal (θ solvent); and if $\chi_{12} > 0.5$, a solvent is considered to be poor [13]. The thermodynamic reason for the formation of macromolecular aggregates is probably incomplete miscibility of polymer fractions of different molecular mass [14]. Therefore, by the formation of aggregates, some fractionating according to molecular mass may proceed. As a result, the structure of the adsorption layer will depend on the ratio between aggregated and nonaggregated macromolecules at the interface with the solid and on their molecular mass distribution.

The experimental data on adsorption from semidiluted and concentrated solutions led to the proposal of the molecular–aggregative mechanism of adsorption described in detail elsewhere [1,2,9].

The peculiarity of polymer adsorption, connected with its aggregative mechanism, consists in the fact that at each concentration of solution, a new state of equilibrium between isolated and aggregated molecules is established. Each point of the isotherm corresponds to another structure of adsorbing entities (from isolated macromolecules up to aggregates of various sizes and their distribution on the surface, depending on solution concentration). During adsorption, both isolated and aggregated molecules pass onto the surface of an adsorbent, with adsorption of aggregates being preferential. In most cases, after adsorption there are no aggregates in solution that appear again some time after establishing the new equilibrium between isolated and aggregated molecules. If, in diluted solutions, the adsorption is higher owing to the poor solvent, for concentrated solutions it is higher owing to the good solvent because of the effect of structure formation in solution on adsorption [8].

The aggregative mechanism of adsorption explains some experimental data on adsorption from semidiluted and concentrated solutions that could not be explained in the framework of traditional concepts of adsorption from a dilute solution. It explains the values of adsorption exceeding those calculated for monomolecular covering of the surface, the small fraction of segments bound with the surface, the dependence of adsorption on the adsorbent–solution ratio, and so forth.

Contrary to the number of studies on adsorption of individual macromolecules from solutions, there are rather few investigations on adsorption of polymer mixtures. This chapter is dedicated to the analysis of the data available on the adsorption of polymer mixtures at the solid interface from semidiluted and concentrated solutions.

II. CRITICAL CONCENTRATIONS FOR SOLUTIONS OF POLYMER MIXTURES

To understand adsorption from solutions of polymer mixtures from the point of view of the molecular–aggregative mechanism, it is very important to establish the regions of the overlapping of macromolecular coils (critical concentrations of overlapping). Unfortunately, up to now, the theory of the behavior of polymer mixtures in solutions is not developed. The analysis of the available data from binary systems (polymer–solvent) [15–21] and a few works on the ternary systems (polymer–polymer–solvent) [22,23] allow one to consider two approaches to the estimation of the critical concentrations C^* for solutions of polymer mixtures.

The first approach is based on the application of the known relations for C^* in binary solutions. C^* is taken to be inversely proportional to the intrinsic viscosity $[\eta]$, with the proportionality coefficient depending on the model of the coils interaction.

Two models for the evaluation of the critical concentration of transitions between dilute and semidilute polymer solution have been proposed [15]. According to the first model, C^* determines the onset of the formation of a continuous network because of overlapping of statistically distributed penetrable spherical macromolecules in solution. In another model, C^* corresponds to the maximum concentration of statistically impenetrable rigid spheres. Both models give the qualitative theoretical background for experimental findings of critical concentrations from the empirical correlation $C^* = 1/[\eta]$.

Table 1 shows the values of C^* for some binary polymer systems to be considered later.

To evaluate C^* in solutions of polymer mixtures, we have used an approach based on the assumption of chaotic distribution and overlapping of macromolecules in solution. According to Roots and Nystrom [24], such an assumption corresponds to the averaged region of the transition from dilute to semidilute solutions that exists in solutions.

The peculiarity of our approach to solutions of polymer mixtures consists of using the empirical correlation $C^* \cong 1/[\eta]$, where the value of $[\eta]$ for one polymer is measured in the "mixed" solvent (second polymer–common solvent) at the condition of variable concentration of the second polymer. This approach was used in

Table 1 Critical Concentration in
Polymer–Solvent Systems

System	C^* (g/100 mL)
PS–CCl$_4$	0.90
PBMA–CCl$_4$	1.05
PCL–CHCl$_3$	7.40
PC–CHCl$_3$	2.20
PS–ethyl acetate	2.20
PDMS–ethyl acetate	9.00
PCL–CH$_2$Cl$_2$	8.00
PS–CH$_2$Cl$_2$	2.30

some of our work [25,26] dedicated to adsorption from solutions of two polymer mixtures: PS ($M_W = 4.4 \times 10^5$)–PBMA ($M_W = 2.7 \times 10^5$)–CCl$_4$ (thermodynamically incompatible pair in solution) and polycarbonate (PC, $M_W = 3.3 \times 10^4$)–poly-ε-caprolactone (PCL, $M_W = 2.0 \times 10^3$)–CHCl$_3$ (thermodynamically compatible pair). For these mixtures, the critical concentrations of overlapping C^* have been determined (Table 2). As is seen, values of C^* for the first polymer in the mixture are higher than with a binary solution and increases with increasing concentration of the second polymer in the "mixed" solvent. Such an effect can be explained by the compression of the macromolecular coils of the first polymer and by diminishing its dimensions in a mixed solvent because of worsening the thermodynamic quality of the "mixed" solvent in comparison to the pure solvent.

The second approach to the estimation of the critical concentration is based on the concept of an "averaged" critical concentration for the mixture, found from the "averaged" intrinsic viscosity according to

$$[\eta]_{mix} = [\eta]_A g_A + [\eta]_B g_B \tag{1}$$

where g_A and g_B are the mass fractions of components in the mixture; $g_A + g_B = 1$. The averaged crossover concentration $C_{mix}^* = 1/[\eta]_{mix}$ is a function of the ratio of concentrations of dissolved molecules of A and B.

Equation (1) is the result of the modification of an equation proposed in Refs. 22 and 23. That equation connects C^* with hydrodynamic volume occupied by molecules of two polymers in solution. In cited works, authors have proposed, for the first time, estimating the critical concentration of the coil overlapping for the mixture of two polymers from the data on C^* of constituent components.

The critical concentrations for some systems were also calculated according

Table 2 Critical Concentrations in Solutions of Polymer Mixtures
(First Method)

"Mixed" solvent concentration, C (g/100 mL)	C* (g/100 mL)	"Mixed" solvent concentration, C (g/100 mL)	C* (g/100 mL)
	PS–PBMA–CCl$_4$ system		
PBMA–"mixed" solvent (PS–CCl$_4$)		PS–"mixed" solvent (PBMA–CCl$_4$)	
0	1.05	0	0.90
0.5	1.66	0.25	1.11
1.0	2.00	0.50	1.25
1.5	2.27	1.00	1.33
2.0	2.94	1.50	2.00
3.0	3.33	2.00	2.50
	PC–PCL–CHCl$_3$ system		
PC–"mixed" solvent (PCL–CHCl$_3$)		PCL–"mixed" solvent (PC–CHCl$_3$)	
0	2.20	0	7.4
1.0	2.20	0.5	13.3
2.0	2.86	1.0	15.4
3.0	3.70	1.5	20.0
4.0	4.50	2.0	22.0
6.0	5.00	2.5	25.0

to the second method [25,26]. These data show that for the PS-PBMA system where values of C^* for each component are close (Table 1), the value of C_{mix}^* only slightly depends on the component ratio in solution. At the same time, for PS–PC and PS–PDMS, a marked increase of C_{mix}^* is observed when the concentrations of PCL or PDMS increase due to worsening of thermodynamic quality of the "mixed" solvent ($[\eta]_{mix}$ diminishes).

Thus, the data on the critical concentration of overlapping show that for all polymer pairs, C^* increases as compared with C^* for binary solutions. In all cases, this value grows with increasing concentration of the "mixed" solvent because of coil compression. Thermodynamic miscibility of polymer components in solution has no effect on the critical concentrations, which are determined preferentially by the presence of the second polymer in solution.

III. AGGREGATION IN SOLUTIONS OF POLYMERS AND THEIR MIXTURES

Equilibrium solutions of polymers contain, besides an isolated macromolecular coil, some supermolecular structures that arise due to aggregation or association of macromolecules. Their formation proceeds both below and above the critical concentration of overlapping and is connected with dependence of the parameter of the thermodynamic interaction polymer–solvent, χ_{12}, on concentration. The interaction of polymer coils leads to the formation of aggregates which can be considered a "swarm" of interacting coils with a definite duration of life. The number of aggregates and of macromolecules entering it depends on the nature of the intermolecular interaction in solution, the nature of the solvent, and the concentration of a solution. Thermodynamic reason for the aggregate formation is partial miscibility of polymer fractions of various molecular masses independent of their chemical nature [14]. Some kind of fractionating can occur by formation of aggregates as a result of the higher solubility of low-molecular-mass fractions [27]. The aggregate formation determines the existence of the different local densities in the bulk of the solution. The density fluctuations determined by the aggregation can be considered the regions of higher concentration of a polymer as compared with the mean density of the solution. These fluctuations exceed the fluctuations typical of liquid as described by the Boltzman distribution [28]. The solution containing aggregates is an equilibrium and stable one-phase system that can become unstable only in the metastable or unstable regions of the phase diagram.

Presently, there are many experimental methods allowing the establishment of the formation of aggregates and evaluation of their number, dimensions, and distribution by size [29–35]. These data served as an experimental basis in formulating the concept of aggregative adsorption (see Introduction).

Meanwhile, there are only scattered data on aggregation in the ternary polymer–polymer–solvent systems [36]. In Refs. 37–39, the aggregation processes on solutions of polymer mixtures have been studied using the turbidity spectrum method that was used earlier for investigation of binary solutions. This method allows the number and dimension of aggregates to be found from the spectrum dependence of the intensity of scattered light in the visible region. Theoretical basis for the method was developed by Heler and Pangonic [40] based on the Mie theory [41] for isotropic spheres. We have used a modification of the method [42]. Some data on the aggregate formation in binary solutions and solutions of polymer mixtures are given in Table 3, where the averaged size of the aggregates of both types is given. The comparison of the results presented in Table 3 shows the different characters of the structure formation in given solutions determined by the thermodynamic parameter of the polymer–polymer interaction. The data were obtained for polymer pairs both miscible and immiscible in solutions.

For the immiscible pair (Table 3), the presence of the second component

Table 3 Dimensions of Aggregates of Incompatible Polymers

	Binary system			Ternary system
	PBMA–CCl$_4$	PS–CCl$_4$		PBMA–PS–CCl$_4$
C (g/100 mL)	r_w (Å)	r_w (Å)	C_{PBMA}/C_{PS} (g/100 mL)	r_w (Å)
0.50	400	350	0.50/3.00	300
0.75	600	450	0.75/2.75	300
1.00	700	500	1.00/2.50	250
1.50	900	550	1.50/2.00	200
2.00	1400	600	2.50/1.00	200
2.50	1900	650	2.75/0.75	250
3.00	2500	700	3.00/0.50	300

leads to the diminishing of aggregate size as compared with binary solutions. From the data given in Refs. 43–48 on the chain dimensions in ternary systems with two immiscible polymers follows that one of the mixture components exerts an essential effect on the dimension of the chain of the second component. Adding one polymer to another sharply changes the thermodynamic environment of another component as shown by the diminishing of the second virial coefficient and inertia radius [43,44]. By the investigation of two ternary systems [43] [I-poly-(β-vinyl naphtalene)–poly(isobutylene)–toluene] and [II-poly(naphtylmethacrylate)–PMMA–benzene] it was found that by increasing the concentration of one component [poly(isobutylene) or poly(naphtyl methacrylate)], the dimensions of the coil of the second component diminish. The dependence of the change of the inertia radius of macromolecules of one polymer in the presence of another is described by the equation

$$\left(\overline{R^2}\right)^{1/2} = KC^{-1/3} \tag{2}$$

where K is a constant and C is the concentration of the second polymer.

The change of the inertia radius of PS as a function of the PMMA concentration in benzene [45] and toluene [43] is described by the exponential dependence with various exponents for different molecular masses of polymers. The decrease of the macromolecular coil size for polymeric components in the ternary systems occurs because of the compression and repulsion of different types of molecules, which leads to diminishing the second virial coefficient and the coefficient of mutual diffusion [43,46,47].

Thus, the diminishing in the aggregate size in the solutions of the mixtures may be explained by the decrease in macromolecular dimensions in the ternary systems. The diminishing of the aggregate size is the result of the worsening of the mixed solvent, in accordance with diminishing the second virial coefficient and intrinsic viscosity [36].

In addition, it is worth noting that the aggregates existing in the thermodynamically incompatible system PS–PBMA–CCl$_4$ are the individual entities; that is, the interpenetrating of the coils of chains of different nature does not take place. This conclusion follows from the data on determining the thermodynamic polymer–polymer interaction parameter and molecular mobility in the mixtures under investigation [36,49]. Studying the ternary system PS–PBMA–CCl$_4$ using the nuclear magnetic resonance (NMR) method of high resolution has shown that in the broad composition interval, the values of δH of the signals corresponding to PS and PBMA are equal to their values in binary solutions (PS–CCl$_4$, PBMA–CCl$_4$). It is evident that the molecular mobility in the solutions of the polymer mixtures is the same as in the binary solution of each component. This fact proves the lack of the common aggregates in the ternary system. This result correlates with the positive values of the parameter of the thermodynamic interaction in the PS–PBMA–CCl$_4$ system. Therefore, in a solution of immiscible polymer pairs, there exist the individual aggregates of macromolecules of each polymer, and their coils and aggregates do not form the common structures.

Another situation was observed for solutions of the mixtures of miscible polymer pairs [50]. The analysis of the experimental data shows that the processes of aggregation in PCL–PC–CHCl$_3$ proceed in two stages. The first stage includes the concentration interval below the critical concentration for PCL (Table 1). When $C_{PCL} < C^*$, the dimensions of the aggregates of PC existing in solution do not differ from the aggregates of PC in the binary solution. At $C_{PCL} > C^*$ (second stage), the aggregate size sharply increases because of the possible formation of the common structures. In reality, using the method of two-dimensional NMR spectroscopy, it was shown [51] that for the thermodynamically compatible system PS–poly(vinyl methyl ether)–deutero-toluene, common structures are formed as a result of the specific interactions between phenyl protons of PS and protons of methoxy groups of poly(vinyl methyl ether) (PVME).

From the comparison of the data for miscible and immiscible polymer pairs in solution, one can conclude that the dimensions of aggregates in solutions are determined by the polymer–polymer thermodynamic interaction parameter. Depending on this parameter, the formation of common aggregates might be possible.

For immiscible polymer pairs, the aggregates of individual components are formed as follows from the data on molecular mobility and interaction parameters. The dimensions of aggregates in the mixture practically do not depend on the initial concentration of components (Table 3) and are much less compared with the aggregate dimensions in binary solutions. Diminishing the aggregate size is explained by the worsening of the thermodynamic quality of the mixed solvent and with collapse of macromolecules of one polymer in the presence of another. The lowest aggregate size is observed at the C^* of each polymer.

For miscible polymer pairs, the common aggregates may be formed, their size increasing with increasing concentration of the components. A sharp increase in the aggregate dimensions occurs at $C \geq C^*$ of each polymer.

The preceding analysis of the conditions of aggregate formation is necessary to interpret the data on adsorption from the ternary systems.

IV. ADSORPTION FROM SOLUTIONS OF POLYMER MIXTURES

A. General Principles

We propose a general approach to the description of the adsorption of polymer mixtures. The solution of two polymers, A and B, in a common solvent can be described as a solution of polymer A in a solution of polymer B and vice versa. In this latter case, the thermodynamic quality of such a mixed solvent will differ from the quality of pure solvent. Adsorption of each polymer will be dependent on the effect of the second polymer on the thermodynamic quality of the mixed solvent. Thus, adsorption of each component will be dependent not only on the affinity of a polymer to the adsorbent surface but on the component ratio, changing the thermodynamic quality of the mixed solvent.

The available experimental data on adsorption of polymer mixtures [52–64] show that, as a rule, pronounced selectivity of adsorption of one of the mixture components occurs by adsorption. In earlier investigations on the adsorption of mixtures, the experimental methods were used, for which the solution concentration was kept constant and only the component ratio was changed. Another experimental approach where the concentration of one polymer in solution is varied while the concentration of the second polymer is kept constant is possible. Finally, it is possible to increase the concentration of both components in a solution at a constant ratio of components. It is evident that neither method can be considered a strict one because each change in the component ratio in the system will lead to the change of the thermodynamic quality of a mixed solvent for each component (i.e., the conditions for adsorption are different for the mixtures of various compositions).

It is evident that different methods will give nonequivalent information about adsorption. At the same time, using them (there is no other way), one can get information about preferential adsorption of one of the mixture components and about the structure of the adsorption layer at the given conditions of experiment.

Generally, considering adsorption of the polymer mixtures in a broad interval of concentration from diluted to concentrated solutions needs to take into account the possible processes of the structure formation in solutions, connected with the coil overlapping and the appearance of the macromolecular aggregates. The values of the critical concentrations of overlapping play a very important role here. As follows from the data given in Tables 1 and 2, values of the critical concentrations C^* in pure solvent differ from values of C^* in mixed solutions. The conditions for the structure formation in solutions are also different for miscible and immiscible polymer pairs in solution, being determined by the thermodynamic interaction parameter χ_{23}.

B. Adsorption Selectivity

First investigations of adsorption from dilute solutions of mixtures had been done by Thies for the immiscible pair PS–PMMA, silica being used as an adsorbent and trichloroethylene as a solvent [52]. The preferential adsorption of PMMA was determined, according to the author's opinion, by the stronger interaction of polar PMMA with the surface has been observed; the presence of PS did not interfere with the adsorption of PMMA.

In Ref. 53, to establish the mutual influence of polymers, the following immiscible pairs have been used: poly(vinylacetate) (PVA)–poly(ethylvinylacetate) (PEVA), poly(dimethyl siloxane) (PDMS)–PDMS with small amount of side vinyl groups and ethyl cellulose–PS, using silica as the adsorbent. From the experimental data, the affinity sequence of a polymer to a surface has been established as follows: PVA > ethyl cellulose > PMMA > PEVA > PS [53].

The study of adsorption for the mixtures of PS and polybutadiene (PB) with corresponding copolymers (di- and tri-block-copolymers of styrene with butadiene) from solution in perchloroethylene has shown [54] that PS is adsorbed preferentially from the mixtures with copolymers, whereas from the mixtures of PB with copolymers, the latter are adsorbed preferentially.

Adsorption on fumed silica from the systems PS–poly(carbonate) (PC)–chloroform and PS–poly(butyl methacrylate) (PBMA)–benzene was studied [55,56] to establish the adsorption effect on the thermodynamic stability of the ternary systems. The thermodynamic stability was estimated from the data on the polymer–polymer interaction parameter calculated from the inverse gas chromatography experiments. It was found that depending on the composition of the ternary system, the selective adsorption of either one (PS) or the other (PC) component occurs from chloroform and only one component (PBMA) is adsorbed from the second system in the whole composition diapason. As a result of the selective adsorption, the region of the thermodynamic miscibility in the ternary system broadens.

The temperature dependence of adsorption is determined mainly by the energetic factors, namely by the parameter of segmental energy of adsorption χ_s. In some works [57–60], the dependence of the selective adsorption from the mixtures on χ_s has been established. The comparison of the values of χ_s obtained from the data on adsorption from binary solutions allows one to predict the selectivity of adsorption of one of the components from the polymer mixture. For example, for the system PVA–PMMA–chloroform by adsorption on the surface of fumed silica, preferential adsorption of PVA should be observed as it has a higher χ_s compared with PMMA. Direct experimental data confirm this supposition.

Using ultraviolet (UV) and infrared (IR) spectroscopy, the competitive and displacement adsorption on nonporous silica has been studied for the mixture of PS and poly(ethylene oxide) (PEO), both having narrow molecular mass distribution (MMD) [61]. For competitive adsorption, when both polymers are in contact with the surface, the preferential adsorption of PEO occurs, adsorption value being proportional to the molecular mass. By displacement adsorption, adding PEO to the adsorption system PS–CCl_4–silica leads to the displacement of prior adsorbed PS. The displacement degree depends on the fraction of PS links directly connected with silanol groups of the adsorbent and on the molecular mass ratio PS to PEO.

Adsorption of the ternary mixture of PS, PMMA, and copolymers of styrene and methyl methacrylate (styrene content 25.5 and 78 vol.%) has been studied in Ref. 62. The polymers with a relatively narrow MMD were used. It was found that by adsorption from the mixture of PS and PMMA, the latter is adsorbed preferentially. By adsorption from the mixture of two copolymers, the preferential adsorption occurs for the copolymer with higher styrene content. For displacement adsorption, it was established that PMMA fully displaces the preliminarily adsorbed PS or the copolymer with a large amount of styrene.

The adsorption kinetics from solutions in CCl_4 has been investigated [63,64] for the mixtures of deuterated PS and *cis*-polyisoprene. It was found that at the beginning of adsorption by the contact of an adsorbent with ternary solution, PS was adsorbed at first and, only after some time, is displaced by *cis*-polyisoprene. The displacement depends on the molecular mass of PS—the higher the PS molecular mass, the lower the adsorption of *cis*-polyisoprene.

The adsorption selectivity from polymer mixtures has been thoroughly studied in the series of our works [65–70]. For the system PBMA–(butaduene-nitrile rubber)–chloroform, it was found [65] that rubber, introduced in the PBMA solution, is not adsorbed in the whole concentration interval, although its presence changes the amount of PBMA adsorbed. The preferential adsorption of epoxy resin (ED-20) from the mixture solutions has been observed for two ternary systems: ED-20–carboxylate rubber–chloroform [66] and ED-20–(polybutadiene rubber)–chloroform [67].

Both from theoretical and practical points of view, establishing the adsorption from the melt of the mixture of two polymers without solvent is of great importance [68]. For the filled systems formed from the melt, adsorption determines the structure of the adsorption layer and the thermodynamic compatibility of the components in the adsorption layer, which is different from compatibility in the bulk [69]. The estimation of the selectivity of adsorption interaction of the blend components in melt with the solid surface has been made from the measurements of molecular mobility using the NMR method [70,71]. For the melt of the thermodynamically incompatible mixture PE–PBMA, the selective adsorption of PBMA was observed.

The most minute investigation of adsorption from solutions of polymer mixtures has been done for the series of thermodynamically incompatible (immiscible) polymer pairs PS–PBMA, PS–PDMS, and miscible PC–PCL [25,72,73].

Adsorption from the solutions of mixtures PS–PBMA [72] was studied for the constant total concentration of polymers in solution at various component ratios, whereas adsorption from PS–PDMS and PC–PCL [25] solutions was studied for the constant concentration of one of the components and alternate concentration of another. The dependence of PDMS adsorption from the mixture with PS [25] in the common solvent has been measured at various PS concentrations below and above the critical concentration of overlapping C^* for PS (Fig. 1). As is seen in the crossover region and above C^*, the passage through a maximum and then diminishing of adsorption is observed both for pure PDMS and for mixtures with various concentrations of PS. The absolute magnitude of adsorption is rather high, which meets the criterion of aggregative adsorption. The diminishing of adsorption at $C > C^*$ also meets this criterion. Increasing the amount of PS in solutions makes the thermodynamic quality worse (see Table 2) and the magnitude of adsorption decreases by preserving the maximum in the crossover region. This regularity is connected with worsening of the mixed solvent quality for PDMS, which leads both to diminishing the coil dimensions and to diminishing the aggregates size (Table 3). This situation is typical of polymer pairs immiscible in solution. In the system under consideration, PDMS has a higher affinity to the adsorbent surface, which determines its high adsorption selectivity. The immiscibility promotes adsorption.

Figure 2 presents isotherms of adsorption of PCL from binary mixtures and adsorption from the mixture at various constant concentrations of PC above and below C^* [23]. It is seen that the isotherm of adsorption of PCL from the binary solution (curve 1) has a maximum in the crossover region (Table 1). Introduction of PC ($C_{PC} < C^*_{PC}$) diminishes the adsorption of PCL (curve 2), although the maximum on the isotherm is preserved in the region of C^* for PCL.

Increasing the amount of PC in the mixture up to $C_{PC} > C^*_{PC}$ diminishes adsorption and simultaneously leads both to the shifts of the maximum on the PCL isotherm to lower concentrations (curves 3 and 4) and to its degeneration. In this system, PCL is adsorbed preferentially. A similar maximum shift is observed if the data are expressed as the ratios A_{PCL}/A_{PC} and A_{PC}/A_{PCL} (Figs. 3a and 3b). These curves depend on the component ratio in solution and show the preferential adsorption of PCL.

The experimental data for the PS–PBMA adsorption show that in the crossover region, the adsorption of PBMA is characterized by the diffuse maximum which shifts to lower concentrations in the presence of PS.

Figure 4a presents the data for this system as a ratio of component adsorp-

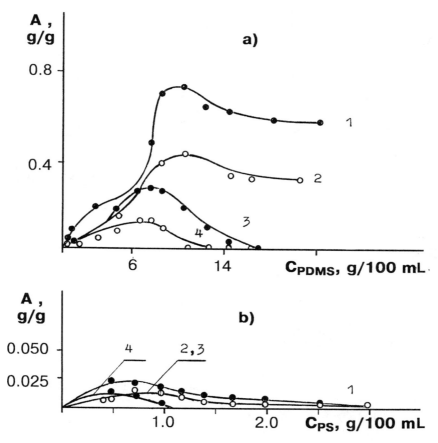

Figure 1 (a) Dependence of PDMS adsorption from a binary solution on C_{PDMS} (1) and from mixtures with PS in ethyl acetate on C_{PDMS} at PS concentrations of 0.5 (2), 1.0 (3), and 2.5 g/100 mL (4). (b) Dependence of PS adsorption from a binary solution on C_{PS} (1) and from mixtures with PDMS in ethyl acetate on C_{PS} at PDMS concentrations of 0.5 (2), 1.0 (3), and 9.0 g/100 mL (4).

tion on the concentration of PBMA in the mixed solvent at preservation of the total concentration of solution 3.0 g/100 mL. It is seen that at any ratio of components, the adsorption of PBMA passes through a minimum and is preferential. A similar minimum (Fig. 4b) is observed for the dependence of the second virial coefficient A_2 for the ternary system PS–PBMA–CCl$_4$. The concentration at the minimum point for a given pair coincides with C^* for each component and C^* for the mixture (Table 2). After passing through C_{mix}^*, the ratio of the adsorption values in-

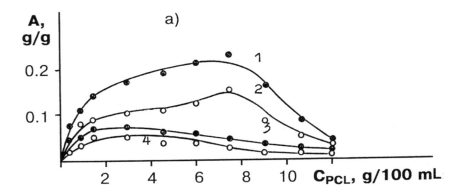

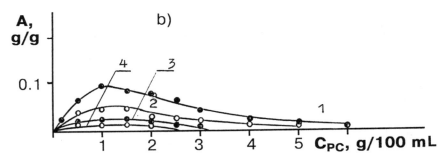

Figure 2 (a) Dependence of PCL adsorption from binary solution on C_{PCL} (1) and from mixtures with PC in methylene chloride on C_{PCL} at PC concentrations of 1.0 (2), 2.3 (3), and 4.0 g/100 mL (4). (b) Dependence of PC adsorption from its individual solution on C_{PC} (1) in methylene chloride on C_{PC} at PCL concentrations of 3.0 (2), 7.4 (3), and 10.0 g/100 mL (4).

creases again. These results indicate the important role of the changing thermodynamic quality of the mixed solvent in adsorption from the mixtures.

In the systems under consideration, the following polymers are less adsorbed: PC for PC–PCL, PS for PDMS–PS, and PS for PBMA–PS mixtures. Adsorption of these polymers is very low and approximately one order less than the adsorption of preferentially adsorbed polymer. The isotherms for less adsorbed polymers at the beginning of adsorption attain the saturation region (Figs. 1b and 2b) and there are no maxima.

From the data discussed follows that both the adsorption magnitude and the isotherm shape of the preferably adsorbed component are the functions of the content of another polymer. The common feature of all isotherms is the appearance of a maximum in the region of C^*. Independent of the miscibility of two polymers in

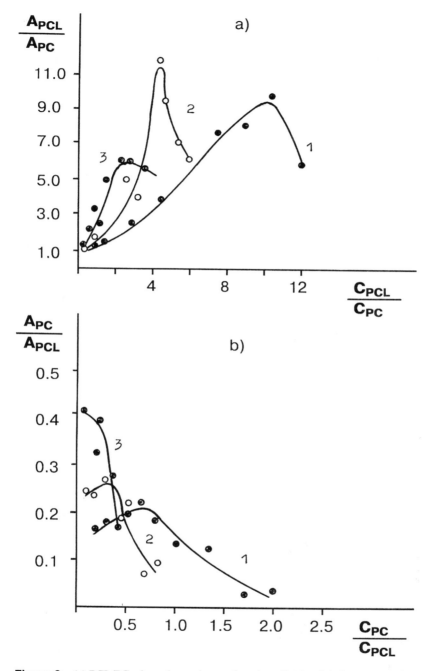

Figure 3 (a) PCL/PC adsorption ratio as a function of ratio of their concentration in solution. PC concentration: 1.0 (1), 2.3 (2), and 4.0 g/100 mL (3). (b) PC/PCL adsorption ratio as a function of ratio of their concentration in solution. PCL concentration: 3.0 (1), 7.4 (2), and 10.0 g/100 mL (3).

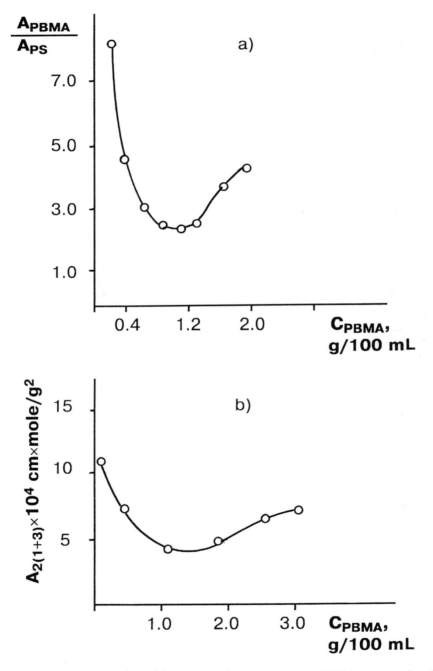

Figure 4 (a) Dependence of the adsorption ratio A_{PBMA}/A_{PS} on PBMA concentration. (b) The second virial coefficient $A_{2(1+3)}$ for a ternary system as a function of the mixed-solvent (PBMA + CCl_4) concentration.

solution (compatible and incompatible polymers), the introduction into the binary polymer solution of the second polymer leads to diminishing of adsorption of a polymer that is adsorbed preferentially. By increasing the concentration of the second polymer component above C^*, the adsorption of preferably adsorbed polymer diminishes much more and the maximum of adsorption is shifted to lower concentrations.

For thermodynamically incompatible systems (PS–PDMS, PBMA–PS), a poor quality of the mixed solvent diminishes the coil dimensions and aggregate size. As a result, the adsorption diminishes. For a compatible system (PCL–PC), adsorption is determined mainly by the formation of the common aggregates, their size and number being dependent on the concentration. Thus, the quality of the mixed solvent determines the conditions of aggregate formation and thus determines adsorption from the mixtures.

The experimental data allow the following general conclusion to reached. There exists an interrelation among polymer adsorption from the mixture, thermodynamic compatibility of components in the solution, change of thermodynamic quality of the mixed solvent, and critical concentration of the crossover for each component and averaged critical concentration for the mixture. It is also evident that the aggregation of macromolecules in the mixture solution, due to the above-mentioned factors, affects the adsorption in the region of concentration above and below C_{av}^*. As a rule, when passing through C^*, the adsorption value drops. This, among the other reasons, may be attributed to the fact that the size and number of macromolecular aggregates also depends on C^* and are not the same in the regions below and above C^*. As far as formation of aggregates is the result of the change in the thermodynamic quality of a solvent, it seems evident that the presence of the second component decreases the thermodynamic quality of the mixed solvent (independent of whether the given pair is compatible in solution or not) and changes the conditions of aggregates formation and their adsorption. High values of adsorption of polymers from pure solvents and from the mixture in common solvent can be explained only from the position of the aggregative mechanism of adsorption. Proceeding from this, the differences in adsorption in the regions below and above the crossover should be connected with the change of aggregation conditions of macromolecules of a different nature in both concentration regimes and under conditions of continuously changing thermodynamic quality.

V. ON THE STRUCTURE OF ADSORPTION LAYERS BY ADSORPTION FROM THE POLYMER MIXTURES

The structure of adsorbing entities by adsorption from the solutions of polymer mixtures determines the features of the adsorption layers. One of the main parameters that characterizes the structure of the adsorption layer is the fraction of bound segments (p) [6,74,75]. The main methods of the evaluation of p are the infrared and NMR spectroscopy [76,77]. Values of p estimated by these methods allow one to make a conclusion about the conformations of adsorbed chains. For adsorption from dilute solutions, it has been established that at a small equilibrium concentration of a solution, the value of p is the magnitude of the order 0.6–0.8. It is explained by the fact that in polymer solutions at $C \ll C^*$ resulting from the chain flexibility and heat motion, there arises a large number of contacts of the polymer chains with the surface (i.e., its flattening). Increasing the solution concentration and the effect of the excluded volume predetermine the change of the conditions of the chain interaction with the surface and transition to the adsorption of the chains in the form of the trains, loops, and tails. Increasing the solution concentration changes the conditions of the interaction and chain conformations, as well as the structure of the adsorption layer. Correspondingly, the value of p diminishes.

The molecular–aggregative mechanism of adsorption in semidiluted and concentrated solutions, as distinct from dilute ones, sharply changes the fraction of segments directly bound with the surface [39]. The fraction of chains in aggregate that directly interact with the surface is much lower. The transition of aggregates onto the adsorbent surface simultaneously increases the amount of the polymer adsorbed and diminishes the fraction of segments interacting with the surface. The continuous change of the conditions for aggregation when the concentration of the solution increases affects the structure of adsorption layers by adsorption from solutions of different concentrations ($C < C^*$ and $C > C^*$).

The interrelation between the adsorption value and p is seen from the comparison of the data for these values. Figure 5 presents the dependencies of p on the solution concentration for adsorption of PC and oligo-ethylene glycol adipinate from solution in dichloroethane. A comparison with adsorption isotherms shows that the maximum on the isotherms corresponds to the minimum fraction of the bound segments [38]. This fact meets the aggregative mechanism of adsorption. Similar dependencies were observed for other binary and ternary systems [10,65,72,78]. For adsorption from the mixtures of PS–PBMA and PBMA–rubber, it was found that by increasing the solution concentration, the value of p changes in a nonmonotonous way. At small concentrations (up to 0.2 g/100 mL), p reaches 0.5. At a higher concentration, the fraction of bound segments diminishes, whereas adsorption increases. Figure 6 presents isotherms of adsorption and fractions of bound segments for PBMA adsorbed from the binary solution and from the solution of its mixture with butadiene–nitrile rubber. As is seen, values of p for adsorption from binary solution and from a mixture solution are different. A comparison of adsorption isotherms shows that when p increases, adsorption decreases and vice versa, both for binary and ternary solutions.

As distinct from the IR method, NMR allows one to determine both the fraction of bound and immobilized segments, P, and molecular mobility directly in the adsorption system (i.e. without separating the adsorbent) [38,78].

The value of P calculated from NMR data differs from the value found from

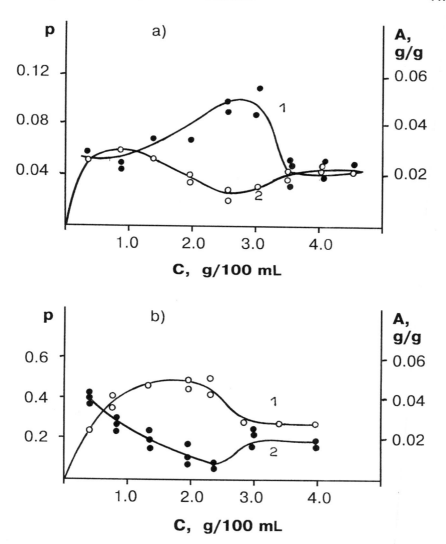

Figure 5 Isotherm of adsorption (1) and fraction of bound segments (2) for the system polycarbonate–fumed silica–dichloroethane (a) and oligoethylene glycoladipinate–fumed silica–dichloroethane (b).

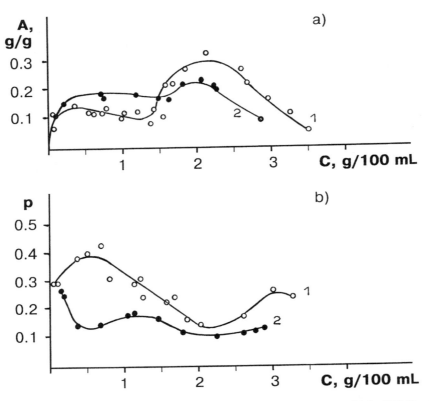

Figure 6 Isotherm of adsorption (a) and fraction of bound segments (b) for PBMA adsorbed from chloroform (1) and from a mixture with rubber (2).

IR spectra. If the latter gives only the segments directly bound with the active center at the surface, NMR gives the fraction of both bound and immobilized segments due to the influence of the surface. For the ternary systems PDMS–PS–ethyl acetate and PC–PCL–CH_2Cl_2, the dependence of P on concentration are given in Figs. 7 and 8. Adsorption isotherms are given in Figs. 1 and 2. It is seen that adsorption is in reverse correlation with P. Maxima of the dependencies of P on concentration coincide for binary solution of PDMS and for mixtures with 0.5 g/100 mL PS. By increasing the amount of PS in the mixture, the extremum on the isotherm of PDMS is shifted to lower concentrations because of structural changes in solution.

Dependencies of the fraction of bound segments for the PCL–PC–CH_2Cl_2 system on solution concentration are given in Figs. 8. With increasing concentration of both components, the value of P decreases linearly, showing a continuous

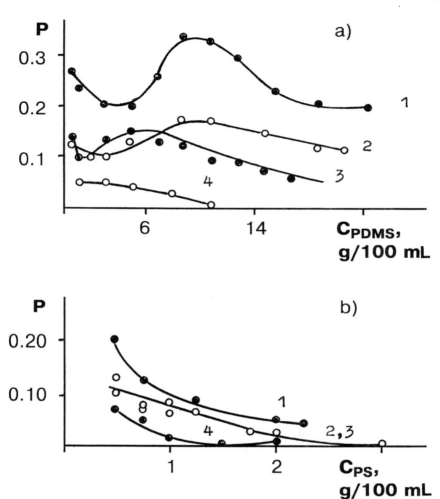

Figure 7 (a) Dependence of P_{PDMS} from its individual solution on C_{PDMS} (1) and from its mixtures with PS in ethyl acetate on C_{PDMS} at PS concentrations of 0.5 (2), 1.0 (3), and 2.5 g/100 mL (4). (b) Dependence of P_{PS} from its individual solution on C_{PS} (1) and from its mixtures with PDMS in ethyl acetate on C_{PS} at PDMS concentrations of 0.5 (2), 1.0 (3), and 9.0 g/100 mL (4).

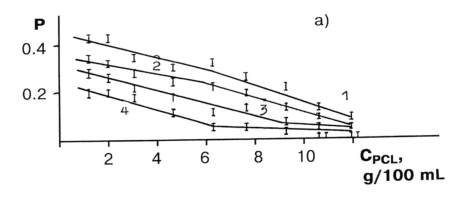

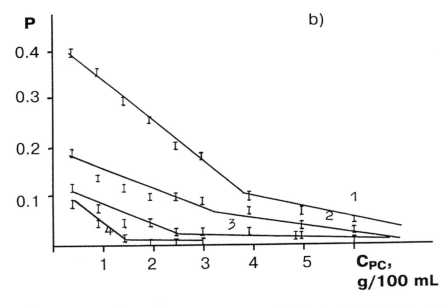

Figure 8 (a) Dependence of P_{PCL} on concentration of PCL for its individual solution (1) and its mixtures with PC in methylene chloride. PC concentrations: 1.0 (2), 2.3 (3), and 4.0 g/100 mL (4). (b) Dependence of P_{PC} on concentration of PC for its individual solution (1) and its mixtures with PCL in methylene chloride. PCL concentrations: 3.0 (2), 7.4 (3), and 10.0 g/100 mL (4).

change in the structure of the adsorption layer. A decreasing P is observed up to crossover point and after it, although the rate of changes (the slope dP/dC) after the crossover point decreases. This fact agrees with the diminishing adsorption at $C > C^*$ (Fig. 2a). Practically, the crossover point corresponds to the limiting value of adsorption [73].

Knowing the adsorption value and fraction of bound segments, one can calculate the number of segments in direct contact with the surface and the thickness of adsorption layer. Such calculations were made for two systems: PDMS–PS–ethyl acetate and PCL–PC–CH$_2$Cl$_2$ [25,73]. Some typical data are presented in Tables 4 and 5.

The thickness of adsorption layer has been calculated according to equation

$$\delta = \frac{A}{\rho S_{ads}} \tag{3}$$

where δ is the thickness of the adsorption layer (cm), ρ is the polymer density in the adsorption layer (g/cm^3), S_{ads} is the specific surface of an adsorbent (cm^2/g), and A is adsorption value (g/g).

For the ternary polymer–polymer–solvent system, the thickness of the adsorption layer was calculated by

$$\delta_{mix} = \frac{A_1/\rho_1 + A_2/\rho_2}{S_{ads}} \tag{4}$$

where δ_{mix} is the thickness of adsorption layer for polymer mixture, and ρ_1 and ρ_2 are the polymer densities in adsorption layer.

The analysis of the experimental data shows that the amount of segments directly interacting with the surface, AP, and the thickness of an adsorption layer have a maximum situated in the region of critical concentrations both for binary and ternary systems. The amount of bound segments for binary system is higher compared with a ternary system. The value of AP for one component diminishes with the increasing concentration of the second one.

Adsorption layers are rather thick, which is typical of aggregative adsorption.

Adsorption bonding leads to the diminishing molecular mobility of adsorbed chains. A change in the adsorption mechanism (i.e. the transition from adsorption of isolated coils to adsorption of aggregates) should change the molecular mobility. Studying the molecular mobility using the NMR method for the system PS–PBMA–CCl$_4$ at various amounts of an adsorbent verified the mutual influence of the mixture components on their ability to adsorption and on molecular mobility [79].

Investigation of adsorption and molecular mobility in the two systems (PC–PCL–CHCl$_3$ and PS–PBMA–CCl$_4$ has shown also the restricted mobility of segments of a polymer that either is not adsorbed at all (PS) or is adsorbed only in a very narrow concentration region [50]. For the PBMA/PS ratio 1:4, the adsorp-

Table 4 Variation of the Thickness of the PS Adsorption Layer on Aerosil and of the Amount of Polymer Directly Contacting the Surface, with Solution Concentration

PS concentration (g/100 mL)	Individual PS solution			Mixture with PDMS (0.5 g/100 mL)			Mixture with PDMS (1.0 g/100 mL)			Mixture with PDMS (9.0 g/100 mL)		
	C_{mix}^* (g/100 mL)	δ (Å)	$A \times P \times 10^3$ (g/g)	C_{mix}^* (g/100 mL)	δ (Å)	$A \times P \times 10^3$ (g/g)	C_{mix}^* (g/100 mL)	δ (Å)	$A \times P \times 10^3$ (g/g)	C_{mix}^* (g/100 mL)	δ (Å)	$A \times P \times 10^3$ (g/g)
0.25	2.2	230	—	4.0	1000	2.00	5.00	1000	1.80	8.0	3600	1.0
0.50		470	5.4		1000	1.40		1000	1.40		3600	0.9
0.75		440	3.6	3.25	1100	1.20	4.05	1070	1.20	7.4	3600	0.5
1.00		390	2.5		1100	1.00		1090	1.00		3000	0.2
1.25		230	1.2	2.8	1200	0.75	3.40	1000	0.75	6.8	—	—
1.50		200	0.9		1000	0.60		1090	0.60		—	—
1.75		190	0.7	2.65	900	0.35	3.05	900	0.35	6.2	—	—
2.00		170	0.5		900	0.25		900	0.25		—	—
2.50		170	—	2.5	900	0.10	2.80	900	0.10	5.5	—	—

Table 5 Variation of the Thickness of the PDMS Adsorption Layer on Aerosil and of Amount of Polymer Directly Contacting the Surface, with Solution Concentration

PDMS concentration (g/100 mL)	Individual solution			Mixture with PS (0.5 g/100 mL)			Mixture with PS (1.0 g/100 mL)			Mixture with PS (2.5 g/100 mL)		
	C_{mix}^{*} (g/100 mL)	δ (Å)	$A \times P \times$ 10^3 (g/g)	C_{mix}^{*} (g/100 mL)	δ (Å)	$A \times P \times$ 10^3 (g/g)	C_{mix}^{*} (g/100 mL)	δ (Å)	$A \times P \times$ 10^3 (g/g)	C_{mix}^{*} (g/100 mL)	δ (Å)	$A \times P \times$ 10^3 (g/g)
2.0	9.0	2,600	34.0	5.7	1,300	5.0	4.5	1,200	5.0	3.4	870	2.0
4.0		3,600	42.0		2,700	14.0		2,600	17.0		1,700	3.0
6.0		4,600	61.0	7.5	4,300	27.0	6.4	4,700	38.0	4.8	2,500	4.2
8.0		8,500	150.0		6,500	53.0		5,500	42.0		2,500	3.5
10.0		13,000	240.0	7.9	8,000	75.0	7.2	4,500	30.0	5.7	870	0.5
12.0		12,000	230.0		8,200	74.0		3,100	16.0		320	—
14.0		11,500	170.0	8.3	7,100	62.0	7.6	1,400	5.4	6.2	—	—
16.0		11,000	150.0		6,600	50.0		1,000	3.0		—	—
18.0		10,000	120.0	8.4	6,300	44.0	7.8	—	—	6.7	—	—

tion of PS from the mixture does not proceed. However, in spite of the lack of PS adsorption, the NMR data show the diminished molecular mobility that is observed in the whole concentration diapason. This effect may be explained by analogy with the data from the mixture of PMMA–PS [80]. For this system, it was shown that a weakly adsorbing polymer may be withheld from the surface or kept nearby for a long time because of the topological interaction with the preferentially adsorbed polymer. A similar result has been obtained [50] by studying adsorption from the mixture of PC with PCL (Fig. 9). In spite of adsorption of PC observed only in a narrow concentration interval of 6.0–9.5 g/100 mL, the restriction of mobility is observed throughout the whole concentration interval 1.0–9.5 g/100 mL. The effects of the restriction of molecular mobility of nonadsorbing polymer were observed both for miscible and immiscible polymer pairs and were explained by the conditions of the aggregate formation depending on the concentration region (above or below C^*) and by the thermodynamic interaction between two polymer components.

Thus, the experimental data on the structure of adsorption layers in the case of adsorption from the polymer mixture show that the latter is determined by adsorption of macromolecular aggregates. These may be aggregates of different macromolecules in the case of immiscible polymers or common aggregates for miscible systems. The transition of aggregates determines the formation of rather thick adsorption layers. The adsorption and topological interactions and restrictions of molecular mobility of polymeric chains lead to such conditions of the layer formation that continuously change with solution concentration.

VI. DYNAMICS OF ESTABLISHING THE ADSORPTION EQUILIBRIUM

As it was stated earlier, the adsorption of polymer mixtures is closely connected with the state of macromolecules in solution, which depends on concentration, temperature, and the quality of the solvent. It is evident that changes in the solution's structure should determine not only the adsorption value and structure of the adsorption layer but also the dynamics of establishing the adsorption equilibrium. Meanwhile, in the literature, there are practically no data on the kinetics of adsorption from solutions of polymer mixtures.

Generally, theoretical analysis of the adsorption from binary mixtures shows [81] that polymer adsorption includes four successive stages:

1. Polymer diffusion to the surface through a stagnant surface layer
2. Initial adsorption of the chain in the flat conformation that results in the formation of a glassy adsorption layer (if the glass transition of the adsorbed polymer is equal or above room temperature)
3. Building of additional macromolecules into the adsorption layer

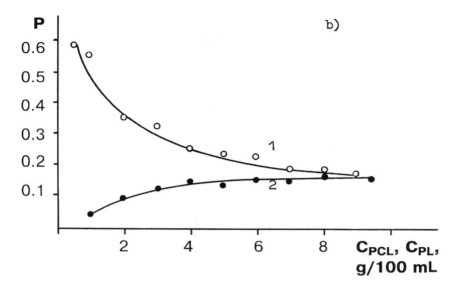

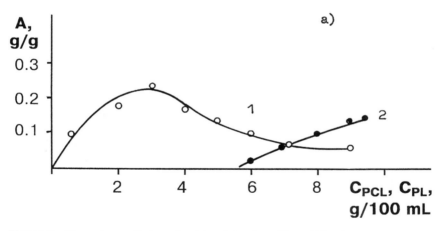

Figure 9 Dependence of adsorption (a) and fraction of immobilized segments (b) on the total polymer solution concentration: PCL in the mixture with PC (1); PC in the mixture with PCL (2).

4. The spreading of the macromolecules into the adsorption layer structure and reaching the equilibrium conformation, if possible.

It was shown that the first two stages proceed relatively quickly and cannot be detected in time by the up-to-date experimental methods. The two following stages take much more time and can be monitored.

Two slow stages of the establishing the adsorption equilibrium have been studied [82] for solutions of polymer mixtures depending on the solution concentration and component ratio. The PS–PBMA–CCl$_4$ system has been chosen that are listed in Table 6. Simultaneously, the kinetics of adsorption was investigated for pure components. Together with adsorbed amount, the fraction of bound segments, P, estimated by NMR method was determined. The influence of various PS concentrations on the kinetics of preferable PBMA adsorption and, at the same time, the dependence of PS adsorption equilibrium establishment time on PBMA concentration were evaluated. Figure 10 presents the kinetic dependencies of adsorption and fraction of the bound segments in system I when $C_{PS} = 0.5$ g/100 mL and the PBMA concentration varies (curves 1, 2, and 3 indicate different PBMA concentrations and curves 4, 5, and 6 indicate different PS concentrations). From Figs. 10a and 10b follows that the establishment of the equilibrium adsorption values and the formation of equilibrium adsorption layer in the system are determined by PBMA concentration. In particular, equilibrium is established most slowly (within 4 h) in mixtures with maximum PBMA concentration (2.5 g/100 mL), whereas equilibrium for PS is established very quickly in all cases (within 15–20 min). Slow attainment of the adsorption equilibrium in the case of $C_{PBMA} = 2.5$ g/100 mL is explained by the fact that the aggregation processes occur in the mixtures. It is supposed that the formation of a large number of PBMA aggregates [25] at $C > C^*$ causes the formation of an entanglement network which prevents the transition of PBMA aggregates and macromolecules onto the adsorbent surface, retarding the time of establishing adsorption equilibrium. This explanation is confirmed by the fact that at the given concentration, PBMA has the lowest P value (curve 3, Fig. 10b) of any of the concentrations (i.e., that a small number of segments are in direct contact with the solid surface).

When the concentration of both polymers are less than the critical value, the

Table 6 Composition of Polymer–Polymer–Solvent Ternary Systems

System PS–PBMA–CCl$_4$	PS concentration (g/100 mL)	PBMA concentration (g/100 mL)
I	0.5 ($C_{PS} < C_{mix}^*$)	0.50; 1.05; 2.50
II	0.9 ($C_{PS} = C_{mix}^*$)	0.50; 1.05; 2.50
III	1.8 ($C_{PS} > C_{mix}^*$)	0.50; 1.05; 2.50

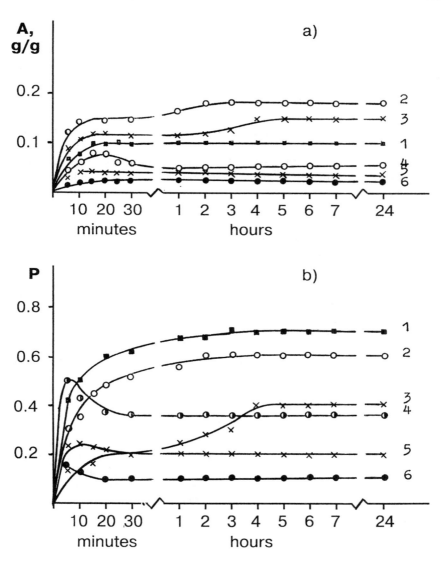

Figure 10 (a) Kinetic dependencies of adsorption for system I: (1) 0.5 g/100 mL PBMA; (2) 1.05 g/100 mL PBMA + 0.5 g/100 mL PS; (3) 2.5 g/100 mL PBMA; (4) 0.5 g/100 mL PS + 0.5 g/100 mL PBMA; (5) 0.5 g/100 mL PS + 1.05 g/100 mL PBMA; (6) 0.5 g/100 mL PS + 2.5 g/100 mL PBMA. (b) *P* dependence on time for system I: curves 1–3—PBMA; curves 4–6–PS.

adsorption equilibrium for PBMA is established quickly (within 15 min) (curve 1, Fig. 10a); a longer time is required (3 h) to form the equilibrium structure of the adsorption layer (curve 1, Fig. 10b). Kinetic dependencies of PS (curve 4, Fig. 10a; curve 4, Fig. 10b) in this case have the maximum in the initial moment of adsorption. A certain increase in adsorption value and in the fraction of bound segments for PS are apparently observed due to the absence of an entanglement network at concentrations below critical, when PS macromolecules are able to contact the adsorbent surface. Later, the displacement adsorption takes place: More polar and mobile PBMA macromolecules can displace some of PS macromolecules from the adsorbent surface. Therefore, the formation of the PBMA adsorption layer takes longer (3h, curve 1, Fig. 10b) than the establishment of an adsorption equilibrium in this system.

At concentration $C_{PBMA} = C^*$, the adsorption equilibrium is established within 1–2 h (curve 2, Fig. 10a), the adsorption values being markedly greater. This may be explained by the peculiarities of the structures formed at these concentrations. It is known that at concentrations equal to C^*, the macromolecular coils touch one another and overlap. The coils are compressed and their dimensions decrease, which is reflected in the P values (curve 2, Fig. 10b) that are between those of the highest (curve 3, Fig. 10b) and the lowest (curve 1, Fig. 10b) PBMA concentrations. If at 0.5 g/100 mL PBMA concentration, the adsorbed macromolecules are chiefly flat ($P = 0.75$) at the transition to the crossover region (curve 2, Fig. 10b), the adsorbed polymer reposes on the surface in the shape of loops and tails ($P = 0.6$). Further increasing the PBMA concentration (2.5 g/100 mL) causes the adsorption of aggregates along with the adsorption of separate macromolecules, resulting in a decrease in the segment number contacting the surface and a decrease in P to 0.4.

Interesting results (Figs. 11a and 11b) were obtained in studies of adsorption kinetics in system II when $C_{PS} = C_{mix}^*(0.9$ g/100 mL). For solutions of mixtures containing 1.05 and 2.5 g/100 mL of PBMA (curves 2 and 3, Fig. 11a), maxima of kinetic dependencies are observed at the initial moment of adsorption; that is, within the first minutes of adsorption, a great number of PBMA macromolecules and aggregates contact the adsorbent surface [25], defining the greater adsorption value which decreases when adsorption equilibrium is attained.

It is with $C \geq C^*$ concentration that aggregation processes are observed due to the contact and overlapping of macromolecular coils which cause the formation of an entanglement network. The introduction of the adsorbent changes the state of equilibrium aggregation in solution, destroying the existing structures of both PBMA and PS and thus enabling the macromolecules of both polymers to contact the adsorbent surface. This results in an increase of the adsorption value and P for polymers at the initial moment of adsorption (maxima on curves 2–6, Figs. 11a and 11b). The destroyed structures are probably restored in the process of establishing the adsorption equilibrium and the maxima on the curves disappear.

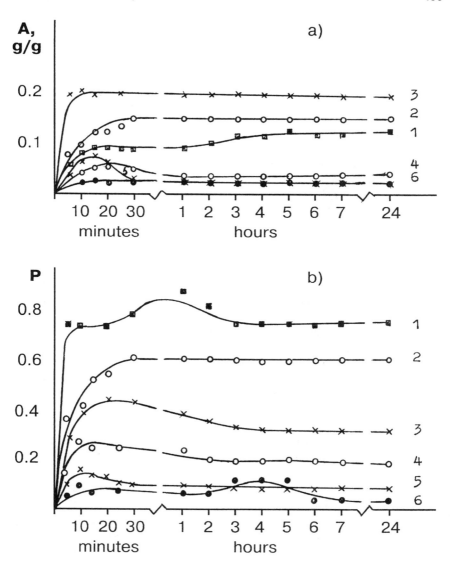

Figure 11 (a) Kinetic dependencies of adsorption for system II: (1) 0.5 g/100 mL PBMA; (2) 1.05 g/100 mL PBMA + 0.9 g/100 mL PS; (3) 2.5 g/100 mL PBMA; (4) 0.9 g/100 mL PS + 0.5 g/100 mL PBMA; (5) 0.9 g/100 mL PS + 1.05 g/100 mL PBMA; (6) 0.90 g/100 mL PS + 2.5 g/100 mL PBMA. (b) P dependence for system II: curves 1–3—PBMA; curves 4–6—PS.

Studies of adsorption kinetics from solutions of system III, where $C_{PS} >$ $C_{mix}*$(1.8 g/100 mL) reveal that the adsorption equilibrium is attained quickly (30 min) for PBMA at concentrations above $C_{mix}*$ When the PBMA concentration is $<C*$, the adsorption equilibrium is attained in 2–4 h. It is possible to suppose that at a concentration $>C*$, the exchange between adsorbed and nonadsorbed PBMA and PS macromolecules is hampered. At a PBMF concentration $<C*$, separate PBMA macromolecules exist that are able to dislodge the PS aggregates from the adsorbent surface due to their mobility. Therefore, the formation of the PBMA adsorption layer requires a long time (2–4 h).

The effects described depend on the nature of an adsorbent. It was shown [82] that using fumed silica modified with diethylene glycol as an adsorbent changes the regularities of kinetics of adsorption to some extent.

To analyze the kinetic curves, the diffusion equation has been applied [81]. This equation gives the relationship between the adsorption and the macromolecules diffusion rate in solution in the form

$$\Gamma_i = 2C\sqrt{Dt/\pi} \tag{5}$$

where Γ_i is the Gibbs surface excess, C is the solution concentration, D is the diffusion coefficient, and t is time.

Figures 12 and 13 show the kinetic curves in terms of the diffusion equation [83]. The results of calculations of the diffusion coefficients are presented in Table 7. As is seen, adsorption proceeds in two stages, with various diffusion coefficients for pure PBMA and PBMA in the mixture with PS. These data show that the introduction of PS into PBMA solutions affects the adsorption kinetics and structure of the adsorbed layer [83].

Thus, establishing the adsorption equilibrium in polymer mixture solutions occurs in two stages—a fast stage and a slow stage; the character of the kinetic dependencies is determined by the concentration of components and the nature of the adsorbent. The factor most influencing the adsorption kinetics is the structure of the solution, which depends on the solution mode (diluted or semidiluted) and on the thermodynamic state of the system. At a concentration above the critical concentration of overlapping, the establishing equilibrium is more prolonged, the attainment of the equilibrium value of adsorption and equilibrium structure of an adsorption layer taking different times. The existence of the entanglement network slows down the adsorption process.

The kinetic curves of adsorption and the time dependence of the fraction of bound segments suggest the existence of the following adsorption stages:

1. Diffusion of macromolecules and aggregates and their bonding with the adsorbent surface
2. Conformation changes in the adsorption layer with the increase in the amount of adsorbed aggregates and isolated macromolecules

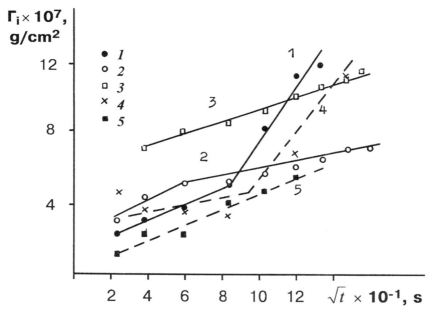

Figure 12 Adsorption of PBMA as a function of the square root from the adsorption time for solutions with the following concentrations: (1) 0.5g/100 mL; (2) 1.05 g/100 mL; (3) 2.5 g/100 mL; (4) solutions of the mixtures with concentrations $C_{PBMA} = 0.5$ g/100 mL and $C_{PS} = 0.5$ g/100 mL; (5) solutions of the mixtures with concentrations $C_{PBMA} = 2.5$ g/100 mL and $C_{PS} = 0.5$ g/100 mL.

3. The process of displacement adsorption (i.e., exchange between the bonded macromolecules and aggregates and those in the solution).

VII. GENERAL CONCLUSION

The experimental data considered in this chapter show that adsorption from solutions of polymer mixtures may be described in the framework of the molecular–aggregative mechanism of adsorption. The main factor governing the adsorption of the mixture is the thermodynamic quality of the solvent. A solution of polymers A and B in common solvent may be considered as the solution of component A in the solution of component B, and as the solution of component B in the solution of component A. Thus, the thermodynamic quality of the mixed solvent will be determined by the thermodynamics of the interaction between the mixture components in solution, and adsorption will be dependent on the miscibility or immiscibility of components in solution. Changing the concentration of

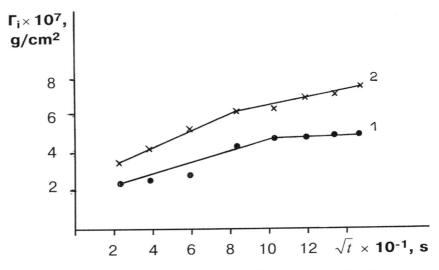

Figure 13 Adsorption of PBMA as a function of the square root from the adsorption time for solutions of mixtures: (1) C_{PBMA} = 2.5 g/100 mL and C_{PS} = 0.9 g/100 mL; (2) C_{PBMA} = 2.5 g/100 mL and C_{PS} = 2.5 g/100 mL.

Table 7 Diffusion Coefficients in Solutions of PBMA and Its Mixtures with PS

Solution of PBMA			Solution PBMA + PS		
PBMA concentration $C \times 10^2$ (g/mL)	Diffusion coefficient D (cm²/s)		PBMA/PS concentration $C \times 10^2$ (g/mL)	Diffusion coefficient D (cm²/s)	
	Stage I	Stage II		Stage I	Stage II
0.5	6.5×10^{-13}	7.5×10^{-13}	0.5/0.5	1.5×10^{-13}	6.0×10^{-12}
1.05	2.1×10^{-13}	4.2×10^{-14}	2.5/0.5	2.5×10^{-14}	2.5×10^{-14}
2.5	2.5×10^{-14}	2.5×10^{-14}	2.5/0.9	7.5×10^{-14}	2.5×10^{-14}
			2.5/2.5	1.5×10^{-14}	1.5×10^{-14}

the component in the solution leads to a continuous change of the thermodynamic quality of the mixed solvent. The latter factor determines the conditions of the formation of macromolecular aggregates of polymers A and B and the possibility of their adsorption. The adsorption of aggregates of polymer components also determines the structure of adsorption layers obtained as a result of adsorption.

REFERENCES

1. YS Lipatov. Colloid Chemistry of Polymers. Amsterdam: Elsevier, 1988.
2. YS Lipatov. Polymer Reinforcement. Toronto: Chem Tec Publ, 1994.
3. VV Klepko, YB Melnichenko, VV Shilov. Polym Sci A36:822–827, 1994.
4. DH Napper, Polymeric Stabilization of Colloidal Dispersions. London: Academic Press, 1983.
5. M Nagasava, ed. Molecular Conformation and Dynamics of Macromolecules in Condensed Systems. Amsterdam: Elsevier, 1986.
6. ThF Tadros, ed. Polymers in Colloid Systems. Amsterdam: Elsevier, 1986.
7. M Cohen Stuart, T Cosgrove, B Vincent. Adv Colloid Interf Sci, 24:143–239, 1986.
8. YS Lipatov, LM Sergeeva. Adsorption of Polymers. New York: Wiley, 1974.
9. YS Lipatov. Prog Colloid Interf Sci 61:12–23, 1976.
10. YS Lipatov, LM Sergeeva. Adv Colloid Interf Sci 6:1–93, 1976.
11. R Simha, J Zakin. J Chem Phys 33:1791–1793, 1960.
12. PG de Gennes. Scaling Concepts in Polymer Physics. Ithaca, NY: Cornell University Press, 1979.
13. AA Tager. Physical Chemistry of Polymers. Moscow: Khimiya, 1978.
14. A Nesterov, Yu Lipatov. Eur Polym J 15:775–780, 1979.
15. JD Wells. J Chem Soc Faraday Trans 80(pt 1):1233–1238, 1984.
16. L Subasini, N Padma, D Manoranjan. J Macromol Sci A19:321–330, 1983.
17. C Nok, A Rudin. Eur Polym J 18:363–366, 1982.
18. S Ioan, BC Simionescu, I Neamtu, CI Simionescu. Polym Commun 27:113–116, 1986.
19. KK Chee. J Macromol Sci B19:257–267, 1981.
20. A Dondos, E Piorri. Polymer 30:1690–1694, 1989.
21. A Shaul, MW Eigene. Angew Macromol Chem 81:75–86, 1979.
22. E Piorri, A Dondos. Eur Polym J 23:347–351, 1987.
23. A Dondos, E Piorri. Makromol Chem 189:1685–1692, 1988.
24. J Roots, B Nystrom. Polymer 20:148–156, 1979.
25. Y Lipatov, T Todosijchuk, V Chornaya. J Colloid Interf Sci 155:283–289, 1993.
26. YS Lipatov, TT Todosijchuk, VN Chornaya. Russian Chem Rev 64:463–469, 1995.
27. SM Lipatov. High-Molecular Compounds—Lyophylic Colloids. Publish Àcad Sci BSSR, Tashkent: 1943.
28. YS Lipatov, VV Shilov. Composite Polymeric Materials. Naukova Dumka, Kiev: 1981, Vol 11, pp 55–69.
29. Z Mzkvikova, E Prakopova. Colloid Polym Sci 265:978–983, 1987.
30. P Cotts. Colloid Polym Sci 265:35–41, 1987.
31. G Kalz. Plaste Kautschuk 35:9–12, 1988.
32. P Cotts. Am Chem Soc Polym Prepr 23:315–321, 1982.
33. H Cackovic, J Springer, FW Weigelt. Prog Colloid Polym Sci 69:134–138, 1984.
34. K Takakazu. J Polym Sci: Polym Phys Ed 18:1685–1695, 1980.
35. GN Timoteeva, VM Avey'yanova. Vysokomol Soed A24:2268–2274, 1982.
36. YS Lipatov, TT Todosijchuk, VN Chornaya, TS Khramova. J Colloid Interf Sci 110:1–8, 1986.
37. YS Lipatov, TT Todosijchuk, LÌ Sergeeva. Vysokomol Soed A14:121–123, 1972.

38. YS Lipatov, LÌ Sergeeva, TT Todosijchuk, TS Khramova. Kolloid Zh 37:280–284, 1975.
39. YS Lipatov. Interfacial Phenomena in Polymers. Naukova Dumka, Kiev: 1980.
40. W Heler, W Pangonic. J Chem Phys 26:498–506, 1957.
41. G Mie. Ann Phys Germany 25:377, 1908.
42. VI Klenin, SYu Shchegolev, VI Lavrushkin. Characteristic Functions of Light Scattering of Dispersed Systems. Saratov University, 1977.
43. YS Lipatov, AE Nesterov. Thermodynamics of Polymer Blends. Lancaster: Tech Publ Com, 1997.
44. C-J Lin, SL Rosen. J Polym Sci: Polym Phys Ed 20:1497–1502, 1982.
45. DB Cotts. J Polym Sci: Polym Phys Ed 21:1381–1388, 1983.
46. R Kuhn, HJ Cantow. Makromol Chem 122:65–81, 1969.
47. MS Kent, M Tirrell, TP Lodge. Polymer 32:314–319, 1991.
48. MS Kent, AF Faldi, M Tirrell, TP Lodge. Macromolecules, 25:4501–4505, 1992.
49. YS Lipatov, VN Chornaya, AE Nesterov, TT Todosijchuk. Polym Bull 12:49–53, 1984.
50. V Chornaya, T Todosijchuk, Y Lipatov. J Colloid Interf Sci 198:201–208, 1988.
51. PA Mirau, H Tanaka, FA Bovey, Macromolecules 21:2929–2933, 1988.
52. C Thies. J Phys Chem 70:3783–3790, 1966.
53. R Botham, C Thies. J Polym Sci C30:369–380, 1970.
54. R Botham, C Thies. J Colloid Interf Sci 45:512–520, 1973.
55. YS Lipatov, AE Nesterov, NP Gudima. Rep Acad Sci Ukraine B 11:44–48, 1985.
56. YS Lipatov, AE Nesterov, TD Ignatova, NP Gudima. Eur Polym J 22:83–87, 1986.
57. AE Nesterov, OT Gritsenko. Polym Sci USSR A 29:1252–1256, 1987.
58. OT Gritsenko, AE Nesterov. Vysokomol Soed B 29:611–615, 1987.
59. OT Gritsenko, AE Nesterov, TS Khramova. Vysokomol Soed A 32:2269–2272, 1990.
60. OT Gritsenko, AE Nesterov. Eur Polym J 27:455–459, 1991.
61. M Kawaguchi, A Sakai, A Takahashi. Macromolecules 19:2952–2955, 1986.
62. M Kawaguchi, K Itoh, A Takahashi. Macromolecules 22:2204–2207, 1989.
63. MH Schneider, S Granick. Macromolecules 27:4714–4720, 1994.
64. MH Schneider, S Granick. Macromolecules 27:4721–4725, 1994.
65. YS Lipatov, G Semenovich, L Sergeeva, L Dubrovina. J Colloid Interf Sci 86:432–435, 1982.
66. GM Semenovich, YS Lipatov, LM Sergeeva, TT Todosijchuk. Polymer Sci A 20:2375–2380, 1978.
67. YS Lipatov, VN Chornaya, TT Todosijchuk, TS Khramova. Ukr Chem J 54:90–93, 1988.
68. YS Lipatov, LM Sergeeva, TT Todosijchuk, AE Nesterov. Rep of Acad Sci USSR 259:11132–1134, 1981.
69. YS Lipatov. Sci Eng Composite Mater 1:35–42, 1995.
70. YS Lipatov, TS Khramova, TT Todosijchuk, EG Gudova. Polymer Sci A 30:443–447, 1988.
71. YS Lipatov, TS Khramova, TT Todosijchuk, EG Gudova. J Colloid Interf Sci 123:143–147, 1988.
72. YS Lipatov, LM Sergeeva, TT Todosijchuk, VN Chornaya. J Colloid Interf Sci 86:437–441, 1982.

73. YS Lipatov, TT Todosijchuk, VN Chornaya. Composite Interf 2:53–69, 1994.
74. M Kawaguchi. Adv Colloid Interf Sci 24:1–41, 1990.
75. A Takahashi, M Kawaguchi. Adv Polym Sci 46:3–65, 1982.
76. RJ Fontana, JR Thomas. J Phys Chem 65:480–490, 1961.
77. ID Robb, R Smith. Polymer 18:500–504, 1977.
78. YS Lipatov, TS Khramova, LM Sergeeva, TT Todosijchuk. Kolloid Zh 39:174–177, 1977.
79. YS Lipatov, TS Khramova, TT Todosijchuk, VN Chornaya. Polym Sci A 28:602–608, 1986.
80. HE Johnson, S Granik. Science 225:966–968, 1982.
81. A Couris, E Gulari. Macromolecules 27:3580–3588, 1994.
82. YS Lipatov, TT Todosijchuk, VN Chornaya. J Colloid Interf Sci 176:361–367, 1995.
83. GM Semenovich, YS Lipatov, TT Todosijchuk, VN Chornaya. J Colloid Interf Sci 184:131–138, 1996.

12

Polymer Adsorption at Oxide Surface

Kunio Esumi
Science University of Tokyo, Tokyo, Japan

I. INTRODUCTION

Studies of interactions between polymers and oxide surfaces have paid attention to the numerous technological and industrial applications as well as the importance of a fundamental standpoint. Several reviews have been reported through theoretical and experimental descriptions [1–10].

Polymers in solution accumulate at an interface, which is called "adsorption," and polymers adsorbed at the interface can take various conformations, such as tail, loop, or train. Such a conformation often affects the stability of colloidal dispersions [11,12]. Although there are several theories on the adsorption of neutral polymers, some work is still needed for the theoretical studies of polyelectrolytes and copolymers. Furthermore, polymer adsorption on oxide surfaces is often influenced by other factors, including surface properties of the oxides [13–15] and some additives such as surfactants.

In this chapter, adsorption characteristics of polymers such as homopolymers and copolymers on oxide surfaces from their aqueous solutions are described, as well as the stability of colloidal dispersion and conformation of polymers adsorbed. The effect of surfactant on the adsorption of polymers on oxide surfaces is also discussed.

II. CHARACTERIZATION OF OXIDE SURFACE

Before discussing the adsorption of a polymer on oxide surfaces from an aqueous solution, it is necessary to provide a description of the oxide–liquid interface.

When an oxide such as Al_2O_3, TiO_2, or Fe_2O_3 is immersed in an aqueous solution, the surface tends to coordinate water molecules and any further dissociation of water molecules leads to a fully hydroxylated surface. Then, the electrical double layer develops, because charged species will migrate across the oxide–water interface, and the oxide surface will acquire a charge with respect to the aqueous phase. In many cases, the electrical double layer plays an important role in the adsorption of polymers and surfactants. The excess charge fixed at the oxide surface is exactly balanced by a diffuse region of equal but opposite charge on the liquid side. The surface charge and the diffuse region constitute the electrical double layer. Figure 1 is a schematic of the structure of the electrical double layer at the solid–water interface. The formation of a charge on an oxide in an aqueous solution occurs by a number of mechanisms, including chemisorption, preferential dissociation of surface ions, selective adsorption of ions, and oxide lattice substitution.

The surface charge of an oxide is determined in part by the pH of the solution ion which it is immersed. For such systems, the H^+ and OH^- ions are the potential-determining ion:

$$\sigma_0 = F(\Gamma_{H^+} - \Gamma_{OH^-})$$

where Γ_{H^+} and Γ_{OH^-} are the adsorption densities of H^+ and OH^-, respectively, σ_0 is the surface charge, and F is the Faraday constant. Two important parameters describing the electrical double layer of an oxide are the point of zero charge (PZC) and the isoelectric point (IEP). The PZC can be defined as the concentration of the potential-determining ion with the surface charge of an oxide $\sigma_0 = 0$, and the IEP as the concentration of potential-determining ion at which the zero potential $\zeta = 0$ can be calculated from electrokinetic measurements.

The surface potential (ψ_0) is determined by the activity of potential-determining ions in solution:

$$\psi_0 = \frac{RT \ln(a^+)}{zF(a^+_{PZC})}$$

where z is the valence of the potential-determining cation in solution, a^+ is the activity of the potential-determining cation in solution, and a^+_{PZC} is its activity when the surface charge is uncharged.

The PZC of an oxide is an essential parameter for describing the electrical layer phenomena at the oxide–water interface. The importance of the PZC is because the surface charge has a major effect on the adsorption of all ions, including polyelectrolytes.

The acid–base behavior of the surface hydroxyl groups can be determined by the electron distribution in the surface O—H bond, which is dependent on the electronegativity of the surface metal atom. In case of a predominantly ionic bond

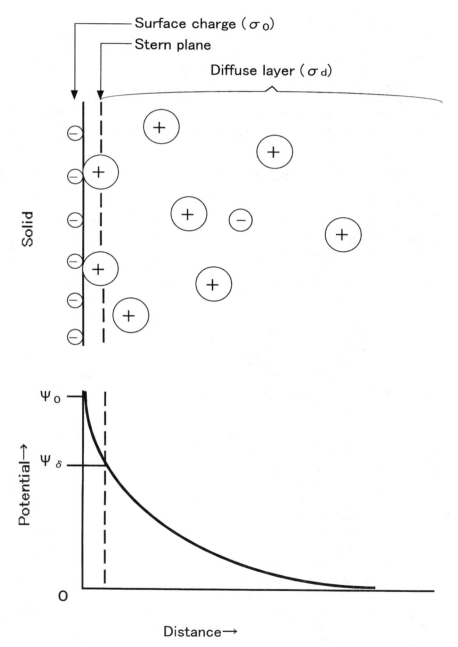

Figure 1 Schematic representation of the structure of the electrical double layer at the solid–water interface.

such as in MgO, the electron pair is close to the oxygen atom, which results in a stronger attraction for proton. However, for an oxide with an increased character of covalent bond such as SiO_2, the proton will not be strongly bound to the oxygen; therefore, this type of hydroxyl group is expected to be acidic. Al_2O_3 has a more covalent character of the Al−O bond than Mg−O, but, at the same time, it is more ionic compared to SiO_2. Thus, SiO_2 has stronger Brønsted acid sites than Al_2O_3. Such an acid–base character of an oxide surface may also play an important role in the adsorption of a polymer, such as nonionic polymer.

III. ADSORPTION OF A HOMOPOLYMER

A. Nonionic Polymer

Adsorption of nonionic homopolymers such as poly(vinyl pyrrolidone) (PVP), poly(ethylene oxide) (PEO), and polyacrylamide (PAM) has been investigated on oxide surfaces such as SiO_2, Al_2O_3, and Fe_2O_3.

Figure 2 shows the adsorption of PVP as a function of molecular weight on SiO_2 [16]. The amount of PVP adsorbed increases with increasing molecular weight. This trend is well correlated with all theories that the adsorbed amount increases with increasing chain length. In addition, changes in the conformation of the polymers adsorbed are an essential aspect of their adsorption behavior. Information on the conformation of adsorbed polymers has been obtained using many techniques, including electron-spin resonance (ESR), nuclear magnetic resonance (NMR), and infrared (IR) [17–19]. Generally, the conformation of adsorbed polymers can be classified into three segments: loops, tails, and trains. The fraction of bound segments (i.e., trains) is usually denoted as p. Figure 3 shows p values as a function of the adsorbed amount of PVP on SiO_2 [20]. In a discussion of ESR results, a suitable spin label must be attached to polymers, in such a way that the labeling does not affect the adsorption behavior. Some evidence has been obtained that the label itself hardly adsorbs on silica, suggesting that the label has no strong effects on the adsorption behavior of PVP [21]. On the other hand, the recent NMR results are reliable, because any solvent signal could be suppressed by an ingenious pulse sequence [22]. It is seen from Fig. 3 that the p value is almost unity at low coverage and decreases to about 0.5 at the highest coverage attained. This indicates that at low coverage, the polymer molecules adsorb in a nearly flat conformation. At higher coverage, the polymer chains adsorbed take a relatively dense loop layer and a few long tails protruding into the solution. Such a decrease in p is in agreement with theoretical results [4].

The adsorption behavior of poly(ethylene oxide) (PEO) on SiO_2 [23] is very similar to that of PVP on SiO_2. In Fig. 4, all adsorption isotherms are of a high-affinity type, where polymers completely adsorb at lower dosages and the adsorbed amount steeply reaches a plateau value. The plateau amount of PEO adsorbed increases with increasing KBr concentration. It is suggested that the concentrated

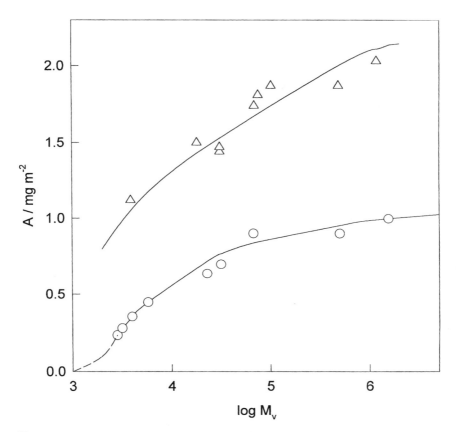

Figure 2 Adsorption isotherms of poly(vinyl pyrrolidone) on SiO_2 at $c = 800$ mg/dm^3 as a function of molecular weight: △—from dioxane; ○, from water. (Reproduced from Ref. 16 with permission of Academic Press, Inc.)

aqueous KBr solution is a poorer solvent for PEO than the dilute aqueous KBr solution. It is very interesting to note [24] that PEO can be appreciably adsorbed on SiO_2, V_2O_5, and MoO_3, but no significant adsorption is observed on oxides such as TiO_2, Fe_2O_3, Al_2O_3, and MgO. The flocculation behavior of PEO-coated particles has been investigated to elucidate the polymer adsorption mechanisms [24–26]. This may suggest that the adsorption of PEO on oxide surfaces is sensitive to the nature of the surface hydroxyl groups and the magnitude of the acidity of the surface hydroxyl groups, as indicated by the point of zero charge of the oxide, is comparable to or more than that of SiO_2 to enhance PEO adsorption.

The adsorption of nonionic poly(acrylamide) (PAM) on a series of oxides has been studied [27]. The adsorption densities of PAM on six different oxides at neu-

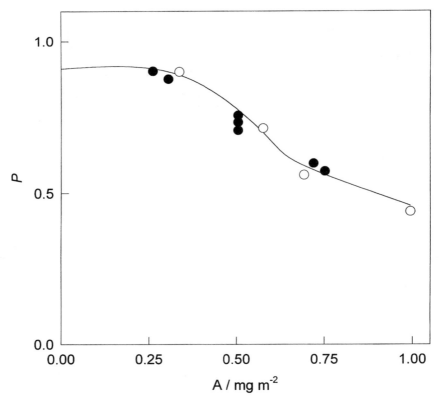

Figure 3 Bound fractions measured with magnetic resonance as function of adsorbed amount of poly(vinyl pyrrolidone) on SiO_2: ●, EPR results; ○, NMR results. (Reproduced from Ref. 20 with permission of Academic Press, Inc.)

tral pH (from 6 to 8) are plotted in Fig. 5. All the isotherms show a steep rise in the low-concentration region followed gradually by a shallower rise with an increase in PAM concentration. In addition, a comparison of the values of PZC with the adsorption densities shows that as the PZC increases, adsorption of PAM on the oxide increases. It has been proposed that the major driving force controlling polymer adsorption is hydrogen-bonding between the electronegative C=O groups on PAM and the proton-donating oxide surface hydroxyls, the positive MOH_2^+, and neutral MOH. Another driving force may be solvation from dissolved species [28].

The adsorption of β-casein at hydrophobic and hydrophilic silica surfaces has been studied by time-resolved ellipsometry [29]. At the hydrophobic surface, the adsorption is fast and the surface is saturated within a relatively short period. On the other hand, the adsorption at the hydrophilic silica surface is much slower.

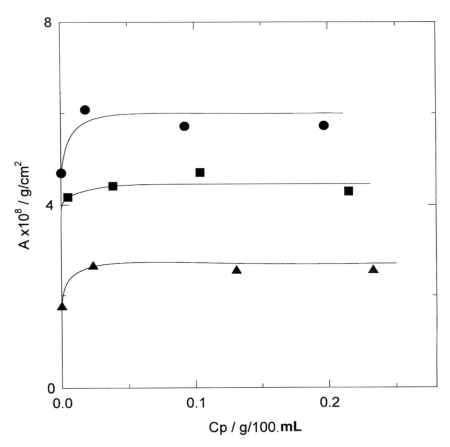

Figure 4   Adsorption isotherms of poly(ethylene oxide) on SiO$_2$: ▲, in pure water; ■, in aqueous 0.01 M KBr solution; ●, in aqueous 0.1 M KBr solution. (Reproduced from Ref. 23 with permission of Academic Press, Inc.)

In addition, although the maximum surface excess at the hydrophilic surface is higher than that measured at the hydrophobic surface, the thickness is slightly smaller. These results indicate that the hydrophobicity of surface affects the conformation of adsorbed layers.

B. Polyelectrolyte

Polyelectrolyte adsorption depends strongly on electrostatic parameters such as the surface charge and the polymer charge, which can depend on both the pH and

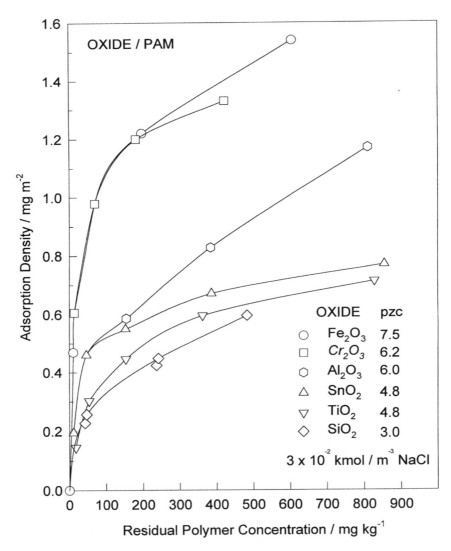

Figure 5 Adsorption densities of poly(acrylamide) on oxides. (Reproduced from Ref. 27 with permission of the American Chemical Society.)

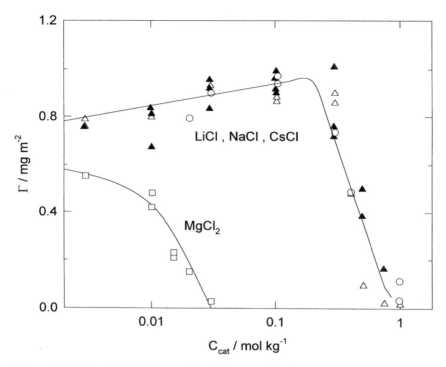

Figure 6 Effect of the ionic strength and the type of cation on the amount of quaternized poly(vinylpyridine) adsorbed onto TiO_2 at pH 8. The cations added are LiCl ($\triangle$), NaCl ($\bigcirc$), CsCl ($\blacktriangle$), and $MgCl_2$ ($\square$). In all cases, a buffer solution with an ionic strength of 0.005 mol/kg was present. The polymer concentration was 10 mg/kg. (Reproduced from Ref. 33 with permission of Academic Press, Inc.)

the ionic strength [30–32]. Figure 6 shows the effect of the addition of salt on the adsorption of quaternized poly(vinylpyridine) (PVP^+) at pH 8 onto TiO_2 as a function of the concentration of several cations [33]. The addition of monovalent cations causes an increase in the adsorbed amount until a maximum is found around $I = 0.2$ mol/kg. At still higher I, the adsorption decreases steeply to zero. No influence of the type of monovalent ion is observed. With the addition of $MgCl_2$, only a decrease in the adsorbed amount is observed. Because the principal interactions are electrostatic, a net electrostatic repulsion arises when overcompensation of the surface charge occurs. The addition of salt screens this replusion, which tends to increase the adsorbed amount. The electrostatic attraction between polymer and surface, however, is also weakened, because the cations compete with the polymer for the negative surface sites. As a result, the adsorbed amount is decreased. On the other hand, the amount of PVP$^+$ adsorbed on SiO$_2$ increases monotonically with ionic strength, suggesting that the adsorption of PVP$^+$ results from both electrostatic and nonelectrostatic interactions. The reversibility of the adsorption of polyelectrolytes on oxides has also been investigated by changing experimental conditions [34]. It is found that polyelectrolyte adsorption on oxides is only partially reversible, because of the strong electrostatic interaction with the surface. This interaction is weakened, and thereby the reversibility enhanced by addition of salt.

Poly(acrylic acid) (PAA) is one of the most popular polyelectrolytes used in many fields. Depending on the pH and electrolyte ion concentration, a variable electrostatic repulsion between ionized carboxyl groups can induce conformational changes in the polymer in solutions. This conformational change plays a major role in the dispersion of metal oxide particles. The variations in the adsorbed PAA conformation under changing pH conditions have been studied by a fluorescence spectroscopic technique using pyrene-labeled poly(acrylic acid) [35,36]. The ratio of the excimer to monomer peak, I_e/I_m, is essentially a measure or index of the coiling/stretching behavior of the polymer: A high ratio is the result of a coiled conformation in the absence of interpolymer interactions. Figure 7 shows the I_e/I_m of PAA at the alumina–liquid interface as a function of final pH. It is clear that the coiled polymer, adsorbed at the solid–liquid interface at a low pH, stretches out when the pH is raised; the initially stretched polymer adsorbed at high pH remains in the stretched form when the pH is lowered. These results can be schematically shown in Fig. 8 [35]. Starting from pH 4, the coiled polymer in solution adsorbs and remains in the same coiled conformation on the positively charged alumina surface. When the pH is raised to the range 5–7, ionization of PAA generates some negative charges in the polymer chain and reduces the extent of coiling. By raising the pH above the PZC of Al$_2$O$_3$, the solid particles become negatively charged like the polymer, and under these conditions, the electrostatic repulsion would cause the polymer to be displaced from the particles. On the other hand, when pH 10 is lowered to pH 4, the polymer is adsorbed wholly in the flat conformation. This conformation change of PAA with pH affects significantly the flocculation properties of Al$_2$O$_3$ (Fig. 9). Adsorption behavior of PAA on TiO$_2$ [37] and Fe$_2$O$_3$ [37,38] is also very similar to that on Al$_2$O$_3$. In the case of SiO$_2$ [39,40], adsorption of PAA occurs to some extent below the PZC of SiO$_2$, but no adsorption is observed above the PZC. However, an enhanced adsorption of PAA is found in the presence of multivalent cations.

IV. ADSORPTION OF A COPOLYMER

Adsorption of poly(acrylonitrile)-co-poly(acrylic acid) (PAN–PAA) on Fe$_2$O$_3$ has been investigated [41,42] to elucidate interactions of carboxylic acids with Fe$_2$O$_3$.

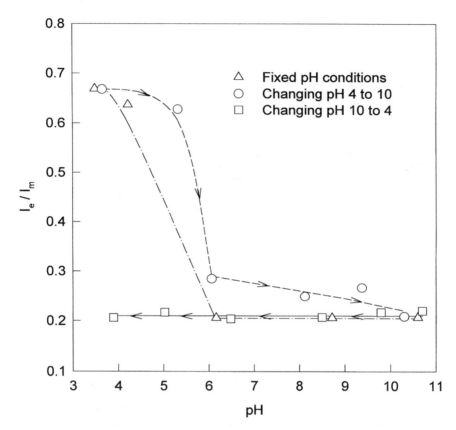

Figure 7 Excimer-to-monomer ratio, I_e/I_m of poly(acrylic acid) (20 ppm) at the Al_2O_3–liquid interface as a function of final pH. (ionic strength = 0.03 M NaCl, S/L = 10 g/200 mL.) (Reproduced from Ref. 35 with permission of Elsevier Science–NL.)

The effect of polymer addition on the stability of iron oxide dispersion is monitored by measuring the particle size distribution and the zeta-potential. It is found that the addition of only small amount of PAN–PAA already increases the mean particle size of Fe_2O_3, which keeps increasing until it reaches a maximum value. Any further addition of PAN–PAA decreases the particle size. Such a flocculation and redispersion process can be explained in terms of a twofold adsorption-layer hypothesis [43]. On the addition of a low concentration of PAN–PAA, the iron oxide particles are discharged and their size increases. As the hydrocarbon chains of PAN–PAA are oriented outward, the particles become hydrophobic and flocculation takes place. On further addition of PAN–PAA, the adsorption of PAN–PAA

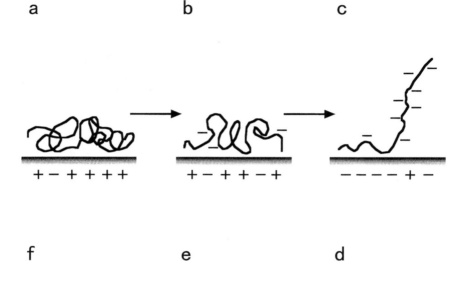

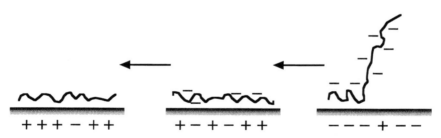

Figure 8 Schematic representation of the variation of polymer conformation at the solid–liquid interface under changing pH conditions. (Reproduced from Ref. 35 with permission of Elsevier Science–NL.)

occurs by a hydrophobic–hydrophobic interaction, resulting in a twofold layer with the ionized groups of the second layer oriented toward the solution. Accordingly, the particles are recharged and redispersion is achieved primarily through electrostatic repulsion. Furthermore, the change in the ratio of PAN and PAA affects the dispersion of Fe_2O_3 significantly.

Another acrylic copolymer of dimethyl diallyl ammonium chloride and acrylic acid (PDMDAAC–AA) has been used for adsorption on SiO_2 [44]. The polymer adsorption depends on the pH and the presence or absence of electrolyte.

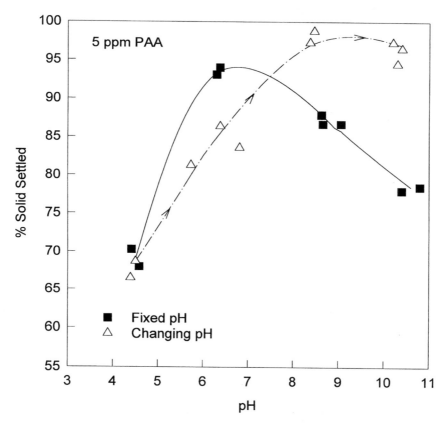

Figure 9 Percent solid settled of alumina suspension with 5 ppm poly(acrylic acid) as a function of pH (fixed and changing pH conditions), (ionic strength = 0.03 M NaCl, S/L = 10 g/200 mL). (Reproduced from Ref. 36 with permission of Elsevier Science–NL.)

In order to estimate conformation of PDMDAC–AA adsorbed on SiO_2 by the (ESR) technique, PDMDAC–AA has been partly spin labeled with 4-amino-2,2,6,6-tetramethylpiperidinoxy [45]. In pure water, the polymer adopts a relatively flat configuration, whereas in the presence of electrolyte, the configuration becomes increasingly more extended as the surface coverage increases. In the absence of the electrolyte, more polymer adsorbs on SiO_2 at pH 3 than at pH 6.5.

Copolymers of acrylamide and acrylate of trimethylaminoethyl chloride have been synthesized by changing the ratio of two monomers [46]. The cationicity τ is expressed by the molar fraction of the acrylate of trimethylaminoethyl chloride in the copolymers. Figure 10 shows an effect of molecular weight of

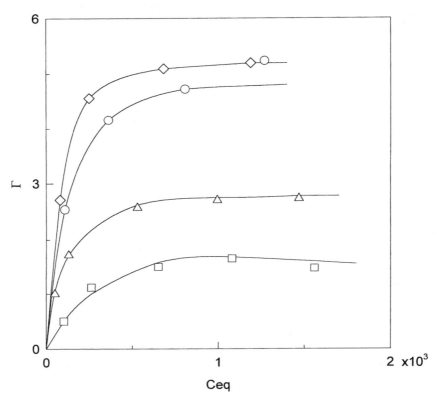

Figure 10 Effect of molecular weight of copolymer of acrylamide and acrylate of trimethylaminoethyl chloride on the adsorption isotherm on SiO_2. The amount of polymer adsorbed, Γ (mg/m²), is plotted versus the polymer concentration in the surrounding medium, C_{eq} (mg/Lr), $\tau = 0.005$. Molecular weights: $\square$, 1.8×10^5; $\triangle$, 7.6×10^5; $\bigcirc$, 1.9×10^6; $\diamondsuit$, 3×10^6. (Reproduced from Ref. 46 with permission of Academic Press, Inc.)

copolymers with $\tau = 0.005$ on the adsorption isotherm on SiO_2. It is interesting that no adsorption of polyacrylamide on SiO_2 is observed, whereas the copolymers of low cationicity are adsorbed even for τ values as low as 0.0025. The amount of the copolymers adsorbed increases with the polymer molecular weight. By this adsorption behavior, it is understood that the amide functions are not adsorption sites for the copolymer and that only a part of the cationic groups leads to a fixation on SiO_2. Using these copolymers, flocculation experiments of SiO_2 have also been examined. It is suggested that the flocculation of silica obeys mainly an electrostatic mechanism for a cationicity > 0.15, whereas for low cationicity

(< 0.05), the major effect is an interparticle bridging. For the intermediate ranges of cationicity, the two effects can be operative simultaneously.

Amphiphilic diblock copolymers have been used more and more in the stabilization of colloidal particles. It can be said that there is a competition between the micellization phenomenon of these copolymers and their adsorption onto solid particles. The adsorption of nonionic copolymers (alkyl polyethylene oxide) of low molecular weights on TiO_2 strongly depends on the ratio R (the relative number of ethylene oxide segments to that of alkyl segments) [47]. It is found that for Synperonic A9 ($R = 1.90$), simple molecules, then spherical micelles, and, finally, lamellar micelles are consecutively adsorbed. For Syneronic A11 ($R = 2.32$), only single molecules are adsorbed, with a tendency to form monomolecular micelles. For Synperonic A20 ($R = 4.23$), only monomolecular micelles are fixed and copolymer A20 behaves as a homopolymer. Furthermore, the importance of the ratio R emphasizes the consideration of the steric stabilization of particle suspensions.

From the adsorption of polystyrene–poly(ethylene oxide) diblock copolymers from aqueous solution to hydrophobic silica, it is found [48] that the surface concentrations attained by three diblock copolymers are independent of molecular weight and are indistinguishable from poly(ethylene oxide) surface concentrations. The hydrodynamic layer thickness is on the order of only a few nanometers for both poly(ethylene oxide) and the diblock copolymer.

The conformation of a copolymer of vinyl pyrrolidone and allylamine adsorbed at the SiO_2/aqueous solution has also been studied by ESR spectroscopy [17]. The polymer changes from a relatively flattened conformation to a more looped one with increasing surface coverage. In addition, the effect on the conformation of changing solvent from pure water to 0.1 M NaCl is small. The time for changes between the flattened and looped conformation is less than 2 min.

V. ADSORPTION OF POLYMER MIXTURES

Although many studies have been concerned with individual adsorption of polymer on surfaces of various materials, competitive adsorption of chemically different polymers on oxides is also important for an understanding of the stability and floccultion of colloidal dispersions.

Figure 11 shows a competitive adsorption of poly(4-vinyl-N-n-propylpyridinium bromide) (PVPP) and poly(ethylene oxide) (PEO) on SiO_2 [23]. The plateau amount of PVPP adsorbed is 1.5 times larger than that of PEO (not shown in the figure.). Therefore, it is expected that PVPP preferentially adsorbs over PEO. However, beyond $C_0 = 0.04$ g/100 mL, the amount of PVPP adsorbed decreases monotonically with increasing C_0, and no adsorption of PVPP is found at $C_0 = 0.07$ g/100 mL. On the other hand, PEO molecules are completely adsorbed

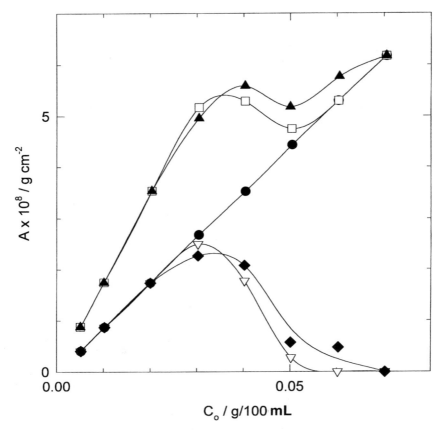

Figure 11   Adsorbed amount of PEO and PVPP as a function of dosage of PEO (or PVPP) C_0 in competitive adsorption of PVPP and PEO: ●, adsorbed amounts of PEO-12 and PEO-860; ◆, adsorbed amount of PVPP; □, total adsorbed amount of PVPP and PEO in PVPP/PEO-12 mixture; ▽, adsorbed amount of PVPP; □, total adsorbed amount of PVPP and PEO in PVPP/PEO-860 mixture. The molecular weights of PEO-12 and PEO-860 were 2.1×10^4 and 8.6×10^5, respectively. (Reproduced from Ref. 23 with permission of Academic Press, Inc.)

over the entire measured range of C_0. Adsorption preference between different polymer species may be caused by the difference in bond strength between the adsorbed segments and the surface sites. For adsorption of PEO to the SiO_2 surface, the predominant interaction can be regarded as hydrogen bonds between ether groups and silanol groups, as mentioned earlier. The interaction between PVPP and the silica surface is mainly due to hydrophobic interaction between siloxane groups and alkyl groups in the PVPP, because hydrogen bonds are not responsible for the adsorption of charged pyridinium ions on SiO_2 in aqueous solution. The contribution of the electrostatic interaction between pyridinium groups and silanol groups is also relatively small because the number of surface silanol groups ionized at pH is small. Accordingly, it can be said that hydrogen bonds between the ether groups in PEO chains and silanol groups, which are stronger than the hydrophobic interactions of PVPP with siloxane groups, may induce the preferential adsorption of PEO molecules over PVPP molecules.

Figure 12 shows a competitive adsorption of poly(dimethyldiallyl ammonium chloride) (PDC) and poly(vinyl pyrrolidone) (PVP) on SiO_2 at pH 6 [49]. In this study, the adsorption of two polymers has been measured as a function of PVP concentration in the presence of feed-constant PDC concentration. In the presence of PDC (0.10 g/dm^3), the amount of PDC adsorbed is almost unchanged with in-

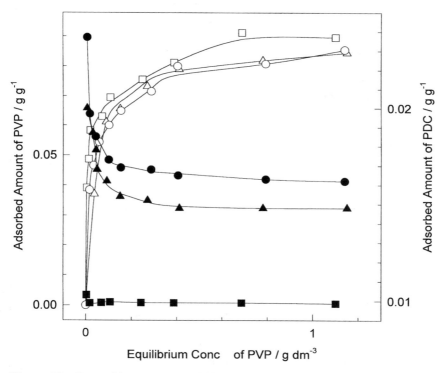

Figure 12 Competitive adsorption of PDC (closed symbols) and PVP (open symbols) on SiO_2 at pH 6. The concentrations of PDC were 0.10 □, 0.20 △, 0.26 g/dm^3 ○. (Reproduced from Ref. 49 with permission of Elsevier Science–NL.)

creasing PVP concentration, whereas that of PVP increases steeply at very low concentrations of PVP. However, at higher feed concentrations of PDC (0.20 and 0.26 g/dm^3), the amounts of PDC adsorbed decrease considerably at very low concentrations of PVP and then reach constant values with a further increase in the PVP concentration, whereas the adsorption of PVP increases with PVP concentration but is lower than that of PVP in the absence of PDC. These results suggest that PVP is preferentially adsorbed on SiO$_2$ rather than PDC, resulting in the replacement of PDC by PVP, in particular at higher feed concentrations of PDC. The competitive adsorption behavior of PDC and PVP on SiO$_2$ has been interpreted by two interactions, including hydrogen-bonding between PVP and the silica surface and the electrostatic interaction between PDC and the silica surface. The conformation of PVP adsorbed through the competitive adsorption of PDC and PVP on SiO$_2$ has been estimated using a spin-labeled PVP. The fraction of train segments, which is called p, is plotted as a function of equilibrium concentration of PVP in Fig. 13. It is seen that in the presence of low feed concentration of PDC (0.10 g/dm^3), the p value is relatively high and decreases gradually with PVP concentration. However, at higher feed concentrations of PDC, the p values decrease sharply with PVP concentration. The p values in the only PVP system decrease from 1 to 0.8 in the concentration region studied. These results demonstrate that PVP is mainly adsorbed in trains at a low feed concentration of PDC, whereas the segments of PVP adsorbed are extended in solution at higher feed concentrations of PDC. In the stability of silica dispersion, SiO$_2$ is completely flocculated through the entire PVP concentration range in the feed concentration of PDC (0.10 g/dm^3), whereas the stability of silica dispersions increases with PVP concentration at higher feed concentrations of PDC. These results suggest that the high stability of silica dispersions is mainly governed by electrostatic and steric repulsion forces at higher feed concentrations of PDC.

Complexation between PVP, PEO, and PAA has been investigated, mainly for biological aspect. The interactive forces responsible for complexation between two chemically different polymers are usually due to electrostatic, hydrogen-bonding, or hydrophobic interactions. Hydrogen-bonding in the PAA–PEO system is the most well established [50,51]. The formation of hydrogen bonds is also considered to be the dominant force in the interaction between PVP and PAA [52,53]. It is found [54] that the association of PAA–PVP is stronger than that of PAA–PEO and the association is affected by the pH. These interactions between PAA–PVA and PAA–PEO have been elucidated through adsorption on alumina particles [54,55]. As already mentioned, no distinct adsorption of PVP or PEO on Al$_2$O$_3$ at various pH's has been observed. In the PAA–PEO system, the adsorption of PEO has not been enhanced with the coadsorption of PAA on Al$_2$O$_3$. However, in the PAA–PVP system, the adsorption of PVP is considerably enhanced with coadsoroption of PAA, and the adsorption behavior is also affected with pH. Figure 14 shows a simultaneous adsorption of PAA and PVP on Al$_2$O$_3$ at feed con-

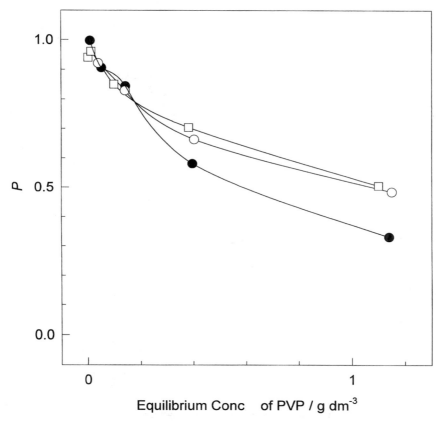

Figure 13 Change in p value for PVP–PDC systems as a function of equilibrium concentration of PVP. The concentrations of PDC were 0.10 □, 0.20 ○, 0.26 g/dm³ ●. (Reproduced from Ref. 49 with permission of Elsevier Science–NL.)

centrations of PAA as well as adsorption of PVP [56]. Adsorption experiments show that PAA completely adsorbs on Al_2O_3 from both of the feed concentrations at pH 5.2 and 7.2, whereas at pH 10.2, PAA adsorbs to some extent on Al_2O_3. It is apparent that the adsorption of PVP is enhanced with the adsorption of PAA at pH 5.2 and 7.2, but no enhancement in the adsorption of PVP is observed at pH 10.2. The reason that the adsorption of PVP is enhanced by the adsorption of PAA is probably due to hydrogen-bonding between the polymers on alumina surface. Because the adsorption of PAA is strong compared to that of PVP, PAA at first would be preferentially adsorbed on alumina surface and, then, a subsequent adsorption of PVP may occur. Also, the complex of PAA–PVP formed in the solution is ad-

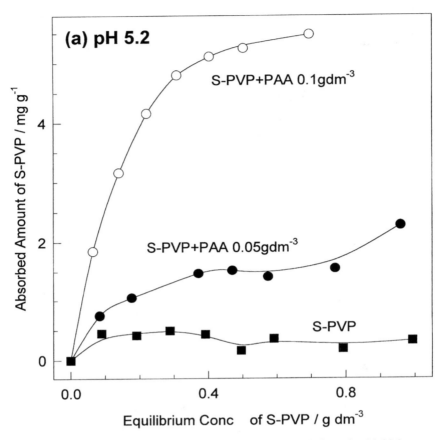

Figure 14 Simultaneous adsorption of PAA and PVP on Al$_2$O$_3$ at fixed initial concentrations of PAA as well as adsorption of PVP. (Reproduced from Ref. 56 with permission of the American Chemical Society.)

sorbed on the alumina surface, orienting the PVP chains to the aqueous solution. With increasing pH, the adsorption of PVP with PAA adsorbed through hydrogen-bonding is prevented by the dissociated PAA adsorbed. At pH 10.2, the adsorption of PVP is hardly observed because all the adsorbed PAA is dissociated, which has no ability of hydrogen-bonding with PVP. Furthermore, in order to estimate the conformation of PAA and PVP adsorbed on Al$_2$O$_3$, both spin-labeled PAA and PVP have been used. The values of p for the spin-labeled PAA and PVP on Al$_2$O$_3$ are plotted in Fig. 15. It is seen that the p values obtained from the adsorption of PAA alone are gradually decreased with increasing PAA concentration as well as

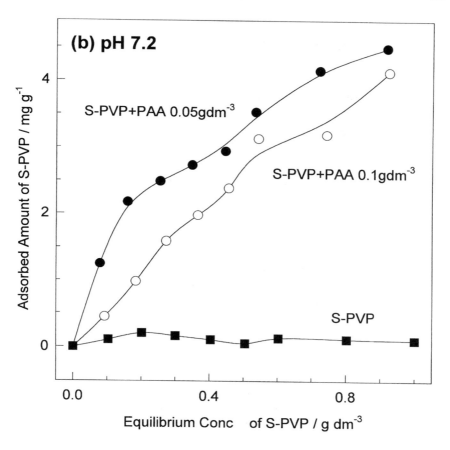

increasing pH, although the values are relatively high, ranging between 0.8 and 0.9. In the simultaneous adsorption of PAA and PVP at both a pH of 5.2 and 7.2, the p values of PAA range between 0.8 and 0.9, whereas those of PVP are very low, ranging between 0.3 and 0.4. These results indicate that, in the simultaneous adsorption of PAA and PVP, PAA molecules are considerably restricted as train segments, whereas PVP molecules are adsorbed with loops or tails and extend to aqueous solution. These results also support the adsorption mechanism of PAA and PVP on Al_2O_3 and that PVP molecules are mainly adsorbed on the PAA-coated alumina at pH's of 5.2 and 7.2. At a pH of 10.2, the p values obtained from both PAA alone and PAA–PVP adsorption are not so different from those at the lower pH. In addition, the dispersion stability of alumina suspensions by ad-

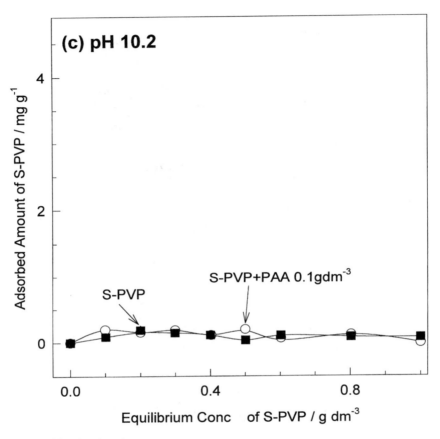

Figure 14 Continued

sorption of PAA and PVP is mainly controlled by electrostatic and steric interactions.

Sequential addition of anionic and cationic polyelectrolytes such as poly(styrene sulfonate) (PSS^-), poly(1-methyl-2-vinylpyridinium) (PVP^+) on SiO_2 and TiO_2 has been performed to form multilayers [57]. Very stable multilayers are formed with a composition ratio of $PVP^+/PSS^- = 3$: The influence of the substrate is restricted to the first few layers and the adsorbed amounts in the higher layers are the same regardless of the type of substrate, the pH, and the sequence of addition.

A competitive adsorption of sodium poly(styrene sulfonate) (PSS), molecular weight 4000 and 1,200,000, on Fe_2O_3, has been studied using size-exclusion chromatography [58]. In the absence of electrolytes, 4000 molecular weight PSS

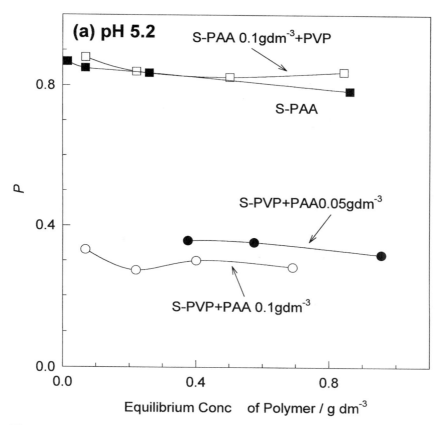

Figure 15 Change in p values for spin-labeled PAA and PVP on Al_2O_3. (Reproduced from Ref. 56 with permission of the American Chemical Society.)

adsorbs more strongly than 1,200,000 primarily because of faster adsorption kinetics. No displacement is observed, probably due to the electrostatic repulsion between adjacent segments. At 0.1 mol/dm³ NaCl, 1,200,000 PSS adsorbs more strongly and is also able to displace the adsorbed 4000 PSS.

VI. EFFECT OF SURFACTANT ON POLYMER ADSORPTION

As already mentioned, adsorption of polymers on oxide surfaces is influenced by many factors such as pH and ionic strength. Furthermore, adsorption of polymer is often varied by addition of surfactants. Until now, interactions between polymers

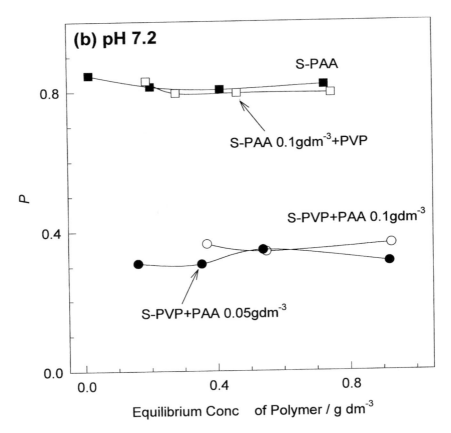

Figure 15 Continued

and surfactants in aqueous solutions have been intensively investigated by many techniques [59–68]. The strength of interactions depends on the kinds of polymer and surfactants. It has been shown that the binding of ionic surfactants to neutral polymers occurs mainly by hydrophobic interactions, whereas a combination of electrostatic and hydrophobic interactions is involved in the binding of oppositely charged polymers and surfactants. A theoretical description of complexation of polymers and surfactants has also been proposed [69–71]. It is expected that aqueous solution properties of polymers and surfactants are greatly reflected in the adsorption of these systems.

Let us start by discussing the adsorption of nonionic polymers and ionic surfactants on oxides, as this combination shows relatively a strong interaction.

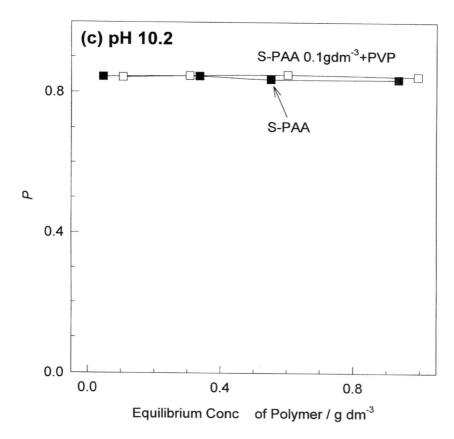

A. Nonionic Polymer–Surfactant

Figure 16 shows the adsorption of PVP and lithium dodecyl sulfate (LiDS) on Al_2O_3 from their mixed aqueous solutions at feed concentrations of PVP (0.2 and 0.5 g/dm³) as a function of LiDS concentration [72]. The amount of PVP adsorbed increases markedly with LiDS concentration, achieves a maximum, and then decreases for both feed concentrations of PVP. On the other hand, the adsorption of LiDS also increases with LiDS concentration and reaches a plateau in the absence of PVP; in the presence of PVP, the amount of LiDS adsorbed is less than that for LiDS only. This result can be explained by a view that PVP–surfactant complexes are formed on Al_2O_3. As it is known that a nonionic polymer and an ionic surfactant can form a polyelectrolytelike complex by hydrophobic interaction in aqueous solution [68], it is postulated that the same interaction takes place and a type

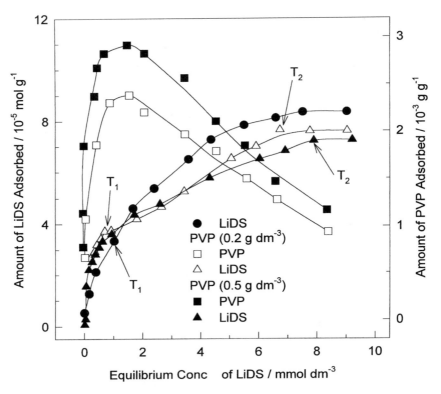

Figure 16 Adsorption of PVP and LiDS on Al_2O_3 from their mixed aqueous solutions containing fixed initial concentrations of PVP (0.2 and 0.5 g/dm^3) as a function of LiDS concentration. (Reproduced from Ref. 72 with permission of the American Chemical Society.)

of surface complex of PVP and surfactant forms at the solid–liquid interface. The anionic surfactant [below the critical micelle concentration (CMC)] adsorbs much faster on the alumina surface than PVP, as a result of its smaller size and the electrostatic attraction force between the negatively charged hydrohpilic groups of the surfactant and the positively charged sites on Al_2O_3. The hydrophobic chain of the surfactant extends toward the aqueous solution. Then, PVP attaches to the hydrophobic chain of the adsorbed layer of the anionic surfactant. This model also applies to surfactant concentrations greater than its CMC, but instead of a bilayer of PVP–surfactant attachments, there might also be patches of hemimicelles to which PVP attaches. As a result, PVP and the surfactant adsorbed on Al_2O_3 can combine with each other by hydrophobic binding, so that more PVP is adsorbed.

With increasing concentration of LiDS, the adsorption of PVP rapidly decreases because PVP–anionic surfactant complexes formed in the bulk are relatively surface inactive and scarcely adsorb at the alumina–solution interface. The formation of a surfactant bilayer on Al_2O_3 may also prevent the formation of the PVP–anionic surfactant complex on Al_2O_3. The formation of PVP–anionic surfactant complexes in aqueous solutions has been confirmed by surface tension [68] and NMR measurements [73]. Using a spin-labeled PVP, the conformation change of PVP with adsorption of an anionic surfactant on Al_2O_3 has also been investigated. Figure 17 shows the plots of p values against LiDS concentration. The values of p steeply increase with increasing LiDS concentration, and then become constant.

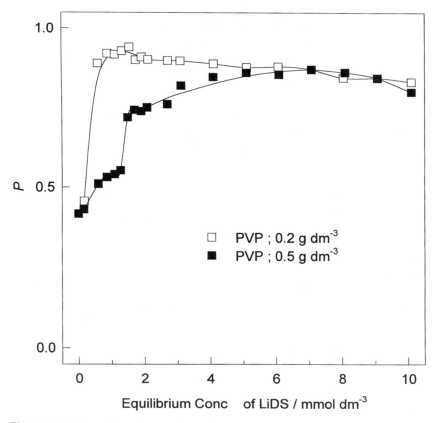

Figure 17 Plots of p values against LiDS for the PVP–LiDS–Al_2O_3 system. The initial polymer concentrations were 0.2 and 0.5 g/dm³. (Reproduced from Ref. 72 with permission of the American Chemical Society.)

This probably occurs because the PVP–surfactant complexes formed by the interaction between PVP and hydrophobic chains of LiDS on Al_2O_3 become more rigid. Then, although the adsorbed amount of PVP decreases remarkably, the values of p remain constant with their segments in trains. The value of p is larger in 0.2 g/dm^3 than in 0.5 g/dm^3 at the maximum PVP adsorption. This difference in p correlates with the bound ratio of PVP and LiDS adsorbed: The bound ratio of LiDS to PVP in the presence of 0.2 g/dm^3 is greater than that in the presence of 0.5 g/dm^3. The adsorption of PVP on Al_2O_3 in the presence of an anionic fluorocarbon surfactant such as lithuim perfluorooctane sulfonate (LiFOS) shows a behavior similar to that of the PVP–LiDS system. However, the p values in the PVP–LiFOS system are higher than those in the PVP–LiDS system, probably due to a rigidity of the fluorocarbon chain of LiFOS. The interaction of PVP with sodium dodecyl sulfate (SDS) on Al_2O_3 has been studied using a calorimetric method [74]. One of the interesting results is that the enthalpy change upon PVP addition to SDS adsorbed on Al_2O_3 is endothermic. This observation would mean that a SDS monolayer is present with the hydrophobic alkyl chain pointing toward the solution. This mechanism is supported by the change of p value for the PVP–LiDS system.

The additive effect of a double-chained anionic surfactant such as sodium bis(2-ethylhexyl) sulfosuccinate (AOT) on the adsorption of PVP on Al_2O_3 has also been studied [75]. The adsorption of PVP is significantly enhanced at a low concentration of AOT, but the amount of PVP adsorbed decreases gradually at higher AOT concentrations. Interestingly, below the CMC of AOT, p values are greater for the PVP–AOT system than for the PVP–LiDS system. These large p values suggest that the PVP–surfactant complexes formed by the interaction between PVP and the hydrophobic chain of AOT on Al_2O_3 become more immobile than those formed between PVP and LiDS. It is likely that because the strength of the interaction between PVP and AOT on Al_2O_3 is stronger than that between PVP and LiDS, the train portion of PVP for the former system is larger than that for the latter system. The adsorption of PVP on Fe_2O_3 is also considerably enhanced in the presence of sodium dodecyl sulfate [76].

It has been shown that PVP adsorption exhibits a strong affinity with SiO_2, different from a weak affinity with Al_2O_3. The effect of anionic surfactant on PVP adsorption on hydrophilic silica is shown in Fig. 18 [77]: The adsorption of PVP increases and shows a maximum, then decreases with increasing LiDS concentration. However, the enhancement in the adsorption of PVP on SiO_2 is small in comparison to that on Al_2O_3. It is very important to confirm the relationship between the polymer conformation adsorbed (values of p) and the thickness of the adsorbed layer (δ). Most frequently, the photon correlation spectroscopy (PCS) technique is used to measure the size of both core particles and adducts, and the thickness of the adsorbed layer is found from the difference between the two. For polydisperse samples of unknown distribution, the technique is less suitable, as it only provides values for the average diameter together with a polydispersity index. Because

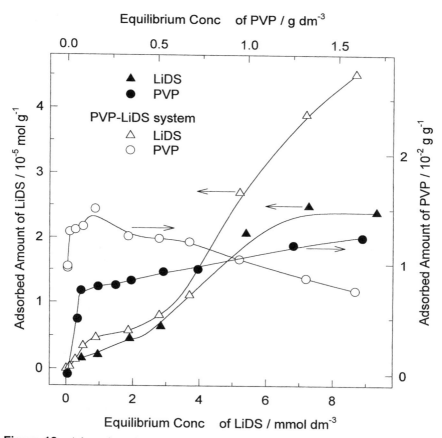

Figure 18 Adsorption of PVP and LiDS on SiO_2 from single and mixed aqueous solutions containing a fixed initial concentration of PVP (1.0 g/dm³) as a function of surfactant equilibrium concentration. (Reproduced from Ref. 77 with permission of Elsevier Science–NL.)

larger particles scatter light more strongly than fine particles, even minor contaminations by particles larger than the average tend to shift the average size toward larger values without noticeably affecting the measured polydispersity. By introducing a sedimentation field-flow-fractionation separation step [78,79] prior to the PCS sizing of the adsorbate, it is possible to improve the accuracy in these size measurements. It is reasonable to assume that δ determined in this process is more reliable. Figure 19 shows the layer thickness and the values of p for the PVP–LiDS–SiO_2 system at a feed concentration of PVP = 1.0 g/dm³. The layer

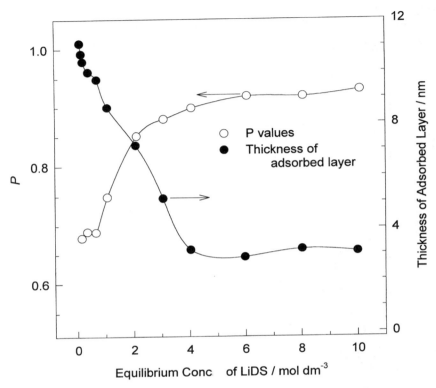

Figure 19 Relationship between the thickness of the adsorbed layer and p values for the PVP–LiDS–SiO$_2$ system. (Reproduced from Ref. 77 with permission of Elsevier Science–NL.)

thickness has been obtained using a sedimentation field-flow fractionation and PCS technique. The adsorption layer thickness of PVP alone on SiO$_2$ (not shown in figure) has suggested that the molecules would be adsorbed flat on the surface in a very low-concentration region, having a thinner layer. When PVP concentration increases, the adsorption layer rapidly becomes thicker because larger loops and tails are formed. The adsorption layer thickness in the range 5–10 nm is known from the literature. Furthermore, from segment density profiles calculated with the Scheutjens–Fleer adsorption theory, the variation of hydrodynamic layer thickness of adsorbed polymer layers with coverage is known to come from long tails that extend far into the solution [80]. Thus, the thicker layer could be attributed to the conformation of polymer adsorbed in the tail. In Fig. 19, a remarkable decrease in the thickness of the adsorbed layer occurs with increasing LiDS and becomes constant, whereas the value of p increases and then becomes constant, indicating that the conformation of PVP adsorbed on SiO_2 significantly affects the thickness of the adsorbed layer. Thus, a good relationship between the layer thickness and the value of p is obtained. Such a relationship is also obtained for the PVP–LiDS–hydrophobic silica system.

The adsorption of hydroethylcellulose and hydrophobically modified hydroethylcellulose and SDS on Al_2O_3 has been investigated [81]. In both systems, the amount of polymers adsorbed increases by twice as much as that without SDS. On the contrary, the adsorption of SDS increases with SDS concentration and reaches a plateau in the absence of a polymer. Further, the addition of hydroethylcellulose hardly alters the amount of SDS adsorbed, but the addition of a hydrophobic polymer reduces the amount of SDS adsorbed to some extent. In the two celluloses systems alone, the values of p are almost zero, indicating that the polymer chains extend considerably in the bulk solution. By the addition of SDS, the values of p are in the range from 0.2 to 0.6, and increase slightly with increasing SDS concentration, where the p values in the hydrophobically modified cellulose are slightly higher than that in the hydroethylcellulose.

The interactions of nonionic polymers and cationic surfactants in aqueous solutions are usually weaker than those between nonionic polymers and anionic surfactants so that the adsorption of nonionic polymers on particles would show a competitive adsorption with cationic surfactant when the polymer and surfactant have affinity to the particle surfaces. Such a case is the PVP–dodecyltrimethylammonium chloride (DTAC)–SiO_2 system [82]. Figure 20 shows the competitive adsorption of PVP and DTAC from their mixed solution in the presence of PVP (0.2, 1.0, and 1.8 g/dm^3). At the highest concentration of PVP studied, the adsorption of PVP decreases gradually with increasing DTAC concentration, whereas the amount of DTAC adsorbed increases, but the magnitude of the increase in the adsorption is lower than that in the absence of PVP. A similar result is obtained at low concentrations of PVP. Thus, the replacement of PVP on SiO_2 by DTAC occurs. Because the interaction between PVP and cationic surfactants in aqueous solutions has been observed to be nonexistent or very weak, in contrast to the case of anionic surfactants, it is readily understood that competitive adsorption between PVP and DTAC on SiO_2 takes place. Figure 21 shows the change in p values as a function of DTAC concentration for the PVP–DTAC–SiO_2 system. The p value decreases gradually with increasing DTAC concentration in the presence of higher concentrations of PVP, whereas p shows an almost constant high value in the presence of a lower PVP concentration. This change in the p value correlates well with the stability of silica dispersion: The dispersion stability of SiO_2 increases with decreasing p value. This suggests that the dispersion stability of SiO_2 is dominated by the steric repulsion force (i.e., the fraction of segments in loops or tails).

The competitive adsorption on SiO_2 between nonionic surfactants and PEO has been investigated as a function of polymer chain length [83]. For low-molec-

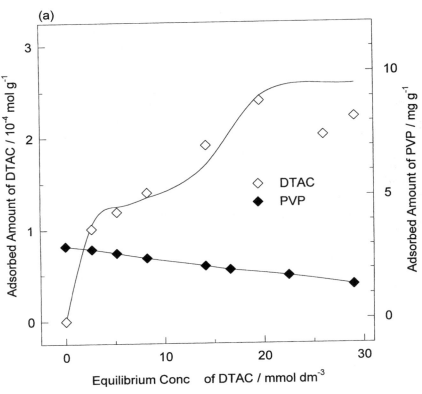

Figure 20 Competitive adsorption of PVP and DTAC on SiO$_2$. The initial polymer concentrations were (a) 0.2, (b) 1.0, and (c) 1.8 g/dm^3. (Reproduced from Ref. 82 with permission of the American Chemical Society.)

ular-weight PEO, a preferential adsorption of the surfactants is observed. However, at higher-molecular-weight PEO, a competitive adsorption between PEO and the surfactant takes place. It is assumed that there is a competition on the silica surface between surfactant surface aggregates whose molecular weight would range between 80,000 and 1,30,000 and adsorbed PEO.

B. Ionic Polymer–Surfactant

The adsorption of anionic polymers such as sodium poly(styrene sulfonate) (PSS) and polyacrylamide including 3 mol% sulfonate groups (PAMS) on oxide surfaces has been influenced by the addition of anionic surfactants. In the case of the PSS–SDS–Al$_2$O$_3$ system [84], the amount of PSS adsorbed decreases with in-

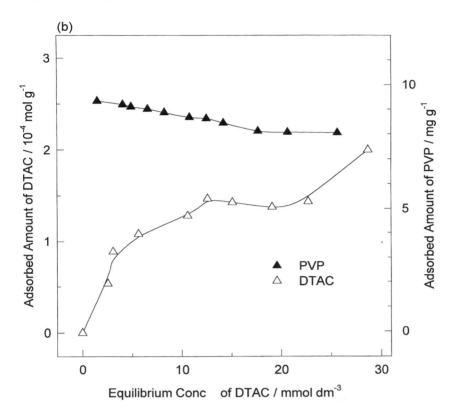

creasing adsorption of SDS, suggesting that the adsorbed PSS is replaced by SDS with increasing SDS concentration. A decrease in the adsorption of PAMS on Fe_2O_3 is also obtained by the addition of sodium dodecylsulfonate [85]. However, no significant change in sodium dodecylsulfonate adsorption on Fe_2O_3 is reported by the adsorption of PAMS.

When oppositely charged polymer and surfactant are mixed in aqueous solutions, precipitation often occurs due to formation of a complex from hydrophobic as well as electrostatic attractive forces. A typical case is a combination of PSS and hexadecyltrimethylammonium chloride (HTAC) [84]. Figure 22 shows the adsorption of PSS and HTAC on positively charged alumina from the mixed solution at a fixed initial concentration of PSS (0.4 g/dm^3). The amount of PSS adsorbed increases gradually and then sharply, accompanied with a similar sharp adsorption of HTAC with increasing HTAC concentration. As there is an electrostatic attraction force between oppositely charged PSS and HTAC in an aqueous solution, and

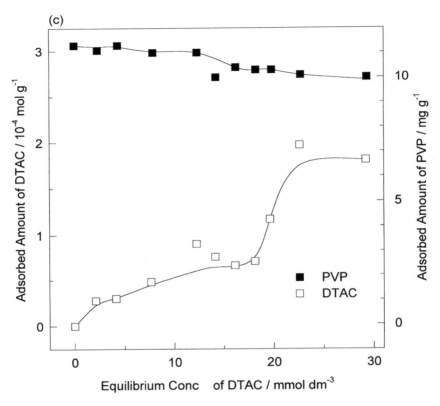

Figure 20 Continued

a hydrophobic force between them, it is reasonable to assume that such a complex forms on the surface of Al_2O_3, resulting in the enhancement of PSS adsorption. A similar coadsorption of PSS and hexadodecyltrimethylammonium bromide (HTAB) on SiO_2 has been studied from mixed solutions by attenuated total reflection techniques [85]. The adsorbed amount of PSS and HTAB is highly dependent on pH and on the order of the addition of both species. Figure 23 is a schematic representation of the proposed surface and solution species formed at various pH's for the sequential addition of HTAB and PSS to SiO_2. At pH 2, although the amount of HTAB adsorbed is small, the hydrophobic nature of the HTAB-coated silica surface facilitates adsorption of PSS. When the solution pH is raised to 7.0, the polyelectrolyte desorbs because of the electrostatic repulsion between the silica surface and PSS. As the solution pH is increased to 9.7, a large rise in the amount of HTAB adsorbed is observed. Surfactant removal from the polyelec-

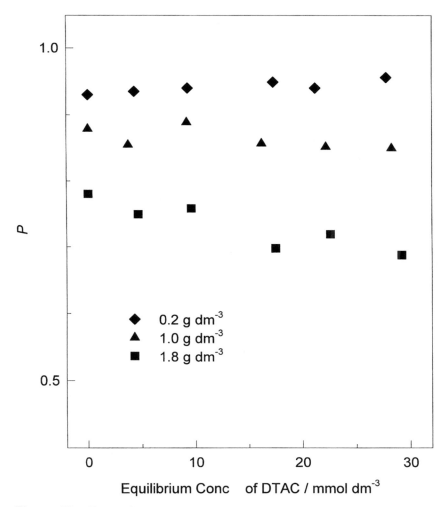

P

Figure 21 Change in p values as a function of DTAC concentration for the PVP–DTAC–SiO$_2$ system. The initial polymer concentrations are indicated by symbols on the figure. (Reproduced from Ref. 82 with permission of the American Chemical Society.)

trolyte to the surface can be understood in terms of the increased electrostatic attraction between HTAB and the highly negatively charged surface at pH 9.7. Anionic polyacrylamide (PAMS) adsorption on positively charged hematite at pH 2.4 has also been investigated by the addition of dodecylamine [86]. At this pH, the amine is fully cationic. The adsorption of PAMS is enhanced compared with that

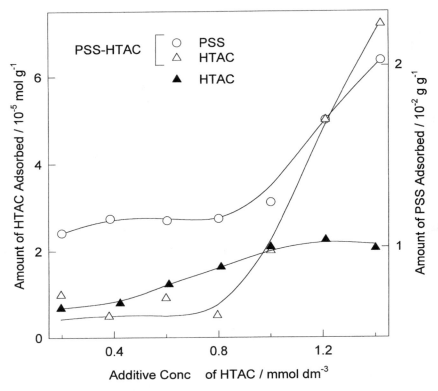

Figure 22   Adsorption of PSS and HTAC on Al$_2$O$_3$ from their mixed aqueous solutions. The fixed concentration of PSS was 0.4 g/dm^3. (Reproduced from Ref. 84 with permission of the American Chemical Society.)

for PAMS alone. This increase is also explained by the associative interaction between the two oppositely charged species.

The adsorption of cationic polymers on oxides has been investigated by the addition of anionic or cationic surfactants. From a study of adsorption of SDS and of JR400 (water-soluble cellulose ether polymer containing quaternary nitogens grafted onto its backbone) on Al$_2$O$_3$ [87], it is found that synergistic adsorption occurs under certain conditions (e.g., at pH 6 and relatively low ratios of SDS to JR400), whereas under different conditions, either component can inhibit the adsorption of the other one. In the case of the simultaneous adsorption of three cationic polymers (JR polymer) and HTAB on SiO$_2$, the polymers are competing with small molecules; the number of adsorbed groups per molecule is an impor-

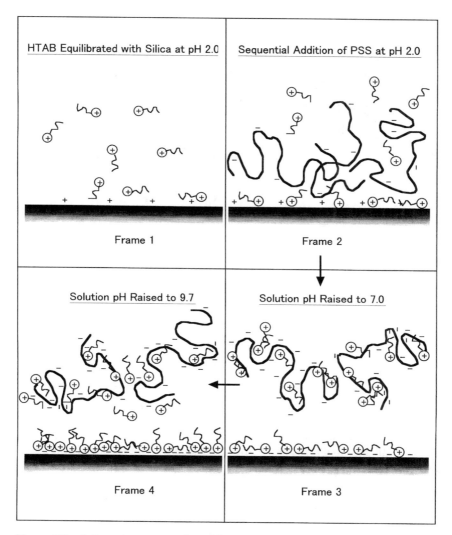

Figure 23 Schematic representation of the proposed surface and solution species formed at pH 2.0, 7.0, and 9.7 for the sequential addition of HTAB (5.5×10^{-5} M) and PSS (100 ppm) to SiO_2. (Reproduced from Ref. 85 with permission of the American Chemical Society.)

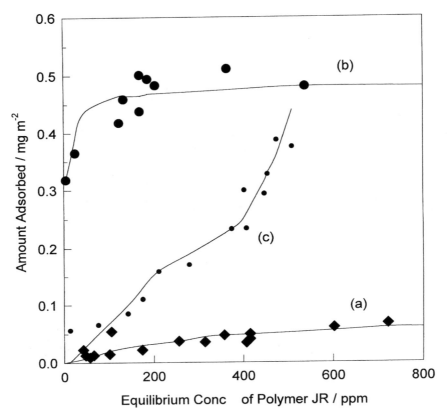

Figure 24   Effect of surfactant on JP polymer adsorption on SiO_2: (a) JR 125; (b) JR 400; (c) JR 30M. (Reproduced from Ref. 87 with permission of The Royal Society of Chemistry.)

tant factor. Here, the order of the molecular weights of the JR polymer is as follows: JR30M > JR400 > JR125. Figure 24 indicates that the diffusion of polymers to the surface before surfactant adsorption is important, as well as the number of adsorbed groups per polymer. The amount of polymer adsorbed increases in the order of JR125 < JR30M < JR400, suggesting that the amount of polymer adsorbed is determined by two factors: (1) the weight of polymer that can diffuse to the surface and (2) the number of segments each polymer chain has on the surface at equilibrium. The larger the polymer, the greater the weight that would be attached to the surface at the start of adsorption, although the slower the diffusion to the surface. In competition with the surfactant, JR125 has the lowest weight adsorbed, probably because of too few contacts with the surface. JR400 may have been adsorbed to a greater extent than JR30M because of its faster diffusion while having sufficient contacts with the surface.

The effect of HTAB on the conformation of the adsorbed copolymer (a copolymer of dimethyldiallyl ammonium chloride and acrylic acid) on SiO_2 has been studied [44]. The ESR spectra obtained indicate that the polymer adsorbed takes a flat conformation at a lower HTAB concentration, whereas at a higher HTAB concentration, the spectra are narrower, suggesting that the polymer adsorbed in the form of loops and tails extending out into solution. When the surfactant concentration increases, HTAB adsorbs on SiO_2 and its competition for the surface sites causes the polymer to adopt a more extended conformation.

VII. CONCLUSIONS

This chapter has described adsorption of polymers on oxide surfaces from their aqueous solutions. The adsorption of polymers is affected by many factors such as the chemical structures of polymer itself and the surface conditions of oxides. Several examples show interesting adsorption behaviors of polymers on oxides, which may be useful for some applications. Although the weight percent of oxide particles in aqueous solutions is relatively low in most of the experimental data shown in this chapter, the adsorption of polymers also significantly alters physicochemical properties of concentrated oxide dispersions. In addition, the use of atomic force microscopy or surface force apparatus can provide new insight for elucidating interactions between polymers and oxide surfaces [88–91].

REFERENCES

1. GJ Fleer, J Lyklema. In: GD Parfitt, CH Rochester, eds. Adsorption from Solution at the Solid/Liquid Interface. London: Academic Press, 1983, pp 153–220.
2. GJ Fleer, JMHM Scheutjens. In: B Dobias, ed. Coagulation and Flocculation. New York: Marcel Dekker, 1993, pp 209–263.
3. A Takahashi, M Kawaguchi. Advances in Polymer Science, Vol. 46, Berlin: Springer-Verlag, 1982, pp 1–65.
4. JMHM Scheutjens, GJ Fleer. J Phys Chem 83:1619, 1979.
5. P-G de Gennes. Rep Prog Phys 32:187, 1969.
6. KM Hong, J Noolandi. Macromolecules 14:727, 1981.
7. CAJ Hoeve, EA Dimarzio, P Peyser. J Chem Phys 42:2558, 1965.
8. E Helfand. J Chem Phys 63:2192, 1974.
9. AG Dahlgen, A Waltermo, E Blomberg, PM Claesson, L Sjostrom, T Akesson, B Jonsson. J Phys Chem 97:11769, 1993.
10. AWP Vermeer, FAM Leermakers, LK Koopal. Langmuir 13:4413, 1997.

11. DH Napper. Polymeric Stabilization of Colloidal Dispersions. London: Academic Press, 1983.

12. J Gregory, In: CA Finch, ed. Chemistry and Technology of Water Soluble Polymers New York: Plenum, 1983, pp 307–320.

13. P Jones, JA Hockey. Trans Faraday Soc 67:2679, 1971.

14. LR Parfitt, RJ Atkinson, RSC Smart. Soil Sci Soc Am J 39:837, 1975.

15. T Hiemstra, WH Van Riemsdijk, GH Bolt. J Colloid Interf Sci 133:91, 1989.

16. MA Cohen Stuart, GJ Fleer, BH Bijsterbosch. J Colloid Interf Sci 90:310, 1982.

17. ID Robb, R Smith. Eur Polym J 10:1005, 1974.

18. KG Barnett, T Cosgrove, B Vincent, DS Sissons, MA Cohen Stuart. Macromolecules 14:1018, 1981.

19. M Korn, E Killmann. J Colloid Interf Sci 76:19, 1980.

20. MA Cohen Stuart, GJ Fleer, BH Bijsterbosch. J Colloid Interf Sci 90:321, 1982.

21. ID Robb, R Smith. Polymer 18:500, 1977.

22. T Cosgrove, KG Barnett. J Magn Res 43:15, 1981.

23. M Kawaguchi, H Kawaguchi, A Takahashi. J Colloid Interf Sci 124:57, 1988.

24. S Mathur, BM Moudgil. J Colloid Interf Sci 196:92, 1997.

25. J Rubio, JA Kitchener. J Colloid Interf Sci 57:132, 1976.

26. BJ Scheiner, GM Wilemon. In: YA Attia, ed. Flocculation in Biotechnology and Separation Systems. Amsterdam: Elsevier Science, 1987, pp 175–186.

27. LT Lee, P Somasundaran. Langmuir 5:854, 1989.

28. GA Parks, PL de Bruyn. J Phys Chem 66:967, 1962.

29. T Kull, T Nylander, F Tiberg, NM Wahlgren. Langmuir 13:5141, 1997.

30. H Tanaka, L Odberg, L Wagberg, T Lindstrom. J Colloid Interf Sci 134:219, 1990.

31. G Durand, F Lafuma, R Audebert. Prog Colloid Polym Sci 266:278, 1988.

32. B Popping, A Deratani, B Sebille, N Desbois, JM Lamarche, A Foissy. Colloids Surf 64:125, 1992.

33. NG Hoogeveen, MA Cohen Stuart, GJ Fleer. J Colloid Interf Sci 182:133, 1996.

34. NG Hoogeveen, MA Cohen Stuart, GJ Fleer. J Colloid Interf Sci 182:146, 1996.

35. KF Tjipangandjara, Yi-B Huang, P Somasundaran, NJ Turro. Colloids Surf 44:229, 1990.

36. KF Tjipangandjara, P Somasundaran. Colloids Surf 55:245, 1991.

37. A Foissy, AE Attar, JM Lamarche. J Colloid Interf Sci 96:275, 1983.

38. CL Tiller, CR O'Melia. Colloids Surf 73:89, 1993.

39. JE Gebhardt, DW Fuerstenau. Colloids Surf 7:221, 1983.

40. GR Joppien. J Phys Chem 82:2210, 1978.

41. K Ogihara, S Tomioka, K Esumi, and K Meguro. J Jpn Soc Colour Mater 55:546, 1982.

42. K Esumi, K Ogihara, K Meguro. Bull Chem Soc Jpn 57:279, 1202, 1984.

43. K Meguro. Nippon Kagaku Zasshi 77:77, 1956.

44. PA Williams, R Harrop, ID Robb. J Colloid Interf Sci 102:548, 1984.

45. MC Cafe and ID Robb. Polymer 20:513, 1979.

46. TK Wang, R Audebert. J Colloid Interf Sci 119:459, 1987.

47. B Siffert, F Badri. Prog Colloid Polym Sci 84:309, 1991.

48. ES Pagac, DC Prieve, Y Solomentsev, RD Tilton, Langmuir 13:2993, 1997.

49. K Esumi and H Matsui. Colloids Surf 80:273, 1993.
50. KL Smith, AE Winslow, DE Petersen. Ind Eng Chem 51:1361, 1959.
51. T Ikawa, K Abe, K Honda, and E Tsuchida. J Polym Sci Polym Chem Ed 13:1505, 1975.
52. Y Osada. J Polym Sci, Polym Chem Ed 17:3485, 1979.
53. IF Pierola, M Caceres, P Caceres, MA Castellanos, J Nunez. Eur Polym J 9:895, 1988.
54. Pradip, C Maltesh, P Somasunsaran. Langmuir 7:2108, 1991.
55. K Ishiduki, K Esumi. J Colloid Interf Sci 185:274, 1997.
56. K Ishiduki, K Esumi. Langmuir 13:1587, 1997.
57. NG Hoogeveen, MA Cohen Stuart, GJ Fleer. Langmuir 12:3675, 1996.
58. R Ramachandran, P Somasundaran. J Colloid Interf Sci 120:184, 1987.
59. ED Goddard. Colloids Surf 19:301, 1986.
60. S Saito. J Colloid Interf Sci 158:77, 1993.
61. S Reekmans, M Gehlen, FC De Schryver, N Boens, MV der Auweraer. Macromolecules, 26:687, 1993.
62. NJ Turro, P-L Kuo. Langmuir 2:438, 1986.
63. DW McQuigg, JI Kaplan, PL Dubin. J Phys Chem 96:1973, 1992.
64. A Carlsson, B Lindman, T Watanabe, and K Shirahama. Langmuir 5:1250, 1989.
65. SM Clegg, PA Williams, P Warren, ID Robb. Langmuir 10:3390, 1994.
66. K Chari, TZ Hossain. J Phys Chem 95:3302, 1991.
67. O Anthony, R Zana. Langmuir 10:4048, 1994.
68. T Nojima, K Esumi, K Meguro. J Am Oil Chem Soc 69:80, 1992.
69. R Nagarajan. Colloids Surf 13:1, 1985.
70. YJ Nikas, D Blankschtein. Langmuir 10:3512, 1994.
71. R Nagarajan. In: K Esumi, M Ueno, eds. Structure–Performance Relationships in Surfactants. New York: Marcel Dekker, 1997, pp 1–81.
72. H Otsuka, K Esumi. Langmuir 10:45, 1994.
73. B Sesta, AL Segre, A D'Aprano, N Proietti. J Phys Chem 101:198, 1997.
74. R Bury, B Desmazieres, C Treiner. Colloids Surf 127:113, 1997.
75. K Esumi, Y Takaku, H Otsuka. Langmuir 10:3250, 1994.
76. CM Ma, C Li. J Colloid Interf Sci 131:485, 1989.
77. H Otsuka, K Esumi, TA Ring, J-T Li, KD Caldwell. Colloids Surf 116:161, 1996.
78. J-T Li, KD Caldwell. Langmuir 7:2034, 1991.
79. JS Tan, DE Butterfield, CL Voycheck, KD Caldwell, J-T Li. Biomaterials 14:823, 1993.
80. MA Cohen Stuart, FHWH Waajen, T Cosgrove, B Vincent, TL Crowley. Macromolecules 17:1825, 1984.
81. Y Yamanaka, K Esumi. Colloids Surf 122:121, 1997.
82. K Esumi, M Oyama. Langmuir 9:2020, 1993.
83. J Ghodbane, R Denoyel. Colloids Surf 127:97, 1997.
84. K Esumi, A Masuda, H Otsuka. Langmuir 9:284, 1993.
85. DJ Neivandt, ML Gee, CP Tripp, ML Hair. Langmuir 13:2519, 1997.
86. BM Moudgil, P Somasundaran. Colloids Surf 13:87, 1985.
87. IM Harrison, J Meadows, ID Robb, PA Williams. J Chem Soc Faraday Trans 91:3919, 1995.

88. JN Israelachvili, GE Adams. J Chem Soc Faraday Trans 1, 4:975, 1978.
89. V Shubin, P Petrov, B Lindman. Colloid Polym Sci 272:1590, 1994.
90. PM Claesson, A Dedinaite, E Blomberg, VG Sergeyev. Ber Bunsenges Phys Chem 100:1008, 1996.
91. PA Maurice. Colloids Surf 107:57, 1996.

13

Preparation of Oxide-Coated Cellulose Fiber

Yoshitaka Gushikem
Universidade Estadual de Campinas, Campinas, São Paulo, Brazil

Eduardo Aparecido Toledo
Universidade Estadual de Maringá, Maringá, Paraná, Brazil

I. INTRODUCTION

Cellulose is the most abundant material on Earth; it is the main constituent of plants, serving to maintain their structure, and it is also found in bacteria, fungi, algae, and even in animals. It is a natural polymer formed by β-ᴅ-glucopyranose linked by $(1 \rightarrow 4)$ glucosidic bonds [1,2]. The monomeric unit is formed by two glucosidic units linked as shown in Fig. 1. The conformational arrangements originated by the $\beta(1 \rightarrow 4)$ bonding and the possible crystalline arrangements may be considered as ideal in forming the fiber structures.

The hydroxyl groups are involved in intramolecular and intermolecular hydrogen-bondings [3,4], where the first is the responsible for the rigidity of the macromolecule and confers a cylindrical structure and the second is the responsible for the parallel and antiparallel alignments of the chains [4–7]. The average bond energy of intramolecular and intermolecular hydrogen-bonding is about 67 kJ/mol. The average energy due to the van der Waals interactions which allow the formation of the fibers is about 36 kJ/mol. Due to the numerous possibilities in forming intermolecular hydrogen bonds, cellulose has five crystalline arrangements: I, II, III, IV, and X [1,2]. Figure 2a shows the unit cell of the crystalline cellulose and Figs. 2b and 2c show the antiparallel and parallel arrangements of cellulose I and II. In the crystalline cellulose I, the chains are arranged parallel, whereas for cellulose regenerated from solution or obtained by mercerization, des-

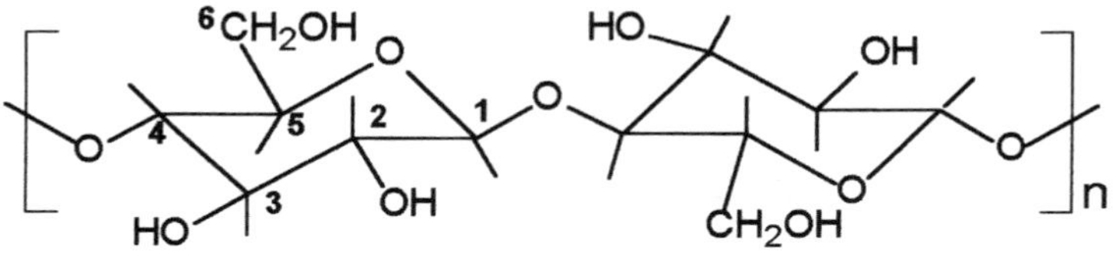

Figure 1   Structure of the cellulose.

ignated as cellulose II, the chains are arranged antiparallel in the crystalline form. The last form is energetically more favorable. The other forms of cellulose are obtained from the form I or II upon divers chemical treatments [8–10]. All the crystalline arrangements have a lamellar morphology which confers a host property to the cellulose and permits it to accept chemical species and form hybrid intercallate composites [11].

The degree of crystallinity varies according from its origin, molecular weight, and the chemical treatment [12]. Celluloses arising from superior organisms have a higher molecular weight and a degree of crystallinity between 30% and 40%, and that arising from inferior organisms such as bacteria have a lower molecular weight and a degree of crystallinity greater than 50%. Because in every case they are semicrystalline, cellulose fibers have crystalline and noncrystalline domains with variable size. Figure 3 is a schematic representation of the cellulose showing the noncrystalline domain region (Fig. 3a), crystalline domain regions (Figs. 3b and 3c), and, the intercrystallite region (Fig. 3d).

In the cellulose polymer, the most reactive groups are the hydroxyl groups, which can be oxidized, esterified, or etherified according with two distinct pathways:

1. Heterogeneous method which consists in reacting a suspension of the polymer with the reagent of interest dissolved in a solvent. In this method, the attack of the reagent occurs preferentially on the noncrystalline regions, which are more acessible. Because the hydroxyl oxygens bonded to C_2 and C_6 are more reactive than those that bonded to C_3 (involved in strong hydrogen-bonding), the former are preferentially attacked [3,13–15].

2. The homogeneous method which consists in reacting the cellulose with the reagent, both in the solution phase; the hydroxyl groups bonded to C_2, C_3, and C_6 behave similarly and, thus, there are no preferential reaction sites.

In the heterogeneous process, the reaction of the cellulose with reagents can be of three types: addition, substitution, and degradation. In the addition process, a swelling-type agent penetrates the cellulose fibers breaking the intermolecular hydrogen bonds separating them [16]. The substitution reaction may occur on the

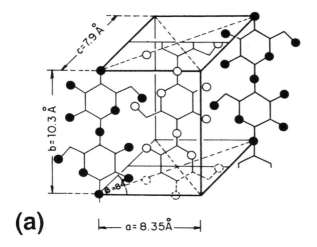

(a)

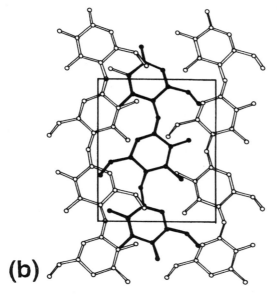

(b)

Figure 2 (a) Crystalline structure of the cellulose; (b) parallel arrangement; (c) antiparallel arrangement.

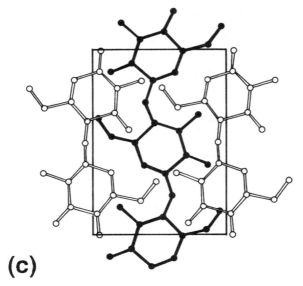

(c)

Figure 2 Continued

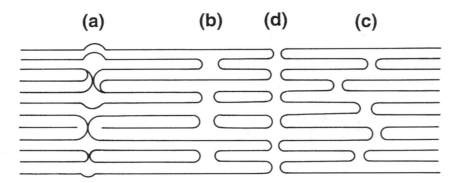

Figure 3 Schematic representation of the cellulose regions: (a) amorphous domain, (b, c) crystalline domains, and (d) intercrystallite region.

monomeric hydroxyl unit, D-glucose, on carbons C_2, C_3, and C_6 by two processes: sterification and etherification. In the degradation process, the bond between carbons C_1 and C_4 of the glucosidic bond is broken, decreasing the size of the chains [17]. In all three reactions, the intramolecular, and intermolecular interactions in the cellulose macrostructure, as mentioned earlier, lower its reactivity and, as consequence, limit its industrial use [18].

Studies of metal oxide particles dispersed in cellulose have also been carried out with the acetylated polymer [19–21]. Depending on the degree of acetylation, the polymer surface becomes more hydrophobic due to the replacement of the hydroxyl groups responsible by the hydrogen-bonding. The polymer becomes soluble in organic solvents and can be molded as fiber or membrane forms and further modified with the metal oxide [22–24]. In the membrane form, the oxide particles are trapped inside the polymer structure and protected against leaching from the matrix when immersed in a solvent; thus, the membrane can be useful in reverse osmosis and ultrafiltration processes [25–27]. In most cases, these acetate membranes do not have a homogeneous cross section i.e., the metal oxide particles are distributed in the interior of the matrix, whereas at the surface of the membrane, they are not detected, as probed by energy-dispersive spectroscopic (EDS) mapping of the fractured membranes [28].

The object of this work is to describe the preparation of the oxide-coated cellulose and some physical and chemical characteristics of the oxide particles dispersed on the polymer as fiber or membrane forms. There are not many examples in the literature in this field despite the enormous potential applications of these materials. Some examples of applications are described in Sec. IX.

II. THE OXIDE-COATING PROCESSES

The experimental methodology of the fiber-coating process varies depending on which form the cellulose is to be obtained: fiber or membrane. In the fiber form, the treatment of cellulose with a precursor reagent can be made in aqueous or nonaqueous solvent. To prepare the membranes, cellulose acetate is normally used because it is very soluble in most of the common organic solvents such as acetone. Generally, two procedures are also used to coat the fibers in the membrane form: (1) Acetate cellulose and the precursor reagents are dissolved in a nonaqueous solvent, molded as a membrane, and followed by the phase-inversion process and (b) the previously prepared membrane is immersed in a solution of the precursor reagent followed by its hydrolysis. The general procedure for each of the two cellulose forms will be discussed.

A. Fiber-Coating Processes

1. Precipitation of the Metal Oxide from an Aqueous Solution

This method of coating consists in immersing the cellulose fiber in an aqueous solution of the metal ion of interest followed by evaporation of the solvent to dryness. The fiber is submitted to a stream of ammonia gas which reacts with the metal ion, and the hydrated oxide is deposited on the fiber surface. Furthermore, it is followed by a careful drying in order to obtain a homogeneously distributed metal oxide particles [29]. Zirconium (IV) oxide deposited on the fiber surfaces by this procedure was reported to be easily detached during the washing stage and only the oxide particles crystallized on the cellulose fiber were stable and resistant to ordinary handling [29]. In this process, the amount of the loaded zirconium oxide on the fiber increased with the initial concentration of the zirconium precursor reagent in the solution phase and the average diameter of the resulting deposited oxide particles increased with the supported amount.

2.   Deposition of the Oxide by Decomposition of the Precursor Reagent

An example of oxide coating by the precursor decomposition process is described in the preparation of the MnO_2-coated cellulose acetate fiber [19]. The precursor reagent in this case is a solution of $KMnO_4$ acidified with HNO_3. The reagent is decomposed to the MnO_2 species which deposit on the fiber surfaces. The amount of supported oxide is described to be about 3 wt% in the various preparation batches, within a 15% deviation. In this process, the oxide-coating layer is more efficiently adhered on the fiber surface and only a negligible release of the oxide metal particles, in aqueous solution, is observed when the composite material is submitted to an ion-exchange reaction experiment.

In both of the procedures described there is no mention of the degree of degradation of the fibers after the chemical treatments.

3.   Preparation from Nonaqueous Solution

As an example of preparation, the procedure to obtain Cel/Nb_2O_5 [30] is described: Fresh sublimed $NbCl_5$ is suspended in dry CCl_4 and a sufficient amount of pure ethanol is added under nitrogen atmosphere. The mixture is refluxed until liberation of HCl gas ceases and complete dissolution of $NbCl_5$ occurs and, at this stage, cellulose is added and the mixture stirred for few hours under dry-nitrogen atmosphere, at room temperature. The mixture is filtered in a Schlenck apparatus under nitrogen atmosphere and washed with CCl_4. The residual solvent is eliminated under vaccum (10^{-4} Torr) and the resulting solid submitted to a NH_3 atmosphere and then exposed to moist air. The solid is then washed with 0.01 M HNO_3 solution and then dried under vaccum (10^{-4} Torr) at room temperature. A similar procedure is used to prepare Cel/ZrO_2 [21] and Cel/TiO_2, where Cel is cellulose [31].

A degradation of the macromolecule may occur in this process of preparation. The mechanism of the degradation in the presence of the Lewis acid reagents, MCl_n, in nonaqueous solvent can be conceived as a coordination of the central atom of the acid with semiacetal, cyclic, or glucosidic oxygen atoms with the for-

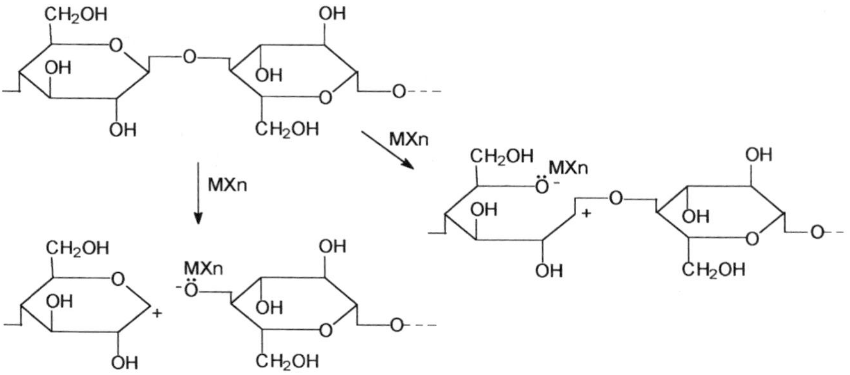

Figure 4 Reaction mechanism of the Lewis acid, MX_n, with cellulose in a nonaqueous solvent.

mation of donor–acceptor bonds involving the metal vacant orbital and the lone-pair electrons of the oxygen atoms. Figure 4 illustrates this reaction mechanism proposed by Sarybaeva et al. [32]. In this mechanism, the Lewis acid MX_n attacks the ring or the glucosidic C—O—C bonds, which are further ruptured. The rupture of the glucosidic bond, in particular, results in the degradation of the macrochain. The formation of the donor–acceptor complex at the first stage of the reaction was evidenced by the infrared spectroscopy data. For instance, the infrared spectra of cellulose/$TiCl_4$ in the 1100–1000-cm^{-1} region was assigned to a vibrational mode associated with the formation of the C—O—Ti bond [33]. It is reasonable to suppose that upon water treatment, the donor–acceptor complex hydrolyzes completely the M—O—C bond with phase separations into degraded cellulose particles and the metal oxide particles dispersed on its surface. Thus, the formation of the donor–acceptor complex precedes the degradation of the cellulose macrochain. The extension of degradation depends on the temperature under which the samples are submitted during the preparation and further drying process.

III. CHANGE OF THE CRYSTALLINITY DEGREE OF THE CELLULOSE

Such studies are made only for cellulose which shows fibers in amorphous and crystalline domains (Fig. 3) and taking into account that those in the former domains are more reactive than those in the crystalline ones. Therefore, the studies normally are not made for cellulose acetate because it is usually obtained as an amorphous form. The extension of degradation of the cellulose structure can be evaluated by measuring the degree of crystallinity (DC) of the untreated cellulose and that of Cel/M_xO_y. The DC can be evaluated from ^{13}C CP-MAS nuclear magnetic resonance (NMR) spectra and by x-ray diffraction patterns of Cel/M_xO_y. In order to illustrate how this is made, one example of each technique is detailed.

Figure 5 shows the ^{13}C CP-MAS NMR for Cel/Nb_2O_5. The assigned bands C_1–C_6 are the resonance peaks of the C atoms of Fig. 1. The areas under the individual bands (dotted lines) were determined by deconvolution of the NMR peaks using the equation $f(ppm) = \Sigma\, G_i(ppm)$, where $G_i(ppm)$ are the Gaussian functions, one for each peak. The degree of crystallinity was calculated by the ratio of the area under the crystalline phase and total area under C_4, where the peaks marked with **a** and **c** refer to the crystalline and amorphous regions of the cellulose [34–37].

Similarly, the DC can be calculated using the x-ray data. Figure 6 shows the x-ray diffraction (XRD) patterns for Cel/ZrO_2. The circles are the experimental points and the solid lines are the fitted curves, where the broad one refers to the diffraction of the amorphous domain of the cellulose. The areas under the x-ray peaks is determined by deconvolution of the x-ray diffractograms peaks by applying the equation $f(2\theta) = \Sigma\, G_i(2\theta) + AM(2\theta) + B(2\theta)$, where $G_i(2\theta)$ are the Gaussian distribution functions, and $AM(2\theta)$ is a broad Gaussian function; for the baseline function $B(2\theta)$, a straight line is assumed. The ratio between the crystalline peaks areas over the total area under the diffractograms gives the crystallinity degree [38]. Table 1 shows the results of DC calculated using both methods for some composites.

It is observed that DC increases as the quantity of M_xO_y loading increases for Cel/ZrO_2, Cel/Nb_2O_5, and Cel/Sb_2O_3. For Cel/ZrO_2, the DC were determined by both XRD and ^{13}C CP-MAS NMR. The absolute values differ, but the relative ones show the same trend (i.e., the DC increases as the metal loading is increased). Although the DC is changed by the coating process, the specific surface areas, S_{BET}, under a practical point of view, remained the same in every case.

The tendency of the DC increase with the metal oxide loading is related to the higher reactivity of the celulose units localized in the amorphous region of the polymer. In the first step of the preparation procedure, the Lewis acid, MX_n, attacks the cellulose units preferentially in region a of Fig. 3, according to the mechanism shown in Fig. 4, and in the second step, the hydrolysis of the donor–acceptor species is followed by a degradation of the cellulose fibrils into smaller units. Evidences of such degradation process is very evident from scanning electron microscopic (SEM) images.

For Cel/TiO_2, the DC increased slightly for sample with 1.2 and 7.0 wt% of the metal oxide loading and decreased slightly for 10 wt% and was more pronounced for 16 wt%. Contrary to that observed for Cel/ZrO_2, Cel/Nb_2O_5, and

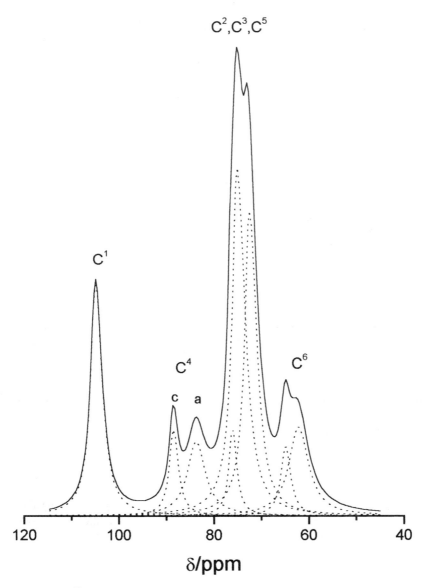

Figure 5 ^{13}C CP-MAS NMR spectra of Cel/Nb$_2$O$_5$ (Nb = 16.0 wt%). (Reproduction from Ref. 30 by permission of Academic Press, Inc.)

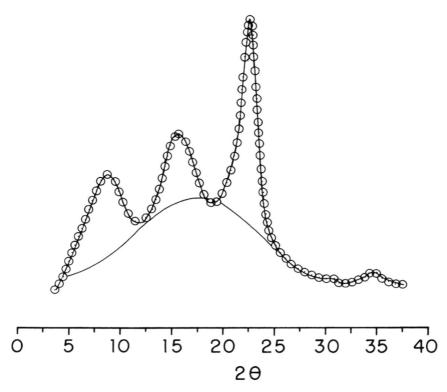

Figure 6   XRD pattern of Cel/ZrO$_2$. The solid lines represent the fitted curves, and the circles represent the experimental points.

Cel/Sb$_2$O$_3$, the specific surface area for Cel/TiO$_2$ increased significantly for samples with metal oxide loadings of 7, 10, and 16 wt%. The main reason is presumably associated with the amorphous and more porous character of the coating TiO$_2$ (i.e., the greater the amorphous character of the oxide, the more it can contribute to increase the area under the broadened XRD peak, given by the solid line in Fig. 6).

IV.   MORPHOLOGY OF THE FIBER SURFACE

The morphology of the oxide layer on the cellulose fiber surface was examined by SEM and the degree of the metal dispersion using a x-ray energy-dispersive spec-

Table 1 Amount of Metal Oxide Loading and Degree of Crystallinity of Cel/M_xO_y

| Cel/M_xO_y | M_xO_y (wt%) | DC (%) | | S_{BET} (m²/g) |
		XRD	¹³C NMR	
Cellulose	—	46	36	1.6
(Ref. 21)				
Cel/TiO_2	1.2	50	—	—
(Ref. 31)	7.0	49	—	19
	10	40	—	25
	16	33	—	46
Cel/ZrO_2	4.0	50	38	—
(Ref. 21)	5.0	57	47	—
Cel/Nb_2O_5	4.5	—	46	4.0
(Ref. 30)	10.2	—	47	4.2
	16.0	—	47	6.4
Cel/Sb_2O_3	2.0	—	41	0.9
(Ref. 39)	4.8	—	49	1.3
	9.2	—	50	1.6

trometer (EDS) for x-ray mapping. The morphology differed depending on the nature of the metal. Islands were formed in the case the of Cel/ZrO_2 composite, the particle dimensions of which varied between 1 and 9 μm, with a predominance of those within 2–3 μm [40]. For Cel/TiO_2 and Cel/Nb_2O_5 composites, a homogeneous and continuous film of TiO_2 or Nb_2O_5 could be observed with no detectable island formation [30,31]. As an example, Fig. 7 shows the SEM images of Cel/TiO_2 with 1.6- (Fig. 7a) and 16-wt% (Fig. 7b) oxide loadings. Figure 7a shows that the fiber surface morphology compared with that of the untreated cellulose surface, within the magnification used, does not present any significant change. However, in case of a higher loading of the oxide, as can be observed in Fig. 7b, it is caused a visible degradation of the microfibrils. In order to certify that the observed layer is constituted predominantly by the metal oxide, its x-ray mapping, using a x-ray EDS, was obtained. The emission lines of Ti$\kappa\alpha$ and Ti$\kappa\beta$ at 4.52 and 4.93 eV, respectively, confirmed that in Fig. 7b, the layer is basically constituted by the TiO_2 species. In every case, these films are amorphous, except in case of Cel/Sb_2O_3, where the oxide is crystalline [39].

The specific surface area changed for Cel/TiO_2 with an oxide loading above 7.0 wt% (Table 1). The amorphous nature and higher porosity of the coating oxide

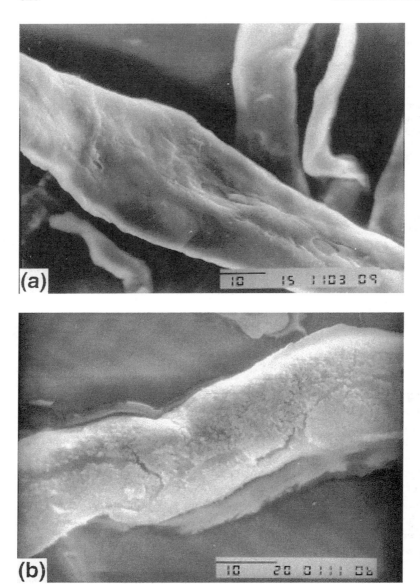

Figure 7 Scanning electron microscopy images of Cel/TiO$_2$: (a) TiO$_2$ = 1.6 wt%; (b) TiO$_2$ = 16 wt%.

may be the responsible for such area increase because for the remaining listed composite materials, such change was not observed (i.e., the specific surface areas were practically the same of that of untreated cellulose).

The oxide in the composite Cel/Sb_2O_3 has a crystalline structure. Figure 8a shows the synchrotron x-ray diffraction patterns for this material having 9.2 wt% of the oxide and is compared with those of crystalline Sb_2O_3 (Fig. 8C), the structure of which is orthorhombic [41]. Figure 8B shows the diffraction patterns of the native cellulose. The synchroton experiments were carried out at the National Synchrotron Radiation Laboratory in Campinas, SP. The XRD patterns using the synchrotron radiation improved the peak resolutions considerably compared with those obtained by using the conventional XRD spectrometer.

The SEM and the EDS mapping images of Sb(III) showed that in this case, despite its crystallinity, the oxide particles are also homogeneously dispersed on the cellulose fibers. No particle agglomerates with detectable sizes, within the magnification used, are observed.

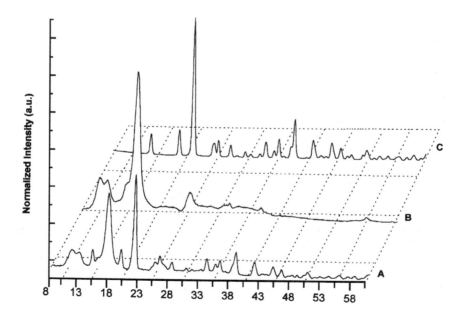

Figure 8 Synchrotron XRD patterns of (A) Cel/Sb_2O_3, (B) untreated cellulose, and (C) Sb_2O_3.

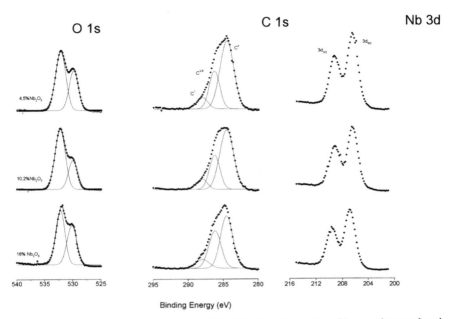

Figure 9 X-ray photoelectron spectra of Cel/Nb$_2$O$_5$. The carbon C1s was deconvoluted into three peaks: C^s is the hydrocarbon peak and C^1 and C^{2-6} correspond to C$_1$ and C$_2$–C$_6$ of Fig. 1. The O1s peak was deconvoluted into two peaks: 532.5 eV due to the cellulose oxygen and 530.1 eV due to the niobium oxygen atom. (Reproduced from Ref. 30, with permission of Academic Press, Inc.)

V. X-RAY PHOTOELECTRON SPECTRA

Figure 9 illustrates the XPS spectra obtained for Cel/Nb$_2$O$_5$. The areas under the binding energy (BE) peaks of O1s and C1s were obtained by the Gaussian deconvolution process as described earlier. Table 2 lists the results of the atomic percentages of oxygen bonded to cellulose matrix and oxygen bonded to the metal ion. The atomic ratios were estimated from the areas under the BE peaks and the Scofield cross section [42].

For Cel/Nb$_2$O$_5$ and Cel/TiO$_2$, the atomic ratios within the experimental error of the technique did not change with the metal oxide loadings. The results indicate that in every case, the surfaces were saturated with the metal oxides, even for the lowest-metal-loaded material, confirming the results obtained by the micrography technique.

For Cel/Sb$_2$O$_3$, estimation of the atomic ratios were carried considering the

Table 2 X-Ray Photoelectron Spectroscopy Data for Cellulose and Cel/M$_x$O$_y$

Coating oxide	M$_x$O$_y$ (wt%)	Metal and oxygen BE (ev)	Atomic ratios[a] (O^i/M)
Cellulose (Ref. 21)	—	532.5 (O1s)	—
Cel/TiO$_2$ (Ref. 31)	7.0	458.5 (Ti 2$p_{3/2}$); 530.5 (Ti O1s)	3.1
	10.0	530.5 (Ti O1s)	3.0
	15.5	530.5 (Ti O1s)	2.9
TiO$_2$ (Refs. 43 and 44)	100	458.5(Ti 2$p_{3/2}$); 530.5(Ti O1s)	—
Cel/Nb$_2$O$_5$ (Ref. 30)	4.5	209.3(Nb 3$d_{3/2}$); 206.7(Nb 3$d_{5/2}$) 530.1 (Nb O1s)	2.2
	10.2	530.1 (Nb O1s)	2.5
	16.0	530.1 (Nb O1s)	2.2
Nb$_2$O$_5$ (Refs. 45–47)	100	210.2(Nb$_{3/2}$); 207.4(Nb$_{5/2}$); 530.7 (Nb O1s)	
Cel/Sb$_2$O$_3$ (Ref. 39)	2.0	540.2 (Sb 3$d_{3/2}$)	2.4
	4.8	539.9 (Sb 3$d_{3/2}$)	3.8
	9.2	540.6 (Sb 3$d_{3/2}$)	5.0
Sb$_2$O$_3$ (Ref. 48)	100	530.3(Sb 3$d_{5/2}$); 539.6(Sb 3$d_{3/2}$); 529.8(Sb O1s)	

[a]O^i is the oxygen bonded to the metal.

O2s and Sb4d BE peaks, in virtue of the overlap between the O1s and Sb 3 $d_{3/2}$ BE peaks (i.e., 529.9 and 530.3 eV, respectively). Therefore, the O^i/M ratio calculated taking into account the O2s and Sb4d binding energy peaks gives a result whose values are only reliable considering the figure magnitude order.

VI. COATING OF THE FIBERS IN MEMBRANES

A. Preparations

General procedures for the preparation of the metal oxide incorporated in a cellulose acetate membrane are described for CelA/Nb$_2$O$_5$ and CelA/ZrO$_2$, illustrating the two procedures cited previously.

Preparation of CelA/Nb$_2$O$_5$ (method a): Cellulose acetate is added to a mixture containing glacial acetic acid and acetone and the mixture is stirred for a few hours and allowed to rest until a viscous syrup is formed. Freshly sublimed

NbCl$_5$ is added to the cellulose acetate syrup and the mixture is stirred under nitrogen atmosphere until complete dissolution of the metal halide, which occurs with the evolution of gaseous HCl. When the evolution of the gas ceases, the mixture is spread over a glass plate with a spacer and the solvent evaporated at room temperature and then immersed in bidistilled water to promote the complete hydrolysis of the metal. The CelA/Sb$_2$O$_3$ composite membrane was prepared following an almost similar procedure.

Preparation of CelA/ZrO$_2$ (method b): In the first step, the casting solution of the membrane is prepared by dissolving celullose acetate in a mixture of acetone and glacial acetic acid and the resulting syrup spread over a glass plate and the thickness controlled by using a spacer. The solvents are evaporated at room temperature and the resulting membrane stored immersed in bidistilled water. At the second step, the membrane, previously dried under air, is first immersed in a propanol solution of zirconium propoxide, Zr(PrO)$_4$, and then in a 10^{-3} M HNO$_3$ solution, in order to hydrolyse the metal alkoxide.

Because of the difference in the experimental procedure, incorporation of the metal oxide in the membrane differed considerably depending on the preparation method used. For instance, for the composites CelA/Nb$_2$O$_5$ and CelA/Sb$_2$O$_3$, metal oxide loadings between 1.1 and 20.9 wt% and 1.8 and 13.4 wt% are achieved, respectively. For CelA/ZrO$_2$, depending on how long time the membrane remained immersed in the solution containing the precursor reagent, a maximum of 0.6 wt% of the metal oxide loading is achieved.

Despite the difference in the metal loadings in the membranes, the SEM image and the mapping of the elements by the EDS technique have shown that the metal oxides in CelA/Nb$_2$O$_5$ and CelA/ZrO$_2$ are uniformly dispersed inside of the matrix [21–30]. A study carried out with CelA/Nb$_2$O$_5$ [28] membrane showed that the band structure which emerged resulting from the increasing interaction of the individual niobium–oxygen complexes for higher metal loadings shifted the maxima of the electronic absorption band treshold to a higher wavelength (i.e., up to 330 nm). For bulk Nb$_2$O$_5$, this band treshold is observed at 410 nm [49]. This observation indicates that the particles are dispersed as monomeric and oligomeric species, even for higher-oxide-loaded membrane.

For CelA/Sb$_2$O$_3$, a distinct phase separation between the membrane and the oxide particles could be observed [39] and the crystalline Sb$_2$O$_3$ phase was distributed along all extension of the matrix, contrary to that observed for other two composite membranes. The migration of the Sb$_2$O$_3$ crystalline particles to the membrane surface occurred with time. For the freshly prepared membrane, no oxide particle on the surface could be observed. Resting the membrane in a pure aqueous solution, the appearance of the crystalline oxide particles at the surface is very clear after 7 days (Fig. 10).

Table 3 presents the results for the celullose acetate membranes CelA/M$_x$O$_y$ with different metal loadings.

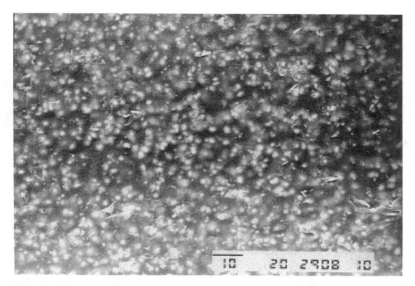

Figure 10 Scanning microscopy image of CelA/Sb$_2$O$_3$ after 7 days of its preparation. Sb$_2$O$_3$ = 13.4 wt%.

Table 3 Amount of Metal Oxide Loading in CelA/M$_x$O$_y$

Samples	M$_x$O$_y$ (wt%)	Nature of the oxide
CelA/Nb$_2$O$_5$	1.1–20.9[a]	Amorphous
CelA/ZrO$_2$	0.16–0.59[b]	Amorphous
CelA/Sb$_2$O$_3$	1.8–13.4[a]	Crystalline

[a]Prepared by mixing the casting solvent and a solution of the precursor metal.
[b]Prepared by the dip-coating method.

VII. METAL–MATRIX INTERACTION IN THE MEMBRANE

The nature of interaction between the metal and matrix is not clear. The infrared band observed at 1560 cm^{-1} has been tentatively assigned as due to the Zr—O—C vibrational mode in the composite membrane of CelA/ZrO$_2$ [23]. Kennedy and Cabral [50] have proposed a hydrogen-bonding interaction between the polymer and the metal oxide instead of Zr—O—C covalent bonding.

The $O1s$ binding energy determined by XPS for CelA/ZrO$_2$ [21] with different metal loadings showed the following oxygen $O1s$ BEs: $O1s^a = 532.2$ eV and $O1s^b = 530.2$ eV, where $O1s^a$ and $O1s^b$ are the oxygen atoms bonded to the cellulose matrix and to the metal ion, respectively. For pure CelA, O_{1s} BE = 532.6 eV, and for ZrO$_2$, $O1s$, BE = 530.1 eV [43]. It is important to note that $O1s^a$ BE in the composite membrane is lower than that observed in the untreated cellulose. This shift to lower energy indicates that in the composite membrane, $O1s^a$ has a higher electronic density than in the untreated membrane. This shift could be assigned to the hydrogen-bonding interaction –OH—OZr– as proposed by Kennedy and Cabral [50].

VIII.   LEWIS ACID SITES

The presence of Lewis and Brønsted acid sites in the incorporated oxide using pyridine as the molecular probe was studied in detail for Cel/Nb$_2$O$_5$ [30] and CelA/Nb$_2$O$_5$ [28] composites with different metal oxide loadings. The absorption bands at about 1590 and 1490 cm^{-1} were assigned as due to the 8a and 19b modes of pyridine adsorbed on the Lewis acid site, and the band at about 1540 cm^{-1} were assigned as due to the 8b mode of pyridine adsorbed on a Brønsted acid site [51–53]. The Brønsted acid sites normally do not resist heat treatment at a temperature above 373 K for either fiber or membrane, because they are formed primarily by the water molecule adsorbed on the hydrated surface oxide [54,55]. The Lewis acid sites are thermally more stable and resistant to heat treatment up to 473 K for both oxides incorporated in membrane or fiber form.

For CelA/Nb$_2$O$_5$, the membrane is particularly interesting to observe in that plotting the integrated area under the band at 1606 cm^{-1} (pyridine 8a mode) against the quantity of loaded Nb$_2$O$_5$ in the membrane, they are proportional, as can be seen in Fig. 11. It means that even for a larger amount of metal oxides in the membrane, no extensive saturation of the Lewis acid sites resulting from the Nb$-$O$-$Nb bond formation occurs. The conclusion is similar of those taken from ultraviolet (UV) electronic absorption bands and from electron miscroscopy of niobium mapping with the EDS image.

IX.   APPLICATIONS

Composite materials are made with the objective of joining the properties of the constituents, and in the particular case of Cel/M$_x$O$_y$ and CelA/M$_x$O$_y$ composites, their use is related to the properties of the incorporated oxides.

The general advantage of using cellulose or cellulose acetate as substrates is related to their low cost, biodegradability, ease of handling, and simplicity of the

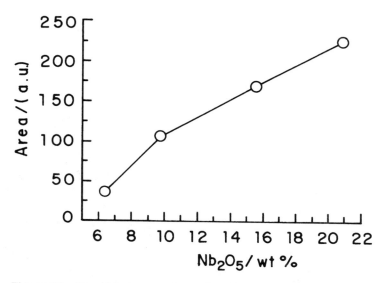

Figure 11 Plot of the integrated area (in arbitrary units) of the band at 1606 cm^{-1} of the pyridine 8a mode against the metal loading.

composite preparations. Cellulose is also a good solvent-resistant substrate and has a reasonable thermal stability over 373 K. This solvent resistance can inhibit its use as a fiber, as a processability step is necessary. However, this restriction is removed using the acetylated cellulose, which is soluble in common organic solvents and permits its easy use in different forms and shapes.

The use of the oxide-coated celluloses in the fiber or other shaped forms, despite their enormous potential for applications, have not been widely explored. The examples found in the recent literature embrace different areas of applications which include ion exchange, enzyme immobilization, reverse osmosis, adsorption of chemical species from solutions, and use in polymer grafting reactions. Applications described in the literature in recent years were collected and details for some of them are presented.

A. TiO$_2$

This oxide is the most used oxide in the preparation of these composites. The fiber of Cel/TiO$_2$ has been used as a packing material of a microcolumn in flow-injection analysis for selective adsorption and preconcentration of the chromium(VI) ion from aqueous solutions [56]. Although Cr$_2$O$_7^{2-}$ is a strong oxydizing agent, it can be adsorbed and desorbed from the column without any detectable decomposition during the experiment. Table 4 shows results of Cr(VI) analysis in the pres-

Table 4 Study of the Interference of Some Ions on the
Adsorption of Cr(VI) by Cel/TiO$_2$

Interferent ion	Ratio of [interf.ion]/[Cr(VI)][a]	Interference factor[b]
Cr^{III}	50	0.93
	10	1.07
Fe^{III}	50	—
	10	0.94
	1	1.00
MoO$_4^{2-}$	50	1.16
	1	0.99
SO$_4^{2-}$	10	0.85
	1	1.00
HPO$_4^{2-}$	10	0.76
	1	0.99

[a]Concentration of [Cr(VI)] = 0.1 ppm.
[b]Interference factor calculated by the ratio of known concentration of
Cr(VI) in the binary mixture and in the pure Cr(VI) solution.

ence of some interferent ions. (the complete list of the adsorbents and details of the
experiments is found in Ref. 56.)

The material is very selective to Cr^{VI} in the presence of Cr^{III} and in the presence of anions which can compete with the adsorptions sites of TiO$_2$. The presence
of a large amount of Cr^{III} does not affect the analysis, which turns the material very
appropriately for chromium (VI) speciation in the presence of Cr(III). When the
interferent ion is Fe^{III}, it can precipitate on the composite surface because of the
hydrolysis of the ion at the solution's pH. Using a flow-injection system, a concentration as low as 0.028 ppm of Cr^{VI} could be detected spiking this ion in natural water [56].

Other applications of the material Cel/TiO$_2$ include immobilization of the
horseradish peroxidase enzyme for oxidation of pyrogalol by the flow-injection
technique [57] and immobilization of urease oxidase, glucose oxidase, L-amino
oxidase, and phosphate adsorption [22]. Photocatalytic reduction of dioxygen to
hydrogen peroxide was described by using the cobalt(III) cyclam [cyclam =
1,4,8,11-tetraazacyclotetradecane] complex adsorbed on oxide-modified cellulose
and irradiated with a xenon lamp [58].

Molded as a membrane, CelA/TiO$_2$, has been employed to entrap the urease
and invertase enzymes [59]. The Michaelis–Menten constant, K_m, and V_{max} was
determined and showed that the first had tendency to increase, an indication that
the affinity of the enzyme to the substrate increased. On the other side, the values

of V_{max} decreased, indicating that the reaction became slower for the immobilized enzyme in comparison with the native one. An advantage was the stability of the material toward the most common solvents, phosphate and electrolyte solutions, over a wide range of pH between 4 and 10.

B. Fe₂O₃

Colloidal Fe_2O_3 dispersed on a cellulose fiber, Cel/Fe_2O_3, has been described as an efficient photosensitizer for grafting acrylamide onto cellulose acetate films [60]. Cel/Fe_2O_3 has also been used in the phosphate nucleation reaction [61] and in the reverse osmosis process for rare earth separation [62].

C. ZrO₂

The cellulose acetate membrane of this oxide has been used to immobilize urease and glucose oxidase enzymes [23]. Highly dispersed oxide in the membrane is an efficient phosphate adsorbent [21].

D. Ag₂O

The porous cellulose acetate membrane $CelA/Ag_2O$ prepared by a counterdiffusion process [63,64] was treated with a $NaBH_4$ solution in order to reduce the oxide to fine particles of metallic silver. The resulting composite material with particle sizes in the nanometer range was used for the detection of trace amounts of uridine and alkaloids such as quinacrine hydrochloride, nicotine, and ellipticine in the concentration range 10^{-5}–10^{-7} M. Detection of these biological molecules was possible by using the surface enhanced Raman spectra (SERS), which was facilitated by the Ag nanoparticles dispersed in the gel network of the polymer.

E. Nb₂O₅

The membrane of cellulose acetate was prepared for immobilization of cobalt (II) metalated hematoporphyrin IX [8,13-bis(1-hydroxyethyl)-3,7,12,17-tetramethyl-21H-porhyne-2,18-dipropionic acid] and used as an electrochemical sensor for dissolved oxygen [65]. The unmetallated porphyrine is easily immobilized on the modified membrane, presumably by the Nb—OOC—bond formation [66] of the two propyonic carboxylic groups of the molecule (Fig. 12). Metallation of the porphyrine ring is made adding it to a metal-ion aqueous solution. An important characteristic is the ability of the material to reduce dissolved oxygen. In order to carry out the experiment, the membrane is spread over a platinum electrode and connected to potentiostat/galvanostat equipment. The electrocatalytic response of the electrode is shown in Fig. 13a, where "Ar in" and "O_2 in" means the electrode re-

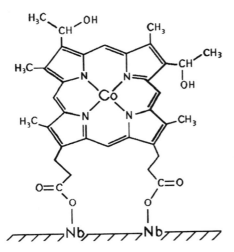

Figure 12 Schematic representation of Co(II) metallated hematoporphyrin IX immobilized on Cel/Nb$_2$O$_5$.

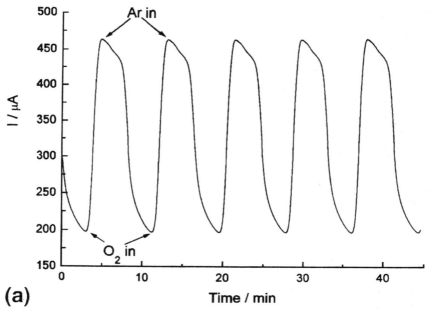

(a)

Figure 13 Electrochemical response of the CelA/Nb$_2$O$_5$ electrode in 1 M KCl supporting electrolyte solution at pH 1: (a) cathodic peak current intensities for alternate cycles in the presence of dissolved oxygen (O$_2$ in) and purged with argon (Ar in); (b) cathodic peak current intensities against the concentration of the dissolved oxygen. (Reproduced from Ref. 65, with permission of the Brazilian Chemical Society.)

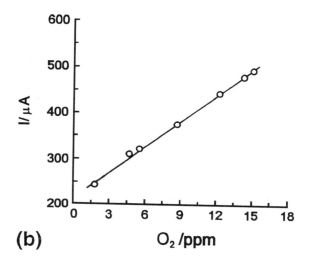

(b)

sponse in presence of argonium or O_2, respectively. The measurements were carried out at −390 mV and the intensity of the cathodic current plotted against the dissolved O_2 concentration showed a linear correlation between 1.5 and 15 ppm (Fig. 13b). Although cellulose acetate is a insulating material, the efficient electrochemical process at the modified electrode surface–solution interface is related to the facility with which the membrane is wet when immersed in the electrolyte solution, and also to the presence of Nb_2O_5 in the matrix, which increases the hydrophylic character of the membrane.

F. MnO$_2$

Cellulose acetate fibers coated with MnO_2, CelA/MnO$_2$, was employed for adsorption of copper(II), zinc(II), and lead(II) ions at very low concentrations [19]. The Langmuir plot for copper(II) and lead(II) adsorption isotherms showed somewhat high adsorption capacities (i.e., 1.7 and 2.0 mmol/g, respectively). Adsorption of zinc was reported to be much higher (i.e., 13 mmol/g, but in this case, this value was interpreted as a result of the formation of a new phase on the oxide surface. Tests revealed that the oxide is the exclusive site of adsorption for CelA/MnO$_2$, indicating that this material as an exceptional adsorbent for removing these metals when they are present as contaminants.

REFERENCES

1. M Yalpani. Studies in Organic Chemistry 36: Polysaccharides, Synthesis, Modifications and Structure/Properties Relations, 1st Ed., Amsterdam: Elsevier Science, 1990, p 17.
2. J Blackwell. In: RM Brown Jr, ed. Cellulose and Other Natural Polymer Systems: Biogenesis, Structure, and Degradation. New York: Plenum Press, 1983, pp 403–428.
3. IG Hurtbise, H Krassig. Anal Chem 32:177, 1960.
4. NV Ivanova, EA Korolenko, EV Korolenko, RG Zhbankov. J Appl Spectrosc 51:847, 1989.
5. MV Shishonok, EV Gert, TI Filanchuk, FN Kaputskii. J Appl Chem–USSR 60:1086, 1987.
6. JJ Sugiyama, J Persson, H Chanzy. Macromolecules 24:2461, 1991.
7. RM Mukhamadeeva, RG Zhbankov, VI Kovalenko, VF Sopin, GN Marchenko. J Appl Spectrosc 52:403, 1990.
8. HJ Marrinan, J Mann. J Polym Sci 21:301, 1956.
9. LE Segal, L Loeb, JJ Creely. J Polym Sci 13:193, 1954.
10. O Ellefsen, N Norman. Chem Abstr 57:14025c, 1962.
11. S Aravindanath, PB Yier, S Sreenivasan. J Appl Polym Sci 46:2239, 1992.
12. P Koiseth, A de Ruvo. In: JA Bristow, PI Kolseth, ed. Paper: Structure and Properties. New York: Marcel Dekker, Inc., 1986, pp 3–26.
13. S Hudson, JA Cuculo. J Macromol Sci—Rev Macromol Chem Phys C18:1, 1980.
14. AN Zhilkin, DA Sukhov, PM Valov. J Appl Spectrosc 2:566, 1990.
15. JJ Cael, KH Gardner, JL Koenig, J Blackwell. J Chem Phys 62:1145, 1975.
16. RA Young, RM Rowell. Cellulose Structure, Modification and Hydrolysis. New York, Wiley, 1986.
17. Y Zhi-li, W Kuo-Ming, M Chein-Fong, F Han-De, Cellulose Chem Technol 16:651, 1982.
18. L Czarnecki, JL White. J Appl Polym Sci 49:1217, 1980.
19. MSB Brandão, F Galembeck. Colloids Surf 48:351, 1990.
20. H Hatayama, T Swabe, Y Kurokawa. J Sol-Gel Sci Technol 7:13, 1996.
21. UP Rodrigues Filho, Y Gushikem, MC Gonçalves, RC Cachichi. Chem Mater 8:1375, 1996.
22. Y Kurokawa, T Sano, H Ohta, Y Nakagawa. Biotechnol Bioeng 42:394, 1993.
23. Y Kurokawa, H Ohta, M Okubo, M Takahashi. Carbon Polym 23:1, 1994.
24. Y Kurokawa. Polym Gels Network 4:153, 1996.
25. I Goossens, A van Haute. Desalination 18:203, 1976.
26. H Finken. Desalination 48:207, 1983.
27. W Doyen, R Leyse, J Mottar, G Waes. Desalination 79:163, 1990.
28. EA Campos, Y Gushikem. J Colloid Interf Sci 193:121, 1997.
29. M Suzuki, T Fujii. In: M Abe, T Kataoka, T Suzuki, ed. New Developments in Ion Exchange. Materials, Fundamentals and Applications, Tokyo: Elsevier, 1991, pp 355–360.
30. EA Campos, Y Gushikem, MC Gonçalves, SC de Castro. J Colloid Interf Sci 180:453, 1996.

31. LRD da Silva, Y Gushikem, MC Gonçalves, UP Rodrigues Filho, SC de Castro. J Appl Polym Sci 58:1669, 1995.
32. RI Sarybaeva, AS Sultankulova, TV Vasilikova, VA Afanasiev. Cellulose Chem Technol 25:199, 1991.
33. VA Afanasev, PI Sarybaeva, AS Sultankulova, TV Vasilikova. Pure Appl Chem 61: 1993, 1989.
34. RH Atalla, JC Gast, DW Sindorf, VJ Bartuska, GE Maciel. J Am Chem Soc 102:3249, 1980.
35. WL Earl, DL Vanderhart. J Am Chem Soc 102:3251, 1980.
36. A Hirai, F Horii, R Kitamaru. Cellulose Chem Technol 24:703, 1990.
37. A Hirai, F Horii, R Kitamaru, W Tsuji. J Appl Polym Sci C, Polym Lett 28:357, 1990.
38. M Kakudo, N Kasai. X-ray Diffraction by Polymers, Tokyo: Kodansha, 1972, pp 359–367.
39. EA Toledo, Y Gushikem. Proceedings of Second International Symposium on Natural Polymers and Composites, Atibaia, Brazil, 1998, pp 103–105.
40. UP Rodrigues Filho, Y Gushikem, FY Fujiwara, SC de Castro, ICL Torriani, LP Cavalcanti. Langmuir 10:4357, 1994.
41. SA Jones, J Fenerty, J Pearce. Thermochim Acta 114:61, 1987.
42. JH Scofield. J Electron Spectrosc Related Phenomena 8:129, 1976.
43. A Dilks. In: JV Dawkins, ed. Developments in Polymer Characterization. London: Applied Science Publications, Ltd., 1980, pp 145–182.
44. M Murata, K Wakino, S Ikeda. J Electron Spectrosc Related Phenomena 6:459, 1975.
45. R Fontaine, R Caillat, L Feve, MJ Guittet. J Electron Spectrosc Related Phenomena 10:349, 1977.
46. F Verpoort, G de Doncker, AR Bossuyt, L Fiermans, and L Verdonck. J Electron Spectrosc Related Phenomena 73:271, 1995.
47. VI Nefedov, D Gati, BF Dzhurinskii, NP Sergushin, YV Salyn. Russ J Inorg Chem 20:1279, 1975.
48. F Garbassi. Surf Interf Anal 2:165, 1980.
49. T Tanaka, H Nojima, H Nakagawa, T Funabiki, S Yoshida. Catal Today 16:297, 1993.
50. JF Kennedy, JMS Cabral. Transition Metal Chem 11:41, 1986.
51. EP Parry. J Catal 2:371, 1963.
52. EV Benvenutti, Y Gushikem, CU Davanzo, SC de Castro, IL Torriani. J Chem Soc Faraday Trans 88:3193, 1992.
53. EV Benvenutti, Y Gushikem, CU Davanzo. Appl Spectrosc 46:1474, 1992.
54. S Denofre, Y Gushikem, SC de Castro, Y Kawano. J Chem Soc Faraday Trans 89:1057, 1993.
55. Y Inoue, H Yamazaki, Y Kimura. Bull Chem Soc Jpn 58:2481, 1985.
56. LRD da Silva, CRM Peixoto, Y Gushikem. Separ Sci Technol 31:1045, 1996.
57. LRD da Silva, Y Gushikem, LT Kubota. Colloids Surf B: Biointerf 6:309, 1996.
58. J Premkumar, R Ramaraj. Radiat Phys Chem 49:115, 1997.
59. Y Ohmori, Y Kurokawa. J Biotechnol 33:205, 1994.
60. S Chatterjee, S Sarkar, SN Bhattacharyya. Polymer 34:1979, 1993.
61. E Dalas. J Crystal Growth 113:140, 1991.
62. Y Kurokawa, K Ueno. J Appl Polym Sci 27:621, 1982.

63. Y Kurokawa, Y Imai, Y Tamai. Analyst 122:941, 1997.
64. H Ishikawa, Y Imai, Y Kurokawa. Vibrat Spectrosc 8:445, 1995.
65. Y Gushikem, EA Campos. J Braz Chem Soc 9:273, 1998.
66. S Denofre, Y Gushikem, CU Davanzo. Eur J Solid State Inorg Chem 28:1295, 1991.

14

The Evaluation of Acid–Base Properties of Polymer Surfaces by Wettability Measurements

Marco Morra
Nobil Bio Ricerche, Villafranca d'Asti, Italy

Claudio Della Volpe and Stefano Siboni
University of Trento, Trento, Italy

I. INTRODUCTION

The characterization and quantification of forces at interfaces is one of the most fundamental efforts in materials surface and interfacial science [1,2]. The ultimate goal of this endeavor is the capability to predict and explain the behavior of materials at interfaces by an analytical approach based on the quantification of interfacial forces, and to design material or surface structures for a specific goal on a rational basis of knowledge of the surface structure–properties relationship.

According to the present status of understanding of interfacial forces in condensed phases, acid–base interactions (the definition of these and related terms is discussed in Sec. II) play an important role in interfacial interactions. The complete recognition of this aspect is a comparatively recent result, probably because the world of interfacial interactions has long been dominated by ideas stemming from works on gases. It is well known that dipoles (either temporary or permanent) give rise to the following interactions, known collectively as van der Waals forces [1,2]: London (dispersion) forces, resulting from the natural oscillation of the electron clouds of the molecules inducing synchronous oscillations in the neighboring molecule, resulting in temporary dipoles; Keesom forces (dipole/dipole) and Debye (dipole/induced dipole) forces. It has long been assumed that nondispersive forces between condensed phases should be ascribed to "polar" interactions, and the role of the "polar" component has been frequently invoked to account for several different adhesive phenomena. It is now well accepted and demonstrated that the permanent dipole contribution to intermolecular forces is negligibly small in condensed phases (see, for instance, Refs. 3–7). The point is that, in condensed phases, the high (as compared to dilute gases) number of nearest neighbors provides conflicting local fields which minimize dipole interactions. Even if the language of intermolecular forces in a condensed phase is dominated by the term "polar," this kind of force plays only a very minor role. The lively discussion on this subject by the late Professor Fowkes is a highly recommended reference [4].

It is a common finding that water wettability or adhesive bonding of apolar polymers such as polyolefins are greatly increased by surface treatments that introduce on the polyolefin chemical moieties containing polar bonds. The dismissal of dipole–dipole interactions as a prime determinant of the enhancement of interfacial interactions requires another explanation for the forces involved. According to the present understanding of intermolecular interactions, the answer is acid–base interactions and, in particular, that specific subset of Lewis acid–base interactions known as hydrogen-bonding [6,7]. Again, Fowkes must be credited for the work he did in clarifying this aspect of interfacial interactions [3–6].

The recognition of the pervasive role of acid–base interactions in the control of the interfacial behavior of condensed phases spurred a great deal of work on this subject and, as underscored by Berg [6], literature on both the fundamental and the practical aspects of acid–base concepts at solid–liquid interfaces has recently experienced a quantum jump in size. Accordingly, the aim of this chapter is not to present a review on acid–base properties of solid and, in particular, of polymer surfaces. The fundamental works of Fowkes [3–5], the collection of papers presented to the Festschrift in honor of Professor Fowkes 75th birthday [8], and the excellent recent review by Berg [6] offer a very rich source of information on this topic. Rather we would like to focus on the interpretation of data stemming from the measurement of acid–base properties of polymer surfaces, in particular, by means of wetting measurements. In order to translate the background and principles contained in the acid–base approach to interfacial interactions into a quantitative and predictive tool for materials and surface design, the correct measurement and interpretation of acid–base properties data are obviously mandatory. Yet, everyday practice and a survey of literature on this topic show that this, indeed, is not the case [9–13]. This is particularly true in the case of approaches to surface acid–base properties based on the wetting measurement which are, as underscored by Jacobash and co-workers, the most frequently used methods to characterize the surface properties of polymers [13]. In this sense, the ultimate goal of this chapter is to promote reflection and discussion and to define some common basis and starting point that could be of help to solve some of the problems that still hinder the full exploitation of the acid–base approach to interfacial interactions.

The chapter is divided-into the following areas: first, some basic concepts and definitions related to acid–base properties are introduced. Then, the basic problem is stated—that is the inconsistency among acid–base properties measured with different methods and the unsatisfactory state of results obtained from wettability measurement. Finally, some discussion on the nature of these difficulties and some suggestion on how to handle them are presented.

II. SOME BASIC CONCEPTS

The most basic concept in the theory of interfacial acid–base interactions is the definition itself. Among the different definitions of acids and bases, the Lewis theory is the most suitable to deal with the chemical nature of polymers, and it is to this theory that all experimental approaches that try to calculate acid–base components of polymer surfaces naturally refer. Strictly speaking, however, it is necessary to remember that the original definition by Lewis describes an acid as any substance that can accept a pair of electrons from a donor substance, whereas a base is any substance capable of furnishing a pair of electrons [14]. In more modern terms [6], the share of both electrons provided by the base is no longer necessary, and a base is defined as any substance that is capable of donating electron density and an acid as any substance capable of accepting electron density [15,16]. It is this broader definition that best describes the kind of acid–base interactions of interest in polymer surface and interfacial science. And, it is within this same definition that the terms "electron donor" and "electron acceptor" are frequently used in the literature as being synonymous with "Lewis base" and "Lewis acid," respectively.

According to the previous definition, acidic sites are all those sites that can act as electron acceptors: metal atoms of organometallic compounds, electrophilic carbons (i.e., those carbon atoms covalently linked to a highly electronegative element, such as oxygen or fluorine), and hydrogen atoms belonging to hydroxyl or carboxyl groups. Lewis bases, on the other hand, are electron donors: atoms containing lone-pair electrons (such as oxygen), or aromatic rings in which the π electrons act as a basic site.

Many compounds contain both acidic and basic sites: Water, for instance, is a self-associated compound because of the balanced basicity of the oxygens and the acidity of hydrogens. The example of water allows one to comment once again on the term "polar": It is obviously correct to define water as a "polar" compound, because of the polarity of the –OH bond. The resulting dipole, however, as measured for instance by the relevant dipole moment, does not contribute to the intermolecular interactions in the condensed phase [3]. Rather, water molecules are polar because they contain electron-rich and electron-poor sites (i.e., basic and acidic sites, respectively). They will interact with neighboring molecules in a

Lewis acid–base way, and this particular acid–base interaction is called hydrogen-bonding. It is well known since 1960 [17] that this bond has preferred bond lengths and bond angles, has a bond strength dependent on the acid strength of the hydrogen donor and on the basic strength of the hydrogen acceptor, and is completely independent of dipole moment. As underscored by Fowkes, bonds involved in hydrogen-bonding are polar, and hydrogen-bonded liquids are often referred to as "polar," but it should not be assumed that they derive their properties from dipoles [3–5]. An interesting recent series of papers by Besseling and co-workers describes a theoretical model that accounts for the properties of water and water at interfaces based on the orientation-dependent properties of water molecules resulting from the presence of electron donor–electron acceptor sites [18–22].

Because of their very nature, acid–base interactions act through a mechanism of complementarity of functions: A compound will exert its acidic (or basic) nature only if a basic (or acidic) site is available, either in a molecule of the same compound (self-associated compounds) or in a different one. This is in sharp contrast to the symmetry expected in models involving interactions between dipoles, where in the interaction between two unlike molecules, i and j, the contribution of i and j are formally of the same kind as the interaction of two i molecules with each other or of two j molecules with each other.

III. THE CALCULATION OF ACID–BASE PROPERTIES FROM WETTING MEASUREMENTS

A. General Introduction

The goal of the calculation of acid–base properties by the wetting measurement is to evaluate the fundamental acid–base properties of solid surfaces by their ability to interact with liquids, as manifested through wetting phenomena. The roots of this approach [3–7] can be found in the splitting of surface free energy into components describing, the contribution of electrodynamic interactions (and overwhelmingly dominated by dispersion forces, as previously discussed), and the acid–base contribution

$$\gamma^{tot} = \gamma^{LW} + \gamma^{AB} \tag{1}$$

where γ^{LW} is the Lifshitz–van der Waals contribution, which includes electrodynamic interactions, and γ^{AB} is the acid–base contribution. Equation (1) basically states that "dispersive" and acid–base interactions are independent. Equation (1) is formally similar to relations used in the dispersive/polar approach, widely diffused before the recognition of the role of acid–base properties. The complementarity of acid–base interactions is clearly reflected in the expression used to describe the acid–base components of solids or liquids surface free energy, developed by

Good–van Oss–Chaudury adopting a combining rule for short-range interactions [7,23–26]

$$\gamma_{L,S} = \gamma_{L,S}^{LW} + 2(\gamma_{L,S}^{+}\gamma_{L,S}^{-})^{1/2} \tag{2}$$

that, combined with the Young equation, results in the following expression for the adhesion work between a liquid and a solid:

$$W_{adh} = \gamma_L(1 + \cos\theta) = 2(\gamma_L^{LW}\gamma_S^{LW})^{1/2} + 2(\gamma_L^{+}\gamma_S^{-})^{1/2} + 2(\gamma_L^{-}\gamma_S^{+})^{1/2} \tag{3}$$

Here, γ^{LW} is the previously defined Lifshitz–van der Waals contribution, γ^- is an electron donor, or Lewis base, and γ^+ is an electron acceptor, or Lewis acid parameter. The subscripts L and S refer to solid and liquid respectively.

The interested reader is referred to the voluminous literature on the subject for the background of Eqs. (2)–(3) [7,23–26]. From the practical point of view, Eq. (3) accounts for the bifunctionalism of liquids in wetting phenomena and allows one to calculate the Lifshitz–van der Waals contribution the electron donor and the electron acceptor parameters from the measurement of the contact angle of (at least) three liquids of known surface free-energy components.

Equation (3), which will be called the GvOC (Good–van Oss–Chaudury) equation in the remainder of this chapter, is, in principle, the tool that allows one to measure the acid–base properties of polymer surfaces, to account for results of interfacial interactions, and to design the target of a given surface modification treatment for a given application in advance. Yet, its practical application shows that it is still very far from this goal. Just to quote a couple of examples, Jacobash and co-workers compare acid–base properties of polypropylene (PP) and flame-treated PP obtained from contact angle and ζ-potential measurement [13]. They conclude that ". . . we can see that the acidic character of the surface is underesti-mated by the approach of van Oss and co-workers. Therefore, we can conclude that ζ-potential measurements give more detailed information regarding the sur-face composition than contact angle measurements." Inverse gas chromatography (IGC) is another technique that has been widely used for the measurement of acid–base properties of polymer surfaces [27–30]. IGC has been applied by Tate and co-workers [12] to measure the acid–base properties of Nylon 66, showing that it ". . . indicates the presence of high-energy acidic sites on the nylon surface." Wetting measurements were performed too, using the GvOC approach, leading to the conclusions that ". . . they are seriously flawed because they show nylon to be basic . . ." [12].

The above are just two examples of the difficulties that are found in the ap-plication of GvOC theory. Surfaces appear overwhelmingly basic when their acid–base properties are measured by the above approach, with little, often negli-gible acidic character. Several explanations have been suggested by Good, van Oss, and Chaudury, including that the occurrence of monopolar basic surfaces could well be a general law for polymer surfaces [24,25].

In addition to the disturbing universality of "monopolar basic surfaces" even in cases, such as polysaccharide surfaces, where electron-acceptor sites are expected to exist (and are detected, for instance, by IGC [6]), the GvOC equation shows some other problems, as follows [11]:

The results strongly depend on the choice of the three solvents used for the measurement.

The unknown of the calculations are the roots of surface tension compo nents and, in some cases, they assume negative values.

This state of the art can be faced in different ways. One can simply ignore these problems and keep using the GvOC equation, taking into account its short-comings, assuming that obtained values, even if not absolute, can be at least used for direct comparison. This approach, however, can be valid only in the short term and, in the authors' views, simply continues the long series of abuses that re-searchers have perpetuated on the contact-angle technique. Born as an equation describing the behavior of an ideal liquid on an ideal surface [31], the Young equa-tion has been stretched and folded to engulf all sorts of contact angle and interac-tion measured on all sorts of surface [32]. With such premises, it does not come as a surprise that it is possible to find in the usually measured, scientific literature, definitions such as "the comedy of errors" [33] or "disposable theories" [34] to de-scribe the status of contact angles and interfacial energetics.

The promises and problems of the wettability approach to acid–base inter-actions are best shown in Ref. 26. This very interesting work shows how acid–base interactions and the GvOC equation can quantitatively account for a large number of complicated interfacial interactions in aqueous media. On the other hand, no clear explanation, or no explanation at all, is given for fundamental quantitative as-pects such as the proper choice of solvents [how to handle the strong dependence of surface free-energy components on the choice of the solvents used for the meas-urement (and why this dependence exists)] and the experimental error and the un-certainty associated with the calculated surface free energy components (why the advancing (or any other) angle should be used among all experimentally accessi-ble angles). The recognition of this general state of the art prompted some authors to denounce shortcomings in the practical applications of existing theories, with the intended goal of promoting an in-depth reflection [9,35], and other to try to solve existing problems borrowing approaches from related fields [11,36,37].

The spirit of the forthcoming sections is to give a closer look at the quanti-tative aspects of the GvOC theory, to understand more about the reasons and meaning of the results it produces, and to try to solve some of its shortcomings. In summary, the goal is to improve the present understanding of the characterization of the acid–base properties of solid surfaces by wetting measurements. The price to be paid for this approach is the introduction of some cumbersome mathematics, but it is the authors' opinion that no price is too high to pay off the debt that the surface science community contracted with contact-angle measurement in a long record of abuses.

B. Calculation of Acid–Base Components According to the Good–Van Oss–Chaudury Theory

From a mathematical point of view, GvOC theory can be classified in the same realm of theories proposed by Drago [38], Taft and co-workers [39], Abraham [40], and others: A thermodynamic potential F_{th}, related to Lewis acid–base properties of two materials X and Y, is expressed as a sum of products of n different coefficients taken pairwise (some of them may be constant, however):

$$F_{th} = \sum_{i,j} X_i Y_j \tag{4}$$

F_{th} may be a free energy or an enthalpy, but this is not mandatory. Thus, it is one of the many linear free-energy relationships (LFER). It has been recently proposed by Douillard [34] that GvOC relations are only approximated; they would be theoretically more acceptable if written in terms of enthalpy. This evaluation can be considered partially correct; however, GvOC theory seems quite satisfactory if compared with other LFER applied to nonenthalpy quantities. (From this point of view, Drago calculations done by using enthalpy appear more correct than other models.) Moreover, it is easy to show that it is possible to write similar relations for both the enthalpy of adhesion and the adhesion work, a Gibbs free-energy quantity. The enthalpy of adhesion can be easily obtained from contact angles measured at different temperatures (at least three, and calculating enthalpy for the central temperature); thus, independent of the correctness of Douillard criticism, the considerations developed in this chapter remain valid. Some other considerations will be developed in a later section.

The basic idea for the computation of free-energy components of a solid consists in measuring the contact angles of L ($L > 3$) appropriately chosen liquids on the test solid. As shown elsewhere [11], to avoid systematic errors, it is convenient to adopt a set of liquids which includes all the typologies of solvents: purely "dispersive," prevalently acid or monopolar acid, prevalently basic or monopolar basic.

Denoting by $\gamma_{l,i}^{LW}$, $\gamma_{l,i}^{+}$, and $\gamma_{l,i}^{-}$ the dispersive, acidic, and basic components, respectively, of the ith liquid, $i = 1, 2, \ldots, L$, and by γ_s^{LW}, γ_s^{+}, and γ_s^{-} the corresponding components of the solid, if $\gamma_{l,i}^{tot}$ is the surface free energy of the ith liquid, the adhesion work W_i^{adh} (which is really a Gibbs free-energy difference) takes the form [slightly modifying the subscripts of Eq. (3)]

$$W_i^{adh} = \gamma_{l,i}^{tot}(1 + \cos\theta_i) = 2(\sqrt{\gamma_{l,i}^{LW}\gamma_s^{LW}} + \sqrt{\gamma_{l,i}^{+}\gamma_s^{-}} + \sqrt{\gamma_{l,i}^{-}\gamma_s^{+}}) \tag{3a}$$

where θ_i is the contact angle of the ith liquid on the solid.

The surface free energy of the same liquid then becomes

$$\gamma_{l,i}^{tot} = \gamma_{l,i}^{LW} + 2\sqrt{\gamma_{l,i}^+ \gamma_{l,i}^-} \tag{2a}$$

and an analogous expression holds for the solid. Whenever the acid–base components of the solvents are known with a sufficient accuracy, Eq. (3a) provides an overdetermined set of linear equations in the variables $\sqrt{\gamma_s^{LW}}$, $\sqrt{\gamma_s^+}$, and $\sqrt{\gamma_s^-}$, which can be solved by standard methods [37,41,42]. The procedure allows one to estimate the standard deviations of the results. As an example, a characterization of PP–EPDM flamed surfaces has been given in Refs. 37 and 42 by using a set of seven solvents with the acid–base components proposed by Good and van Oss. Note that it is possible to use a reduced set of liquids, at least three; in this case, which is the most common, the choice of the liquids becomes increasingly critical, because the square roots of the polymer components are obtained by solving a linear set of three equations in three variables and serious problems of ill-conditioning may arise. This is a rather typical occurrence if no attention is paid to the choice of the liquid triplet. If A denotes the 3×3 matrix whose rows are the coefficients $\sqrt{\gamma_s^{LW}}$, $\sqrt{\gamma_s^+}$, and $\sqrt{\gamma_s^-}$ of each liquid, a standard estimate to the conditioning of the set is provided by the so-called condition number $C_n = \|A\| \cdot \|A^{-1}\|$, where $\|A\| = \Sigma_{i,j=1}^3 \|A_{i,j}\|$; the larger C_n is, the stronger is the sensitivity of solutions to round-off or data errors (the relative error of the solution is estimated by multiplying by C_n the relative error of the wetting data). As an example, for the water–glycerol–bromonaphthalene triplet, the condition number turns out to be about 5.98, whereas $C_n = 7.54$ in the case for water–diiodomethane–formamide. It is noticeable that bromoform, dimethyl sulfoxide, and glycerol lead to the unacceptably large value $C_n = 28.6$, clearly because of the odd choice of the triplet which includes no dispersive liquid [11].

However, the determination of the acid–base components of liquids is not trivial and it requires rather arbitrary assumptions about the nature of the surfaces involved, or complex experimental procedures difficult to be reproduced and managed, particularly for the check and estimate of the errors. This is the case for the values calculated by Good and van Oss, achieved by employing, alternatively, some solid surfaces "reasonably" monopolar or liquids encased in gels [26]. That is why it is useful to introduce a more general approach, as follows.

In general, as the set (3a) is overdetermined, it is expected that it has no exact solution; this is even truer if the experimental errors and the approximate nature of the acid–base theory are taken into account.

We can then choose a set of S solids, each denoted with the index $j = 1, 2, \dots, S$, and introduce a larger set of equations containing the following relationships for every liquid i or solid j:

$$\gamma_{l,i}^{tot} = \gamma_{l,i}^{LW} + 2\sqrt{\gamma_{l,i}^+ \gamma_{l,i}^-}, \qquad \gamma_{s,j}^{tot} = \gamma_{s,j}^{LW} + 2\sqrt{\gamma_{s,j}^+ \gamma_{s,j}^-} \tag{2b}$$

together with the adhesion work equalities for each liquid–solid pair:

$$W_{i,j}^{adh} = \gamma_{l,i}^{tot}(1 + \cos\theta_{i,j}) = 2(\sqrt{\gamma_{l,i}^{LW}\gamma_{s,j}^{LW}} + \sqrt{\gamma_{l,i}^{+}\gamma_{s,j}^{-}} + \sqrt{\gamma_{l,i}^{-}\gamma_{s,j}^{+}}) \tag{3b}$$

where θ_{ij} is the contact angle of the ith solvent on the jth solid and a further index j distinguishes the acid–base components of different solids.

Whenever the number of solvents and solids is large enough, the result is an overdetermined set of $L + S + LS$ nonlinear equations in the $3L + 4S$ variables:

$$\sqrt{\gamma_{l,i}^{LW}}, \quad \sqrt{\gamma_{l,i}^{+}}, \quad \sqrt{\gamma_{l,i}^{-}}, \quad i = 1, 2, \ldots, L$$

$$\sqrt{\gamma_{s,j}^{tot}}, \quad \sqrt{\gamma_{s,j}^{LW}}, \quad \sqrt{\gamma_{s,j}^{+}}, \quad \sqrt{\gamma_{s,j}^{-}}, \quad j = 1, 2, \ldots, S \tag{5}$$

and if no further physical information can be invoked, we are led to a purely formal optimization problem consisting in Eqs. (2b) and (3b) to be solved with respect to the variables (5). Note that set (5) does not include the total surface free energies $\gamma_{l,i}^{tot}$ of the liquids, as they can be directly measured and it seems more reasonable to treat them as given parameters. In contrast, no commonly accepted experimental technique is available for solids, so that each $\gamma_{s,j}^{tot}$ must be taken as an unknown.

It is worthy of note that a similar nonlinear approach has been developed also in the case of two-component theory by Erbil and Meriç [43]. Some of their results will be commented on later in this chapter.

Now, suppose that the optimization problem admits a best-fit solution (which is mathematically nontrivial). If it is also unique (and the acid–base model is correct), we can be confident that the physical meaning given to the components γ^{LW}, γ^{+}, and γ^{-} is essentially correct and useful for a direct description of surface interactions. But if uniqueness of the best-fit solution fails, a serious problem arises about the interpretation of acid–base components, because of the indeterminacy involved by the best-fit procedure. To overcome these difficulties, we only have two alternatives:

1. On the one hand, we can decide to reject any physico-chemical interpretation of γ^{LW}, γ^{+}, and γ^{-}, and to state consistently that the acid–base components are significant only when inserted into Eq. (3b) or (2b) to compute adhesion work or surface free energies. This choice has been originally done by GvOC and implies the impossibility of comparing acidic and basic components of the same material; no definition of "strength" was proposed by GvOC for acids and bases. Most importantly, the arbitrary choice of taking the acidic and basic components of water as equal produces as an obvious effect that it is not possible to compare the acidic and basic components of the same material and to gauge their respective contributes to the total interaction. The only possible comparison is that between the acidic or basic component of a certain material with the corresponding one of a different material [11]. Unfor-

tunately, this fact has not been sufficiently emphasized and the current use of GvOC theory in literature is often incorrect; many authors compare directly the magnitude of the acidic and basic components of the same material, but they use the original scale and coefficients of GvOC theory, with the obvious consequence of obtaining very low acidic components.

2. On the other hand, if the amount of indeterminacy of the best-fit solution is large enough, we can hope to harness it so as to define an appropriate scale of acid–base components, suitable for comparison with other and more direct scales of acid–base strength.

In the case of the GvOC model, it can be shown that the solvability of the best-fit problem actually implies the existence of an infinite number of solutions, a feature of the mathematical form of the model which does not depend on the best-fit algorithm adopted. A specific solution must be selected by means of some conventionally assigned components of reference solvents, in a way which recalls Drago's calibration procedure for his model equations about the enthalpies of adduct formation in gas-phase or poorly solvating media [38].

Among the infinite and formally equivalent choices, we can define, for instance, a scale based on a realistic ratio γ^+/γ^- for a reference solvent (as in the Abraham scale, where γ^+/γ^- is 5.5 for water); such a scale would also make possible the comparison of acid *and* base components of different solids and liquids.

Although the previous procedure seems difficult to carry out, owing to the general disagreement in the values of contact angles for nonpolar and basic solvents on common polymers and to an undoubted lack of data about acid solvents, it constitutes the most correct way to apply GvOC theory.

It is useful to compare our approach to the similar one proposed by Drago. The two methods share some features, such as the use of best-fit solutions to describe concisely a large amount of experimental data, the lack of uniqueness of these best-fit solutions because of the existence of appropriate infinite groups of linear transformations which leave the model equations invariant, or the possibility of defining different scales. However, Drago distinguishes "acidic" and "basic" solvents (electron acceptors and donors), each characterized by *two* variables in such a way that the enthalpy of adduct formation for any acceptor–donor pair is written as

$$-\Delta H = E_A E_B + C_A C_B = D_A^T D_B, \qquad D_A = \begin{pmatrix} E_A \\ C_A \end{pmatrix}, \qquad D_B = \begin{pmatrix} E_B \\ C_B \end{pmatrix} \qquad (6)$$

where the subscripts A and B indicate acceptor and donor, respectively, and E and C represent electrostatic and covalent contributions, respectively (the superscript T stands for "transpose"). In contrast, acid–base theories do not classify molecules into purely acid or purely basic: Each liquid or solid is described in terms of the three components γ^{LW}, γ^+, and γ^-, which appear inside the appropriate expressions (2b) and (3b). These relationships can also be rewritten in the matrix form

$$\gamma_{l,i}^{tot} = X_i^T R X_i, \qquad \gamma_{s,j}^{tot} = Y_j^T R Y_j, \qquad \tfrac{1}{2}(1 + \cos\theta_{ij})\gamma_{l,i}^{tot} = X_i^T R Y_j \qquad (7)$$

if we associate to each liquid (or solid) a column vector X_i (respectively, Y_j) of components $\sqrt{\gamma^{LW}}$, $\sqrt{\gamma^+}$, and $\sqrt{\gamma^-}$ and introduce an appropriate constant matrix R. Therefore, although matrix formalism is applicable to both cases, the situations are rather different. The matrix R and the vectors X_i and Y_j for each i and j are as follows:

$$X_i = \begin{pmatrix} \sqrt{\gamma_{l,i}^{LW}} \\ \sqrt{\gamma_{l,i}^+} \\ \sqrt{\gamma_{l,i}^-} \end{pmatrix}, \qquad Y_j = \begin{pmatrix} \sqrt{\gamma_{s,j}^{LW}} \\ \sqrt{\gamma_{s,j}^+} \\ \sqrt{\gamma_{s,j}^-} \end{pmatrix}, \qquad R = \begin{pmatrix} 1 & 0 & 0 \\ 0 & 0 & 1 \\ 0 & 1 & 0 \end{pmatrix} \qquad (8)$$

Given any 2×2 real nonsingular matrix A, Drago's theory allows one to redefine the acceptor and donor components by means of the linear transformations

$$D_A \rightarrow A D_A, \qquad D_B \rightarrow (A^{-1})^T D_B \qquad (9)$$

which trivially leave Eq. (6) invariant. The use of a different transformation for acidic and basic components makes sense only because acceptors and donors are a priori recognized and classified, so that there is no reason to impose that the linear transformation rule must be the same for both.

On the contrary, acid–base theories treat any molecule formally in the same way and a unique transformation has to be applied to all the chemicals, by means of an appropriate nonsingular matrix C:

$$X_i \rightarrow C X_i, \qquad Y_j \rightarrow C Y_j \qquad (10)$$

which must also ensure the invariance of Eqs. (7). The transformation matrix C satisfies the required condition if and only if

$$C^T R C = R \qquad (11)$$

and belongs to a subgroup of transformations much smaller than the general group of linear nonsingular transformations available for the Drago model. The subgroup, isomorphic to the orthogonal group $O(2, 1; \mathbb{R})$, includes a three-parameter family of nonsingular matrices C and, therefore, provides an infinite set of linear transformations (10). The invariance of Eqs. (7)–(10) means the invariance of any objective function one may use to compute best-fit solutions and, consequently, applying transformation (10) to a best-fit solution provides a best-fit solution as well.

IV. SOME COMPARISON WITH AND SOME REFLECTION ON THE LITERATURE

A. Modeling Enthalpy or Free Energy?

It would be useful to show the relationship between this point of view and that expressed by Fowkes [5]; the relations proposed by GvOC represent the free energy of the interface or, better, the work of the adhesion, which is a difference of free energies in the formation of the interface. Fowkes expresses the work of adhesion in term of the enthalpy of the same process; obviously, this cannot be done without remembering the thermodynamic relation between these two quantities in an isothermal process:

$$\Delta G = \Delta H - T\Delta S \tag{12}$$

To suppose, as Fowkes does, that ΔG can be expressed as the product of the enthalpy for a certain coefficient f is a "strong" hypothesis whose coherence with the experimental context must be proved and, generally speaking, is false [34]. Moreover, to choose that the f value is about 1 would correspond to the fact that the entropy variation during the interface formation is null; this is possible in some cases, but cannot be considered a general conclusion at all.

It is possible to express either the Gibbs free energy or the enthalpy of the interface formation process using an equation such as LFER. Obviously, this choice has different consequences on the mathematical form of the remaining quantity.

In particular, if one chooses to introduce a LFER equation for the Gibbs free energy, according to GvOC the expression of enthalpy of adhesion becomes

$$\Delta H = \Delta G + T\Delta S = \Delta G - T\frac{\partial \Delta G}{\partial T} \tag{13a}$$

$$\Delta H_{adh} = W_{adh} - T\frac{\partial W_{adh}}{\partial T}$$

$$= \gamma_l(1 + \cos\theta) - T\left\{(1 + \cos\theta)\frac{\partial \gamma_l}{\partial T} + \gamma_l\frac{\partial(\cos\theta)}{\partial T}\right\} \tag{13b}$$

and substituting Eq. (3b), one obtains

$$\Delta H_{adh} = 2(\sqrt{\gamma_{l,i}^{LW}\gamma_{s,j}^{LW}} + \sqrt{\gamma_{l,i}^{+}\gamma_{s,j}^{-}} + \sqrt{\gamma_{l,i}^{-}\gamma_{s,j}^{+}})$$

$$- 2T\left\{\frac{\partial(\sqrt{\gamma_{l,i}^{LW}\gamma_{s,j}^{LW}} + \sqrt{\gamma_{l,i}^{+}\gamma_{s,j}^{-}} + \sqrt{\gamma_{l,i}^{-}\gamma_{s,j}^{+}})}{\partial T}\right\} \tag{13c}$$

which can be simplified to

$$\Delta H_{adh} = 2\left\{\left(\sqrt{\gamma_{l,i}^{LW}\gamma_{s,j}^{LW}} - T\frac{\partial\sqrt{\gamma_{l,i}^{LW}\gamma_{s,j}^{LW}}}{\partial T}\right) + \left(\sqrt{\gamma_{l,i}^{+}\gamma_{s,j}^{-}} - T\frac{\partial\sqrt{\gamma_{l,i}^{+}\gamma_{s,j}^{-}}}{\partial T}\right)\right.$$
$$\left. + \left(\sqrt{\gamma_{l,i}^{-}\gamma_{s,j}^{+}} - T\frac{\partial\sqrt{\gamma_{l,i}^{-}\gamma_{s,j}^{+}}}{\partial T}\right)\right\} \tag{13d}$$

and, finally, the three terms in curly brackets (in Eq. 13d) can be considered as corresponding to LW (the first term) and to the acid–base component (the last two terms) of the total enthalpy of adhesion:

$$\Delta H_{adh} = \{\Delta H_{adh}^{LW} + \Delta H_{adh}^{AB}\} \tag{14}$$

From this equation, one can deduce that by knowing the values of the acid–base components of adhesion work and their trend with the temperature, it is possible to calculate some corresponding enthalpic quantities.

It is worthy of note that although the equivalence (or a similar one for the acid–base components)

$$\sqrt{\gamma_{l,i}^{LW}\gamma_{s,j}^{LW}} - T\frac{\partial\sqrt{\gamma_{l,i}^{LW}\gamma_{s,j}^{LW}}}{\partial T} = \Delta H_{adh}^{LW} \tag{15}$$

is legitimate from a thermodynamic and dimensional point of view, the situation appears different for terms of the form $\sqrt{\gamma_{s,j}^{LW}} - T\,(\partial\sqrt{\gamma_{s,j}^{LW}}/\partial T)$. They are neither enthalpic quantities nor square roots of enthalpic quantities, even if their mathematical forms appear similar.

However, it remains still valid that knowing the single acid–base components of adhesion work and their trend with the temperature, it is also possible to calculate the quantities as $\gamma_{s,j}^{LW} - T\,(\partial\gamma_{s,j}^{LW}/\partial T)$, which can be considered the true acid–base components of the enthalpy of a material, but they are not explicitly present in the previous equations.

Vice versa, if one chooses to express the enthalpy by an equation of the LFER form, by proposing the equivalence (which becomes so a starting point, fully equivalent to the GvOC proposal)

$$\Delta H_{adh} = 2(\sqrt{h_{l,i}^{LW}h_{s}^{LW}} + \sqrt{h_{l,i}^{+}h_{s}^{-}} + \sqrt{h_{l,i}^{-}h_{s}^{+}}) \tag{16a}$$

and using the Gibbs–Helmholtz equation, one can then obtain

$$\frac{d\Delta W_{adh}/T}{dT} = -\frac{\Delta H_{adh}}{T^2} = -\frac{2(\sqrt{h_{l,i}^{LW}h_{s}^{LW}} + \sqrt{h_{l,i}^{+}h_{s}^{-}} + \sqrt{h_{l,i}^{-}h_{s}^{+}})}{T^2} \tag{16b}$$

where the left-hand side can be obtained from contact-angle measurements at different temperatures, and it represents the known term of an equation in the unknowns h's. An overdetermined set of these equations can be solved with the same methods as described for the adhesion work and its acid–base components.

It is noticeable that, again, some apparently obvious relationships are not valid; for example, although it is true that $h^{LW} = \gamma^{LW} - T(\partial\gamma^{LW}/\partial T)$, the situation is different for the acid (or the base) component: $h^+ \neq \gamma^+ - T(\partial\gamma^+/\partial T)$.

In both cases, starting from a free-energy term or from an enthalpy one, one needs to use measurements of contact angles at different temperatures to express the remaining function correctly (i.e., the acid–base components of surface enthalpy or of surface free energy).

Considering correctly the thermodynamic relationships, there is no contradiction between these two approaches. The problem is only to have good experimental data, otherwise the numerical calculation of derivatives is a really discouraging task.

B. A Comparison with Other Proposals to Modify the GvOC Theory

The article by Lee [44] in 1996 has been the first one in which a correlation among the coefficients of solvatochromic scale and the coefficients of GvOC theory has been suggested. However, Della Volpe and Siboni (DVS) were not aware of Lee's article at the time they published their article in 1997 [11], nor did any reviewer note that the article was not cited among the quoted references. Both articles consider the same correlation, but from different points of view.

In the article by Lee, the correlation is presented as an "unexpected result" whereas in the DVS article, this correlation depends on the analysis of the mathematical relations of the Taft–Abraham and GvOC theories. The individualization of the exact functional correspondence between the two mathematical expressions is very important and is widely developed by means of a matrix formalism; this highlights nicely the peculiarities of GvOC theory with respect to other LFER applications.

A second important conceptual difference, with marked consequences on the numerical estimates of coefficients, is that Lee considered the components γ^+ and γ^- to be directly related to the parameters α and β by Taft and others. Now, the LFER express the interaction surface free energy of two materials (or better, the logarithm of an equilibrium constant, which is proportional to a free energy but actually adimensional) as a linear combination of products; in these products, each coefficient belongs to one of the two materials. In the case of GvOC theory, the adhesion work or the surface free energy of a liquid or a solid depends on the product of the *roots* of acidic and basic components, not on the product of the components themselves. The correlation which can be established between the two theories must take into account this element: The correspondence is not between γ^+ and α, or γ^- and β, as proposed by Lee, but between $\sqrt{\gamma^+}$ and α, or $\sqrt{\gamma^-}$ and β. This changes the ratio between the water components in a significant way and the value 1.8 suggested by Lee is probably too low; a value of $(1.8)^2 = 3.2$ is more realistic.

In the DVS article, two values of α/β were considered, 6.5 and 5.5; for the first of them, proposed by Taft and co-workers [39], the ratio of roots was 1.17/0.18, with a small and strongly uncertain denominator; the correct value of the ratio should then be $(6.5)^2 \approx 42$, a value which appears too high. The most recent and experimentally supported of the two values is 5.5, calculated from the data of Abraham [40]; in fact, in that article, the values of the ratio between the acid and base components of water α/β was 0.82/0.35 = 2.34 and must be elevated to the second power, becoming 5.5. For these reasons, some calculation based on the suggested ratio of 6.5 were proposed in Ref. 11, but if this value also corresponds numerically to a particular α/β ratio, the correct general approach is $\gamma^+/\gamma^- = (\alpha/\beta)^2$.

Actually, it is not known what the best γ^+/γ^- ratio for water at 20°C is; this will be the subject of the following discussion and of future measurements. It is certainly greater than 1, and the interval 3.2–5.5 can be considered a reasonable guess.

It is worth noting that in common pure water, a fast equilibrium (in about 10–20 min) with atmospheric carbon dioxide is established; the final pH is about 5.5–6, which corresponds to a strong excess of hydrogen ions, about 900 times more concentrated than OH^- ions; this can be the origin of the measured prevalently acid character of common pure water. Measurements of the contact angles of ultrapure water (at pH 7 and with a very low conductance, about 0.05 μS) on common polymers are not available in literature nor would they probably be useful for practical purposes, because water is commonly in equilibrium with carbon dioxide in all the systems of interest, particularly in biological systems. However, for scientific reasons this kind of measurement would be very interesting and could even show that ultrapure water is really a Lewis base before it is "acidified" by carbonic acid formation.

As a third important difference, in the Lee article, no attention is paid to other mathematical aspects of the theory: the ill-conditioning of the set of equations obtained using only three solvents, the limited significance of results calculated only from advancing contact angles, and so forth.

In the DVS article, it has been widely shown that the use of only three *polar* liquids for the calculation of the LW, acidic, and basic components of surface free energy strongly enhances the ill-conditioned equation set with the worst results. A correct choice is the use of a wide (greater than three) and *proper* (well-balanced) set of monopolar or mainly acidic, monopolar or mainly basic, and mainly dispersive liquids. As a first and limited example of this more correct method and also of the sensitivity it is possible to obtain from the use of receding angles in the analysis of activated surfaces, a recent article [37] describes the characterization of PP–EDPM surfaces by seven liquids and the calculation of surface free energies from advancing and receding contact angles.

The most urgent task for the surface community is now to organize, as indicated in Ref. 11, a roundrobin to measure advancing and receding contact angles for a set of the most common liquids and (polymer) solids; the set developed in this way can be used to test GvOC or other similar theories.

Finally, the results by Lee and DVS agree from a qualitative point of view; there is a general increase of the acidic components and a decrease of the basic ones. In the case of liquids, this implies that, as a general result, the chemical nature of each liquid is better reflected by its components.

In the case of polymers, acidic components appear greater than before, but never predominant; polyvinylchloride (PVC) is not acidic, and polyvinylfluoride (PVF) is a weak acid for DVS, but not for Lee.

In the literature, there is also another set of articles (by Qin and Chang) [45–48] in which the problem of acid–base theories has been dealt with by new and original ideas. Qin and Chang introduce as variables the square roots of acid and base components, accept explicitly that they can be positive or negative, and consider that a material is acidic or basic if *both* of its components have a certain sign—positive for acids and negative for bases. They propose, therefore, a new and different set of coefficients for the test liquids and recalculate the components of some polymers accordingly.

From our point of view, this approach is not generally wrong: It corresponds to the choice of a particular scale, among the infinite possible ones, and the properties of the corresponding equation set are not different from those found for the original GvOC model. Indeed, the Qin and Chang expressions for the surface free energy and the adhesion work of a liquid L and a solid S are respectively [46]

$$\gamma^L = \tfrac{1}{2}(P_L^d)^2 - P_L^a P_L^b; \qquad \gamma^S = \tfrac{1}{2}(P_S^d)^2 - P_S^a P_S^b;$$

$$W^{\text{adh}} = P_S^d P_L^d - (P_S^a P_L^b + P_L^a P_S^b)$$

and can be reduced easily to the GvOC form (3a) and (2b) by means of the substitutions

$$P^d \rightarrow \sqrt{2\gamma^{LW}}, \qquad P^a \rightarrow \sqrt{2\gamma^+}, \qquad P^b \rightarrow -\sqrt{2\gamma^-}$$

We tried to apply the procedure of Qin and Chang to the set of data previously described [46], comparing these results with a scale where all the coefficients are chosen as positive; the conclusion is that all the parameters, but the sign, are the same. No real advance is actually achieved.

The problem is the actual acceptability of the values proposed as a reference. Unfortunately, the values for water (and also for other liquids) chosen by those authors have been published, but *never* justified. They were presented at a congress in 1992 [45], reported in two subsequent papers in 1995 [46] and 1996 [47], but never fully explained or justified; in fact, the corresponding paper appears as "submitted" to the *Journal of Adhesion Science and Technology* from about 1996. Thus, we cannot judge the procedure used by the authors, but only the obtained parameters. The latter do not seem coherent with the accepted view of the properties of the test liquids; in fact, water appears to be more basic than acid, and methylene iodide and bromonaphthalene are both described as fully basic.

From this point of view, the scale adopted by Qin and Chang is in the same situation as the original GvOC one; it is not possible to compare the acidic and basic properties of the same material, but only the acidic (or basic) properties of different ones. It is an arbitrary scale and cannot be justified.

C. Some Considerations About the Comparison with Other Techniques: The Role of Receding Angles

The last point we would like to discuss in this section is that a wider and proper set of advancing and receding contact angles is necessary, possibly obtained as a result of a round-robin, in order to reduce errors. The use of receding contact angles to calculate the surface tension of solids could seem scandalous, but it is not the case.

The calculation of surface free-energy components of solids from receding contact angles is, in principle, as valid as that done from advancing ones or from values considered as equilibrium ones, because they all really correspond to metastable states.

For the sake of clarity, one could call the values calculated from receding angles the "apparent" surface free-energy components. This term would emphasize that the original contact angle utilized in the calculation is not an equilibrium one; note, however, that this is always the case. In general, the "surface science community" uses sessile static contact angles, considering them as equilibrium ones, or simply prefers to use surface tension calculated from advancing contact angles. However, no accepted experimental method exists to measure the true "equilibrium" value of a contact angle, a method able to distinguish it from the other infinite number of metastable values [49]. Thus, all calculated values of the surface free energy of solids available in literature are "apparent," unless they are calculated on (nearly) ideal surfaces, which is very seldom the case. This conclusion may be difficult to accept, although it is a simple truth. Moreover, in the case of heterogeneous surfaces, one must remember the important results obtained by Johnson and Dettre [50,51]; both advancing and receding contact angles are representative; their values give us a greater insight in the real surface properties; and the advancing contact angle is better correlated with the low-energy portion and the receding contact angle with the high energy portion of the surface.

If the two "apparent" contact angles provide a deeper description of the surface than the "Young contact angle" does, why not trying to extend this reasoning to the calculation of surface free energy?

In two recent articles, Jacobash and co-workers [13,52] strongly support the idea that the contact-angle measurements are not able to evaluate the acidic properties of the flame-treated materials. Those articles contain very good contact-angle data of flame- and plasma-treated thermoplastic polyolefins (TPO) surfaces in three liquids, measured by a sessile technique (which permits to obtain the so-called "recently advancing" and "recently receding" contact angles). The authors calculated acid–base surface free-energy components by GvOC theory and compared them with the results of x-ray photoelectron spectroscopy (XPS) and ζ-potential analysis. They found that, in contrast to the other two techniques, contact angles did not reveal the acidic component. Is there a reason?

The experimental results of Jacobash and co-workers are in good agreement with published data and their presentation is unquestionable; however, some criticism can be raised of the calculation procedure, in the light of the previous sections, namely:

Only advancing angles are used for calculations.
Only three liquids are used.
The acidic and basic components of every material are directly compared.

This procedure is the commonly used approach to the GvOC theory, but it contains three important errors, as previously described:

Advancing angles are mostly indicative of the low-surface-free-energy portion of a surface, so that they are not necessarily representative of the flame-treated or plasma-treated portion of the surface [49–51].
The small number of liquids employed enhances the ill-conditioning of the GvOC equation set and produces a significant deviation of the results.
The use of the liquid coefficients originally proposed by GvOC implies the impossibility of comparing acidic and basic components of the same material; only the acidic or basic component of a certain material with the corresponding one of a different material can be compared.

If the Jacobash data are treated according to these criteria, it is possible to show that (at least in those cases where the advancing contact angles are significantly modified by the flame or plasma treatment) the acidic component of a flame-treated PP surface is as enhanced as the basic one with respect to the corresponding value of the untreated material. In other cases, however, the advancing contact angles are not sufficiently modified by the process to correctly show the high-free-energy portion of the surface; in those cases, the GvOC theory "appears" to fail.

The use of the receding angle as a better index of high-energy surfaces and particularly of the acidic (or basic) character of a surface is presented in Ref. 13 by

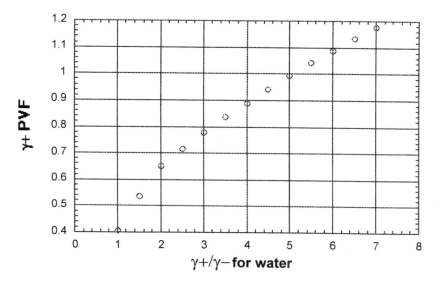

Figure 1 The acidic component of PVF computed by the best-fit method, as a function of the γ^+/γ^- ratio for water. The dispersive component of water is 22 mN/m.

the same authors; they consider the general advantages offered by the receding contact angles of polypropylene–ethylenvinylacetate (PP–EVA) blends in water to estimate the acidic character of these surfaces, but they do not take this idea to the point of calculating (apparent) surface free-energy components from receding contact angles.

V. THE PROBLEM OF SCALES

A. Some New Ideas About the Choice of a Correct Scale of Acid–Base Components for Solids

We analyzed the dependence of the acidic and basic components of solids and liquids on the numeric value of the chosen γ^+/γ^- ratio for water by using the set of values already introduced in Ref. 11. This is a wide set, containing 10 liquids and 14 solid polymers; it was already described and used in the quoted reference and was obtained by collecting the data from international literature. Looking at the graph of Fig. 1, one can see, for instance, that the acidic component of PVF (polyvinyl

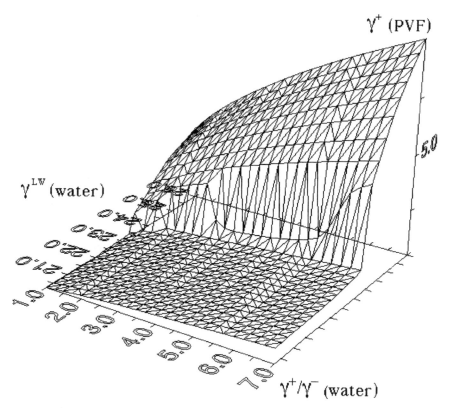

Figure 2   A plot of the acidic component of PVF calculated by the DVS best-fit method versus the γ^+/γ^- ratio and the value of γ^{LW} for water.

fluoride) increases as the water γ^+/γ^- ratio grows. This is generally true for all the acidic components; the basic components decrease, but the variations are not as large as one would expect.

We decided to explore the domain of the variables more extensively, changing all the elements of the water reference scale; in Figs. 2 and 3, the variation of the acidic component of PVF and of the basic component of DMSO versus the γ^+/γ^- ratio and the value of γ^{LW} for water are plotted. From these graphs, it is possible to appreciate an unexpected result. For a value of the dispersive water component of about 23–24 mN/m, there is a strong nonlinear variation of all the components; data refer to PVF, but the situation is similar for other polymers. For the DMSO base component, this nonlinear jump is present at about 26–27 mN/m; thus, in the interval 23–27 mN/m, there is the optimum of the solution.

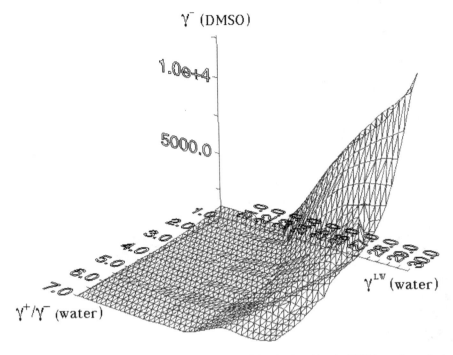

Figure 3 A plot of the basic component of DMSO calculated by the DVS best-fit method versus the γ^+/γ^- ratio and the value of γ^{LW} for water.

The global trend of the mathematical function whose minimum gave the co-efficient values is shown in Fig. 4; this function, which we will call the merit function, is minimized to solve the GvOC overdetermined equation set. It contains all the data and coefficients and depends on 71 unknowns, 3 of which are the acid–base components of water. By assuming a surface tension of water of 72.8 mN/m, these components are uniquely fixed by assigning the values of γ^{LW} and γ^+/γ^- in an appropriate interval for each; the minimum of the merit function with respect to the remaining 68 variables, at the given γ^{LW}, γ^+, and γ^- of water, is then calculated and the value of that minimum is plotted. The resulting plot is displayed in Fig. 4. As one can see, the minimum of the global merit function becomes slightly lower with the increase of the γ^+/γ^- ratio for water, but the most important thing is that the minimum is at a value of the water dispersive component of 26.25, completely different from that commonly used, 21.8. This suggests taking a closer look at the origin of this widely accepted value.

The value of 21.8 was proposed by Fowkes in 1964 [53]; in that article, he

Merit function

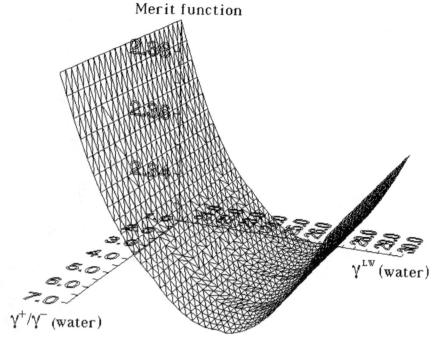

Figure 4 An illustration of the merit function trend. The function depends on the acid–base components of water and on 68 other variables, components of various liquids and polymers. By assuming a surface tension of 72.8 mN/m, water components are uniquely fixed by assigning the values of γ^{LW} and γ^+/γ^- in an appropriate interval for each; the minimum of the merit function with respect to the remaining 68 variables, at the given γ^{LW}, γ^+, and γ^- of water, is then calculated and the value of that minimum plotted.

reported measurements of the interfacial energy of water at the interface with completely dispersive liquids, with a very low scattering of data.

In the same article, he also reported on the interfacial energy of water on completely dispersive solids, but in this case, the numerical values were not reported but only plotted on a graph (Fig. 6 of Ref. 53) at a large scale, showing that they occupied a wide interval, between 15 and 30 mN/m. It should be very difficult to recalculate the original values from the figure, so we tried to use other data of the literature to calculate the LW component of water from interactions of water with a purely dispersive solid whose surface tension was calculated independently, using the contact angle of a purely dispersive liquid.

We used Eq. (3b) in the suitable form, as reported in the following. The LW component of the solid was the first to be calculated from the contact angle of

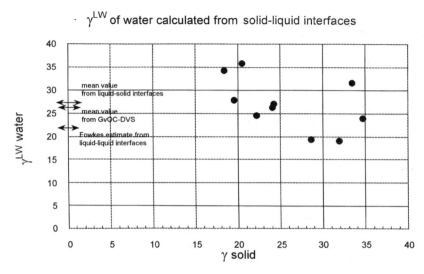

Figure 5 Estimates of the dispersive component of water by data of solid–liquid interfaces. The values on abscissas are the surface free energies of the solids employed for the calculation. The data sets were collected from Refs. 41, 49, and 54–56. The mean values by Fowkes, GvOC, and DVS [11] are shown for comparison.

purely dispersive liquids on a purely dispersive solid and the LW component of water was the second to be calculated from its contact angle on the same solid:

$$\gamma_s^{LW} = \frac{\gamma_l(1 + \cos\theta)^2}{4}; \qquad \gamma_l^{LW} = \frac{\gamma_l^2(1 + \cos\theta)^2}{4\gamma_s^{LW}}$$

It is possible to find many useful collected data in the literature; we found 10 useful sets, reported in Refs. 41, 49, and 54–56. The results are shown in Fig. 5; one can see that the estimates are very different, with a large scattering but with a mean value greater than that reported for the liquid–liquid interaction, about 27.1 ± 5.7. In contrast, the value obtained by Fowkes for the liquid–liquid interaction was 21.8 ± 0.2.

Is it conceivable that a liquid–liquid interaction is different from a liquid–solid one, even if only dispersive interactions are present? It is not easy to answer to this question, but, in principle, an affirmative answer cannot be excluded. For example, one can consider that for a solid polymer, at a temperature higher than its T_g, molecular motions required to adapt its structure to the force field generated by the water molecules are much more difficult than for a liquid; moreover the geometrical shape of the interface is different, because the flattest polymer surface cannot be as flat as a liquid.

It is worth noting that a similar result, with a dispersive component for water higher than that commonly accepted, has been obtained by Erbil and co-workers [57] using a nonlinear approach to the calculation of two-component models. Unfortunately Erbil's conclusion (i.e., the presence of a specific contribution in each liquid–solid couple) introduces another ad hoc coefficient and is not useful in the case of three-component models, because the corresponding set of equations does not admit solutions.

If these new coefficients are applied as a reference scale for water, 26.25 for the LW component, and 48.5 and 11.2 for the acidic and basic components, respectively, with a relative ratio of 4.35 for the γ^+/γ^- ratio exactly in the middle point of the interval 3.2–5.5, previously considered as a possibly acceptable interval, we obtain the results reported in Table 1.

It is possible to observe that results are now relatively acceptable; liquids and solids are acidic or basic, the acid components are significant and greater than basic ones, with a considerable respect of the chemical intuition. PVF or Nylon 66 are effectively acid surfaces, whereas PMMA remains strongly basic; among liquids, DMSO is the most basic one, whereas the bromoform, pyrrole, water, and glycerol have a strong acid content.

It is interesting to note that whenever the square root of the acid component is negative, the standard deviation of γ^+ is greater than γ^+ itself except in one case. The case of PBT is probably caused by the quality of the original data; in fact, those data were collected in the laboratory of one of the authors analyzing the surface of a blood leukodepletion filter made by nonwoven fabric, a material whose surface is very hard to analyze by the Wilhelmy microbalance; a systematic error was probably present in those data, so that they were comparable among them, but not with others.

Once again, however, it is not expected that these calculations provide "the correct components"; much work is necessary to obtain good and acceptable contact angles as starting points for a calculation of the acid–base coefficients, a work which can be completed only if many laboratories work together in an international round-robin. This is true also for the data necessary to set the scale; that is, to obtain (1) advancing and receding contact angles of dispersive liquids on dispersive solids and (2) advancing and receding contact angles of water on the same solids to set the dispersive component of water from liquid–solid interfaces with a more acceptable standard deviation.

Only if this global work does not give acceptable results should the "surface community" reconsider completely its ideas about solids surfaces and the utility of contact angles.

From the above discussion, it is easy to suggest that one of the main limits of the data already obtained is the lack of contact angles in monopolar or strong acidic liquids; water is the most powerful acid used up to now. In other terms, in a set where a strong hydrogen-bond acceptor such as the aprotic solvent DMSO is

Table 1 Lifshitz–Van der Waals and Acid–Base Components of Some Liquids and Solids Calculated with a Reference Scale for Water: $\gamma^{LW} = 26.25$, $\gamma^+ = 48.5$, and $\gamma^- = 11.2$ by GvOC theory; SD = ± 2 mN/m

Wlw	5.12	26.2	P3FEt	5.04	25.4
W+	6.97	48.5	P3FElw	4.80	23.1
W–	3.34	11.2	P3FE+	1.34	1.79
Glw	5.92	35.0	P3FE–	0.856	0.733
G+	5.28	27.8	PVDFt	5.52	30.5
G–	2.71	7.33	PVDFlw	5.35	28.6
FAlw	5.96	35.5	PVDF+	3.24	10.5
FA+	3.35	11.3	PVDF–	0.290	0.0839
FA–	3.36	11.3	PVFt	6.07	36.8
MIlw	7.13	50.8	PVFlw	5.84	34.1
MI+	0.00	0.00	PVF+	2.87	8.25
MI–	0.00	0.00	PVF–	0.476	0.227
GElw	5.82	33.9	PVDCt	6.35	40.3
GE+	0.983	0.966	PVDClw	6.36	40.4
GE–	7.18	51.6	PVDC+	–0.0431	0.00186
BNlw	6.66	44.4	PVDC–	1.62	2.63
BN+	0.00	0.00	PVCt	6.29	39.5
BN–	0.00	0.00	PVClw	6.25	39.1
DMSOlw	5.68	32.3	PVC+	0.239	0.0572
DMSO+	0.193	0.0373	PVC–	0.940	0.883
DMSO–	27.6	763	PSt	6.18	38.2
BFlw	5.30	28.1	PSlw	6.20	38.4
BF+	4.46	19.9	PS+	–0.162	0.0261
BF–	1.50	2.24	PS–	0.468	0.219
PYlw	5.09	25.9	PMMAt	6.14	37.7
PY+	3.29	10.8	PMMAlw	6.12	37.4
PY–	1.63	2.66	PMMA+	0.104	0.0108
HDlw	5.25	27.6	PMMA–	1.49	2.23
HD+	0.00	0.00	PCTFEt	4.68	21.9
HD–	0.00	0.00	PCTFElw	5.30	28.1
PTFEt	4.49	20.1	PCTFE+	–1.57	2.46
PTFElw	4.49	20.1	PCTFE–	1.97	3.89
PTFE+	0.00	0.00	PA66t	6.54	42.8
PTFE–	0.00	0.00	PA66lw	6.11	37.4
PPt	5.32	28.3	PA66+	1.73	2.99
PPlw	5.32	28.3	PA66–	1.57	2.46
PP+	0.00	0.00	PETt	6.19	38.4
PP–	0.00	0.00	PETlw	6.21	38.5
PEt	5.85	34.2	PET+	–0.0545	0.00297
PElw	5.85	34.2	PET–	1.61	2.58
PE+	0.00	0.00	PBTt	0.774	0.599
PE–	0.00	0.00	PBTlw	6.19	38.4
			PBT+	–5.15	26.5
			PBT–	3.67	13.4

Abbreviations: WA = water; GL = glycerol; FA = formamide; MI = diiodomethane; GE = ethylenglycol; BN = bromonaphthalene; DMSO = dimethylsulfoxide; BF = bromoform; PY = pyrrole; HD = hexadecane; PTFE = polytetrafluoroethylene; PP = polypropylene; P3FE = polytrifluoroethylene; PVDF = polyvinylidene fluoride; PVF = polyvinylfluoride; PVDC = polyvinylidene chloride; PVC = polyvinyl chloride; PS = polystyrene; PMMA = polymethyl metacrylate; PCTFE = polychloro-trifluoroethylene; PA66 = polyhexamethylene adipamide; PET = polyethylene terephtalate; PBT = polybutylene terephtalate.

The square root (first numerical column) and the actual value (second numerical column) of the Lifshitz–van der Waals, acid and base components of liquids and solids are indicated by the symbols lw, +, and – respectively. In the case of solids, the symbol t indicates the total surface free energy.

present, we also need a strong hydrogen-bond donor. The main problem is to find liquids with both a high acid or hydrogen-bond donor ability and a high value of the total surface tension, because of the need to have contact angles different from zero on common polymers, without significant swelling or reaction effects. Some candidates can be found among those shown in Table 2. Halogen derivatives and nitroderivatives of phenol and cresol can also be explored.

A very intriguing idea is the use of hydrogen peroxide; its surface tension is similar to that of water, about 76 mN/m, and it is estimated to be about 10^6 times less basic than water [58]!

B. Some Mathematical Aspects of the Definition of Scales

Let us tackle in more detail the problem of scale definition. Suppose that a satisfactory best-fit solution has been found for a given set of donors and acceptors. Our goal is to fix the components of one or more chemicals in order to remove any invariance property and select a particular scale.

The four entries of Drago's matrix A in Eq. (9) are essentially arbitrary, the only constraint being det $A \neq 0$. Thus, we simply have to compute the parameters E and C for an acceptor and a donor and to give them some conventional values; the set of four equations imposed by the transformation rule will finally determine the four required entries. Some care is needed in the choice of the reference chemicals to obtain a set of four independent equations.

The definition of a specific scale for acid–base theories seems a more delicate task. If we suppose the components of a particular solvent to have been calculated, another set of γ^{LW}, γ^+, γ^- for the same chemical can be obtained through a

Table 2 Monopolar or Strong Acidic Liquids Suitable for Contact-Angle Measurements on Polymers

Liquid	Surface tension at room temp.	Acid solvatochromic	Basic solvatochromic
H_2SO_4 (98.5%)	55.1	(?)	0 (?)
HNO_3 (98.8%)	42.7	(?)	0 (?)
Formic acid	37.6	0.75	0.38
Monochloroacetic acid	35.4	0.74	0.36
Dichloroacetic acid	35.4	0.90	0.27
Pyrrole	36.6	0.41	0.29
Bromoform	41.53	0.15	0.06
Phenol	40.9	0.60	0.30
Water	72.8	0.82 (1.17)	0.35 (0.18)

Note: Data from Abraham [40] and those in parentheses from Taft and co-workers [39]. Water data are reported for comparison.

transformation of the invariance group. As $O(2, 1; \mathbb{R})$ essentially depends on a triple of parameters, we expect that whenever three components of some reference solvents are specified, the global best-fit solution is uniquely determined and the scale selected. Of course, we have not only to check that the three imposed conditions are independent, but also that the related matrix C satisfies Eq. (11). The invariance of $X^T R X$ obviously provides a necessary condition to accomplish the last requirement.

As a less restrictive constraint, we may impose that the acidic and basic components γ^+ and γ^- of a liquid are equal, without assigning a specific value to them. In this case, we introduce a unique condition and the scale is not completely fixed (we note here that GvOC theory in its present form makes this unsatisfactory choice). It is noticeable, however, that a solvent whose components γ^+ and γ^- are both nonzero certainly admits an invariance transformation C leading to $\gamma^+ = \gamma^-$, because any diagonal matrix with diagonal elements $(1, e^\zeta, e^{-\zeta})$, $\forall \zeta \in \mathbb{R}$, belongs to $O(2, 1; \mathbb{R})$ [11]. For an appropriate value of ζ, such a matrix expresses the condition imposed in GvOC theory for water, although it remains only a particular case of the most general requirement given in Eq. (11).

We emphasize once again that the reference solvent and components must be taken "judiciously," because a wrong choice could be inconsistent with the experimental data. The previous analysis is based on the simple observation that any best-fit algorithm must minimize some (objective) function of the rests:

$$\Delta_{l,i} = X_i^T R X_i - \gamma_{l,i}^{tot}, \qquad \Delta_{s,j} = Y_j^T R Y_j - \gamma_{s,j}^{tot}$$
$$\Delta_{ij} = X_i^T R Y_j - \tfrac{1}{2}(1 + \cos \theta_{ij})\gamma_{l,i}^{tot} \tag{17}$$

where we have introduced the column vectors

$$X_i = \begin{pmatrix} \sqrt{\gamma_{l,i}^{LW}} \\ \sqrt{\gamma_{l,i}^+} \\ \sqrt{\gamma_{l,i}^-} \end{pmatrix}, \quad Y_j = \begin{pmatrix} \sqrt{\gamma_{s,j}^{LW}} \\ \sqrt{\gamma_{s,j}^+} \\ \sqrt{\gamma_{s,j}^-} \end{pmatrix}, \quad \forall i = 1,\ldots,L; \forall j = 1,\ldots,S \tag{18}$$

and the symmetric orthogonal matrix

$$R = \begin{pmatrix} 1 & 0 & 0 \\ 0 & 0 & 1 \\ 0 & 1 & 0 \end{pmatrix}$$

The main result we can prove is the following:

Proposition. *Let A_i, $i = 1, 2, \ldots, L$, and B_j, $j = 1, 2, \ldots, S$, be some (real or complex) nonsingular matrices such that the substitutions $X_i \to A_i X_i$ and $Y_j \to B_j Y_j$ leave expressions [Eq. (17)] invariant. Then, necessarily,*

$$A_i = B_j = C, \quad \forall i = 1, 2, \ldots, L; j = 1, 2, \ldots, S,$$

where C is any matrix satisfying $C^T R C = R$.

Proof. Insert the previous linear transformations into Eqs. (17) and use invariance to obtain the relationships

$$A_i^T R A_i = R, \qquad B_j^T R B_j = R, \qquad A_i^T R B_j = R, \quad \forall i, j.$$

From the first equation, we deduce that $A_i^T R = R A_i^{-1}$, which when inserted into the third leads to the equality $R A_i^{-1} B_j = R$. As $R^{-1} = R$, we obtain the identity $A_i^{-1} B_j = I$ and the result follows from the arbitrariness of i and j. The common matrix C actually turns out to be nonsingular, because all the above equalities take the same form $C^T R C = R$, which implies $[\det(C)]^2 = 1$.

Now, simple algebraic manipulations allow us to show that the set of all the complex or real matrices C satisfying $C^T R C = R$ is actually nonempty, that it forms a group with respect to the usual matrix product, and that such a group is isomorphic to the orthogonal group $O(2, 1; \mathbb{C})$ [11]. Its intersection with the group $Gl(3, \mathbb{R})$ of real 3×3 invertible matrices is isomorphic to the orthogonal group $O(2, 1; \mathbb{R})$.

VI. CONCLUSIONS

In the beginning of this chapter, it was stated that the main problems arising from the application of the GvOC theory to the calculation of Lewis acid–base properties of polymer surfaces from contact angle data are the following:

All surfaces are overwhelmingly basic.
Results strongly depend on the choice of the three liquids used for contact-angle measurement.
Calculations yield, sometimes, negative values for unknowns, despite the fact that the latter are the roots of acid–base parameters.

Throughout this chapter, it was possible to account for these problems, namely:

The Lewis base, or electron-donor component, is, in absolute terms, much greater than the Lewis acid or electron-acceptor component because of the reference values for water chosen in the original GvOC theory. According to this theory, in its original form, the only meaningful approach is to compare acidic components with other acidic components of different materials, and the same for basic components. The direct comparison of the magnitude of the acidic component with the basic one of the same materials has no meaning. This explains some discrepancies between results obtained from wettability and IGC measurements [6,12] and, most of all, accounts for the unrealistic (if gauged by common chemico-physical knowledge) complete predominance of monopolar basic surfaces stemming from the application of the GvOC theory. It has also be shown, as an example and without any pretension to be right, a reference scale for water which is able to solve this problem; however, much work is necessary to obtain data suitable for a definitive solution of the "scale problem."

The strong dependence of the value of the components on the three liquids used is a result of the ill-conditioning of the related set of equations, an intrinsic and purely mathematical feature which cannot be completely cured by any realistic improvement of experimental accuracy. To reduce or eliminate the effect, one only needs a proper set of liquids representative of all kinds of different solvents.

As for negative values, whenever a large and proper set of solvents is employed and the contact-angle data are suitably treated, the number of negative coefficients reduces in a significant way. Moreover, in many cases, the standard deviation of a negative component is typically greater than the estimate itself, so that the actual value may be zero. Negative coefficients appear then as a simple consequence of measurement uncertainty. We cannot exclude, however, that in some cases they could have a different origin.

Beside these results, we hope that a certain consciousness inched its way through the mathematics of this chapter: There are many factors, not always considered in everyday practice, that come into play in the calculation of surface free-energy components from contact-angle measurement. Some of them have been frequently invoked and are directly related to the experiment itself, that is whether it is better to use the advancing, the receding, or any other contact angle. Some of them are more subtle and related to more general aspects, such as the effects on the results of the mathematical nature of the theory, the choice of the proper scale of acid/base components, and the definition of scales. No wonder that results obtained neglecting these aspects are tactfully deemed "theoretically interesting, but not always useful" [59]. There is no hope to obtain good surface free-energy data from contact-angle measurements or, more in general, to evaluate whether contact-angle measurement and related theories can yield useful surface free-energy data, unless all these aspects are taken into due account.

ACKNOWLEDGMENT

The authors acknowledge the financial support of CNR: Materiali Speciali per Tecnologie Avanzate II.

REFERENCES

1. AW Adamson. Physical Chemistry of Surfaces. 5th ed. New York: Wiley, 1990.
2. JN Israelachvili. Intermolecular and Surface Forces. 2nd ed. London: Academic Press, 1992.
3. FM Fowkes, In: KL Mittal, ed. Physicochemical Aspects of Polymer Surfaces. New York: Plenum Press, 1983, Vol. 2.
4. FM Fowkes. In: JD Andrade, ed. Surface and Interfacial Aspects of Biomedical Polymers. New York: Plenum Press, 1985, Vol. 1, Chap. 9.
5. FM Fowkes. J Adhesion Sci Technol 1:7–27, 1987.
6. JC Berg, In: JC Berg, ed. Wettability. New York: Marcel Dekker, 1993, Chap. 2.
7. RJ Good, MK Chaudury. In: LH Lee, ed. Fundamentals of Adhesion. New York: Plenum Press, 1991, Chap. 3.
8. KL Mittal, HR Anderson Jr, eds. Acid–Base Interactions: Relevance to Adhesion Science and Technology. Utrecht: VSP, 1991.
9. M Morra. J Colloid Interf Sci 182:312–314, 1996.
10. LH Lee. Langmuir 12:1681–1687, 1996.
11. C Della Volpe, S Siboni. J Colloid Interf Sci 195:121–136, 1997.
12. ML Tate, YK Kamath, SP Wesson, SB Ruetsch. J Colloid Interf Sci 17:579–588, 1996.
13. HJ Jacobasch, K Grundke, S Schneider, F Simon. J Adhes 48:57–73, 1995.
14. GN Lewis. Valence and the Structure of Atoms and Molecules. New York: The Chemical Catalog Co., 1923.
15. WB Jensen. J Adhes Sci Technol 5:1–7, 1991.
16. SR Cain. J Adhes Sci Technol 5:71–80, 1991.
17. GC Pimentel, AL McClellan. The Hydrogen Bond. San Francisco: Freeman, 1960.
18. NAM Besseling. Statistical Thermodynamics of Molecules with Orientation-Dependent Interactions in Homogeneous and Heterogeneous Systems. PhD thesis, Wageningen Agricultural University, Wageningen, The Netherlands, 1993.
19. NAM Besseling, JMHM Scheutjens. J Phys Chem 98:11597–11609, 1994.
20. NAM Besseling. J Phys Chem 98: 11610–11622, 1994.
21. NAM Besseling, J Lyklema. J Pure Appl Chem 67:881–888, 1995.
22. NAM Besseling. Langmuir 13:2109–2112, 1997.
23. CJ van Oss, RJ Good, MK Chaudhury. J Protein Chem 5:385–402 1986.
24. CJ van Oss, MK Chaudhury, RJ Good. Adv Colloid Interf Sci 28:35–60, 1987.
25. RJ Good, CJ van Oss. In: ME Schrader, G Loed, eds. Modern Approach to Wettability: Theory and Application. New York: Plenum Press, 1991, Chap. 1.
26. CJ van Oss. Interfacial Forces in Aqueous Media. New York: Marcel Dekker, 1994.
27. M Sidqi, G Ligner, J Jagiello, H Balard, E Papirer. Chromatographia 28:588–596, 1989.
28. VI Bogillo, A Voelkel. Polymer 36:3503–3510, 1995.
29. A Voelkel, E Andrzejewska, R Maga, M Andrzejewski. Polymer 37:455–462, 1996.
30. E Andrzejewska, A Voelkel, M Andrzejewski, R Maga. Polymer 37:4333–4344, 1996.
31. T Young. Phil Trans 95:65, 1805.
32. M Lampin, R Warocquier-Clerout, C Legris, M Degrange, MF Sigot-Luizard. J Biomed Mater Res 36:99–108, 1997.
33. JF Padday. In: E Matijevic, ed. Surface and Colloid Science. New York: Wiley, 1969.

34. JM Douillard. J Colloid Interf Sci 188:511–514, 1997.
35. M Morra, C Cassinelli. J Biomater Sci Polymer Ed 9:55–74, 1997.
36. LH Lee. Langmuir 12:1681–1687, 1996.
37. C Della Volpe, A Deimichei, T Riccò. J Adhes Sci Technol 12:1141–1180 (1998).
38. RS Drago, Struct Bond 15:73–139, 1973.
39. MJ Kamlet, JM Abboud, MH Abraham, RW Taft. J Org Chem 48:2877–2891 1983.
40. MH Abraham. Chem Soc Rev 22:73–83, 1993.
41. A Deimichei, Thesis, University of Trento, Trento, 1996
42. WH Press, BP Flannery, SA Teukolsky, WT Vetterling. Numerical Recipes. Cambridge: Cambridge University Press, 1989.
43. HY Erbil, RA Meriç. Colloids Surf 33:85–97, 1988.
44. L-H Lee. Langmuir 12(6):1681–1687, 1996.
45. WV Chang, X Qin. Paper presented at ACS San Francisco Meeting, San Francisco, California April 1992; J Adhes Sci Technol (submitted).
46. X Qin, WV Chang. J Adhes Sci Technol 9:823–841, 1995.
47. X Qin, WV Chang. J Adhes Sci Technol 10:963–989, 1996.
48. WV Chang, X Qin. J Adhes Sci Technol (submitted).
49. S Wu. Polymer Interface and Adhesion. New York: Marcel Dekker, 1982.
50. RE Johnson, Jr, RH Dettre. J Phys Chem 68:1744, 1964.
51. RE Johnson, Jr, RH Dettre. In: E. Matijevic, ed., Surface and Colloid Science Vol. 2. New York: Wiley, 1969.
52. K Grundke, H-J Jacobasch, F Simon, S Schneider. J Adhes Sci Technol 9:327–350, 1995.
53. FM Fowkes. Ind Eng Chem 56:40–52, 1964.
54. GEH Hellwig, AW Neumann. Proceedings, Internat. Congress on Surface Activity, Section B, p. 687, Barcelona, 1968.
55. B Janczuk, T Bialopiotrowicz. J Colloid Interf Sci 140:362–372, 1990.
56. AW Neumann, D Renzow. Z Phys Chem 11:68–73, 1969.
57. HY Erbil, RA Meriç. Colloids Surf 33:85–97, 1988.
58. FA Cotton, G Wilkinson. Advanced Inorganic Chemistry. New York: Wiley, 1966.
59. RE Baier, AE Meyer. In: JL Brash, P Wojciechowski, eds. Interfacial Phenomena and Bioproducts. New York: Marcel Dekker, 1996, p 94.

Index